49 181 $ 66,-

125,40

Nuclear and Particle Physics at Intermediate Energies

NATO ADVANCED STUDY INSTITUTES SERIES

A series of edited volumes comprising multifaceted studies of contemporary scientific issues by some of the best scientific minds in the world, assembled in cooperation with NATO Scientific Affairs Division.

Series B: Physics

RECENT VOLUMES IN THIS SERIES

Volume 7 – Low-Dimensional Cooperative Phenomena
edited by H. J. Keller

Volume 8 – Optical Properties of Ions in Solids
edited by Baldassare Di Bartolo

Volume 9 – Electronic Structure of Polymers and Molecular Crystals
edited by Jean-Marie André and János Ladik

Volume 10 – Progress in Electro-Optics
edited by Ezio Camatini

Volume 11 – Fluctuations, Instabilities, and Phase Transitions
edited by Tormod Riste

Volume 12 – Spectroscopy of the Excited State
edited by Baldassare Di Bartolo

Volume 13 – Weak and Electromagnetic Interactions at High Energies (Parts A and B)
edited by Maurice Levy, Jean-Lous Basdevant, David Speiser, and Raymond Gastmans

Volume 14 – Physics of Nonmetallic Thin Films
edited by C.H.S. Dupuy and A. Cachard

Volume 15 – Nuclear and Particle Physics at Intermediate Energies
edited by J. B. Warren

Volume 16 – Structure and Reactivity of Metal Surfaces
edited by E. G. Derouane and A. A. Lucas

The series is published by an international board of publishers in conjunction with NATO Scientific Affairs Division

A	**Life Sciences**	Plenum Publishing Corporation
B	**Physics**	New York and London
C	**Mathematical and Physical Sciences**	D. Reidel Publishing Company Dordrecht and Boston
D	**Behavioral and Social Sciences**	Sijthoff International Publishing Company Leiden
E	**Applied Sciences**	Noordhoff International Publishing Leiden

Nuclear and Particle Physics at Intermediate Energies

Edited by
J. B. Warren
University of British Columbia
Vancouver, B.C.
Canada

PLENUM PRESS • NEW YORK AND LONDON
Published in cooperation with NATO Scientific Affairs Division

Library of Congress Cataloging in Publication Data

Brentwood Summer Institute on Nuclear Physics at Intermediate Energies, Brentwood College. 1975.
Nuclear and particle physics at intermediate energies.

(Nato advanced study institutes series: Series B, Physics; v. 15)
"Lectures presented at the Brentwood Summer Institute on Nuclear Physics at Intermediate Energies held at Brentwood College, Victoria, Canada, June 23-July 2, 1975."
"Organized by the TRIUMF group of universities.
"Published in cooperation with NATO Scientific Affairs Division."
Includes bibliographical references and index.
1. Particles (Nuclear physics)—Addresses, essays, lectures, I. Warren, John Bernard, 1914- II. North Atlantic Treaty Organization. Scientific Affairs Division. III. Title. IV. Series.

QC793.28.B73 1975 539.7'21 76-8270
ISBN 0-306-35715-1

Lectures presented at the Brentwood Summer Institute on Nuclear and Particle Physics at Intermediate Energies held at Brentwood College, Victoria, Canada, June 23-July 2, 1975

A Division of Plenum Publishing Corporation
227 West 17th Street, New York, N.Y. 10011

United Kingdom edition published by Plenum Press, London
A Division of Plenum Publishing Company, Ltd.
Davis House (4th Floor), 8 Scrubs Lane, Harlesden, London NW10 6SE, England

Printed in the United States of America

Preface

The Brentwood Summer Institute on Nuclear and Particle Physics at Intermediate Energies was the second of its kind organised by the TRIUMF group of Universities, the first taking place at Banff in 1970. With the advent of initial beams at the new meson facilities at LAMPF, SIN, NEVIS, CERN S.C. and TRIUMF it was an eminently suitable time for an in-depth study of some of the science which will be possible when these accelerators achieve their design intensities in proton and meson beams.

The organizing committee, comprising:-

G.A. Beer	Univ. of Victoria	J.M. Cameron	Univ. of Alberta
J.M. McMillan	U.B.C.	D.F. Measday	U.B.C.
R.M. Pearce	Univ. of Victoria	J.E.D. Pearson	U.B.C.
	J.B. Warren	U.B.C.	

wishes to acknowledge the financial support provided by the North Atlantic Treaty Organisation, the National Research Council of Canada, and Atomic Energy of Canada Ltd., without which the Institute could not have been held. Also we wish to acknowledge the helpful advice of the Scientific Committee of NATO and of Dr. T. Kester, Secretary of this Committee.

Many persons from the University of Victoria and the University of British Columbia helped with the local arrangements and we are grateful to them and particularly to the staff of Brentwood College who made the stay of the participants such a pleasant one.

For the preparation of these Proceedings we are indebted to many persons, particuarly to Mrs. Lilian Ratcliffe and Mrs. Hilary Prior. The Proceedings contain texts of the advertised series of lectures and of a number of seminars given by attendees. No attempt was made to record the lively discussions which took place in the question periods but participants were asked to submit written versions of their questions and comments if they so desired and these are included.

While "trifles make perfection and perfection is no trifle", a compromise is necessary between blemishes in the text and publication date. It is hoped that the reader will be satisfied with the result.

J. B. Warren

Contents

MUON PHYSICS

H. Primakoff

University of Pennsylvania

Philadelphia, Pennsylvania, U.S.A.

1. INTRODUCTION

The muon is only one of the now numerous known elementary particles and yet it possesses an interest which is perhaps greater than that characterizing most of the others. The basic reasons for this special interest in the muon may be enumerated as follows:

1) The muon (μ) was the first elementary particle discovered which was found to be "superfluous", in fact the first of a class of "superfluons". By this is meant that the muon is unnecessary for the understanding of all the usual phenomena of molecular, atomic, and nuclear physics. In fact, all the usual molecular and atomic phenomena can be adequately described by supposing that electromagnetic radiation is made up of photons (γ) while matter consists of electrons (e) and nuclei grouped under the influence of electromagnetic-type interactions into atoms and molecules; moreover the nuclei themselves are composed of protons (p) and neutrons (n) and all the usual nuclear phenomena can be reasonably well understood on the basis of such a proton-neutron model if, in addition, one introduces the various mesons ($\pi, \eta, \eta', \rho, \omega, \phi$ ....) required to transmit the strong-interaction-type nuclear forces, and the neutrino (ν_e) required for the interpretation of the weak-interaction-type nuclear β-decay ($n \to p + e^- + \bar{\nu}_e$, $H^3 \to He^3 + e^- + \bar{\nu}_e$ ....) and π-meson (pion) β-decay ($\pi^+ \to \pi^0 + e^+ + \nu_e$).

The existence of a world containing only the elementary particles listed above, namely the photon (γ), leptons (e, ν_e) and the hadrons (p, n, mesons), with the couplings between them the same as in our world, would consistute a relatively tidy universe from the point of view of elementary-particle physics. In such a relatively tidy universe, the charged pion would decay predominantly via the channel: $\pi^+ \rightarrow e^+ + \nu_e$ with a lifetime 2.6024 x 10^{-8}sec./ 1.24 x 10^{-4} = 2.10 x 10^{-4}sec. [1]. Unfortunately this relatively tidy universe is, at least on the elementary particle level, very different from ours; in fact, many elementary particles other than γ; e, ν_e; p,n,π, η, η', ρ, ω, ϕ, ... are present. These additional elementary particles, the so-called "superfluons" : μ, ν_μ, the strange hadrons (Λ, Σ, Ξ, ..., K, K^*, ...),... no doubt appear "superfluous" from the point of view of ordinary nuclear and atomic physics only because we lack a deeper understanding of their role, but some such terminology is nevertheless expedient at present. The muon is a particularly interesting "superfluon" because, apart from the neutretto (ν_μ), it is the only "superfluous" lepton, [2] and because its non-zero charge, intermediate-sized mass, and relatively long lifetime render it especially accessible to experimental investigation and to employment as a probe of nuclear, atomic, and solid-state properties.

2) The second basic property of the muon which is of fundamental interest is the unique relationship between the muon and the electron: on the one hand, the electromagnetic and weak interactions of μ and e appear to be essentially identical, and neither exhibits any strong interactions; on the other hand, their masses differ very considerably, viz.: $(m_\mu - m_e)/m_e = 205.769$. Usually, mass differences between the elementary particles are associated with differences in internal quantum numbers carrying dynamical significance. This certainly seems to be true of mass differences between particles other than μ or e - for example the mass difference between π^+ and π^0 is attributed to a difference in their electric charges (1 vs. 0) and therefore to their different electromagnetic self interactions. One can indeed speculate that the μ-e mass difference: $(m_\mu - m_e)/m_e = 205.769$ is also somehow caused by the difference in the electromagnetic self interactions of the μ and the e (in spite of the equality of their electric charges) but no satisfactory quantitative development of this concept has ever been given. A curious and possibly significant fact in this connection is: $\frac{3}{2}(1/\alpha) = 3/2\ (137.036) = 205.554$.

3) A more practical consideration regarding muons, or rather muons as compared to electrons, involves the fact that, because of the very different muon and electron masses, parallel processes such as

$$\mu^- + p \rightarrow \nu_\mu + n \quad ,$$

$$e^- + p \to \nu_e + n \quad ,$$

$$|\vec{p}_\mu| \approx |\vec{p}_e| << m_\mu$$

occur at different values of momentum transfer, i.e.

$$(p_\mu - p_{\nu_\mu})^2 = (p_n - p_p)^2 = -m_\mu^2 + 2(E_\mu E_{\nu_\mu} - p_\mu \cdot p_{\nu_\mu}) \cong m_\mu^2$$

$$(p_e - p_{\nu_e})^2 = (p_n - p_p)^2 = -m_e^2 + 2(E_e E_{\nu_e} - \vec{p}_e \cdot \vec{p}_{\nu_e}) \cong 2 E_e^2 - m_e^2$$

$$= 2\vec{p}_e^{\,2} + m_e^2$$

Assuming that the various primitive interactions of μ and of e with all other elementary particles are the same if the momentum transfers are the same, we can, by a study of parallel processes, determine vertex functions or form factors describing the primitive interactions for at least two different values of momentum transfer.

We proceed to discuss briefly the physical parameters of a muon in slowly varying gravitational and electromagnetic fields. The numerical values of these physical parameters are

$$m_\mu = (206.76922 \pm 0.00041) m_e$$

$$s_\mu = \frac{1}{2}\hbar$$

$$e_\mu = (1.000000 \pm 0.000002) e_e$$

$$\frac{\mu_\mu}{\frac{e_e \hbar}{2 m_e c}} = \left(\frac{e_\mu}{e_e}\right)\left(\frac{m_e}{m_\mu}\right)\left(\frac{g_\mu}{2}\right) = (4.8419497 \pm 0.0000095) \times 10^{-3};$$

$$\frac{g_\mu}{2} = 1.00116616 \pm 0.00000031$$

As regards the experimental determination of these numerical values we recall that the muon magnetic moment is given by a measurement of the muon spin precession frequency in a known magnetic field while the muon magnetic moment anomaly $\frac{g_\mu - 2}{2}$ is deduced from a study of the change in angle between the spin and the momentum vectors of muons stored in a known magnetic field. Thus, a very accurate experimental value is available for $\frac{e_\mu}{e_e}\frac{m_e}{m_\mu}$. Further, an extremely precise value is available for the triplet ↔ singlet transition frequency in the ground state of muonium $[\mu^+ e^-]$

$$\nu_{\text{trip.}\leftrightarrow\text{sing.}} = \left(\frac{e_\mu}{e_e}\right)^4 \left(\frac{16}{3}\right)\alpha^2 c \ (\text{Ryd.}) \frac{\left(\frac{m_e}{m_\mu}\right)\left(\frac{g_\mu}{2}\right)\left(\frac{g_e}{2}\right)}{(1 + m_e/m_\mu)^3}$$

$$\times \left(1 + \varepsilon_1 + \varepsilon_2 \cdot \frac{m_e}{m_\mu}\right)$$

$$= (4463302.5 \pm 1.6) \times 10^3 \sec^{-1} \quad ;$$

$$\alpha^{-1} = (e_e{}^2/\hbar c)^{-1} = 137.03602 \pm 0.00021 \quad ,$$

$$c = (2.99792462 \pm 0.00000018) \times 10^{10} \text{ cm/sec.} \quad ,$$

$$\text{Ryd.} = (1.09737312 \pm 0.00000011) \times 10^5 \text{ cm}^{-1} \quad ,$$

$$\frac{g_e}{2} = \frac{\mu_e}{\frac{e_e \hbar}{2m_e c}} = 1.001159657 \pm 0.000000035 \quad ,$$

$$\varepsilon_1 \equiv \left(1 + \frac{\alpha}{2\pi}\right)^{-1} \left[(\ell n2-1)\alpha^2 - \left(\frac{8}{3\pi}\ \ell n\alpha(\ell n\alpha - \ell n4 + \frac{281}{480}\right) + \frac{18.4 \pm 5.0}{\pi}\right)\alpha^3\right],$$

$$\varepsilon_2 \equiv \left(1 + \frac{\alpha}{2\pi}\right)^{-1}\left[\left(\frac{3}{\pi}\ \ell n\left(\frac{m_e}{m_\mu}\right)\right)\alpha - \left(\frac{9}{2}\ \ell n\alpha\right)\alpha^2\right] \cong \left(1 + \frac{\alpha}{2\pi}\right)^{-1}$$

$$\times \quad \left[\left(\frac{3}{\pi}\ \ell n\left(\frac{2}{3}\ \alpha\right)\right)\alpha - \left(\frac{9}{2}\ \ell n\alpha\right)\alpha^2\right]$$

Thus, a very accurate value is also available for

$$\left(\frac{e_\mu}{e_e}\right)^4 \frac{(m_e/m_\mu)}{(1+(m_e/m_\mu))^3}\left(1 + \varepsilon_1 + \varepsilon_2 \cdot \frac{m_e}{m_\mu}\right)$$ and this, together with the previously mentioned very accurate value for $\left(\frac{e_\mu}{e_\mu}\right)\left(\frac{m_e}{m_\mu}\right)$, allows the determination of numerical values for $\frac{e_\mu}{e_e}$ and $\frac{m_\mu}{m_e}$ separately. We also recall that s_μ is given as $\frac{1}{2}\hbar$ from the analysis of the fine structure of muonic atoms and, again, from the analysis of the just discussed hyperfine structure of muonium; this, on the basis of the conventional spin-statistics theorem, identifies the muon as a fermion. The fermion identification is supported by agreement of fermion-type pair-production cross sections with experiment in the case of $\gamma + Z \to \mu^+ + \mu^- + Z$ and $\mu^+ + Z \to \mu^+ + \mu^+ + \mu^- + Z$ (precision $\approx$ 10%). However, a completely unequivocal test of muon statistics can only be obtained by a detailed examination of states containing two or more muons of the same sign of charge, as for example, in the process $\mu^+ + Z \to \mu^+ + \mu^+ + \mu^- + Z$. In such a process, the momentum and spin correlation of the two μ^+ would depend on the muon statistics and, in particular, one could search for $\mu^+\mu^+$ events which violate the exclusion principle, i.e. $\mu^+\mu^+$ events with $S_{\mu^+\mu^+} = 1$, $L_{\mu^+\mu^+} = 0,2,4,\ldots$ and $S_{\mu^+\mu^+} = 0$, $L_{\mu^+\mu^+} = 1,3,5,\ldots$.

In concluding our brief discussion of the physical parameters of a muon we summarize the essential features of the behavior of a muon in what are effectively very rapidly varying electromagnetic fields, i.e. of the behavior of a muon in 1) high momentum-transfer elastic, shallow inelastic, and deep inelastic scattering from a proton or a neutron ($\mu^\pm + p,n \to \mu^\pm + p,n$; $\mu^\pm + p,n \to \mu^\pm + p,n$ + mesons), 2) wide-angle photoproduction of muon pairs ($\gamma + Z \to \mu^+ + \mu^- + Z$), 3) high-energy large-angle bremsstrahlung of muons ($\mu^\pm + Z \to \mu^\pm + \gamma + Z$), and 4) conversion of electron pairs into muon pairs ($e^+ + e^- \to \mu^+ + \mu^-$). In all of these processes the charge and current distributions within the muon appear identical within experimental accuracy to those within the electron, i.e. appear essentially point-like. More quantitatively, the root-mean-square radius associated with the charge (or current) distribution within the muon can be estimated on the basis of these experiments as less than 5×10^{-15} cm.

We proceed to the discussion of the physical parameters of a muon which are characteristic of its weak interactions. In particular we shall treat the elementary-particle aspects of the decay of a muon and of the capture of a muon by a proton

$$\mu^+ \to e^+ + \nu_e + \bar{\nu}_\mu \, , \; \mu^- \to e^- + \bar{\nu}_e + \nu_\mu$$

and

$$\mu^- + p \to \nu_\mu + n \, .$$

2. GENERALITIES REGARDING THE WEAK INTERACTIONS

The Hamiltonian (density) of the leptonic weak interactions is given, in lowest order, at least approximately by a bilinear expression in the appropriate leptonic weak currents; we shall suppose that each of these leptonic weak currents transform as a linear combination of a polar-vector (V) current and an axial-vector (A) current under space-time translations, rotations and inversions, there being no need, within available experimental precision, to invoke the presence of scalar (S), pseudo-scalar (P), and tensor (T) leptonic weak currents (see, however, below). Similarly, the Hamiltonian (density) of the semileptonic strangeness-preserving ($\Delta S = 0$) weak interactions is given, in lowest order, at least approximately by a bilinear expression in the appropriate leptonic and hadronic {V,A} weak currents. Thus, in lowest order,

$$H_{\text{lept}:\mu\leftrightarrow e}(x) = -\frac{G}{\sqrt{2}}\{\ell_\alpha(x;\nu_\mu,\mu^-)\ell_\alpha(x;e^-,\nu_e) + \text{herm. conj.}\}$$

$$H_{\text{semilept}:\Delta S=0}(x) = -\frac{G\cos\theta_C}{\sqrt{2}}\{[\ell_\alpha(x;\nu_\mu,\mu^-)+\ell_\alpha(x;\nu_e,e^-)] \times h_\alpha^{(-),\Delta S=0}(x) + \text{herm. conj.}\} \qquad (1)$$

where G is the weak-interaction coupling constant (see Eq. (18) below) and θ_C is the Cabibbo angle (see Eq. (26) et seq. below). Further, the leptonic weak currents $\ell_\alpha(x;a.b)$ are given explicitly in terms of the corresponding lepton field operators by

$$\ell_\alpha(x;a,b) = i(\psi_a^\dagger\gamma_4\gamma_\alpha(1 + \gamma_5)\psi_b) = \ell_\alpha^\dagger(x;b,a)(1-2\delta_{\alpha 4}) \qquad (2)$$

while the $\Delta S = 0$ hadronic weak currents $h_\alpha^{(-),\Delta S=0}(x)$

$$h_\alpha^{(-),\Delta S=0}(x) = V_\alpha^{(-)}(x) + A_\alpha^{(-)}(x) = \{h_\alpha^{(+),\Delta S=0}(x)\}^\dagger(1-2\delta_{\alpha 4})$$

$$=(\{V_\alpha^{(+)}(x)\}^\dagger + \{A_\alpha^{(+)}(x)\}^\dagger)(1-2\delta_{\alpha 4}) \qquad (3)$$

need not be given explicitly in terms of presumed fundamental hadron-field (e.g. quark-field) operators but rather can be specified directly in terms of the isospin currents $I_\alpha^{(\mp)}(x) = I_\alpha^{(1)}(x) \mp iI_\alpha^{(2)}(x)$ and the pion-source current $J_{\pi^\pm}(x)$ as follows

$$V_\alpha^{(\mp)}(x) = I_\alpha^{(\mp)}(x) \quad , \quad \frac{\partial V_\alpha^{(\mp)}(x)}{\partial_\alpha} = 0$$

$$\left[\int V_0^{(+)}(\vec{x},t)d\vec{x}, \int V_0^{(-)}(\vec{y},t)d\vec{y}\right]_- = \left[I^{(+)}(t) \, , \, I^{(-)}(t)\right]_- = 2I^{(3)}(t)$$

$$CV_\alpha^{(\mp)}(x)C^{-1} = -V_\alpha^{(\pm)}(x) \quad , \quad GV_\alpha^{(\mp)}(x)G^{-1} \equiv \left(Ce^{i\pi I^{(2)}}\right)V_\alpha^{(\mp)}(x)\left(Ce^{i\pi I^{(2)}}\right)^{-1}$$

$$= V_\alpha^{(\mp)}(x) \qquad (4)$$

where $I^{(\mp)}(t) \equiv \int I_0^{(\mp)}(\vec{x},t)d\vec{x}$, $C \equiv$ particle-antiparticle conjugation operator, and

$$A_\alpha^{(\mp)}(x) = a_\pi m_\pi^{\,3}\left(-\frac{\partial}{\partial x}\cdot\frac{\partial}{\partial x} + m_\pi^{\,2}\right)^{-1}\left(\frac{\partial}{\partial x}\cdot\frac{\partial}{\partial x}\right)^{-1}\frac{\partial}{\partial x_\alpha}\, J_{\pi^\pm}(x)$$

$$= a_\pi m_\pi^{\,3}\left(\frac{\partial}{\partial x}\cdot\frac{\partial}{\partial x}\right)^{-1}\frac{\partial}{\partial x_\alpha}\,\Phi_{\pi^\pm}(x)$$

$$\frac{\partial A_\alpha^{(\mp)}(x)}{\partial x_\alpha} = a_\pi m_\pi^{\,3}\left(-\frac{\partial}{\partial x}\cdot\frac{\partial}{\partial x} + m_\pi^{\,2}\right)^{-1} J_{\pi^\pm}(x) = a_\pi m_\pi^{\,3}\,\Phi_{\pi^\pm}(x)$$

$$a_\pi = m_\pi^{-3}\,\frac{\langle vac|\frac{\partial A_\alpha^{(-)}(0)}{\partial x_\alpha}|\,\pi^+\rangle}{\langle vac|\,\Phi_{\pi^+}(0)\,|\,\pi^+\rangle} = 0.94 \pm 0.01 \text{ (see Eq. (26) et seq. below)}$$

$$\langle F(p+q)|\frac{\partial A_\alpha^{(\mp)}(0)}{\partial x_\alpha}|I(q)\rangle = \frac{a_\pi m_\pi^3}{q^2+m_\pi^2}\langle F(p+q)|J_{\pi^\pm}(0)|I(p)\rangle$$

$$= \frac{a_\pi m_\pi^3}{q^2+m_\pi^2}\, g_{\pi IF}(q^2)(U_F^\dagger(p+q)KU_I(q))$$

$$\left|[g_{\pi IF}(q^2) - g_{\pi IF}(-m_\pi^2)]/g_{\pi IF}(-m_\pi^2)\right| \cong \left|\{[g_{\pi IF}(0) - g_{\pi IF}(-m_\pi^2)]/g_{\pi IF}(-m_\pi^2)\}(1+\frac{q^2}{m_\pi^2})\right|$$

$$<<1 \text{ for } -m_\pi^2 \leqq q^2 \lessapprox m_\pi^2$$

$$[I^{(\pm),5}(t), I^{(\pm)}(t)]_- = 0 \quad , \quad [I^{(+),5}(t), I^{(-),5}(t)]_- = 2I^{(3)}(t)$$

$$CA_\alpha^{(\mp)}(x)C^{-1} = A_\alpha^{(\pm)}(x) \ , \ GA_\alpha^{(\mp)}(x)G^{-1} \equiv (Ce^{i\pi I^{(2)}})A_\alpha^{(\mp)}(x)(Ce^{i\pi I^{(2)}})^{-1}$$

$$= -A_\alpha^{(\mp)}(x) \qquad (5)$$

where $\Phi_{\pi^\pm}(x) \equiv$ pion-field operator; $|I(p)\rangle$ and $|F(p+q)\rangle$ are, respectively, hadron states of four-momentum p and p+q; $u_I(p)$, $u_F(p+q)$ are the corresponding hadron "spinors" describing the "center-of-mass" motion of I and F; K is an appropriate kinematic-type pseudoscalar isovector operator acting on $u_I(p)$ or $u_F(p+q)$; $g_{\pi IF}(q^2)$ is the vertex function associated with the $\pi^- + I \to F$ vertex; and $I^{(\mp),5}(t) \equiv \int A_0^{(\mp)}(\vec{x},t)d\vec{x}$.

We emphasize that the values of $[I^{(+)}(t),I^{(-)}(t)]_-$ and of $[I^{(\pm),5}(t),I^{(\pm)}(t)]_-$ are a consequence of the fact that $I^{(\pm)}(t)$ is a component of isospin and $I^{(\pm),5}(x)$ $(\sim J_{\pi^\pm}(\vec{x},t))$ is an isovector; on the other hand, the assignment of a definite value to $[I^{(+),5}(t), I^{(-),5}(t)]_-$ is an additional and crucial assumption. Eqs. (4) and (5) constitute the quantitative formulation of the conserved vector current hypothesis for $V_\alpha^{(\mp)}(\vec{x},t)$, which identifies the hadronic $\Delta S=0$ vector weak current with the isospin current (CVC), the partially conserved axial-vector current hypothesis for $A_\alpha^{(\mp)}(\vec{x},t)$, which is characterized essentially by the assumption of the relatively slow and approximately linear variation of $g_{\pi IF}(q^2)$ with q^2 in the range $-m_\pi^2 \leqq q^2 \lessapprox m_\pi^2$ (PCAC), and the current-algebra-type hypothesis for the equal-time commutator $[I^{(+),5}(t),I^{(-),5}(t)]_-$, which fixes the scale of $A_\alpha^{(\pm)}(\vec{x},t)$ (CAC). We note that Eq. (1) is consistent with muon-electron universality in the weak interactions, this universality being upset only if $\ell_\alpha(x;\nu_\mu,\mu^-)$ on the right-hand side is multiplied by a constant different from unity. We also note that the leptonic and hadronic weak currents in Eqs. (1)-(5) are singly charged, but emphasize that the theory can be easily generalized to describe the possible presence of neutral (or doubly charged) leptonic and hadronic weak currents. In fact, such a generalization appears to be required in view of the recent experimental discovery of neutral weak currents [3] but does not affect (at least not to lowest order in G) the description of the various charge-exchange leptonic and semileptonic weak processes such as $\mu^+ \rightarrow e^+ + \nu_e + \bar{\nu}_\mu$ and $\mu^- + p \rightarrow \nu_\mu + n$.

Eqs. (1)-(5) imply the validity of the law of conservation of total muon-family leptonic number

$$(L_\mu)_{tot} = \sum_j (L_\mu)_j = \text{const.}$$

$$L_\mu = \begin{cases} 1 & \text{for } \mu^-, \nu_\mu \\ -1 & \text{for } \mu^+, \bar{\nu}_\mu \end{cases} \tag{6}$$

(provided $m_{\nu_\mu} = 0$) and the validity of the law of conservation of total electron-family leptonic number

$$(L_e)_{tot} = \sum_j (L_e)_j = \text{const.}$$

$$L_e = \begin{cases} 1 & \text{for } e^-, \nu_e \\ -1 & \text{for } e^+, \bar{\nu}_e \end{cases} \tag{7}$$

(provided $m_{\nu_e} = 0$). These laws of conservation are analogous to the experimentally far better established laws of conservation of total baryonic number and total electric charge and forbid reactions such as

$$\mu^\pm \rightarrow e^\pm + \gamma,\ e^\pm + \gamma + \gamma,\ \ldots$$

$$\mu^\pm \rightarrow e^\pm + e^+ + e^-$$

$$[Z,A] \rightarrow [Z+2,A] + e^- + e^-$$

$$\mu^- + [Z,A] \rightarrow e^+ + [Z-2,A]$$

$$K^{\pm} \to \pi^{\pm} + \mu^{\pm} + e^{\mp},\ \pi^{\mp} + \mu^{\pm} + \mu^{\pm},\ \pi^{\mp} + e^{\pm} + e^{\pm},\ \pi^{\mp} + \mu^{\pm} + e^{\pm}$$

$$\nu_\mu + n \to e^- + p$$

$$\nu_\mu + p \to \mu^+ + n$$

$$\nu_\mu + p \to e^+ + n$$

$$\nu_e + n \to \mu^- + p$$

$$\nu_e + p \to e^+ + n$$

$$\nu_e + p \to \mu^+ + n \tag{8}$$

It is however possible that, e.g., $\{(L_\mu)_{tot} + (L_e)_{tot}\}$ is exactly conserved while $(L_\mu)_{tot}$ and $(L_e)_{tot}$ are separately conserved only to some approximation - in this case $K^{\pm} \to \pi^{\mp} + \mu^{\pm} + \mu^{\pm}$, $\pi^{\mp} + e^{\pm} + e^{\pm}$, and $\pi^{\mp} + \mu^{\pm} + e^{\pm}$ would still be forbidden while $K^{\pm} \to \pi^{\pm} + \mu^{\pm} + e^{\mp}$ could proceed albeit at a much reduced rate (relative to $K^+ \to \pi^0 + \mu^+ + \nu_\mu$). Observationally, no evidence exists in favor of any of the reactions listed in Eq. (8) and rather low limits have been set experimentally for their branching ratios

(e.g., $\dfrac{\text{Rate } [\mu^+ \to e^+ + \gamma]}{\text{Rate } [\mu^+ \to e^+ + \nu_e + \bar{\nu}_\mu]} < 2 \times 10^{-8}$). We also note (again provided $m_{\nu_\mu} = 0$, $m_{\nu_e} = 0$) that Eqs. (1)-(5) imply

$$|\nu_e(\vec{p};\lambda)\rangle = |\nu_e(\vec{p}; -\tfrac{1}{2})\rangle$$

$$|\nu_\mu(\vec{p};\lambda)\rangle = |\nu_\mu(\vec{p}; -\tfrac{1}{2})\rangle$$

$$|\bar{\nu}_e(\vec{p};\lambda)\rangle = |\bar{\nu}_e(\vec{p}; \tfrac{1}{2})\rangle$$

$$|\bar{\nu}_\mu(\vec{p};\lambda)\rangle = |\bar{\nu}_\mu(\vec{p}; \tfrac{1}{2})\rangle \tag{9}$$

so that a one-to-one correspondence exists between the helicity (λ) and the leptonic number of a neutrino or an antineutrino state. In view of the fact that states such as $|\nu_e(\vec{p};\tfrac{1}{2})\rangle$, etc., are not admitted in the massless-neutrino case by these equations, one can, without a loss of generality and with a gain in economy, take $\tilde{\psi}^{\dagger}_{\nu_e}(x) = \psi_{\nu_e}(x)$, $\tilde{\psi}^{\dagger}_{\nu_\mu}(x) = \psi_{\nu_\mu}(x)$ (Majorana field operators) so that

$\psi_{\bar{\nu}_e}(x) = \psi_{\nu_e}(x)$, $\psi_{\bar{\nu}_\mu}(x) = \psi_{\nu_\mu}(x)$ and the only physical distinction between a neutrino and an antineutrino (of either the electron or the muon family) resides in their opposite helicity.

3. MUON DECAY

We now discuss muon decay: $\mu^+ \to e^+ + \nu_e + \bar{\nu}_\mu$ and $\mu^- \to e^- + \bar{\nu}_e + \nu_\mu$ on the basis of Eqs. (1), (2). We begin by setting down the expressions for the probability per unit time for the emission of an electron (e^-) with energy between $E_e = (|\vec{p}_e|^2 + m_e^2)^{1/2}$ and $E_e + dE_e$ and in a direction making an angle between θ_e and $\theta_e + d\theta_e$ with the muon (μ^-) polarization $\vec{\mathcal{P}}_\mu$ $(0 \leq |\vec{\mathcal{P}}_\mu| \leq 1)$

$$P(E_e,\theta_e)\, dE_e\, 2\pi \sin\theta_e\, d\theta_e = \sum_{\nu_\mu,\bar{\nu}_e,e^-\,\text{spins}} \int 2\pi \left| \left\langle \nu_\mu, \bar{\nu}_e, e^- \right| \int \mathcal{H}_{\text{lept}:\mu\to e}(\vec{x},0)\, d\vec{x} \left| \mu^- \right\rangle \right|^2$$

$$\times\ \delta\left(E_e + |\vec{p}_{\bar{\nu}_e}| + |\vec{p}_e + \vec{p}_{\bar{\nu}_e}| - m_\mu\right) \times \frac{d\vec{p}_{\bar{\nu}_e}}{(2\pi)^3}\, \frac{|\vec{p}_e| E_e\, dE_e\, 2\pi \sin\theta_e\, d\theta_e}{(2\pi)^3}$$

$$= \left[P_1(E_e) + |\vec{\mathcal{P}}_\mu| \cos\theta_e\, P_2(E_e) \right] 2\pi \sin\theta_e\, d\theta_e\, dE_e \tag{10}$$

with [4]

$$P_1(E_e) = \left(\frac{G^2 m_\mu |\vec{p}_e| E_e}{12\pi^4} \right) \left[3\left(1 + \rho' \frac{m_e}{E_e}\right)\left((E_e)_{max.} - E_e\right) + 2\rho\left(\frac{4}{3}E_e - (E_e)_{max.} - \frac{1}{3}\frac{m_e^2}{E_e}\right) \right] \tag{11}$$

$$P_2(E_e) = \left(\frac{G^2 m_\mu |\vec{p}_e| E_e}{12\pi^4} \right) \left(\frac{|\vec{p}_e|}{E_e} \right) \rho'' \left[\left((E_e)_{max.} - E_e\right) + 2\rho'''\left(\frac{4}{3}E_e - (E_e)_{max.} - \frac{1}{3}\frac{m_e^2}{E_e}\right) \right] \tag{12}$$

In Eqs. (10)-(12) m_{ν_μ} and m_{ν_e} have been taken = 0; $m_e \leq E_e \leq (E_e)_{max} = \frac{1}{2} m_\mu \left(1 + \frac{m_e^2}{m_\mu^2}\right)$; $\vec{P}_\mu = |\vec{P}_\mu| \hat{p}_\mu$ for μ^- from $\pi^- \to \mu^- + \bar{\nu}_\mu$ with $|\vec{P}_\mu| < 1$ only because of depolarization effects that occur between the instant of emission of the μ^- and the instant of its decay; finally the parameters ρ, ρ', ρ'', ρ''' are given by

$$\rho = \frac{3}{4}, \quad \rho' = 0, \quad \rho'' = -1, \quad \rho''' = \frac{3}{4} \tag{13}$$

The motivation of introducing the parameters ρ, ρ', ρ'', ρ''' into Eqs. (11), (12), rather than writing their numerical values directly as given in Eq. (13), lies in the fact that a large class of modifications of the $H_{\text{lept}:\mu\leftrightarrow e}$ of Eqs. (1), (2) results in expressions $P_1(E_e)$ and $P_2(E_e)$ which differ from those in Eqs. (11), (12) only in the numerical values of these parameters (see below).

The expressions for $P_1(E_e)$ and $P_2(E_e)$ in Eqs. (11) and (12) are subject to an electromagnetic radiative correction arising essentially from the emission and reabsorption of a virtual photon by the μ^- and by the e^- and from the exchange of a virtual photon between the μ^- and the e^-; this results in the following replacements in Eqs. (11) and (12)

$$P_1(E_e) \longrightarrow P_1(E_e)\left[1 + \frac{\alpha}{2\pi} f_1\left(\frac{E_e}{(E_e)_{max}}\right)\right] \equiv P_1'(E_e)$$

$$P_2(E_e) \longrightarrow P_2(E_e)\left[1 + \frac{\alpha}{2\pi} f_2\left(\frac{E_e}{(E_e)_{max}}\right)\right] \equiv P_2'(E_e) \tag{14}$$

where $f_1\left(\frac{E_e}{(E_e)_{max}}\right)$ and $f_2\left(\frac{E_e}{(E_e)_{max}}\right)$ are rather complicated functions of their argument [4] which however assume simple forms when $\left(\frac{E_e}{(E_e)_{max}}\right)$ is close to 1, viz. [5]

$$f_1\left(\frac{E_e}{(E_e)_{max}}\right) \cong f_2\left(\frac{E_e}{(E_e)_{max}}\right) \cong 2\left(\ln\left[\frac{m_\mu}{m_e}\right] - 1\right) \ln\left[1 - \frac{E_e}{(E_e)_{max}}\right] + \left(\frac{3}{2}\ln\left[\frac{m_\mu}{m_e}\right] - \frac{\pi^2}{3} + \frac{1}{2}\right) \tag{15}$$

Eqs. (10), (11), (13)-(15) yield the predicted electron energy spectrum, and Eqs. (10), (12), (13)-(15) yield the predicted energy dependence of the electron directional asymmetry. As is most easily seen on the basis of CPT invariance, all of the above relations hold for $\mu^+ \to e^+ + \nu_e + \nu_\mu$ provided that $\cos\theta_{e^-} = \hat{p}_{e^-} \cdot \hat{\mathcal{P}}_{\mu^-} = \hat{p}_{e^-} \cdot \hat{p}_{\mu^-}$ is replaced by $-\cos\theta_{e^+} = -\hat{p}_{e^+} \cdot \hat{\mathcal{P}}_{\mu^+} = \hat{p}_{e^+} \cdot \hat{p}_{\mu^+}$. Thus, since $\rho'' = -1$ and $\rho''' = \frac{3}{4}$, both the e^- and the e^+ with E_e close to $(E_e)_{max}$ tend to be emitted in a backward direction relative to the incoming μ^- or μ^+. Further the muon decay rate is given by

$$\Gamma(\mu^\pm \to e^\pm + \nu_e(\bar{\nu}_e) + \bar{\nu}_\mu(\nu_\mu)) = \int_{m_e}^{(E_e)_{max}} \int_0^\pi \left(P_1'(E_e) \mp |\vec{\mathcal{P}}_\mu| \cos\theta_e P_2'(E_e)\right) \times dE_e \, 2\pi \sin\theta_e \, d\theta_e$$

$$= \frac{(Gm_\mu^2)^2 m_\mu}{192\pi^3} \left(1 + f\left(\frac{m_e}{m_\mu}\right)\right)\left(1 + \frac{\alpha}{2\pi} \Delta_{rad}(\mu^+ \to e^+ + \nu_e + \nu_\mu)\right)$$

$$f\left(\frac{m_e}{m_\mu}\right) \cong -8\left(\frac{m_e}{m_\mu}\right)^2, \qquad \Delta_{rad}(\mu^+ \to e^+ + \nu_e + \bar{\nu}_\mu) = \frac{25}{4} - \pi^2 \tag{16}$$

which, on comparison with the most recent value of the muon lifetime [6,7]

$$\tau(\mu^+ \to e^+ + \nu_e + \bar{\nu}_\mu)|_{exp} = \left\{\Gamma(\mu^+ \to e^+ + \nu_e + \bar{\nu}_\mu)|_{exp}\right\}^{-1}$$
$$= (2.20026 \pm 0.00081) \times 10^{-6} \text{ sec} \tag{17}$$

yields

$$G = (1.43481 \pm 0.00026) \times 10^{-49} \text{ erg cm}^3$$
$$= 1.026 \times 10^{-5}(m_p c^2)\left(\frac{\hbar}{m_p c}\right)^3$$
$$\xrightarrow[h=1,c=1]{} 1.026 \times \frac{10^{-5}}{m_p^2} \tag{18}$$

We now proceed to consider the consequences with regard to muon decay of some a priori conceivable modifications of the $\mathcal{H}_{lept:\mu\leftrightarrow e}(x)$ of Eqs. (1) and (2). Experimental verification of any of these consequences would clearly be of great importance in the further development of our understanding of the weak interactions.

Thus, suppose that $H_{\text{lept}:\mu\leftrightarrow e}(x)$ retains the {leptonic weak current} × {leptonic weak current} form of Eq. (1) but with

$$\ell_\alpha(x;a,b) = i\left(\psi_a^\dagger(x)\gamma_4\gamma_\alpha\left\{\frac{(1+\gamma_5)+\eta(1-\gamma_5)}{\sqrt{1+|\eta|^2}}\right\}\psi_b(x)\right) \qquad (19)$$

where η is a parameter [8]. With the $H_{\text{lept}:\mu\leftrightarrow e}(x)$ of Eqs. (1) and (19), Eqs. (10)-(13) are modified only to the extent that now

$$\rho = \frac{3}{4}\left(1 - \frac{2|\eta|^2}{(1+|\eta|^2)^2}\right),\quad \rho' = 0,\quad \rho'' = -\left(\frac{1-|\eta|^2}{1+|\eta|^2}\right),\quad \rho''' = \frac{3}{4} \qquad (20)$$

It is to be emphasized that ρ' vanishes for any value of η, i.e. for any value of the $\frac{A}{V}$ ratio $\left(\frac{1-\eta}{1+\eta}\right)$ which enters in the $\ell_\alpha(x;a,b)$ of Eq. (19). In fact, a nonvanishing value of ρ' is obtained only if one includes within $H_{\text{lept}:\mu\leftrightarrow e}(x)$ terms of the form $\{$non- {V-A} leptonic weak current$\}$ × $\{$non-{V,A} leptonic weak current$\}$ such as

$$\left\{H_{\text{lept}:\mu\leftrightarrow e}(x)\right\}_{S,P} = -\frac{G}{\sqrt{2}}\left\{2\kappa\left(\psi_{\nu_\mu}^\dagger(x)(1+\gamma_5)\gamma_4\psi_{\mu^-}(x)\right)\right.$$
$$\left.\times\left(\psi_{e^-}^\dagger(x)\gamma_4(1+\gamma_5)\psi_{\nu_e}(x)\right) + \text{herm. conj.}\right\} \qquad (21)$$

where κ is another parameter [9]; with this $\left\{H_{\text{lept}:\mu\leftrightarrow e}(x)\right\}_{S,P}$ Eqs. (10)-(13) are again modified only to the extent that

$$\rho = \frac{3}{4},\quad \rho' = -\frac{\frac{1}{2}(\kappa+\kappa^*)}{1+|\kappa|^2},\quad \rho'' = -\left(\frac{1-|\kappa|^2}{1+|\kappa|^2}\right),\quad \rho''' = \frac{3}{4} \qquad (22)$$

so that any experimental upper limit on ρ' and/or on $|\rho''-(-1)|$ will serve to delimit the value of κ and hence the value of any contribution to $H_{\text{lept}:\mu\leftrightarrow e}(x)$ from non-{V,A} leptonic weak currents. We also note that the magnitudes of the longitudinal, perpendicular, and transverse polarizations of the $e^\mp$ are given by

(for $|(E_e)_{max} - E_e)/(E_e)_{max}| << 1$, $\hat{p}_e \cdot \hat{\mathcal{P}}_\mu = 0$, and any $|\vec{\mathcal{P}}_\mu|$)

$$\left\langle\frac{\vec{S}_{e^\mp}\cdot\hat{p}_{e^\mp}}{S_{e^\mp}}\right\rangle = \begin{cases} \mp 1 & : \ \rho,\ \rho',\ \rho'',\ \rho''' \text{ as in Eq. (13)} \\ \mp\left(\frac{1-|\eta|^2}{1+|\eta|^2}\right) & : \ \rho,\ \rho',\ \rho'',\ \rho''' \text{ as in Eq. (20)} \\ \mp\left(\frac{1-|\kappa|^2}{1+|\kappa|^2}\right) & : \ \rho,\ \rho',\ \rho'',\ \rho''' \text{ as in Eq. (22)} \end{cases}$$

$$\left\langle \frac{\vec{S}_{e^{\mp}} \cdot \hat{P}_{\mu^{\mp}}}{S_{e^{\mp}}} \right\rangle = \begin{cases} 0 & : \rho,\ \rho',\ \rho'',\ \rho''' \text{ as in Eq. (13)} \\ 0 & : \rho,\ \rho',\ \rho'',\ \rho''' \text{ as in Eq. (20)} \\ |\vec{P}_{\mu^{\mp}}| \left(\frac{\kappa+\kappa^*}{1+|\kappa|^2} \right) & : \rho,\ \rho',\ \rho'',\ \rho''' \text{ as in Eq. (22)} \end{cases}$$

$$\left\langle \frac{\vec{S}_{e^{\mp}} \cdot \hat{P}_{e^{\mp}} \times \hat{P}_{\mu^{\mp}}}{S_{e^{\mp}}} \right\rangle = \begin{cases} 0 & : \rho,\ \rho',\ \rho'',\ \rho''' \text{ as in Eq. (13)} \\ 0 & : \rho,\ \rho',\ \rho'',\ \rho''' \text{ as in Eq. (20)} \\ |\vec{P}_{\mu^{\mp}}| \left(\frac{-i(\kappa-\kappa^*)}{1+|\kappa|^2} \right) & : \rho,\ \rho',\ \rho'',\ \rho''' \text{ as in Eq. (22)} \end{cases} \qquad (23)$$

We emphasize that the observation of a nonvanishing $\left\langle \frac{\vec{S}_{e^{\mp}} \cdot \hat{P}_{e^{\mp}} \times \hat{P}_{\mu^{\mp}}}{S_{e^{\mp}}} \right\rangle$ for the $e^{\mp}$ from $\mu^{\mp}$ decay would be of the greatest importance since such a nonvanishing $\left\langle \frac{\vec{S}_{e^{\mp}} \cdot \hat{P}_{e^{\mp}} \times \hat{P}_{\mu^{\mp}}}{S_{e^{\mp}}} \right\rangle$ would imply a violation of T invariance and of CP invariance (e.g. by the $\left\{ H_{\text{lept}}(x)_{\mu\leftrightarrow e} \right\}_{S,P}$ of Eq. (21)).

We now record the available experimental values of ρ, ρ', ρ'', ρ''' and $\left\langle \frac{\vec{S}_{e^{\mp}} \cdot \hat{P}_{e^{\mp}}}{S_{e^{\mp}}} \right\rangle$ – there are no measurements as yet of $\left\langle \frac{\vec{S}_{e^{\mp}} \cdot \hat{P}_{\mu^{\mp}}}{S_{e^{\mp}}} \right\rangle$ and $\left\langle \frac{\vec{S}_{e^{\mp}} \cdot \hat{P}_{e^{\mp}} \times \hat{P}_{\mu^{\mp}}}{S_{e^{\mp}}} \right\rangle$. We have [10]

$$\rho|_{\exp} = 0.752 \pm 0.003, \qquad \rho'|_{\exp} = -(0.12 \pm 0.21)$$

$$\rho''|_{\exp} = -(0.972 \pm 0.013), \quad \rho'''|_{\exp} = 0.755 \pm 0.009$$

$$\left\langle \frac{\vec{S}_{e^{\mp}} \cdot \hat{P}_{e^{\mp}}}{S_{e^{\mp}}} \right\rangle \Bigg|_{\exp} = \mp \left(1.00 \begin{smallmatrix} +0.00 \\ -0.13 \end{smallmatrix} \right) \qquad (24)$$

in agreement with the corresponding theoretical values in Eqs. (13) and (23). Comparison of these experimental values with the theoretical values in Eqs. (20), (22), and (23) yields the following upper limits on the parameters η and κ (at momentum transfers $< m_\mu$)

$$|\eta| < 0.03, \qquad |\kappa| < 0.15 \qquad (25)$$

the limit on $|\kappa|$ implying the possibility of a value of $|\rho'|$ as large as 0.15 and of values of $\left|\left\langle\frac{\vec{S}_{e^{\mp}}\cdot\hat{P}_{\mu^{\mp}}}{S_{e^{\mp}}}\right\rangle\right|$ and of $\left|\left\langle\frac{\vec{S}_{e^{\mp}}\cdot\hat{P}_{e^{\mp}}\times\hat{P}_{\mu^{\mp}}}{S_{e^{\mp}}}\right\rangle\right|$ as large as 0.3. As a result, we suggest that serious consideration be given to the practicalities of a search for $e^{\mp}$ transverse (and perpendicular) polarization and that further efforts be made to obtain precise data on the low energy end of the $e^{\mp}$ spectrum [11].

4. MUON CAPTURE

We proceed to discuss the process of muon capture by a proton: $\mu^- + p \to \nu_\mu + n$. As a preliminary we treat the process of pion decay: $\pi^\pm \to \mu^\pm + \nu_\mu(\bar{\nu}_\mu)$ - all the $\mu^\pm$ so far carefully studied in muon decay and in muon capture have originated in pion decay. We have on the basis of Eqs. (1)-(5) the following expression for the pion decay rate:

$$\Gamma(\pi^- \to \mu^- + \bar{\nu}_\mu) = \Gamma(\pi^+ \to \mu^+ + \nu_\mu) = 2\pi\left|\left\langle \nu_\mu \mu^+ \right| \int \mathcal{H}_{\text{semilept}:\Delta S=0}(\vec{x},0)\,d\vec{x} \left| \pi^+ \right\rangle\right|^2$$

$$\times\left(\frac{4\pi(E_{\nu_\mu})^2}{(2\pi)^3(1+E_{\nu_\mu}/E_\mu)}\right)\left(1+\frac{\alpha}{2\pi}\Delta_{rad}(\pi^+\to\mu^++\nu_\mu)\right)$$

$$= 2\pi\left(\frac{G\cos\theta_c}{\sqrt{2}}\right)^2\left|\left(u^\dagger_{\nu_\mu}\gamma_4\gamma_\alpha(1+\gamma_5)u^*_{\mu^+}\right)\left\langle \text{vac}\right|A^{(-)}_\alpha(0)\left|\pi^+\right\rangle\right|^2$$

$$\times\left(\frac{4\pi(E_{\nu_\mu})^2}{(2\pi)^3(1+E_{\nu_\mu}/E_\mu)}\right)\left(1+\frac{\alpha}{2\pi}\Delta_{rad}(\pi^+\to\mu^++\nu_\mu)\right)$$

$$= \frac{1}{8\pi}\left(G\cos\theta_c m_\mu m_\pi\right)^2 m_\pi a_\pi^2\left(1-\frac{m_\mu^2}{m_\pi^2}\right)^2\left(1+\frac{\alpha}{2\pi}\Delta_{rad}(\pi^+\to\mu^++\nu_\mu)\right) \tag{26}$$

where $E_{\nu_\mu} = \dfrac{m_\pi^2-m_\mu^2}{2m_\pi}$, $E_\mu = \dfrac{m_\pi^2+m_\mu^2}{2m_\pi}$,

$\cos\theta_C = (0.990 \pm 0.002)\Big/\left[1+\frac{\alpha}{2\pi}\Delta_{rad}(^{26m}A\ell \to {}^{26}Mg + e^+ + \nu_e)\right]^{1/2}$, G as in Eq. (18), $a_\pi = a_\pi(q^2 \equiv [p_{vac}-p_\pi]^2 = -m_\pi^2)$ is the {pion$\leftrightarrow$vacuum} axial form factor introduced in Eq. (5), $\Delta_{rad}(\pi^+ \to \mu^+ + \nu_\mu)$ and $\Delta_{rad}(^{26m}Al \to {}^{26}Mg + e^+ + \nu_e)$ are the electromagnetic radiative correction factors for the indicated weak processes, and the numerical value of $\cos\theta_C[1 + \frac{\alpha}{2\pi}\Delta_{rad}(^{26m}Al \to {}^{26}Mg + e^+ + \nu_e)]^{\frac{1}{2}}$ is obtained from $\Gamma(^{26m}Al \to {}^{26}Mg + e^+ + \nu_e)$, the G of Eq. (18), and CVC. Comparison of $\Gamma(\pi^+ \to \mu^+ + \nu_\mu)|_{Eq.(26)}$ with $\Gamma(\pi^+ \to \mu^+ + \nu_\mu)|_{exp} = \{(2.6024 \pm 0.0024) \times 10^{-8}\ \mathrm{sec}\}^{-1}$ yields, as quoted in Eq. (5)

$$|a_\pi| = (0.930 \pm 0.001)\left[\frac{1 + \frac{\alpha}{2\pi}\Delta_{rad}(^{26m}Al \to {}^{26}Mg + e^+ + \nu_e)}{1 + \frac{\alpha}{2\pi}\Delta_{rad}(\pi^+ \to \mu^+ + \nu_\mu)}\right]^{\frac{1}{2}}$$

$$= 0.94 \pm 0.01$$

if we estimate somewhat model-dependently [12] the ratio of the electromagnetic radiative correction factors. We also emphasize the importance of the process $\pi^+ \to \mu^+ + \nu_\mu + \gamma$. Here, while it is known [10,11a] that $\left[\frac{\Gamma(\pi^+ \to \mu^+ + \nu_\mu + \gamma)}{\Gamma(\pi^+ \to \mu^+ + \nu_\mu)}\right]\Bigg|_{exp} = (1.24 \pm 0.25) \times 10^{-4}$, no study has ever been made of the photon-muon coincidence spectrum with an extrapolatory determination of the photon and muon energy end-points (corresponding to the pion rest-frame configuration $\vec{p}_{\nu_\mu} \to 0$, $\vec{p}_\gamma \to -\vec{p}_\mu$). A study of this type would be of considerable interest since the sum of the photon and muon endpoint energies

$$(E_\gamma)_{max} + (E_\mu)_{max} = \frac{(m_\pi - m_{\nu_\mu})^2 - m_\mu^2}{2(m_\pi - m_{\nu_\mu})} + \frac{(m_\pi - m_{\nu_\mu})^2 + m_\mu^2}{2(m_\pi - m_{\nu_\mu})}$$

$$\cong m_\pi\left(1 - \frac{m_{\nu_\mu}}{m_\pi}\right) \qquad (27)$$

depends linearly on the small quantity m_{ν_μ}/m_π (we recall that $(m_{\nu_\mu}/m_\pi)|_{exp} < 5 \times 10^{-3}$). Such a <u>linear</u> dependence is characteristic of a 3-particle final state and is to be contrasted with the <u>quadratic</u> dependence of the muon energy on m_{ν_μ}/m_π in the 2-particle final state of $\pi^+ \to \mu^+ + \nu_\mu$ (where $E_\mu = \left(\frac{m_\pi^2 + m_\mu^2}{2m_\pi}\right)\left(1 - \left(\frac{m_\pi^2}{m_\pi^2 + m_\mu^2}\right)\left(\frac{m_{\nu_\mu}}{m_\pi}\right)^2\right)$).

Having done with these preliminaries, we commence on the treatment of the capture of a muon by a proton. From Eqs. (1)-(5), the capture rate is given by

$$\Gamma(\mu^- + p \to \nu_\mu + n) = 2\pi \int \left| \left\langle n\,\nu_\mu \left| \int \mathcal{H}_{\text{semilept}:\Delta S=0}(\vec{x},0)\, d\vec{x} \right| p\,\mu^- \right\rangle \right|^2 \left(\frac{4\pi E_{\nu_\mu}^2}{(2\pi)^3 (1 + E_{\nu_\mu}/E_n)} \right) \frac{d\hat{p}_{\nu_\mu}}{4\pi}$$

$$\times \left(1 + \frac{\alpha}{2\pi} \Delta_{\text{rad}}(\mu^- + p \to \nu_\mu + n)\right)$$

$$= 2\pi \left(\frac{G\cos\theta_C}{\sqrt{2}} \right)^2 \frac{[\alpha m_\mu (1 + m_\mu/m_p)^{-1}]^3}{\pi} \int \left| \left(u_{\nu_\mu}^\dagger \gamma_4 \gamma_\alpha (1+\gamma_5) u_{\mu^-}\right) \left\langle n \left| V_\alpha^{(-)}(0) + A_\alpha^{(-)}(0) \right| p \right\rangle \right|^2$$

$$\times \left(\frac{4\pi E_{\nu_\mu}^2}{(2\pi)^3 (1 + E_{\nu_\mu}/E_n)} \right) \frac{d\hat{p}_{\nu_\mu}}{4\pi} \left(1 + \frac{\alpha}{2\pi} \Delta_{\text{rad}}(\mu^- + p \to \nu_\mu + n)\right)$$

$$= 2\pi \left(\frac{G\cos\theta_C}{\sqrt{2}} \right)^2 \frac{[\alpha m_\mu (1 + m_\mu/m_p)^{-1}]^3}{\pi} \int \left| \left(u_{\nu_\mu}^\dagger \gamma_4 \gamma_\alpha (1+\gamma_5) u_{\mu^-}\right) \right.$$

$$\times \left(u_n^\dagger \gamma_4 \left[\gamma_\alpha F_V(q^2) - \frac{\sigma_{\alpha\beta} q_\beta}{2m_p} F_M(q^2) + \frac{i(m_p+m_n)}{m_\pi^2} q_\alpha F_S(q^2) + \gamma_\alpha \gamma_5 F_A(q^2) - \frac{\sigma_{\alpha\beta} q_\beta \gamma_5}{2m_p} F_E(q^2) \right. \right.$$

$$\left. \left. + \frac{i(m_p+m_n)}{m_\pi^2} q_\alpha \gamma_5 F_P(q^2) \right] u_p \right) \right|^2 \left(\frac{4\pi E_{\nu_\mu}^2}{(2\pi)^3 (1 + E_{\nu_\mu}/E_n)} \right) \frac{d\hat{p}_{\nu_\mu}}{4\pi} \left(1 + \frac{\alpha}{2\pi} \Delta_{\text{rad}}(\mu^- + p \to \nu_\mu + n)\right) \tag{28}$$

where $\sigma_{\alpha\beta} \equiv (2i)^{-1}(\gamma_\alpha\gamma_\beta - \gamma_\beta\gamma_\alpha)$, $E_{\nu_\mu} = (m_\mu + m_p) - E_n = (m_\mu + m_p) - (m_n + m_\mu^2/2m_n + \ldots)$, $q^2 = (p_n - p_p)^2 = (p_\mu - p_{\nu_\mu})^2 = -m_\mu^2 + 2m_\mu E_{\nu_\mu} = m_\mu^2(1 - m_\mu/m_n + \ldots)$, with G as in Eq. (18) and $\cos\theta_C$ as in Eq. (26) et seq. Here, $[\alpha m_\mu(1 + m_\mu(m_p)^{-1}]^3/\pi$ is the muon-proton coincidence probability density appropriate to the muon's 1s orbit about the proton in the $[\mu^- p]$ atom (with neglect of the small and oppositely directed effects of the vacuum-polarization and finite-size corrections to the proton's Coulomb potential), and $F_V(q^2)$, $F_M(q^2)$, $F_S(q^2)$, $F_A(q^2)$, $F_E(q^2)$ and $F_P(q^2)$ are the polar, weak-magnetism, scalar, axial, weak-electricity or pseudotensor, and pseudoscalar proton$\leftrightarrow$neutron form factors – the values of these form factors as functions of q^2 summarize in a Lorentz-invariant way the dependence of $\langle n|V_\alpha^{(-)}(0) + A_\alpha^{(-)}(0)|p\rangle$ on the (squared) momentum transfer q^2.

To evaluate the form factors we proceed as follows. We assume (I) that $|n\rangle$ and $|p\rangle$ are isopure $[|n\rangle = -\exp(i\pi I^{(2)})|p\rangle$, $|p\rangle = \exp(i\pi I^{(2)})|n\rangle]$ and (II) that the hadronic isovector electromagnetic current = $I_\alpha^{(3)}(x)$. Then assumption (I), together with the CVC-implied condition $GV_\alpha^{(\mp)}(x)G = V_\alpha^{(\mp)}(x)$ [Eq. (4)] or, equivalently, the CVC condition $\frac{\partial V_\alpha^{(\mp)}(x)}{\partial x_\alpha} = 0$ [Eq. (4)], yields (for all q^2)

$$F_S(q^2) = 0 \tag{29}$$

while assumptions (I) and (II), together with the CVC-specification $V_\alpha^{(\mp)}(x) = I_\alpha^{(\mp)}(x)$ [Eq. (4)], give

$$F_V(q^2) = F^p_{Dirac}(q^2) - F^n_{Dirac}(q^2) = \left(1+\frac{q^2}{4m^2}\right)^{-1}\left[\left(G^p_{ec}(q^2) - G^n_{ec}(q^2)\right) + \frac{q^2}{4m^2}\left(G^p_{mm}(q^2) - G^n_{mm}(q^2)\right)\right]$$

$$F_M(q^2) = F^p_{Pauli}(q^2) - F^n_{Pauli}(q^2) = \left(1+\frac{q^2}{4m^2}\right)^{-1}\left[\left(G^p_{mm}(q^2) - G^n_{mm}(q^2)\right) - \left(G^p_{ec}(q^2) - G^n_{ec}(q^2)\right)\right]$$

$$F_V(0) = \left(G^p_{ec}(0) - G^n_{ec}(0)\right) = 1$$

$$F_M(0) = \left(G^p_{mm}(0) - G^n_{mm}(0)\right) - \left(G^p_{ec}(0) - G^n_{ec}(0)\right) = \mu_p - \mu_n - 1$$

$$\left.\frac{dG^n_{ec}(q^2)}{d(q^2)}\right|_{q^2=0} = -\frac{1}{4m_n^2}G^n_{mm}(0) = -\frac{1}{4m_n^2}\mu_n \tag{30}$$

with $\mu_p = 2.793$ and $\mu_n = -1.913$ the magnetic moments of the proton and the neutron, and $m \equiv \frac{1}{2}(m_p + m_n)$. Here the explicit dependence on q^2 of the electron charge (ec) and the magnetic moment (mm) form factors of the proton and the neutron is obtained, respectively, from the analysis of experimental data on the elastic scattering of electrons in hydrogen and on the quasielastic and elastic scattering of electrons in deuterium. This explicit dependence yields, for q^2 appropriate to $\mu^- + p \to \nu_\mu + n$, i.e., for $q^2 = m_\mu^2(1 - m_\mu/m_n + \ldots) = 0.88m_\mu^2$

$$\begin{aligned} F_V(0.88m_\mu^2) &= F_V(0)\left(1 - q^2 \times 2.31(\mathrm{GeV})^{-2} + \ldots\right)_{q^2=0.88m_\mu^2} \\ &= F_V(0) \times 0.977 \end{aligned} \tag{31}$$

$$F_M(0.88m_\mu^2) = F_M(0)(1 - q^2 \times 2.96(\text{GeV})^{-2} + \ldots)_{q^2=0.88m_\mu^2}$$

$$= F_M(0) \times 0.971 \qquad (32)$$

Further, the above assumption (I), together with the PCAC-implied condition $GA_\alpha^{(\mp)}(x)G = -A_\alpha^{(\mp)}(x)$ [Eq. (5)], yields (for all q^2)

$$F_E(q^2) = 0 \qquad (33)$$

while the PCAC-condition $\frac{\partial A_\alpha^{(\mp)}(x)}{\partial x_\alpha} = a_\pi m_\pi^3 \Phi_{\pi^\pm}(x)$ [Eq. (5)] gives (again for all q^2)

$$F_A(q^2) + \frac{q^2}{m_\pi^2} F_P(q^2) = \frac{a_\pi g_{\pi pn}(q^2)\left(\frac{m_\pi}{m_p+m_n}\right)}{1 + q^2/m_\pi^2} \equiv \frac{a_\pi f_{\pi pn}(q^2)}{1 + q^2/m_\pi^2}$$

$$F_A(0) = a_\pi f_{\pi pn}(0)$$

$$\lim_{q^2\to -m_\pi^2}\left\{\left(1 + \frac{q^2}{m_\pi^2}\right)\left(F_A(q^2) + \frac{q^2}{m_\pi^2} F_P(q^2)\right)\right\} = -\lim_{q^2\to -m}\left\{\left(1 + \frac{q^2}{m_\pi^2}\right)F_P(q^2)\right\}$$

$$= a_\pi f_{\pi pn}(-m_\pi^2) \qquad (34)$$

or, rearranging Eq. (34)

$$F_P(q^2) = -\left(\frac{F_A(q^2)}{1+q^2/m_\pi^2}\right)\left[1 + \frac{m_\pi^2}{q^2}\left(1 - \frac{(f_{\pi pn}(q^2)/f_{\pi pn}(0))}{F_A(q^2)/F_A(0)}\right)\right] \qquad (35)$$

The second of Eqs. (34) constitutes the Goldberger-Treiman relation while the third of Eqs. (34) in the approximate form $F_P(q^2) \cong -a_\pi f_{\pi pn}(-m_\pi^2)/(1 + q^2/m_\pi^2)$ was first given by Wolfenstein.

From the recent relatively precise determinations of $\Gamma(n \to p + e^- + \bar{\nu}_e)\big|_{exp}$, the G of Eq. (18), and CVC $[F_V(0) = 1]$ one obtains

$$\left\{(\cos\theta_C)^2(1 + 3(F_A(0))^2\ (1 + \frac{\alpha}{2\pi}\Delta_{rad}(n \to p + e^- + \bar{\nu}_e))\right\}$$

$$= (0.990 \pm 0.002)^2(1 + 3(F_A(0)^2)\left[\frac{1 + \frac{\alpha}{2\pi}\Delta_{rad}(n\to p+e^-+\bar{\nu}_e)}{1 + \frac{\alpha}{2\pi}\Delta_{rad}(^{26m}Al\to {}^{26}Mg+e^++\nu_e)}\right]$$

$$= 5.52 \pm 0.12 \qquad (36)$$

whence, with $\left[\dfrac{1 + \frac{\alpha}{2\pi}\,\Delta_{rad}(n \to p + e^- + \bar{\nu}_e)}{1 + \frac{\alpha}{2\pi}\,\Delta_{rad}(^{26m}A\ell \to {}^{26}Mg + e^+ + \nu_e)}\right]$ estimated as [12] 1.004 ± 0.001, one gets

$$|F_A(0)| = F_A(0) = 1.24 \pm 0.01, \quad f_{\pi pn}(0) = F_A(0)/a_\pi = 1.32 \pm 0.02 \tag{37}$$

This experimental value of $|F_A(0)|$ is close to that deduced theoretically from CAC, PCAC and the experimental values (for the various available pion energies) of $\{\sigma(\pi^- + p \to all) - \sigma(\pi^+ + p \to all)\}$ (Adler-Weissberger sum rule). The sign of $F_A(0)$ and so of $a_\pi = F_A(0)/f_{\pi pn}(0)$ $[f_{\pi pn}(0) = |f_{\pi pn}(0)|]$ is fixed by the good agreement between the $\Gamma(\mu^- + p \to \nu_\mu + n{:}S_{\mu^- p} = 0)|_{th}$ of Eq. (48) and the $\Gamma(\mu^- + p \to \nu_\mu + n{:}S_{\mu^- p} = 0)|_{exp}$ of Eq. (51) [see below]. Further, from an analysis of experimental data on the "quasi-elastic scattering" of (muon-family) neutrinos in deuterium - $\nu_\mu + d \to \mu^- + p + [p]_{spectator}$ - one has [13]

$$\frac{F_A(q^2)}{F_A(0)} = \frac{1}{[1 + q^2/((0.95 \pm 0.12)\mathrm{GeV})^2]^2}$$

$$= 1 - q^2 \times 2.22(\mathrm{GeV})^{-2} + \ldots \tag{38}$$

while, from the dispersion-theoretic analysis of experimental data on pion-proton elastic scattering, one can extract

$$f_{\pi pn}(-m_\pi^2) = \sqrt{2}\,\sqrt{4\pi}\,[0.0790 \pm 0.0010]^{\frac{1}{2}} = 1.41 \pm 0.01 \tag{39a}$$

so that, using Eqs. (5), (37) and (39a)

$$f_{\pi pn}(q^2) = f_{\pi pn}(-m_\pi^2) + \left(f_{\pi pn}(0) - f_{\pi pn}(-m_\pi^2)\right)\left(1 + \frac{q^2}{m_\pi^2}\right)$$

$$= 1.41 - 0.09\left(1 + \frac{q^2}{m_\pi^2}\right)$$

$$f_{\pi pn}(0.88m_\mu^2) = 1.28 = f_{\pi pn}(0) \times 0.97 \tag{39b}$$

Thus from Eqs. (38), (35) and (39b)

$$F_A(0.88m_\mu^2) = F_A(0) \times 0.978 \tag{40}$$

$$F_P(0.88m_\mu^2) = -F_A(0) \times 0.666$$

$$\left(\frac{m_\mu(m_p + m_n)}{m_\pi^2}\right) F_P(0.88m_\mu^2) = -F_A(0) \times 6.78 \qquad (41)$$

Finally, suppose one adjoins to our hadronic $\Delta S=0$ axial-vector, weak current, which satisfies $GA_\alpha^{(\mp)}(x)G^{-1} = -A_\alpha^{(\mp)}(x)$ [Eq. (5)], another hadronic $\Delta S=0$ axial-vector weak current $A^{(\mp)'}(x)$, which like $A_\alpha^{(\mp)}(x)$ is an isovector but which satisfies $GA_\alpha^{(\mp)'}(x)G^{-1} = A_\alpha^{(\mp)'}(x)$ [$A_\alpha^{(\mp)}(x)$ and $A_\alpha^{(\mp)'}(x)$ are called, respectively, "first-class" and "second-class" currents]. Then, one can prove that insofar as $|n>$ and $|p>$ are isopure $A_\alpha^{(\mp)'}(x)$ does not contribute to $F_A(q^2)$ and $F_P(q^2)$, but, in general, does produce a nonvanishing $F_E(q^2)$, which moreover is real (for $q^2 > -m_{S-C}^2$) if $A_\alpha^{(\mp)'}(x)$ is normal under time reversal [i.e. $TA_\alpha^{(\mp)'}(x)T^{-1} = -\{A_\alpha^{(\mp)'}(\vec{x},-t)\}^\dagger$]; on the other hand, if $A_\alpha^{(\mp)'}(x)$ is abnormal under time reversal [i.e. $TA_\alpha^{(\mp)'}(x)T^{-1} = \{A_\alpha^{(\mp)'}(\vec{x},-t)\}^\dagger$], $F_E(q^2)$ is imaginary. Thus, since $F_V(0.88m_\mu^2)$, $F_M(0.88m_\mu^2)$, $F_S(0.88m_\mu^2)$, $F_A(0.88m_\mu^2)$, and $F_P(0.88m_\mu^2)$ are specified in Eqs. (30)-(32), (29), (37), (40), and (41), comparison of $\Gamma(\mu^- + p \to \nu_\mu + n)$ as calculated from Eq. (28) with $\Gamma(\mu^- + p \to \nu_\mu + n)|_{exp}$ should set a limit on $F_E(0.88m_\mu^2)$ and so indicate the extent to which the "second-class" current $A_\alpha^{(\mp)'}(x)$ can possibly be present (see below).

We now proceed with the calculation of

$$M(\mu^- + p \to \nu_\mu + n) \equiv \left(u_{\nu_\mu}^\dagger \gamma_4\gamma_\alpha(1+\gamma_5)u_{\mu^-}\right)\left(u_n^\dagger\gamma_4\left[\gamma_\alpha F_V(q^2) - \frac{\sigma_{\alpha\beta}q_\beta}{2m_p} F_M(q^2)\right.\right.$$
$$+ \gamma_\alpha\gamma_5 F_A(q^2) - \frac{\sigma_{\alpha\beta}q_\beta\gamma_5}{2m_p} F_E(q^2)$$
$$\left.\left. + \frac{i(m_p+m_n)}{m_\pi^2} q_\alpha\gamma_5 F_P(q^2)\right] u_p\right) \qquad (42)$$

for $q^2 = 0.88m_\mu^2$ where we have allowed for the existence of a non-vanishing $F_E(q^2)$. Then, noting that $\vec{p}_n = -\vec{p}_{\nu_\mu} = -E_{\nu_\mu}\hat{p}_{\nu_\mu}$, $\vec{p}_p = -\vec{p}_\mu \approx -(\alpha m_\mu)\hat{p}_\mu$, and that apart from terms in $|M(\mu^- + p \to \nu_\mu + n)|^2$ of order α^2 one can take $\vec{p}_\mu = 0$, we obtain

$$M(\mu^- + p \to \nu_\mu + n) = v_{\nu_\mu}^\dagger v_n^\dagger H_{eff}(\mu^- + p \to \nu_\mu + n)v_{\mu^-} v_p$$

$$H_{eff}(\mu^- + p \to \nu_\mu + n) \equiv \frac{1}{\sqrt{2}}\left(\frac{E_n+m_n}{2E_n}\right)^{1/2} (1 - \vec{\sigma}_L \cdot \hat{p}_{\nu_\mu})$$
$$\times (G_V + G_A\vec{\sigma}_L \cdot \vec{\sigma}_N + G_P\vec{\sigma}_N \cdot \hat{p}_{\nu_\mu}) \qquad (43)$$

where $\frac{1}{2}\vec{\sigma}_L$ and $\frac{1}{2}\vec{\sigma}_N$ are (two-by-two matrix) spin operators which work, respectively, on the lepton (two-component) spinors v_{ν_μ}, v_{μ^-} and on the nucleon (two-component) spinors v_n, v_p. The effective lepton-nucleon coupling constants G_V, G_A and G_P are given by

$$G_V = F_V(0.88m_\mu^2)\left(1+\frac{E_{\nu_\mu}}{E_n+m_n}\right) - F_M(0.88m_\mu^2)\left(\frac{m_\mu}{2m_p}\right)\left(\frac{E_{\nu_\mu}}{E_n+m_n}\right) = 1.02 \pm 0.002$$

$$G_A = -F_A(0.88m_\mu^2) - \left[F_V(0.88m_\mu^2) + F_M(0.88m_\mu^2)\left(\frac{E_n+m_n}{2m_p}\right)\right]\left(\frac{E_{\nu_\mu}}{E_n+m_n}\right)$$

$$-F_E(0.88m_\mu^2)\left[\frac{m_\mu-E_{\nu_\mu}}{2m_p} - \frac{E_{\nu_\mu}^2}{2m_p(E_n+m_n)}\right] + F_M(0.88m_\mu^2)\left[\left(\frac{m_\mu-E_{\nu_\mu}}{2m_p}\right)\left(\frac{E_{\nu_\mu}}{E_n+m_n}\right)\right]$$

$$= -(1.46 \pm 0.02) - \left[F_E(0.88m_\mu^2)\right] \times (4\times 10^{-4})$$

$$G_P = \left[\left(\frac{m_\mu(m_p+m_n)}{m_\pi^2}\right)F_P(0.88m_\mu^2) + F_A(0.88m_\mu^2) + F_E(0.88m_\mu^2)\left(\frac{E_n+m_n}{2m_p}\right)\right.$$

$$\left.-F_V(0.88m_\mu^2) - F_M(0.88m_\mu^2)\left(\frac{E_n+m_n}{2m_p}\right)\right]\left(\frac{E_{\nu_\mu}}{E_n+m_n}\right)$$

$$+F_E(0.88m_\mu^2)\left(\frac{E_{\nu_\mu}^2}{2m_p(E_n+m_n)}\right) + F_M(0.88m_\mu^2)\left[\left(\frac{m_\mu-E_{\nu_\mu}}{2m_p}\right)\left(\frac{E_{\nu_\mu}}{E_n+m_n}\right)\right]$$

$$= -(0.63 \pm 0.01) + \left[F_E(0.88m_\mu^2)\right] \times 0.056 \tag{44}$$

where we have used Eqs. (30)-(32), (37), (40), (41) and (28) for specification of the numerical values. The expression for $M(\mu^- + p \to \nu_\mu + n)$ or $H_{eff}(\mu^- + p \to \nu_\mu + n)$ in Eqs. (43), (44) exhibits clearly the (enormous) "hyperfine effect" in the capture rate of a 1s-orbit muon by a proton. Thus, with $\{v_{\mu^-}v_p\}_{S_{\mu^-p}=1}$ and $\{v_{\mu^-}v_p\}_{S_{\mu^-p}=0}$ the spin-triplet and spin-singlet hyperfine substates of the 1s-orbit ground state of the $[\mu^-p]$ atom, we have, substituting Eqs. (42)-(44) into Eq. (38)

$$\Gamma(\mu^- + p \to \nu_\mu + n : S_{\mu\text{-}p}) = \left\{ \left[\frac{1 + \alpha/2\pi\, \Delta_{rad}(\mu^- + p \to \nu + n)}{1 + \alpha/2\pi\, \Delta_{rad}(^{26m}Al \to {}^{26}Mg + e^+ + \nu_e)} \right] (0.990 \pm 0.002)^2 \right.$$

$$\times \left(\frac{(Gm_\mu^2)^2 m_\mu}{2\pi} \right) \left(\frac{\alpha^3}{\pi} \left[1 + \frac{m_\mu}{m_p} \right]^{-3} \right) \left(\frac{(E_{\nu_\mu}/m_\mu)^2}{(1 + E_{\nu_\mu}/E_n)} \right) \left(\frac{E_n + m_n}{2E_n} \right) \Bigg\}$$

$$\times \left\{ \left[G_V^2 + 3G_A^2 - G_A(G_P + G_P^*) + |G_P|^2 \right] \right.$$

$$\left. + \langle \vec{\sigma}_L \cdot \vec{\sigma}_N \rangle_{S_{\mu^- p}} \left[2G_V G_A - 2G_A^2 - \frac{1}{3} G_V (G_P + G_P^*) + \frac{2}{3} G_A (G_P + G_P^*) \right] \right\}$$

$$\langle \vec{\sigma}_L \cdot \vec{\sigma}_N \rangle_{S_{\mu^- p}} \equiv \left(\left\{ \nu_{\mu^-} \cdot \nu_p \right\}^\dagger_{S_{\mu^- p}} \vec{\sigma}_L \cdot \vec{\sigma}_N \left\{ \nu_{\mu^-} \cdot \nu_p \right\}_{S_{\mu^- p}} \right)$$

$$= 2 \left[S_{\mu^- p}(S_{\mu^- p} + 1) - \frac{3}{2} \right] = \begin{cases} 1, & S_{\mu^- p} = 1 \\ -3, & S_{\mu^- p} = 0 \end{cases} \qquad (45)$$

whence, assuming henceforth that $F_E(0.88\ m_\mu^2)$ and so G_P are real, and using the numerical values of G_V, G_A, G_P in Eq. (44),

$$\frac{\Gamma(\mu^- + p \to \nu_\mu + n : S_{\mu^- p} = 0)}{\Gamma(\mu^- + p \to \nu_\mu + n : S_{\mu^- p} = 1)} = \frac{(G_V - 3G_A + G_P)^2}{(G_V + G_A)^2 - 2/3(G_V + G_A)\, G_P + G_P^2}$$

$$= (56.0 \pm 2.8) \frac{(1 + 0.012 F_E(0.88 m_\mu^2))^2}{(1 - 0.13 F_E(0.88 m_\mu^2) + 0.0077 [F_E(0.88 m_\mu^2)]^2)}$$

$$= \begin{cases} 41.3 \pm 2.1, & F_E(0.88 m_\mu^2) = -2 \\ 56.0 \pm 2.8, & F_E(0.88 m_\mu^2 = 0 \\ 76.5 \pm 3.8, & F_E(0.88 m_\mu^2) = 2 \end{cases} \qquad (46)$$

Finally, again estimating somewhat model dependently [12] the ratio of the electromagnetic radiative correction factors as

$$\left[\frac{1 + \alpha/2\pi\, \Delta_{rad}(\mu^- + p \to \nu_\mu + n)}{1 + \alpha/2\pi\, \Delta_{rad}(^{26m}Al \to {}^{26}Mg + e^+ + \nu_e)} \right] = 0.98 \pm 0.01 \qquad (47)$$

and substituting Eqs. (47), (44), and (18) into Eq. (45), we obtain

$$\Gamma(\mu^- + p \to \nu_\mu + n : S_{\mu^- p} = 0) = \left[(29.2 \pm 0.3)\,\mathrm{sec}^{-1}\right]\left(G_V - 3G_A + G_P\right)^2$$

$$= \left[(664 \pm 20)\,\mathrm{sec}^{-1}\right]\left(1 + 0.012 F_E(0.88 m_\mu^2)\right)^2$$

$$= \begin{cases} (631 \pm 20)\,\mathrm{sec}^{-1}, & F_E(0.88 m_\mu^2) = -2 \\ (664 \pm 20)\,\mathrm{sec}^{-1}, & F_E(0.88 m_\mu^2) = 0 \\ (697 \pm 20)\,\mathrm{sec}^{-1}, & F_E(0.88 m_\mu^2) = 2 \end{cases} \quad (48)$$

$$\Gamma(\mu^- + p \to \nu_\mu + n : S_{\mu^- p} = 1) = \left[(29.2 \pm 0.3)\,\mathrm{sec}^{-1}\right]\left((G_V + G_A)^2 - 2(G_V + G_A)\frac{1}{3} G_P + G_P^2\right)$$

$$= \left[(11.9 \pm 0.7)\,\mathrm{sec}^{-1}\right]\left(1 - 0.13 F_E(0.88 m_\mu^2) + 0.0077\left[F_E(0.88 m_\mu^2)\right]^2\right)$$

$$= \begin{cases} (15.3 \pm 0.9)\,\mathrm{sec}^{-1}, & F_E(0.88 m_\mu^2) = -2 \\ (11.9 \pm 0.7)\,\mathrm{sec}^{-1}, & F_E(0.88 m_\mu^2) = 0 \\ (9.1 \pm 0.6)\,\mathrm{sec}^{-1}, & F_E(0.88 m_\mu^2) = 2 \end{cases} \quad (49)$$

$$\Gamma(\mu^- + p \to \nu_\mu + n : \mathrm{stat.}) = \frac{3}{4}\Gamma(\mu^- + p \to \nu_\mu + n : S_{\mu^- p} = 1) + \frac{1}{4}\Gamma(\mu^- + p \to \nu_\mu + n : S_{\mu^- p} = 0)$$

$$= \left[(29.2 \pm 0.3)\,\mathrm{sec}^{-1}\right]\left(G_V^2 + 3G_A^2 - 2G_A G_P + G_P^2\right)$$

$$= \begin{cases} (169 \pm 5)\,\mathrm{sec}^{-1}, & F_E(0.88 m_\mu^2) = -2 \\ (175 \pm 5)\,\mathrm{sec}^{-1}, & F_E(0.88 m_\mu^2) = 0 \\ (181 \pm 5)\,\mathrm{sec}^{-1}, & F_E(0.88 m_\mu^2) = 2 \end{cases} \quad (50)$$

Experimentally, one has [14]

$$\Gamma(\mu^- + p \to \nu_\mu + n : S_{\mu - p} = 0)\big|_{\mathrm{exp}} = (651 \pm 57)\,\mathrm{sec}^{-1} \quad (51)$$

so that values of $F_E(0.88\ m_\mu^2) > 2$ and < -4 are excluded; alternatively, taking $F_E(0.88\ m_\mu^2) = 0$ but considering $[f_{\pi pn}(0.88m_\mu^2)/f_{\pi pn}(0)]$ in Eq. (35) for $F_P(0.88m_\mu^2)$ as a free parameter, values of $[f_{\pi pn}(0.88m_\mu^2/f_{\pi pn}(0)] > 1.1$ and < 0.7 are excluded. This corresponds to the exclusion of values of $\left\{-\left(\frac{m_\mu(m_p+m_n)}{m_\mu^2}\right)\frac{F_P(0.88m_\mu^2)}{F_A(0)}\right\}$ that are > 5 and < 11 [Eqs. (44), (35), (40), (37), and (30)-(32)]. Equivalently, one can say that the CVC, PCAC (and hence no "second-class" current) theoretical prediction of $\Gamma(\mu^- + p \to \nu_\mu + n\colon S_{\mu^-p} = 0)$, i.e., Eq. (48) for $F_E(0.88m_\mu^2) = 0$ (which uses $\left(\frac{f_{\pi pn}(0.88m_\mu^2)}{f_{\pi pn}(0)}\right) = 0.97$, $\left\{-\left(\frac{m_\mu(m_p+m_n)}{m_\mu^2}\right)\frac{F_P(0.88m_\mu^2)}{F_A(0)}\right\} = 6.78$ [Eqs. (39b) and 41)]), is in good agreement, within the overall uncertainty, with the $\Gamma(\mu^- + p \to \nu_\mu + n\colon S_{\mu^-p} = 0)|_{exp}$ of Eq. (51). The value of $\Gamma(\mu^- + p \to \nu_\mu + n\colon S_{\mu^-p} = 0)|_{exp}$ given in Eq. (51) is found when muons stop in isotopically (and chemically) pure medium-density gaseous hydrogen where

$$\frac{\Gamma(\{[e^-p] + \text{spin-triplet}[\mu^-p]\}\to\{e^- + p + \text{spin-singlet}[\mu^-p]\})}{\Gamma(\mu^- \to e^- + \bar{\nu}_e + \nu_\mu)} >> 1$$

Future experiments in which muons stop in isotopically (and chemically) pure low-density gaseous hydrogen where

$$\frac{\Gamma(\{[e^-p] + \text{spin-triplet}[\mu^-p]\} \to \{e^- + p + \text{spin-singlet}[\mu^-p]\})}{\Gamma(\mu^- \to e^- + \bar{\nu}_e + \nu_\mu)} << 1$$

will permit determination of the $\Gamma(\mu^- + p \to \nu_\mu + n\colon\text{stat.})$ of Eq. (50).

To continue, we calculate the capture rate of a muon by a proton in a $[p\mu^-p]$ molecule-ion; such $[p\mu^-p]$ molecule-ions are eventually formed when muons stop in isotopically (and chemically) pure high-density gaseous hydrogen or isotopically (and chemically) pure liquid hydrogen where

$$\frac{\Gamma(\{[e^-p] + \text{spin-singlet}[\mu^-p]\}\to\{e^- + [p\mu^-p]\}}{\Gamma(\mu^- \to e^- + \bar{\nu}_e + \nu_\mu)} >> 1$$

A priori, at the moment of muon capture, the $[p\mu^-p]$ molecule-ion can have $S_{p\mu-p} = 1/2$ and $S_{pp} = 0$: para-$[p\mu^-p]$ ($L_{pp} = 0,2,\ldots$) or $S_{p\mu-p} = 1/2$ or $3/2$ and $S_{pp} = 1$: ortho-$[p\mu^-p]$ ($L_{pp} = 1,3,\ldots$) so that

$$\langle\vec{\sigma}_L\cdot\vec{\sigma}_N\rangle_{S_{p\mu p},S_{pp}} = 2\left\langle\frac{1}{2}\vec{\sigma}_\mu\cdot\left(\frac{1}{2}\vec{\sigma}_{p_1}+\frac{1}{2}\vec{\sigma}_{p_2}\right)\right\rangle_{S_{p\mu p},S_{pp}}$$

$$= \left\{S_{p\mu p}(S_{p\mu p}+1)-\frac{3}{4}-S_{pp}(S_{pp}+1)\right\}$$

$$\langle\vec{\sigma}_L\cdot\vec{\sigma}_N\rangle_{S_{p\mu p}=\frac{3}{2},S_{pp}=1} = 1 = \langle\vec{\sigma}_L\cdot\vec{\sigma}_N\rangle_{S_{\mu p}=1}$$

$$\langle\vec{\sigma}_L\cdot\vec{\sigma}_N\rangle_{S_{p\mu p}=\frac{1}{2},S_{pp}=1} = -2 = \frac{1}{4}\langle\vec{\sigma}_L\cdot\vec{\sigma}_N\rangle_{S_{\mu p}=1}+\frac{3}{4}\langle\vec{\sigma}_L\cdot\vec{\sigma}_N\rangle_{S_{\mu p}=0}$$

$$\langle\vec{\sigma}_L\cdot\vec{\sigma}_N\rangle_{S_{p\mu p}=\frac{1}{2},S_{pp}=0} = 0 = \frac{3}{4}\langle\vec{\sigma}_L\cdot\vec{\sigma}_N\rangle_{S_{\mu p}=1}+\frac{1}{4}\langle\vec{\sigma}_L\cdot\vec{\sigma}_N\rangle_{S_{\mu p}=0} \quad (52)$$

Thus, using Eq. (45) with $\langle\vec{\sigma}_L\cdot\vec{\sigma}_N\rangle_{S_{\mu^-p}}$ in that equation replaced by $\langle\vec{\sigma}_L\cdot\vec{\sigma}_N\rangle_{S_{p\mu^-p},S_{pp}}$ and with a muon-proton coincidence probability appropriate to $[p\mu^-p]$ we have

$$\Gamma(\mu^-+p\rightarrow\nu_\mu+n:S_{p\mu p}=\tfrac{3}{2},S_{pp}=1) = 2\gamma_{ortho.}\Gamma(\mu^-+p\rightarrow\nu_\mu+n:S_{\mu p}=1) \quad (53)$$

$$\Gamma(\mu^-+p\rightarrow\nu_\mu+n:S_{p\mu p}=\tfrac{1}{2},S_{pp}=1) = 2\gamma_{ortho.}\left\{\frac{1}{4}\Gamma(\mu^-+p\rightarrow\nu_\mu+n:S_{\mu p}=1)+\frac{3}{4}\Gamma(\mu^-+p\rightarrow\nu_\mu+n:S_{\mu p}=0)\right\} \quad (54)$$

$$\Gamma(\mu^-+p\rightarrow\nu_\mu+n:S_{p\mu p}=\tfrac{1}{2},S_{pp}=0) = 2\gamma_{para.}\left\{\frac{3}{4}\Gamma(\mu^-+p\rightarrow\nu_\mu+n:S_{\mu p}=1)+\frac{1}{4}\Gamma(\mu^-+p\rightarrow\nu_\mu+n:S_{\mu p}=0)\right\} \quad (55)$$

where

$$2\gamma_{ortho,para} = \frac{2\left[\left(\Psi_{p\mu p:ortho,para}(|\vec{r}_\mu-\vec{r}_{p_1}|=0,|\vec{r}_\mu-\vec{r}_{p_2}|=|\vec{r}_{p_1}-\vec{r}_{p_2}|)\right)^2\right]_{av.over|\vec{r}_{p_1}-\vec{r}_{p_2}|}}{\left|\psi_{\mu^-p}(|\vec{r}_\mu-\vec{r}_p|=0)\right|^2}$$

$$= \begin{cases} 1.01\pm0.01\ , & \text{ortho} \\ 1.15\pm0.01\ , & \text{para} \end{cases} \quad (56)$$

is obtained on the basis of a rather elaborate variational-type calculation of $\psi_{p\mu^-p:ortho,para}(|r_\mu - r_{p1}|,|\vec{r}_\mu - \vec{r}_{p2}|)$.

Now, as already mentioned, $S_{\mu^-p} = 0$ just before the attachment of the $[\mu^-p]$ to the other p and, since the spin-flipping magnetic forces within the $[p\mu^-p]$ are relatively weak, $S_{p\mu^-p} = 1/2$ to a good approximation not only at the moment of $[p\mu^-p]$ formation but also at the subsequent moment of muon capture by one of the $[p\mu^-p]$ protons. Further, the process

$$\{[e^-p] + \text{spin-singlet}[\mu^-p]\} \rightarrow e^- + \text{ortho} - [p\mu^-p]\}$$

proceeds via an "electric-dipole" collisional transition, while the process

$$\{[e^-p] + \text{spin-singlet}[\mu^-p] \rightarrow e^- + \text{para} - [p\mu^-p]\}$$

proceeds via a far less probable (factor $\approx 10^4$) "electric-monopole" collisional transition; in addition [15]

$$\frac{\Gamma(\text{ortho} - [p\mu^-p] \rightarrow \text{para} - [p\mu^-p])}{\Gamma(\mu^- \rightarrow e^- + \bar{\nu}_e + \nu_\mu)} \ll 1$$

Thus, to a high degree of approximation, the rate of muon capture in the $[p\mu^-p]$ molecule is appropriate to the $S_{p\mu^-p} = {}^1/2$, $S_{pp} = 1$ configuration, and is given, using Eqs. (56), (54), (48), and (49) by

$$\Gamma(\mu^- + p \rightarrow \nu_\mu + n:\ S_{p\mu^-p} = \tfrac{1}{2},\ S_{pp} = 1)$$

$$= \begin{cases} (483 \pm 20)\text{sec}^{-1} & F_E(0.88m_\mu^2) = -2 \\ (506 \pm 20)\text{sec}^{-1} & F_E(0.88m_\mu^2) = 0 \\ (530 \pm 20)\text{sec}^{-1} & F_E(0.88m_\mu^2) = 2 \end{cases} \qquad (57a)$$

These theoretical values are to be compared with the experimental values [16]

$$\Gamma(\mu^- + p \rightarrow \nu_\mu + n:\ S_{p\mu^-p} = \tfrac{1}{2},\ S_{pp} = 1)\Big|_{exp}$$

$$= \begin{cases} (515 \pm 85)\text{sec}^{-1} \\ (464 \pm 42)\text{sec}^{-1} \end{cases} \qquad (57b)$$

where in the experiment one detects only "late-arriving" neutrons, i.e. essentially only those neutrons whose parent muons and protons have had time to form $[p\mu^-p]$. Comparison of Eq. (57a) with Eq. (57b) again exhibits good agreement, within the overall existing uncertainty, between the CVC, PCAC theoretical prediction and experiment. However, since this overall existing uncertainty is some 10%, one would very much welcome increased precision in the experimental $[p\mu^- p]$ capture rate quoted in Eq. (57b) and particularly in the experimental $[\mu^- p]$ capture rate quoted in Eq. (51). Increased precision in the experimental value of $\Gamma(n \to p + e^- + \bar{\nu}_e)$ would also be helpful since such increased precision would decrease the uncertainty in $F_A(0)$ [Eq. (37)], this last uncertainty being the chief contributor to the total uncertainty in $\Gamma(\mu^- + p \to \nu_\mu + n: S_{\mu^- p} = 0)|_{th}$ [Eq. (48)] and to the total uncertainty in $\Gamma(\mu^- + p \to \nu_\mu + n: S_{p\mu^- p} = \frac{1}{2}, S_{pp} = 1)|_{th}$ [Eq. (57a)].

We proceed to report on the process of radiative muon capture by a proton: $\mu + p \to \nu_\mu + n + \gamma$ [17]. The process $\mu^- + p \to \nu_\mu + n + \gamma$ is of particular interest since it permits determination of $\Gamma(\mu^- + p \to \nu_\mu + n: S_{\mu^- p} = 1)$ [Eq. (49)] in contradistinction to the previously treated process $\mu^- + p \to \nu_\mu + n$, which as we have seen, yields at best (by a suitable combination of muon-capture rate measurements in gaseous hydrogen at various low and medium densities) the ratio

$$\left[\frac{\Gamma(\mu^- + p \to \nu_\mu + n\text{: stat.})}{\Gamma(\mu^- + p \to \nu_\mu + n: S_{\mu^- p} = 0)}\right] = \frac{1}{4}\left[\frac{3\Gamma(\mu^- + p \to \nu_\mu + n: S_{\mu^- p} = 1)}{\Gamma(\mu^- + p \to \nu_\mu + n: S_{\mu^- p} = 0)} + 1\right]$$

The numerical disparity between $\Gamma(\mu^- + p \to \nu_\mu + n: S_{\mu^- p} = 1)$ and $\Gamma(\mu^- + p \to \nu_\mu + n: S_{\mu^- p} = 0)$ [Eq. (46) or Eqs. (49) and (48)] then implies that the ratio of $[\Gamma(\mu^- + p \to \nu_\mu + n\text{:stat.})/\Gamma(\mu^- + p \to \nu_\mu + n: S_{\mu^- p} = 0)]$ must be measured with high precision. to obtain a moderate precision in the ratio $[\Gamma(\mu^- + p \to \nu_\mu + n: S_{\mu^- p} = 1)/\Gamma(\mu^- + p \to \nu_\mu + n: S_{\mu^- p} = 0)]$. On the other hand, if the muons stop in medium-density gaseous hydrogen, the process $\mu^- + p \to \nu_\mu + n + \gamma$ will, as discussed above, originate from spin-singlet $[\mu^- p]$, and, since ν_μ and n are uncharged and $S_\gamma = 1$, will proceed predominantly via the muon internal bremsstrahlung (IB) mechanism

$$\begin{aligned} \text{spin-singlet}[\mu^- p]_{1s\text{-orbit}} &\to \text{spin-triplet}[\mu^- p]_{Ns\text{-orbit}} + \gamma \\ &\to \nu_\mu + n + \gamma\,, \\ &\qquad N = 1,2,3,\ldots \end{aligned} \tag{58}$$

This mechanism yields

$$\begin{aligned} \left[\frac{\Gamma(\mu^- + p \to \nu_\mu + n + \gamma : S_{\mu^- p} = 0)}{\Gamma(\mu^- + p \to \nu_\mu + n: S_{\mu^- p} = 0)}\right] &\approx \frac{\alpha}{\pi}\int_0^{m_\mu}\left[\frac{(m_\mu - E_\gamma)^2}{m_\mu^2}\,\frac{E_\gamma^2}{m_\mu^2}\,\frac{dE_\gamma}{E_\gamma}\right]\left[\frac{\Gamma(\mu^- + p \to \nu_\mu + n: S_{\mu^- p} = 1)}{\Gamma(\mu^- + p \to \nu_\mu + n: S_{\mu^- p} = 0)}\right] \\ &= \frac{\alpha}{12\pi}\left[\frac{\Gamma(\mu^- + p \to \nu_\mu + n: S_{\mu^- p} = 1)}{\Gamma(\mu^- + p \to \nu_\mu + n: S_{\mu^- p} = 0)}\right] \\ &= 4 \times 10^{-6} \end{aligned} \tag{59}$$

where

$$P(E_\gamma)dE_\gamma \equiv \left\{12\,\frac{(m_\mu - E_\gamma)^2}{m_\mu^2}\,\frac{E_\gamma^2}{m_\mu^2}\,\frac{dE_\gamma}{E_\gamma}\right\}$$

is the normalized photon energy spectrum and where the exact calculation [17], which includes the proton and neutron IB and the structure-dependent non-IB contributions, replaces 4×10^{-6} by 8×10^{-6}. Of course, the anticipated smallness of $[\Gamma(\mu^- + p \to \nu_\mu + n + \gamma{:}S_{\mu^- p} = 0)/\Gamma(\mu^- \to e^- + \overline{\nu}_e + \nu_\mu)] \approx 10^{-8}$ [Eqs. (59), (48), and (17)] will render difficult any precise measurement of $[\Gamma(\mu^- + p \to \nu_\mu + n + \gamma{:}S_{\mu^- p} = 0)/\Gamma(\mu^- + p \to \nu_\mu + n{:}S_{\mu^- p} = 0)]$ and so, any subsequent extraction of $[\Gamma(\mu^- + p \to \nu_\mu + n{:}S_{\mu^- p} = 1)/\ (\mu^- + p \to \nu_\mu + n{:}S_{\mu^- p} = 0)]$. However, the prospect of even a moderately accurate determination of the $F_E(0.88m_\mu^2)$-sensitive quantity $[\Gamma(\mu^- + p \to \nu_\mu + n{:}S_{\mu^- p} = 1)/\Gamma(\mu^- + p \to \nu_\mu + n{:}S_{\mu^- p} = 0)]$ [Eq. (46)] should encourage a serious attempt to observe radiative muon capture in medium-density gaseous hydrogen. In fact, such an attempt seems at the moment to be more attractive than ever because of the very recent experimental evidence in favor of the existence of "second-class"-current effects in nuclear β-decay [18].

FOOTNOTES AND REFERENCES

1. The actual π^+ lifetime is 2.6024×10^{-8} sec. while the actual $(\pi^+ \to e^+ + \nu_e)/(\pi^+ \to \mu^+ + \nu_\mu)$ branching ratio is 1.24×10^{-4}.

2. All the other "superfluons" are strange hadrons except for the newly discovered ψ and ψ' mesons which are non-strange.

3. A. Rousset, AIP Conference Proceedings No. 22, Neutrinos, Philadelphia, 1974, AIP, New York (1974).

4. See, e.g., R.E. Marshak, Riazuddin, and C.P. Ryan, "Theory of Weak Interactions in Particle Physics", Wiley, New York (1969).

5. When $\left(\frac{E}{(E_e)_{max}}\right)$ is so close to 1 that $\frac{\alpha}{2\pi} f_{1,2}\left(\frac{E_e}{(E_e)_{max}}\right)$ is no longer << 1 we must exponentiate $1 + \frac{\alpha}{2\pi} f_{1,2}\left(\frac{E_e}{(E_e)_{max}}\right)$ which, using Eq. (15), then becomes equal to

$$e^{\frac{\alpha}{2\pi}\left(\frac{3}{2}\ln\left[\frac{m_\mu}{m_e}\right] - \frac{\pi^2}{3} + \frac{1}{2}\right)}\left(1 - \frac{E_e}{(E_e)_{max}}\right)^{\frac{\alpha}{2\pi}\,2\left(\ln\frac{m_\mu}{m_e} - 1\right)} = 0.994\left(1 - \left(\frac{E_e}{(E_e)_{max}}\right)\right)^{0.010}$$

6. R.W. Williams and D.W. Williams, Phys. Rev. D6, 737 (1972).

7. We note that, strictly speaking, both the $\Gamma(\mu^+ \to e^+ + \nu_e + \overline{\nu}_\mu)|_{exp}$ of Eq. (17) and the $\Gamma(\mu^+ \to e^+ + \overline{\nu}_e + \nu_\mu)|_{th}$ of Eq. (16) correspond to the determination of the total decay rate of the μ^+ and therefore include the (relatively small) contribution of the rate of $\mu^+ \to e^+ + \nu_e + \overline{\nu}_\mu + \gamma$ with any possible $\vec{p}_\gamma$.

8. If $\eta \neq \eta^*$ the $H_{\text{lept}:\mu\leftrightarrow e}^{(x)}$ of Eqs. (1) and (19) violates T invariance and CP invariance (but still conserves CPT invariance). Also, η need not be treated as a constant parameter but may be considered as a function of $(-\frac{\partial}{\partial x}\cdot\frac{\partial}{\partial x})$ so that one has

$$\frac{\eta\left(-\frac{\partial}{\partial x}\cdot\frac{\partial}{\partial x}\right)}{\sqrt{1+\left|\eta -\left(\frac{\partial}{\partial x}\cdot\frac{\partial}{\partial x}\right)\right|^2}}\; i\left(\psi_a^\dagger(x)\gamma_4\gamma_\alpha(1-\gamma_5)\psi_b(x)\right)$$

9. The remarks made about η in the preceding footnote can also be made about κ.

10. See Particle Data Group, Rev. Mod. Phys. 45, S1 (1973).

11. As may be seen from Eq. (11), the term involving ρ' is most important when $m_e/E_e \to 1$.

11a. Also, strictly speaking, both the quoted $\Gamma(\pi^+ \to \mu^+ + \nu_\mu)|_{exp}$ and the $\Gamma(\pi^+ \to \mu^+ + \nu_\mu|_{th}$ of Eq. (26) include the (relatively small) contribution of the rate of $\pi^+ \to \mu^+ + \nu + \gamma$.

12. See, e.g., G. Källen, Springer Tracts Mod. Phys. 46, 67 (1968).

13. W.A. Mann et al., Phys. Rev. Letters 31, 844 (1973).

14. A. Alberigi Quaranta et al., Phys. Rev. 177, 2115 (1969).

15. S. Weinberg, Phys. Rev. Letters 4, 585 (1960).

16. E.J. Bleser et al., Phys. Rev. Letters 8, 288 (1962); J.E. Rothberg et al., Phys. Rev. 132, 2664 (1963).

17. G.I. Opat, Phys. Rev. 134, B428 (1964).

18. K. Sugimoto, I. Tanihata, and J. Goring, Phys. Rev. Letters 34, 1533 (1975) and F. Calaprice, Princeton University, Seminar at the University of Pennsylvania. These investigators report data on the electron-energy dependence of the

decay asymmetry of β-rays emitted from oriented nuclei (B^{12}, N^{12}, and N^{19}, respectively) which appears to require introduction of a "second-class"-current contribution.

NUCLEAR MUON-CAPTURE

J.P. Deutsch

Universite Catholique de Louvain

Louvain-La-Neuve, Belgium

NUCLEAR MUON-CAPTURE

We do not attempt in these two lectures to develop a coherent formalism of nuclear muon-capture, nor do we drive for completeness in the description of what the study of these processes may teach us about the basic interaction or the nuclear structure. The formalism may be consulted in ref. 1 (impulse approximation) or ref. 2 (elementary particle approach). Information on the basic interaction was reviewed in ref. 3 and 4, the nuclear structure aspects being covered in refs. 3, 4 and 5. Some of the topics will be also taken up in the lectures of Professor Primakoff.

We shall attempt to follow up in these lectures some of the new research lines which may open up in nuclear muon-capture at the new research facilities.

More specifically, we shall address ourselves to the following subjects:

1) Muon-capture as a tool of nuclear spectroscopy: an example.
2) How similar are isobar analogue states: a test of the "elementary particle" approach.
3) How to disentangle the vector- and axial-strength in muon capture?
4) Are nuclei built up from nucleons only?: the nuclear renormalization of the coupling constants.
   4.1. The vector-current and CVC.
   4.2. The axial-current (g_A, g_P).

5) What about second-class currents?
6) Electron-muon universality: helicity of the muon-neutrino.
7) T-violation and muon capture.

Though some of these questions are coupled by experiment, they unfortunately cannot be covered completely in two lectures: we hope however to open discussions to be pursued informally after the talks.

1) Muon-Capture as a Tool of Nuclear Spectroscopy: An Example

The experiment we consider was performed and published by the Louvain group many years ago [6]. We recall, in this written version of our talk, the main features of it only. Partial muon-capture rates were measured in ^{11}B on transitions leading to ^{11}Be bound states by an activation method and high-resolution Ge(Li) spectroscopy. The de-excitation gamma-ray was found to be Doppler broadened and its appearance decreased more slowly in time than that of the muon-decay electrons. This could be accounted for by a $F_+ \rightarrow F_-$ conversion process [7] from the upper to the lower hyperfine levels of the $^{11}B - \mu$ system and a preferential capture from the lower hyperfine level to the gamma-unstable ^{11}B final state. These observations allowed us to infer spin and parity assignments to the ^{11}Be states reached in the muon-capture process and to verify a theoretical conjecture [8] on the inversion of shell-model states in this region. This example illustrates the services, though unconventional, muon-capture may render to nuclear spectroscopy specifically when the (Z-1, N+1)-neighbour of a stable (Z,N)-nucleus can not be reached by nuclear reactions on stable targets. In this connection, it is amusing to recall a later study of the ^{11}Be states [9] performed using the $^{10}Be(d,p)^{11}Be$ reaction on the very unconventional target of ^{10}B extracted from neutron-irradiated carbon.

The discussion of the experiment gave us the opportunity to elaborate, in the lectures, on the observation of Doppler-broadened gamma line-shapes [10] and hyperfine-conversion [7], observational techniques we shall have to rely upon in the discussion of further topics.

2) How Similar are Isobar Analogue States? : Test of the Elementary Particle Approach

It is standard procedure [2] to compute the axial form factor $F_A(q^2 \simeq m_\mu^2)$ of an "allowed" $\Delta J = 1$ muon-capture process between the $|i\rangle$ ground state of a (Z,N)-nucleus and the $|f\rangle$-state of its (Z-1,N+1)-neighbour using the ft-value of the $|f\rangle \rightarrow |i\rangle$ beta-decay

[$\rightarrow F_A(0)$] and the q^2 dependence of the backward electron scattering $|i\rangle \rightarrow |f'\rangle$ [$\rightarrow F_A(q^2)/F_A(0)$], where $|f'\rangle$ is the isobar analogue state of $|f\rangle$ in the (Z,N)-nucleus. A somewhat similar approach fits the parameters of a model function of $|i\rangle$ and $|f'\rangle$ considering electromagnetic data on and between these states and computes, on this basis, the $|f\rangle \rightarrow |i\rangle$ beta-decay rate and the $|i\rangle \rightarrow |f\rangle$ muon-capture axial form factor [11]. [This method requires, of course, the assumption of some potential e.g. harmonic oscillator and a truncation of the basis in which the wave-functions are expanded: a limitation which may render the conclusions dubious, in principle at least.]

Both approaches imply that: a) the exchange corrections to the axial form factors of the three processes (electron scattering, beta-decay, muon-capture) are identical and b) the wave-functions of the isobaric analogue states $|f\rangle$ and $|f'\rangle$ are really similar. Assumption b) was questioned, in particular, by the authors of ref. [12] who noted that because of differences in binding, the radial behaviour of $|f\rangle$ and $|f'\rangle$ should be different, in principle. This difference may have a serious (though as yet uncomputed) influence on the correctness of muon-capture computations like the ones outlined above, especially if the binding differences are important.

Such may be the case for the ^{6}Li $\rightarrow$ ^{6}He(g.s.) transition, ^{6}He(g.s.) being bound by 964 keV but its analog, ^{6}Li(3.56 MeV), by only 136 keV [12]. Our doubts on the validity of the above-mentioned procedure [11] are perhaps strengthened by the discrepancy between experiment [13] and a computation performed on the same basis [14] on the ^{6}Li(γ,π^+)^{6}He(g.s.) reaction which involves in good approximation the same form factor $F_A(q^2)$ as muon-capture. It was observed that to relax some of the restrictive conditions in the computations of ref.[13](harmonic oscillator radial wave-function) seems to remove part of the discrepancy [15].

The question may be solved by a precision measurement of the partial muon-capture rate, known to a moderate accuracy only [16], which is not subject to some of the difficulties inherent to photoproduction. Fortunately, due to the super-allowed nature of the transition (ft = 802 s), the rate is rather insensitive to the somewhat dubious induced pseudoscalar coupling and so constitutes a good measurement of $F_A(q^2 \simeq m_\mu^2)$ [17].

A final warning: the rate computations assume statistical population of the F = 1/2 and F = 3/2 hyperfine levels at the moment of the capture. This seems to hold with sufficient accuracy in this case [18,19].

3) How (and why) to Distentangle the Vector and Axial Strength in Muon-Capture.

The vector form factor $F_V(q^2)$ is highly hindered in the "allowed" muon-capture of light nuclei because of isospin selection rules. This will not be the case any more in $\Pi_i\Pi_f = -$, $\Delta J = 1$ transitions like the muon-capture in ^{16}O to the $^{16}N(1^-$; 396 keV) level, in which the vector matrix element M_V $[\int r]$ is of the same order as the axial one M_A $[\int \sigma \times r]$.

The ratio M_V/M_A is rather model dependent: $\sim$.9 to $\sim$1.2 in the Migdal theory [20] (for spherical ^{16}O, respectively 2p-2h and 4p-4h admixtures included), $\sim$.45 with more conventional configuration-mixed wave-functions [21]. The accuracy claimed for the Migdal prediction is only about 20%; the difference between the M_V/M_A ratios predicted by the two approaches is, however, so huge, that even a crude measurement of it would allow us to test the virtues of the Migdal approach to this field.

How can we measure this ratio? It is easy to realize the $^{16}N(1^-$, 396 keV) $\rightarrow$ $^{16}N(0^-$, 120 keV) gamma-ray will not be emitted isotropically referred to the neutrino-momentum as quantization-axis and that the anisotropy will be a measure of $r = M_V/M_A$ [22]. With the wave-functions of ref.[23] the correlation-coefficient A_2 in the correlation function, $W(\theta_{\nu\gamma}) \simeq 1 + A_2P_2(\cos\theta_{\nu\gamma})$, turns out to be about 0.25 with $\partial A_2/\partial r \simeq 1$ [24] ! With modern high-resolution detectors, the Doppler broadened line would have about three times the "natural" width, which should allow a measurement of A_2 with the method discussed in ref. [10] well within the accuracy required to distinguish the Migdal prediction from the conventional shell-model approach.

A final warning: the life-time of the $^{16}N(1^-$, 396 keV) level is $\sim$ 40 ps so care should be taken to avoid slowing of the recoiling nucleus before the gamma emission.

4) Are Nuclei Built Up from Nucleons Only? : The Nuclear Renormalization of Coupling-Constants

Are nuclei built up from nucleons only? In other words, can we compute the muon-capture observables in nuclei from the weak coupling constants of the free nucleon and some wave-functions of the nucleus in terms of single nucleon coordinates? Can we forget about the mesonic degrees of freedom?

What we measure is a product of coupling constants and matrix elements, so the answer to our question would require a reliable knowledge of these nuclear matrix elements, i.e. of the nuclear wave-functions. (The mesonic degrees of freedom are, of course, hidden behind the potential which gives rise to the nuclear wave-functions). Though the task to compute nuclear matrix elements with reliability is by no means achieved, much attention was given recently to the question whether the coupling constants keep their free-nucleon value in nuclei, i.e. to the nuclear renormalization of coupling constant due to mesonic effects [25,26,27]. Beyond the nuclear renormalization problem one should remember also the well-known efforts to explain the renormalization of the nucleon weak coupling constants g_A and g_P compared to the "bare" value g_V and 0 they would have in absence of strong interactions, i.e. the mesonic effects [28,29].

4.1. The vector-current and CVC. Let us recall the CVC-hypothesis which requires [30] that the vector and "weak-magnetism" form factors $F_V(q^2)$ and $F_M(q^2)$ in the weak hadronic current($\langle f|V_\lambda|i\rangle \equiv \langle f|F_V(q^2)\gamma_\lambda + F_M(q^2)\sigma_{\mu\lambda}q_\mu\ |i\rangle$) be identical (up to the bare coupling-constant ratio G/e and Clebsch-Gordon coefficients) to the isobar-analog iso-vector electro-magnetic form factors. This means that mesonic contributions to the two currents should be identical.

A model independent test of this hypothesis on $F_V(q^2)$ would be achieved by the muon-capture experiment discussed in section 3 comparing the matrix element M_V obtained to the corresponding one obtained in electron scattering [31]. A closer control on $F_M(q^2)$ could be obtained from a more precise knowledge of the partial capture rate $^{12}C(0^+) \rightarrow {}^{12}B(g.s.\ ;\ 1^+)$ [32].

4.2. The axial current. Let us neglect first in the axial-vector weak hadronic current($\langle f|\ A_\lambda\ |\ i\rangle \equiv \langle f|F_A(q^2)\gamma_\lambda\gamma_5 - iF_P(q^2)q_\lambda\gamma_5 - F_T(q^2)\sigma_{\mu\lambda}q_\mu\gamma_5|i\rangle$)the induced tensor term $F_T(q^2)$. This term is of the "second class" [33]: it transforms under $G \equiv Ge^{i\pi T_2}$ oppositely to $F_A(q^2)$ and so should be absent as the strong interaction is invariant under G-transformation.

Let us turn first to the mesonic effects on $F_A(q^2)$. A renormalization of the coupling constant $g_A(q^2)$ of about 0.8 is expected in nuclear matter for $0 < q^2 < m_\mu^2$ [$\tilde{g}_A/g_{A,free}$: 0.75 - 0.8] [25,26,27], but it is not clear what survives of this renormalization in real nuclei [27,34]. In beta-decay of light nuclei there is a slight indication of some downward renormalization of $g_A(0)$ [35]. In muon-capture, $g_A(q^2 \simeq m_\mu^2)M_A$ could be measured by partial rate determinations with rather good accuracy if the $F_P(q^2)$ contribution is small, such as in the strong M1 transitions of the 1p shell [36] and heavier nuclei [5]. The test requires, of course, a reliable knowledge of the matrix element M_A. The suggestion to extract it from the form factor of the analog gamma transition [37] works only if the exchange effects on these form factors can be neglected.

We shall comment in next section on the value of $C_P(q^2) = m_\mu F_P(q^2)/F_A(0) = 10.0 \pm 1.6$ found in muon capture by the nucleon [38] and compare it to $C_P(q^2)_{th.} = 7.4$ [29]. The renormalization of C_P in nuclei was considered by many authors; one of the most recent ones obtains in nuclear matter $r \equiv \tilde{C}_P/C_{P,free} \simeq 0.44$ [27] around $q^2 \simeq m^2$.

The methods to measure $\tilde{C}_P$ in nuclear muon-capture were discussed in ref. 3 and we shall expand this point in the oral version of the lectures. Let us recall however the list of observables we discussed: a) ratio of partial muon-capture rates (^{16}O), b) radiative muon capture (^{40}Ca), c) ratio of partial muon-capture rates from hyperfine levels of a $J \neq 0$ nucleus (^{11}B) and d) neutrino-gamma directional correlation (^{28}Si). We should add to this list two more recent approaches: e) the measurement of the average ^{12}B(g.s.) polarization in the capture of polarized muons by ^{12}C [39] and f) the comparison of the partial muon-capture rate $^{16}O \rightarrow {}^{16}N(0^-)$ and its inverse beta-decay rate measured recently [40].

The results for C_P and r are:

a) $C_P = 10.8 \pm 1.0$ [41], $r = 1.1 \pm 0.2$;
b) $13 \leq C_P \leq 18$ [42], $1.1 \leq r \leq 2.1$†;
c) $C_P \leq 12$ [43], $r \leq 1.4$;
d) $C_P = 5 \pm 8$ [44], $.3 \leq r \leq .9$ or $-7 \leq C_P \leq 1$, $-8 \leq r \leq .1$ according to the author of ref. [45] who re-analyzed the data of ref. [44];
e) $C_P = 10^{+5}_{-4}$, [39]* $.5 \leq r \leq 1.8$;
f) $13 \leq C_P \leq 20$ [40], $1.1 \leq r \leq 2.4$ values changing to $8 \leq C_P \leq 12$, $.7 \leq r \leq 1.4$ if one considers Coulomb corrections in the induced terms of the $^{16}N(0^-) \rightarrow {}^{16}O$ beta-decay [46].

Considering these values and leaving aside the dubious result of approach d) it is fair to say that there is no compelling evidence as yet for $r \neq 1$, i.e. for a renormalization of C_P in nuclei. In particular, there is no indication, in the light nuclei studied, of the strong quenching predicted for nuclear matter. More accurate data are needed, in heavy nuclei if possible.

---

†Note that this measurement of $C_P(q^2)$ for $q^2 \neq m_\mu^2$ may not be directly compared to the other ones in view of the possible fast variation of $C_P(q^2)$ with q^2 [27].

*$C_P = 12 \pm 5 \rightarrow C_P = 10^{+5}_{-4}$: private communication from Professor Grenacs.

5) What About Second-Class Currents?

The first indication for the presence of a second-class form factor $F_T(q^2)$ in weak interactions was inferred from mirror beta-decay rate asymmetries. It was recognized, however, that these indications were dubious because nuclear-structure induced asymmetries were hard to assess. For a discussion on these points, see ref. [47] and the references cited therein. It was recognized also that beta-decay correlation coefficients and contributions to muon capture were free of these uncertain asymmetries [48]. (I do not enter here in the discussion of ref. [48] of how off-shell effects affect the expression of the axial current and how the observables we mentioned, related to the time part of the current, are free of some difficulties which handicap the space part of the current). Unfortunately, the eventual second-class coupling, $C_T(q^2) \equiv 2M_P F_T(q^2)/F_A(0)$, is always linked to $C_P(q^2)$ and we can measure in muon capture their sum $C_P(q^2) + C_T(q^2)$ only.

In hydrogen $(C_P(q^2) + C_T(q^2))_{exp} = 10.0 \pm 1.6$ [38] and $(C_P(q^2))_{th} \simeq 7.4$ [29], indicating $(C_T(q^2 \simeq m^2))_{nucleon} = 2.6 \pm 1.6 \pm$?. The question mark reflecting the inaccuracy of $(C_P(q^2))_{th}$ (cf. the last reference [29]). So, there may be a slight indication of second-class currents in nucleon muon-capture*.

In nuclei the situation is more confused. Preliminary results on beta-decay correlation experiments indicate

$C_T(q^2 \simeq 0) \simeq -10$ [49] (A=19),

$C_T(q^2 \simeq 0) = -3.5 \pm 1.3$ [50] (A=12),

$C_T(q^2 \simeq 0) = 3.8 \pm 0.9$ [51] (A=12), and

$-3 \leq C_T(q^2 \simeq 0) \leq 0$ [52] (A=20).

C_T differences from one nucleus to the other can only be accommodated if we assume contributions from the ω-meson [47]. This contribution may induce variations depending on the values taken by a two-nucleon matrix element which is hard to compute [47].

These possible variations (if confirmed in beta-decay) will render, of course, precise measurement of $C_P(q^2 \simeq m^2)$ extremely difficult.

Let us now briefly come to the last two topics pertaining to somewhat more "fundamental" aspects of the basic muon-capture interaction.

---

*Refer, however, on this point to the last lecture of Professor Primakoff.

6) Electron-Muon Universality; Helicity of the Muon-Neutrino

The neutrino being left-handed in beta-decay the hypothesis of muon-electron universality requires the same left-handedness for the muon-neutrino emitted in ($\pi \to \mu$)-decay, μ-decay and muon-capture.

Some early measurements on π-decay (to be described in the lectures) [53] are in favour of the muon-neutrino left-handedness with 20% - 30% accuracy. In muon decay, neutrino left-handedness is assured by $\rho = 0.75$ (favoured anti-parallel emission of the electron and the two neutrinos) and the complete longitudinal polarization of the decay positron, as it can be seen easily by an angular-momentum/momentum scheme of the reaction.

In muon-capture, only the measurement of the neutron longitudinal polarization in $\mu^- + p \to \nu + n$ is free of nuclear structure uncertainty. In nuclear muon-capture one has to measure (a) the momentum of the neutrino ("easily" inferred from the recoil direction) and (b) its spin direction. This latter quantity is linked to the spin of the captured muon via the multipoles of the transition operators ($L \neq 0$ for the induced terms) whose amplitudes have to be known.

If one insists in choosing only between the helicity ± 1 (two-component neutrino), then the measurement of the average polarization of the ^{12}B recoil we mentioned [39] allows one to choose the helicity - 1, in agreement with the hypothesis of universality.

7) T-Violation and Muon-Capture

It was noted by Professor Primakoff some time ago [54] that our most accurate check of time-reversal invariance in weak interactions was performed within a supermultiplet ($\Delta I = 0$) and that one could concoct a T-violating weak interaction which would show up only in $\Delta I \neq 0$ transitions. No corresponding tests were performed up to now; let us see what can be learned on this point from muon-capture.

One advantage of muon-capture compared to beta-decay is the absence of electromagnetic final-state interactions which may simulate small T-odd correlations. These correlations were considered by the author of ref. [55]; we choose for illustration the $(\bar{k}_\nu x \bar{\sigma}_\mu) \cdot J_f$ -correlation between the neutrino momentum, the muon spin and the spin of the final nuclear state.

Attention was called recently to the possible use of double correlation experiments [56]: T-conservation implies a relationship between the average polarization (P_{av}) and the longitudinal polarization (P_L) of nuclei produced in the capture of polarized muons. In the case of the $^{12}C \rightarrow {}^{12}B$(g.s.) transition P_{av} is already measured [39] and a measurement of P_L is under way by the same physicists. Similar tests can be performed comparing the recoil asymmetry and the alignment of the nucleus produced [57].

It may be noticed, that the $^{12}C \rightarrow {}^{12}B$(g.s.) transition proceeds predominately by the axial form factor, so the test measures the relative phase of g_A and g_P. If one is interested in the relative phase of g_A and g_V, it may be possible to combine recoil asymmetry and alignment measurements on the $^{16}O \rightarrow {}^{16}N(1)$ transition considered in Section 4.1.

We are indebted to Professor L. Grenacs and Dr. N. Mukhopadhyay for many useful discussions.

REFERENCES

[1] A. Fujii and H. Primakoff, Nuovo Cimento 12 (1959) 327 and H. Primakoff, Rev. Mod. Phys. 31 (1959) 319.
[2] C.W. Kim and H. Primakoff, Phys. Rev. 140 (1965) B566.
[3] M. Morita: "Beta Decay and Muon Capture", W.A. Benjamin, Inc. Reading, Mass., 1973.
[4] J. Deutsch: "Muon Capture", in Proceedings of the SIN Spring School on Weak Interactions and Nuclear Structure, SIN, 1972.
[5] H. Uberall: "Study of Nuclear Structure by Muon Capture", Springer Tracts in Modern Physics, 71, pp. 1-38, Springer-Verlag, Berlin, Heidelberg, New York, 1974 and P. Springer: "Emission of Particles Following Muon Capture in Intermediate and Heavy Nuclei", ibid. pp. 39-87.
[6] J.P. Deutsch et al., Phys. Letters 28B (1968) 179 and Nuovo Cimento 52B (1967) 557.
[7] R. Winston and V.L. Telegdi, Phys. Rev. Letters 7 (1961) 104 and R. Winston, Phys. Rev. 129 (1969) 2766.
[8] I. Talmi and U. Unna, Phys. Rev. Letters 4 (1960) 469.
[9] D.R. Goosman and R.W. Kavanagh, Phys. Rev. C1 (1970) 1939.
[10] L. Grenacs et al., Nucl. Instr. and Methods 58 (1968) 164.
[11] T.W. Donnelly and J.D. Walecka, Phys. Letters 44B (1973) 330.
[12] F. Cannata and C. Werntz, Phys. Rev. C9 (1974) 782.
[13] J. Deutsch et al., Phys. Rev. Letters 33 (1974) 316.
[14] K. Koch and T.W. Donnelly, Nucl. Phys. B64 (1973) 478.
[15] J.C. Bergström et al., Saskatoon preprint, 1975.
[16] J.P. Deutsch et al., Phys. Letters 26B (1968) 315.
[17] N.C. Mukhopadhyay, Phys. Letters 40B (1972) 157.

[18] L. Hambro, N.C. Mukhopadhyay, SIN-Report PR-75-006, 1975.
[19] D. Favart et al., Phys. Rev. Letters 25 (1970) 1348.
[20] M. Rho, Phys. Rev. 161 (1967) 955 (set c) and private communication see also ref. 24.
[21] V. Gillet, D.A. Jenkins, Phys. Rev. B140 (1965) 32.
[22] Method suggested by Professor L. Grenacs.
[23] J. Gillet, N. Vinh Mau, Nucl. Phys. 54 (1964) 321.
[24] Private communication from Prof. V. Devanathan and Dr. P.R. Subramanian (see also Madras University Report No. Ph./31/1974, to appear in Annals of Physics).
[25] M. Ericson, A. Figureau and C. Thévenet, Phys. Letters 45B (1973) 19.
[26] K. Ohta and M. Wakamatsu, Phys. Letters B51 (1974) 325 and Nucl. Phys. A234 (1974) 445.
[27] M. Rho, Nucl. Phys. A231 (1974) 493 and private communication.
[28] S.L. Adler, Phys. Rev. 140B (1965) 736 and W.I. Weisberger, Phys. Rev., 143 (1966) 1302.
[29] Kabir, Zeitschrift für Phys. 191 (1966) 447.
P. Pascual, CERN Th. Preprint 1081 (1969).
L. Wolfenstein in "Proc. Conf. on High Energy Physics and Nuclear Structure", Columbia, 1969 (ed. S. Devons) (Plenum Press, New York, 1970).
[30] e.g. T.D. Lee and C.S. Wu, Ann. Rev. Nuclear Sci. 15 (1965) 381.
[31] J.P. Deutsch and L. Frenacs, 4th Intern. Conf. High Energy Physics and Nuclear Structure, Dubna, USSR, 1971.
[32] L.L. Foldy and J.D. Walecka, Phys. Rev. 140 (1965) B1339.
[33] S. Weinberg, Phys. Rev. 112 (1958) 1375.
[34] M. Ericson, J. Delorme, "Saclay Meeting on Mesonic Effects in Nuclei", 1975 (ed. Cl. Schuhl), Saclay Report, 1975.
[35] D.H. Wilkinson, Phys. Rev. C7 (1973) 930; Nucl. Phys. A209 (1973) 470; Nucl. Phys. A255 (1974) 365.
[36] N.C. Mukhopadhyay, Phys. Letters 45B (1973) 309.
[37] A suggestion of N.C. Mukhopadhyay.
[38] World average quoted by E. Zavattini "Muon Capture" in "Muon Physics" (ed. V.W. Hughes and C.S. Wu), CERN Preprint, 1972.
[39] A. Possoz et al., Phys. Letters 50B (1974) 438.
[40] L. Palffy et al., Phys. Rev. Letters 34 (1975) 212.
[41] J.P. Deutsch et al., Phys. Letters 29B (1969) 66.
[42] L.M. Rosenstein and I.S. Hammerman, Phys. Rev. CB (1973) 603.
[43] J.P. Deutsch et al., 4th Intern. Conf. High Energy Physics and Nuclear Structure, Dubna, USSR, 1971 (revisited !)
[44] G.H. Miller et al., Phys. Rev. Letters 29 (1972) 1194.
[45] S. Ciechanowicz, Wroclaw preprint no. 320 1975.
[46] Private communication from Professor C.W. Kim and A. Bottino et al., Phys. Rev. 11C (1975) 991.
[47] D.H. Wilkinson, Physics Letters 48B (1974) 169 and refs. cited therein.
[48] J. Delorme and M. Rho, Nucl. Phys. B34 (1971) 317.

[49] F. Calaprice et al., Princeton Progress Report 1974, quoted by M. Rho, "Saclay Meeting on Mesonic Effects in Nuclei" (ed. Cl. Schuhl), Saclay Report, 1975.
[50] K. Sugimoto et al., OULNS-preprint 75-3, 1975 and M. Morita, H. Ohtsubo, contributed to 6th Intern. Conf. on High Energy Physics and Nucl. Str., Santa Fe, 1975.
[51] M. Steels et al., contributed to the 6th Intern. Conf. on High Energy Physics and Nucl. Str., Santa Fe, 1975.
[52] N. Rolin et al., Louvain University Report IPC-N-7503, 1975.
[53] A.I. Alikhanov et al., Sov. Journ. Nucl. Phys. JETP 11 (1960) 1380.
G. Backenstoss et al., Phys. Rev. Letters 6 (1961) 415.
M. Bardon et al., Phys. Rev. Letters 7 (1961) 23.
[54] C.W. Kim and H. Primakoff, Phys. Rev. 180 (1969) 1502.
[55] Z. Oziewicz, Acta Phys. Pol. 31 (1967) 501.
[56] J. Bernabeu, Phys. Letters 55B (1975) 315.
[57] Private communication from Prof. Devanathan and Dr. J. Bernabeu.

RECENT MUON PHYSICS AT SREL

John R. Kane

College of William and Mary

Williamsburg, Virginia, U.S.A.

As most here know, the SREL facility in Virginia is a 600 MeV synchrocyclotron which has steadily produced both an extracted proton beam and various meson beams since 1967. The muon channel facility shown in Figure 1 presently delivers beams of high duty cycle to the Meson Cave with the following intensities: backward μ^- - 3 x 10^5 sec^{-1} and backward μ^+ - 1 x 10^5 sec^{-1}. During the coming month of July the channel will be rotated in order to bring it closer to the machine. Monte Carlo simulations indicate that the muon levels will thereby increase by about 50% because of the improved acceptance.

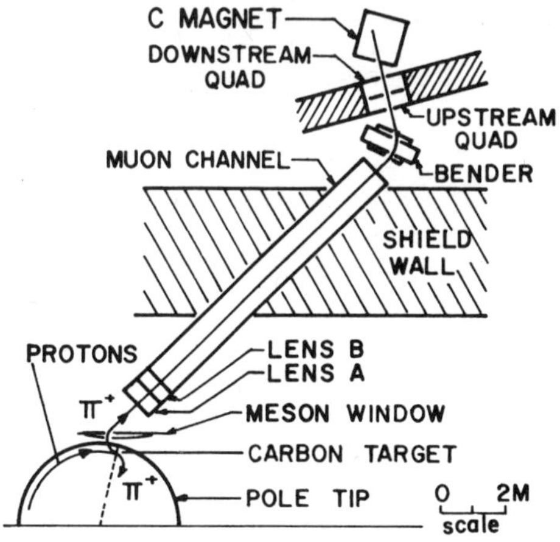

Fig. 1

Experimental Area

In this talk I shall limit myself to a description of current and future activity in the muon program at SREL. In so doing I will not touch upon a number of scattering experiments which utilize the extracted proton beam and internally and externally produced pions.

In Table 1 I have sorted the muon experiments of the past year into what I feel is a natural set of categories. As you can see all areas of muon physics are well represented. I shall make comments on all of these experiments in sequence, emphasizing those with which I am most familiar.

TABLE 1

MUON PHYSICS AT SREL

Classification	Experiments	Institutions
Quantum Electrodynamics	1.Formation of $(\alpha\mu^-)e^-$	Yale-Heidelberg
	2.QED Effects in μ atoms	Carleton-Chicago Nat. Research Council of Canada
	3.Muonium hfs Interval	Chicago-U.of Calif.
V-A Weak Interactions	Radiative Muon Capture (RMC) in ^{40}Ca	William & Mary
Nuclear Charge Structure	Nuclear Charge Parameters in 6 Hg Isotopes	Cal. Tech. William & Mary Wyoming SIN
μ^+ Solid State Studies	μ^+ Precession in Ferromagnets, Superconductors, Spin Glass Materials	William & Mary Bell Labs.
Weak Neutral Currents	Survey of 2γ Process in $2S_{1/2} \rightarrow 1S_{1/2}$ μ atom Transitions	Carleton William & Mary NRC Canada
Lepton Conservation Law	Muonium in Vacuum Production Studies for $M \rightarrow \bar{M}$	Maryland William & Mary

1. QUANTUM ELECTRODYNAMICS (QED)

Within the past year three SREL experiments have had as their objective the study of QED effects in muonic systems. Most recently Hughes et. al. [1] have used the Larmor precession method to demonstrate for the first time the formation of a polarized one-electron muonic helium atom $(\alpha\mu^-)e^-$. They achieved this by first demonstrating free precession for the $(\alpha\mu^-)^+$ ion in pure helium gas at 7 and 14 atmospheres, and then by observing the muonium-like hyperfine precession frequency of the $(\alpha\mu^-)e^-$ atom which resulted from the addition of 1.2% Xe to the helium. With pure helium the free signal was observed to be $A_\mu(\alpha\mu^-) = (1.24 \pm 0.22)\%$ (65% beam polarization), while upon the addition of the charge-exchanging impurity the signal was $A(\alpha\mu^-e^-) = (0.53 \pm 0.09)\%$. This loss of signal amplitude is consistent with the retention of polarization in the F = 1 hyperfine state.

This development should make it possible to perform precision measurements of the hyperfine structure interval (hfs) $\Delta\nu$ and the Zeeman effect for this heavy muonium-like system. It is felt that, despite the structure of the $\alpha\mu^-$ core, muonium resonance methods offer the promise of precise determinations for the μ^- mass and magnetic moment. This will permit a test of CPT invariance for $\mu^\pm$ properties where the data is derived from a common method of measurement. The approximate theoretical value for $\Delta\nu(\alpha\mu^-e^-)$ is 4494.1 MHz. This differs from $\Delta\nu(\mu^+e^-) = 4463.32$ MHz mainly due to a different reduced mass factor and the structure of the $(\alpha\mu^-)^+$ nucleus.

In a second QED-type experiment the Ottawa - Chicago group made a careful series of muonic X-ray energy measurements in the range from 100 keV to 450 keV. This range was selected to minimize various muon-nucleus and muon-electron effects such as nuclear finite size, nuclear polarization, and electron screening. As a result they were sensitive mainly to the vacuum polarization corrections to the Dirac energy value.

It is well known that an earlier experiment by this group at Chicago [2], and a similar measurement by another group at CERN [3] resulted in values which deviated with existing theory by as much as six standard deviations for μ - Ba and μ - Pb transitions. Since then modifications in the theory together with a recent shift in the reference line of ^{198}Au at 412 keV have adjusted predictions until the original Chicago measurements now fall within one standard deviation of prediction below 350 keV and to within one and one half standard deviations in the 440 keV region of Ba and Pb. In this year's run at SREL, this group has obtained one standard deviation agreement with the latest prediction for essentially the same set of target samples. This experiment is scheduled to run again in the Fall.

A high field measurement of the muonium hfs interval was attempted by Telegdi et. al. during the past year, but it appears that experimental difficulties may have limited the value of that measurement. A second effort is expected by this group in the near future.

The same group had successfully completed a "zero" field $\Delta\nu$ measurement two years before at SREL. In that work [4] they applied the high precision method of double pulsed microwave resonance at low pressure, and obtained a value of $\Delta\nu(0) = 4463.3013\ (40)$ MHz which when combined with earlier Chicago data gave a value of $\Delta\nu(0) = 4463.3012\ (23)$ (0.5 ppm). This should be compared to the recent measurement by Hughes et. al. [5] of $\Delta\nu(0) = 4463.3011\ (16)$ (0.36 ppm) MHz.

2. V-A WEAK INTERACTIONS

In the area of V-A interactions a William and Mary group is in the process of measuring the rate for the radiative muon capture process in ^{40}Ca. As is well known the basic muon capture process (MC) is $\mu^- + p \rightarrow n + \nu_\mu$, while the radiative muon capture process (RMC) is $\mu^- + p \rightarrow n + \nu_\mu + \gamma$. The RMC process is of substantial interest because of its sensitivity to the induced pseudoscalar coupling constant g_P. Of course the basic RMC process is quite difficult because the following ratio of rates must be multiplied:

$(\Lambda_{MC}/\Lambda_{TOT})_P \sim 10^{-3}$ and $(\Lambda_{RMC}/\Lambda_{MC})_P \sim 2 \times 10^{-4}$. For this reason the medium-Z nucleus of ^{40}Ca for which $(\Lambda_{MC}/\Lambda_{TOT})_{Ca} \sim 0.85$ and $(\Lambda_{RMC}/\Lambda_{MC})_{Ca} \sim 2 \times 10^{-4}$ has served as the target in most RMC experiments.

An early measurement by Conversi et. al. [6] gave for the ratio of coupling strengths g_P/g_A a value of $+ 13.3 \pm 2.7$. More recently Rosenstein and Hammerman [7] obtained $g_P/g_A \simeq 5.9 \pm 5$ after correcting for the fact that 45% of their high energy neutral events were neutrons. Here it should be pointed out that a very recent treatment of the last experiments' data by Ohta [8] is able to produce agreement with the Goldberger-Treiman prediction of $g_P/g_A \sim 7$ by accounting for the influence of the N^* resonance upon RMC.

In the present experiment at SREL an attempt is being made to avoid ambiguity in the γ telescope between high energy neutrons

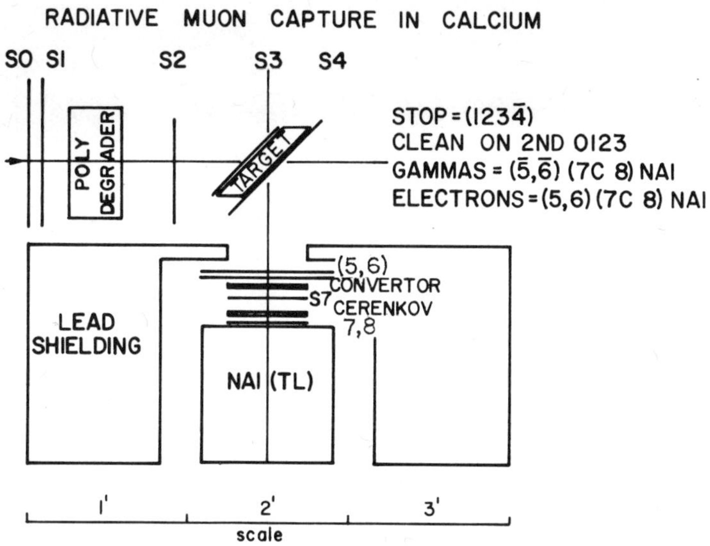

Fig. 2

and gamma rays by requiring that the gamma ray convert before entering the NaI crystal. Figure 2 shows the counter arrangement. The telescope consists of two veto counters, an 11% Pb converter, followed by two scintillators, a plastic Cerenkov counter and the 10" dia. NaI crystal. The ability of this system to detect Panofsky gamma rays is shown in Figure 3. As yet results are not available on this work.

It is hoped that a measurement of the asymmetry of photons from the RMC process can be combined with that of the photon spectrum in the next running period. To date the only previous measurement of the asymmetry coefficient α [9] is 2.5 standard deviations away from that calculated by Rood and Tolhock [10].

3. NUCLEAR CHARGE STRUCTURE

A measurement has recently been made at SREL of the nuclear charge parameters (mean square radius and deformation) for six separated Hg isotopes. The object was to chart the evolution of these parameters versus neutron number in the region of doubly magic ^{208}Pb. Data were collected simultaneously for all six

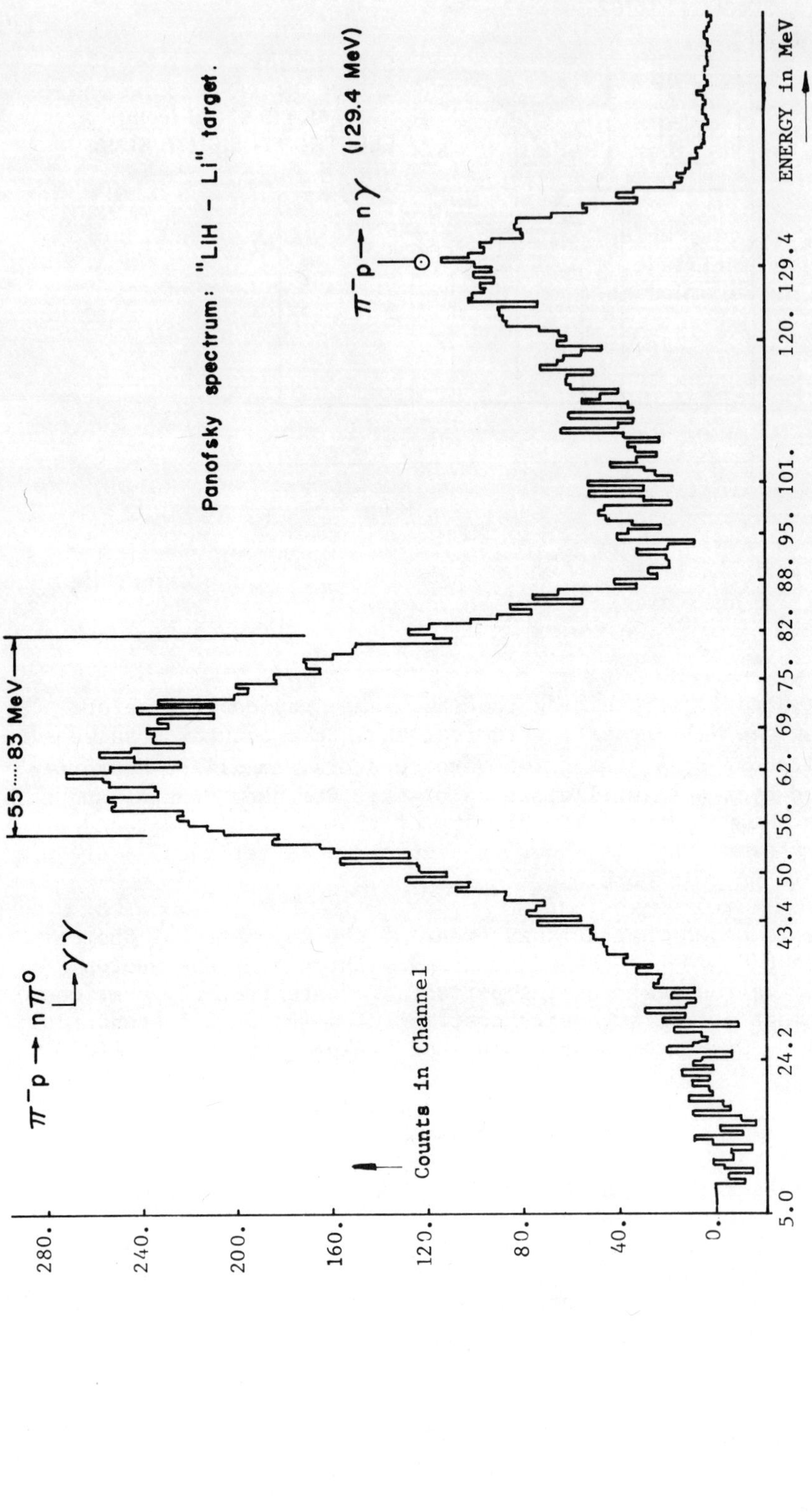

Panofsky spectrum: "LIH - Li" target.
$\pi^- p \to n\gamma$ (129.4 MeV)
$\pi^- p \to n\pi^0$
$\hookrightarrow \gamma\gamma$
55 .... 83 MeV
Counts in Channel
ENERGY in MeV
5.0
24.2
43.4
50.
56.
62.
69.
75.
82.
88.
95.
101.
120.
129.4
0.
40.
80.
120.
160.
200.
240.
280.

Fig. 3

isotopes, covering the low energy region with a 3 cc intrinsic Ge detector and the high energy region with an 85 cc Ge(Li) detector.

4. μ^+ SOLID STATE STUDIES

An active program of systematically probing solid state properties with positive muons is well established at the muon channel facility. The technique involves implanting polarized μ^+ in solid targets which are subject to a uniform external field. The signal amplitude, frequency, and relaxation of precessing μ^+ moments is recorded in terms of the observed μ^+ decay asymmetry pattern. A steady investigation of the internal fields of ferromagnetic materials is being made for Co, Ni, and Fe [11], and is to be extended to the rare earths. In these studies signals are recorded as a function of temperature above and below the Curie point. Measurements will also be made for intermetallic compounds such as Fe_3Si, while a search for interesting muonium systems will continue. Finally the collaboration of Bell Laboratories and William and Mary has also succeeded in detecting the feature of vortex structure in the magnetic field distribution of Type II superconductors [12]. As we have learned from Dr. Schenck this area of μMR has remarkable versatility for probing solid state properties, and so we anticipate a number of additional important developments from this work.

5. WEAK NEUTRAL CURRENTS

It has been proposed in a number of papers [13,14,15] that a neutral weak current may express itself in a parity-violating manner for the $2S_{1/2}$ - $2P_{1/2}$ level system of low Z muonic atoms. In particular it has been shown by Bernaben, Ericson, and Jarlskog [14] that for Li and Be this parity-violating effect will be as large as 10% in terms of an angular correlation between the spin of the polarized μ^- in the 2S state and the direction of a low yield M1 photon from 2S → 1S. In order to sketch their argument we shall make use of the level diagram below. In the region of Li and Be the $2S_{1/2}$ - $2P_{1/2}$ level separation ΔE is at a minimum value as the finite nuclear size effect for the 2S state is nearly balanced by the opposite shift due to vacuum polarization. The existence of a parity-violating term in the hamiltonian for neutral currents will generate an admixture $\zeta = \frac{\langle 2P_{1/2}|H_{PV}|2S_{1/2}\rangle}{\Delta E}$ of one nearly degenerate state with another of opposite parity. Thus, subject to this interaction the $2S_{1/2}$ state becomes $|2S_{1/2}\rangle+\zeta|2P_{1/2}\rangle$ and the $2P_{1/2}$ state becomes $|2P_{1/2}\rangle-\zeta|2S_{1/2}\rangle$.

The metastable 2S state is depleted by a two photon E1

emission, and occasionally by a small amplitude M1+ζE1 single photon. It happens that the admixed E1 amplitude, ζE1, is comparable to small amplitude M1 leading to a single photon with circular polarization, or directional correlation with the muon spin. Complications result if nearby electrons are able to promote strong depletion of the 2S state via the Auger process, or if Stark mixing to the 2P state further reduces the 2S population. Our first step in this experiment is to find a low Z material which is basically free of such complications. We have begun to look for evidence of the two photon process as a measure of the 2S state population and hope to determine its rate $\Gamma_{2\gamma}$. This is done by operating large Ge detectors on both sides of the target in fast coincidence with one another. Since the 2 photon emission is slow for low Z materials, evidence of the process will consist of a fast Ge1 - Ge2 coincidence which is delayed relative to the muon signal and has nearly the expected $\Gamma_{2\gamma}$ rate. I can show data taken recently in BeH_2 which indicates coincidences, although this plot corresponds to <u>all</u> times relative to the muon stop. Figure 4 is a scatter plot of coincident E_1 and E_2 detector energies in which it is possible to pinpoint coincidence between a K_α event in one and an L_α event in the other or a K_α in one and an L_β in the other. We, of course, must eventually concentrate upon $E_1 + E_2 = K_\alpha - \Delta E$ events which are delayed relative to the muon stop. While the present plot is not restricted to delayed events, it is interesting to note here however, that this $E_1 + E_2 \sim K_\alpha$ line is populated by coincident events in which the Ge escape energy for a K_α into one detector is captured by the other detector. We should have some estimate regarding the feasibility of the eventual angular correlation study in the low Z region in the near future. The angular correlation work itself would necessitate a μ^- beam of high intensity.

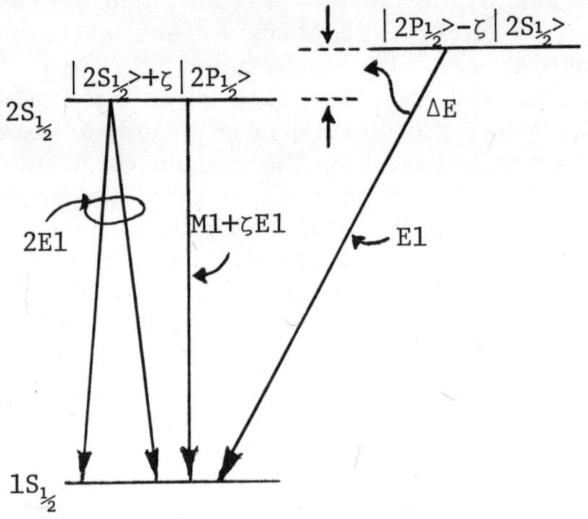

LEVEL DIAGRAM REFERRED TO ON PREVIOUS PAGE

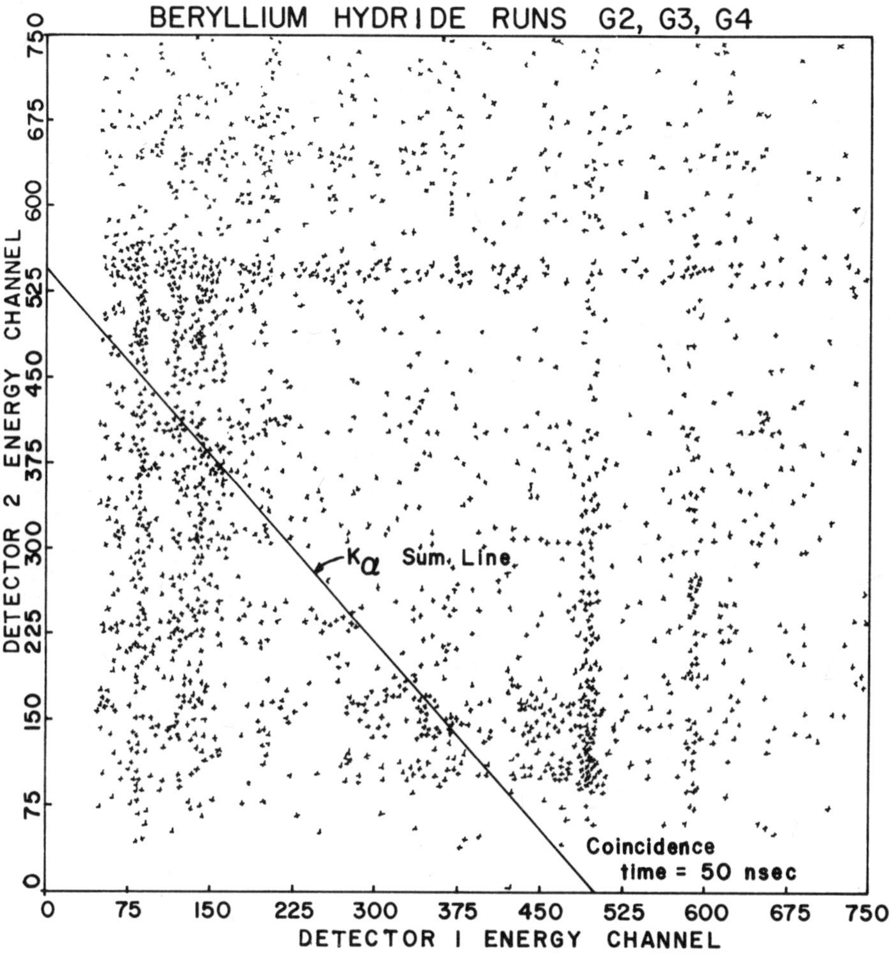
BERYLLIUM HYDRIDE RUNS G2, G3, G4
DETECTOR 2 ENERGY CHANNEL
DETECTOR I ENERGY CHANNEL
0
75
150
225
300
375
450
525
600
675
750
Kα Sum Line
Coincidence time = 50 nsec

Fig. 4

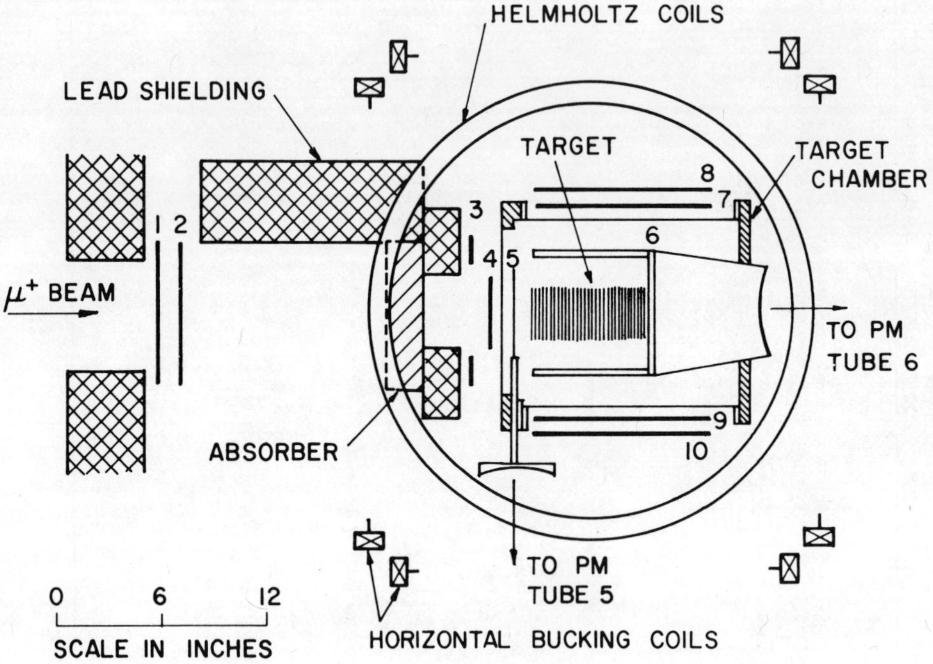

Fig. 5. Top view of the experimental set up

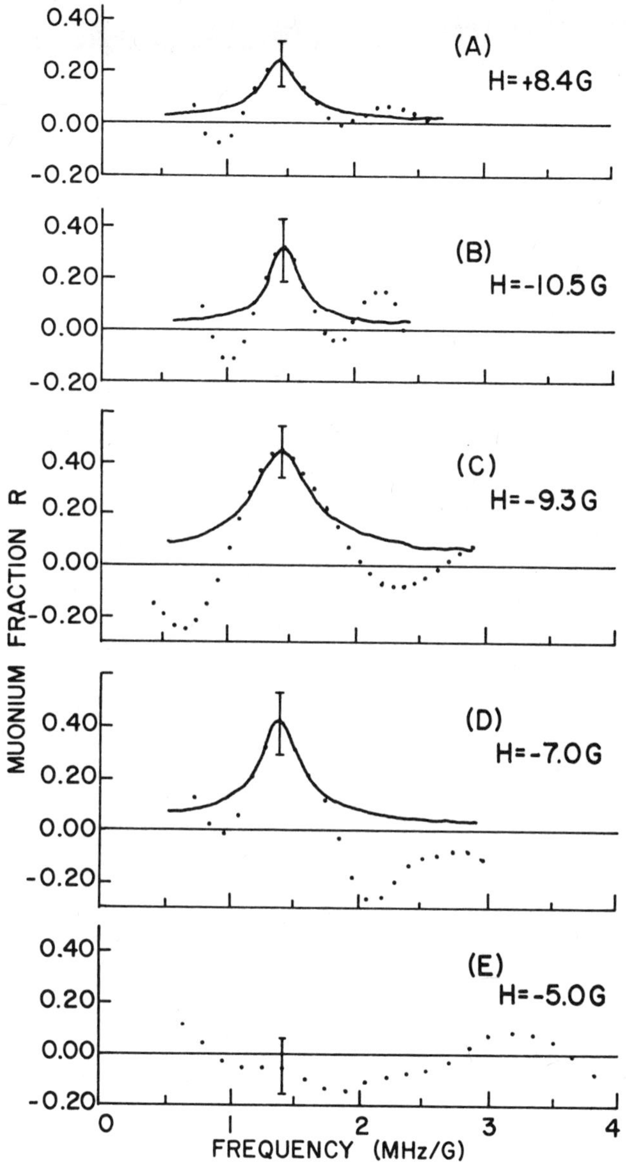

Fig. 6 Results of Fourier frequency analyses

6. LEPTON CONSERVATION LAW

In the remaining time I can only make a few brief remarks about our efforts to prepare for a search for the process: muonium $M(\mu^+e^-) \rightleftarrows$ antimuonium $\bar{M}(\mu^-e^+)$. Pontecorvo [16] first considered this transition while Feinberg and Weinberg [17] stressed its significance as a test of the nature of the lepton conservation law. Basically this process is forbidden by the familiar additive form of lepton number conservation, but allowed by the multiplicative form which preserves muon number parity.

In order to prevent strong quenching of the process by perturbing external fields which would break the energy degeneracy of a $M - \bar{M}$ system, the M atom should spend its time in a free space environment. We have taken Feinberg and Weinberg's suggestion in preparing a target of 200 thin metal plates which are separated by vacuum drift space. Our "plates" are 1000 Å thick Au foils, spaced at 1 mm intervals along the beam line. This configuration is shown in Figure 5. I should stress that our work at SREL is primarily an effort to develop a way to generate M atoms which quickly find their way to a vacuum region. Unambiguous evidence of this mechanism has been looked for in terms of a characteristic muonium signal. The present state of our data, shown in Figure 6, is suggestive of muonium precessing at the frequency of $\sim$ 1.4 MHz/G. Because this effect must be well established before attempting the $M - \bar{M}$ process we shall soon make further studies of muonium production which include a 10" dia. NaI as part of the decay positron telescope. This will permit us to study asymmetry amplitude as a function of Michel energy. Additionally we hope to improve our sensitivity to μ^+ which stop in the region of the target foils. Given the success of this investigation, the final search for $M \rightarrow \bar{M}$ will require higher μ^+ intensity.

REFERENCES

[1] P.A. Sonder et al., Phys. Rev. Lett. 34 (1975) 1417.

[2] M.S. Dixit et al., Phys. Rev. Lett. 27 (1971) 878.

[3] H.K. Watter et al., Phys. Lett. 40B (1972) 197.

[4] D. Favart et al., Phys. Rev. Lett. 27 (1971) 1336.

[5] D.E. Casperson et. al., Abs. of Cont. Papers, LAMPF Conf. (1975) 145.

[6] M. Conversi et al., Phys. Rev. 136 (1964) B1077.

[7] L.M. Rosenstein and I.S. Hammerman, Phys. Rev. C, 8, (1973) 603.

[8] K. Ohta, Phys. Rev. Lett. 33 (1975) 1507.

[9] L. DiLella et al., Phys. Rev. Lett. 27 (1971) 830.

[10] H.P.C. Rood and H.A. Tolhock, Nucl. Phys. 70 (1965) 658.

[11] M.L.G. Foy et al., Phys. Rev. Lett. 30 (1973) 1064.

[12] A.T. Fiory et al., Phys. Rev. Lett. 33 (1974) 969.

[13] A.N. Moskalev, Sov. JETP Lett. 19 (1974) 141.

[14] J. Bernaben et al., Phys. Lett. 50B (1974) 467.

[15] G. Feinberg and M.Y. Chen, Phys. Rev. D10 (1974) 190.

[16] B. Pontecorvo, Zhur. Eksp. i Teoret. Fig. 33 (1957) 549.

[17] G. Feinberg and S. Weinberg, Phys. Rev. 123 (1961) 1439.

INVARIANCE PRINCIPLES

P.K. Kabir

University of Virginia

Department of Physics, Charlottesville, Va., U.S.A.

INVARIANCE PRINCIPLES

The important role of invariance principles in physical theory was recognized only relatively recently, after the development of analytical dynamics. In classical mechanics, the equations of motion are completely solved if one knows all the constants of motion, therefore the problem of solving the equations of motion is the same as the problem of finding all the "angle" variables ϕ_k, viz those coordinates which the Hamiltonian is independent of, since the canonically conjugate "action" variables J_k are constant in time. Since the J_k are also the generators of infinitesimal canonical transformations which induce changes of the ϕ_k alone, recognition of all the invariance transformations of the Hamiltonian, and identification of the corresponding generators, is equivalent to a complete solution of the equations of motion. In quantum theory, the situation is a little more complex in that there is a limitation of principle in the extent to which one can specify the state of a dynamical system, and therefore *a fortiori*, in the manner in which one can trace its evolution in time. The corresponding statement is that, if one can find a maximal set of commuting operators, each of which commutes with the Hamiltonian, then the states which are simultaneous eigenstates of all those operators have a particularly simple time dependence. They form a complete set of "stationary" states and it is convenient, in describing the time-development of an arbitrary state, to express that state (more precisely the state vector corresponding to that state) in terms of this complete set. In contrast to the classical situation, however, the relation of symmetry transformations to conservation laws is very direct and eigenstates of the operator corresponding to the generator of a symmetry transformation suffer only a phase change under the corresponding coordinate transformation, which means that all observable properties of the state remain unchanged [1].

This brings us to a point which was apparently first clearly brought out by H. Poincaré and P. Curie, which is that it is necessary to have some departure from symmetry in order to recognize the existence of that symmetry. If, for example, a physical system were in every respect invariant with respect to a certain symmetry transformation S, it would be impossible in principle to discover the existence of that symmetry, because there would be no way of knowing whether such a transformation had indeed been made. Perhaps this is why, as artists have known for long, a slightly asymmetric arrangement is far more attractive to the eye than a too regular geometrical symmetry. Wigner has emphasized that our recognition of space-time symmetries, for example, rests on the fact that our knowledge of physical systems can be separated into factors which depend on initial boundary conditions and factors which depend on physical laws [2]. This relies on the hypothesis that the physical systems that we study may, with sufficient accuracy, be regarded as uninfluenced by their surroundings. The different environments in which we place two such "identical" systems allows us to distinguish between them while, at the same time, they are both regarded as being sufficiently isolated that their physical behavior is not affected by the difference of their situations. In seeking invariance principles, we are concerned with discovering the symmetries of the physical laws governing those systems; the initial conditions - at least at the microscopic level - appear to be far too complicated for us to find any order in them. If detailed comparison of two such systems should reveal that a certain postulated symmetry is not exactly obeyed, a possible way of reconciling the result with the hypothetical symmetry is to attribute the observed asymmetry to the influence of the environment. A well-known illustration of this is provided by the phenomenon of ferromagnetism [3], where the hypothesis of rotational invariance for the interaction between the elementary magnets is not contradicted by the occurence of a preferred magnetization axis: the choice of that direction is dictated by initial conditions, e.g. the presence of an arbitrarily small background magnetic field. This is the analogy which is used in the currently fashionable theories of "spontaneous" symmetry-breaking. It should be noted, however, that the analogy is not perfect. In the case of the ferromagnet, the direction of magnetization is attributed to conditions external to the magnet. If we take a large assembly of ferromagnetic atoms and cool it below the Curie temperature, in a region where we have excluded external magnetic fields to the best of our ability, we find that the atoms adopt the ferromagnetic phase in domains, with the axis of magnetization being different in different domains. If there are a sufficiently large number of domains, rotational symmetry will be restored because there will be no preferred direction of magnetization for an average domain. The present theories of spontaneous symmetry breaking correspond to the presence of a single ferromagnetic domain of infinite extent; the analogs of domain boundaries have not yet received a satisfactory interpretation. At the present state, therefore, theories of spontaneous symmetry breaking cannot be distinguished physically from theories in which the symmetry is broken by the physical laws themselves. This is a somewhat unsatisfactory situation compared to the idealization of being able to isolate physical systems, which we had assumed earlier. The previous approach was consistent with the philosophy of successive approximations. At a certain level of precision, one could regard a system as being completely isolated; higher accuracy would reveal finer details whose interpretation would require inclusion of some effects of the surroundings, even greater precision would require consideration of even more remote influences and so on, ad infinitum. As presently formulated, the philosophy of the theory of spontaneous symmetry breaking, on the other hand, seems to require knowledge of the whole in order that we may describe any part. Since the theory is equivalent to one in which there is no influence of the environment, and the symmetry is broken by the physical laws themselves, it is still possible to proceed by successive approximations according to the hierarchy of the various approximate symmetries which are present, but in that case, one may well ask what has been gained by the hypothesis of "spontaneous" symmetry breaking?

Finally, it should be remarked that the greatest utility of an approach based on invariance principles is when the basic laws are unknown or, if they are known, their mathematical structure is too complex to admit of ready solution. If the laws and their solutions are known, there is no need for an explicit declaration of the invariance properties which will necessarily be built into those solutions. This is well-illustrated by the example of the conservation laws of classical mechanics which were discovered long before their relation to various invariance properties of Lagrange's equations [4].

INTERNAL SYMMETRIES

In this section, we shall review our present knowledge of the internal structure of hadrons, in particular of protons and neutrons, which is derived largely from symmetry considerations.

The approximate equivalence of neutrons and protons with respect to nuclear forces finds its mathematical expression in the hypothesis of charge-independence or isospin-invariance. Neutrons and protons are viewed as two distinct states of a fundamental entity called the nucleon, distinguished by a two-valued internal coordinate which, by analogy with electron spin, is taken to be the z-component of a "spin" in a fictitious space called isospin space. Then, charge-independence of nuclear forces is assured by the postulate that all directions in isospin space are equivalent as far as nucleon-nucleon interactions are concerned. Another way of stating this hypothesis is to say that, except for electromagnetic interactions which distinguish the charged proton from the electrically neutral neutron, it makes no difference, from the point of view of nuclear interactions, if we redefine the two independent nucleon states as nucleons with isospin "up" and "down" along any other direction in isospin space. From the mathematical representation of electron spin, we know that such a redefinition corresponds to a unitary transformation on the original "isospin-up" and "isospin-down" states. Thus, the hypothesis of charge-independence is the hypothesis that neutrons and protons are unitarily equivalent with respect to nuclear forces. If we factorize out those unitary transformations which represent a common phase-transformation of neutron and proton states, the remaining symmetry is called SU(2) for obvious reasons. As can be seen from the analogy with electron spin, invariance under isospin transformations implies the conservation of isospin, defined as the generator of rotations in that space. All observed deviations from this unitary symmetry, such as the neutron-proton mass-difference, can be attributed to the effect of electromagnetic interactions which do not respect that symmetry.

After Yukawa proposed the meson theory of nuclear forces, it was realized that charge-independent nuclear forces would result only if there were neutral mesons in addition to charged mesons. One will obtain nuclear forces invariant under isospin rotation if the meson-nucleon interaction is itself invariant under isospin rotation. This can be assured by taking the meson-nucleon interaction to have the form of the (iso)scalar product of the nucleon isospin density with an isovector meson field whose quanta would form an isotriplet of mesons with positive, negative, and zero charge*. The $\pi^{\pm},\pi^{0}$ which were subsequently discovered comprise such an isotriplet and their coupling to nucleons conforms to the predicted relations. The $\pi^{\pm}-\pi^{0}$ mass-difference is of the same order as the neutron-proton mass-difference and may also be regarded as an electromagnetic effect. Electromagnetic interactions

---

*The form of the coupling between the isodoublet nucleons and isotriplet mesons can also be deduced from the condition that the mass-degeneracy of nucleons and of mesons persist after including self-energies. See Ref. 5.

single out a preferred direction in isospin space, electric charge being related to the z-component of isospin by

$$Q = I_3 + B/2 \tag{1}$$

where B is the baryon number or baryon charge, carried by nucleons but not by mesons. After the discovery of strange particles, it was realized that the concept of isospin should be extended to them also and that a consistent description of strange particle behavior could be obtained by generalizing Eq. (1) to read

$$Q = I_3 + (B+S)/2 \tag{2}$$

where the "strangeness" S (zero for π-mesons and nucleons) is conserved in all strong and electromagnetic interactions.* This scheme, proposed by Gell-Mann and by Nishijima, has been verified with respect to all its implications. It was found that, in addition to the neutron and proton states, there are six other nearby baryon states with the same spin and parity, stable except for weak decays with a typical lifetime of 10^{-10} sec - comprising an isosinglet Λ^0 (with mass 1115 MeV/c^2), and an isotriplet $(\Sigma^-,\Sigma^0,\Sigma^+)$ [with mean mass 1190 MeV/c^2] with $S= -1$, and an $S= -2$ isodoublet (Ξ^-,Ξ^0) with mean mass 1320 MeV/c^2, see Fig. 1. The mass-differences between the members of an isomultiplet are of the same order as the neutron-proton mass difference and consistent with being of electromagnetic origin. Similarly, the pseudoscalar meson triplet (π^-,π^0,π^+) with mean mass 140 MeV/c^2 has two isodoublets of strange partners: (K^0,K^+) with mean mass about 500 MeV/c^2, and their corresponding antiparticles $(K^-,\bar{K}^0)$**. The occurrence of these other particle states associated with the new "strangeness" degree of freedom strongly suggested the existence of a symmetry higher than isospin-invariance arising from the unitary equivalence of two objects, or SU(2). The most fruitful suggestion, made by Sakata, was to regard all the observed particles as compounds of three fundamental components, which he chose to be the neutron, proton and the Λ^0 particle, and their antiparticles. Fermi and Yang had already shown that the observed $\pi^\pm$, and π^0 mesons could be regarded as bound nucleon-antinucleon states. By adding the Λ^0 to the list of basic constituents, the Fermi-Yang picture could be

---

*With integer values of B and S, one must impose the restriction $(-1)^{B+S} = (-1)^{2I}$ to avoid the occurrence of half-electronic charges.

**Because strangeness is not exactly conserved, the distinction of K^0 and $\bar{K}^0$ is not absolute, leading to the beautiful phenomena associated with K^0-$\bar{K}^0$ mixing.

S						Mass in MeV/c^2
-2		Ξ^-		Ξ^0		1320
			Σ^0			1190
-1	Σ^-				Σ^+	
			Λ^0			1115
0		n		p		940
I_3	-1		0		+1	

Fig. 1. Baryon octet with $J^P = \frac{1}{2}^+$.

S					
1		$K^0 = (\bar{\Lambda} n)$		$K^+ = (\bar{\Lambda} p)$	
0	$\pi^- = (\bar{p} n)$		π^0 $\frac{(\bar{p}p - \bar{n}n)}{\sqrt{2}}$		$\pi^+ = (-p\bar{n})$
-1		$K^- = (\Lambda \bar{p})$		$\bar{K}^0 = (-\bar{\Lambda} n)$	

Fig. 2. Pseudoscalar mesons according to the Sakata model.

extended to include strange mesons as well, and Fig. 2 shows the composition of the observed pseudoscalar mesons according to the Sakata model. As in the Fermi-Yang model, the question arises why there should not be a I = 0 pseudoscalar bound state of nucleon and antinucleon, represented by the isospin wavefunction $(p\bar{p} + n\bar{n})/\sqrt{2}$ in addition to the neutral component π^0 of the $\underline{I} = 1$ bound states, which has the isospin wavefunction $(p\bar{p} - n\bar{n})/\sqrt{2}$. Such a state, named the η meson, was indeed found, but it has a mass of about 550 MeV/c^2 and decays to $\gamma\gamma$ and 3π states with a lifetime of the order of 10^{-18} sec. Including the η, one has an octuplet of pseudoscalar mesons all of which are accounted for by the Sakata model. The $^3S_1$ bound states of the same fundamental constituents comprise a similar octuplet of vector mesons, whose quantum numbers correspond exactly to the ρ, ω, and K^* states subsequently discovered. A ninth vector meson ϕ found later, as well as a corresponding pseudoscalar meson η', can be regarded as $^3S_1$ and $^1S_0$ $(\Lambda\bar{\Lambda})$ bound states. Ikeda, Ogawa, and Ohnuki made the further hypothesis that the unitary equivalence of neutron and proton postulated by isospin-invariance should be extended to include the third fundamental constituent of the Sakata model, the Λ^0, thus enlarging the symmetry group to SU(3). This SU(3) symmetry can only be approximate since the Λ^0 mass differs significantly from the nucleon mass. Also, in the limit of exact SU(3) symmetry, all eight members of the pseudoscalar meson octet should be degenerate and, likewise, the vector meson octet should all have a common mass. The relatively large mass-difference between K and π mesons confirms the approximate nature of SU(3) symmetry. Unlike the case of the n-p or $\pi^{\pm}-\pi^0$ mass-differences, which could be attributed to electromagnetic effects, we do not know the source of this symmetry-breaking, which remains one of the major problems of particle physics.

Despite the success of the Sakata model in explaining and predicting the observed mesons, it was not able to account for the observed baryons in any simple or natural way. This was achieved by Gell-Mann and Ne'eman [6] who retained the hypothesis of SU(3) symmetry but assigned the observed baryons (Fig. 1) directly to an octuplet representation of SU(3), which may or may not be built up from more basic triplets. According to this view, none of the baryons (including n, p, Λ^0) is a member of such a basic triplet. The situation is not unlike that which occurred in the quantum theory of angular momentum where only states with integral angular momentum were first considered before the discovery of electron spin. Although the latter came from analysis of experimental data, it is conceivable that the occurrence of half-integral angular momenta could have been predicted from the mathematical theory of angular momentum and its intimate relation with SU(2) symmetry. At the same time, conservation of angular momentum, viz. the existence of the underlying symmetry does not <u>require</u> the existence of half-integral angular momenta. Similarly, it is quite possible to have SU(3) symmetry without having its spinorial representations realized in Nature.

The occurrence of octets in SU(3) symmetry can be easily understood. It is convenient to represent an arbitrary unitary transformation U as exp (iH) where H is Hermitian. For the most general unitary transformation on three objects, we need the most general 3x3 Hermitian matrix. This can be written as a linear combination of the 9 linearly independent 3x3 Hermitian matrices, which may be chosen as the unit matrix and the 8 generalizations λ_j of Pauli's σ matrices. Then

$$U = \exp\{i[(\mathrm{Tr}H)II + \tilde{H}\}$$

where $\tilde{H}$, which is traceless, is completely specified by the 8 real coefficients $\xi_j : \tilde{H} = \sum_{j=1}^{8} \xi_j \lambda_j$. Thus $\tilde{H}$ is characterized by a real 8-dimensional vector in the so-called adjoint space and it can be easily verified that under a unitary change of basis states, $\tilde{H}$ indeed transforms as a vector in the 8-dimensional space. In the Gell-Mann Ne'eman assignment, the octet of baryons as well as the meson octets are assumed to correspond to such vector representations.

Although, as we have noted, the existence of SU(3) symmetry does not require that basic triplets, corresponding to the fundamental representation of SU(3), actually occur in Nature, it is instructive to inquire what they might be. A scheme proposed by Gell-Mann and by Zweig possesses the particular appeal of simplicity while explaining many otherwise ununderstood features of hadron structure [7]. They proposed that there is indeed a basic triplet of building-blocks, corresponding to the fundamental representation of SU(3), with the isospin and strangeness quantum numbers of the original Sakata triplet n, p, Λ^0 although physically distinct from those particles. To build both baryons and mesons from these basic units, it is necessary to assign them half-integral spin and simplicity dictates the choice of spin-½. The success of the Sakata model in accounting for the observed mesons is carried over in this model. To have the baryons as bound states of the basic triplets, their spin-½ requires an odd number of constituents. The lowest possible choice is 3 and it is indeed possible to have an octet representation of SU(3) from a compound of three triplet objects. Since SU(3) invariance requires conservation of I_3 and S, we expect each of the quantities on the R.H.S. of Eq. (2) to be additively conserved, as is the quantity on the L.H.S. It is therefore reasonable to require Eq. (2) to apply to the basic triplet also. Calling the members of the basic triplet n, p, λ (to distinguish them from the physical n, p, Λ^0 states) we see that n

and λ have the same value of $I_3 + S/2$ and therefore the same electric charge, say q. Then, again according to Eq. (2), p has a charge which is one unit higher, $Q_p = q + 1$. If we now assume that the mean charge of the triplet is zero, which is equivalent to the hypothesis that the average energy of the triplet does not change* if a trio of such particles is placed in an external electric field, then $q = -1/3$ is the charge carried by n and λ while p carries charge $2/3$. Basic triplets with such fractional electric charges have been named "quarks" by Gell-Mann. Since three quarks make up a baryon, each quark must carry baryon number $1/3$. From the quarks n,p,λ and their corresponding antiquarks $\bar{n},\bar{p},\bar{\lambda}$ we can make up all the known hadrons. The observed breaking of SU(3) symmetry can be qualitatively understood by taking the λ quark to be somewhat more massive than the other two. For example, the proton and the neutron can be thought of as ppn and nnp compounds respectively. In the lowest state of relative motion, only the intrinsic spins of the quarks contribute to the total angular momentum so the quark spins must be coupled to add to ½. If all the quark spins were parallel, one would obtain a $3/2^+$ state, which explains the observed decuplet of baryon states with $J^P = 3/2^+$ which has been seen. Fig. 3 shows the possible quark configurations of such $3/2^+$ states. These correspond exactly to the quartet of Δ states (whose doubly charged component Δ^{++} was the first resonance observed in the πp system), the $I = 1$, $S = -1$ Y_1^* resonances in the $\Lambda\pi$ system, the $I = ½$, $S = -2$ Ξ^* resonances, and the famous isosinglet triply strange Ω^-, whose existence was in fact predicted on the basis of the other resonances.

Now let us suppose that, starting from a quark configuration in which all three quarks have spin "up", we reverse one of the spins. If this is one of the quark configurations at the corners of the triangle in Fig. 3, e.g. ppp we simply obtain another arrangement of the same quarks with $m = ½$ instead of $m = 3/2$. Since all the quarks are identical, there is only one such state and since a spin-$3/2$ particle must have states with $m = ½, -½, -3/2$ in addition to $m = 3/2$, this is just the $m = ½$ state of the $3/2^+$ Δ^{++} found earlier. When we consider npp, however, we have two independent $m = ½$ states depending on whether a n quark or a p quark has its spin flipped. From these, we can form one linear combination which is the $m = ½$ component of Δ^+, but the other (orthogonal) combination must correspond to a $j = ½^+$ state and therefore can be identified with the proton. Similarly, we get $j = ½^+$ states for each of the quark combinations shown in Fig. 3, excepting those at the corner positions and the centre position, where we find two new states because any one of the three quarks $np\lambda$ could have its spin reversed.

---

*The fact that this condition is not satisfied for the basic (n,p) doublet of SU(2) may be taken as an indication of a higher symmetry for hadrons.

Such an arrangement of $\frac{1}{2}^+$ states corresponds exactly to the octet of baryons shown in Fig. 1. Thus the simplest bound states of three quarks chosen from (n,p,λ) account exactly for the observed octuplet of $\frac{1}{2}^+$ baryons and the decuplet of $\frac{3}{2}^+$ baryons. The meson octets are now simply quark-antiquark bound states without orbital motion: the singlet combination of spins represents the pseudoscalar mesons while the triplet combination corresponds to the vector mesons.

In addition to the meson and baryon states which we have mentioned, many other resonances with higher angular angular momentum, and both positive and negative parities, have been discovered. These can also be explained within the quark model by taking account of the possibility that the quarks may take up orbital angular momentum. The lowest excitation, corresponding to one unit of angular momentum, should lead to states of opposite parity and for the baryons, it is found that in addition to octuplets and decuplets, SU(3) singlet states also become possible. Odd parity baryon resonances have been found with $j = \frac{1}{2}, \frac{3}{2}, \frac{5}{2}$ and can be assigned in all cases to SU(3) multiplets with one of these dimensionalities. In many cases, several of the SU(3) partners of nucleon resonances have been identified. Higher even-parity states are explained in terms of two units of orbital excitation, and so on. The important point is that all the known baryon states can be thought of as "molecules" of three quarks, and similarly the mesons as quark-antiquark bound states. Furthermore, it is remarkable that not a single "exotic" state has been found, that cannot be thought of in this way. Examples of these would be any meson with double charge or strangeness, or a baryon state with $S = +1$.

The quark model has other remarkable successes. Since the "stable" baryons are assumed to be three-quark states without orbital motion, their magnetic moments must arise from the magnetic moments of the quarks. If it is assumed that the magnetic moments of the various quarks are proportional to their charges, i.e. that the mean energy of a triplet does not change in an external magnetic field, then our earlier assumption that the baryon spin is determined by the coupling of the quark spins leads directly to predictions for magnetic moments of all the baryons in terms of a basic quark magnetic moment, which we shall take to be -1 for the n and λ quarks. Then the p quark is required to have magnetic moment +2 and the calculated values of the magnetic moments of the members of the baryon octet are shown in Table I and compared with the experimental values expressed in nuclear magnetons. It will be seen that a choice of the quark unit a little smaller than one nuclear magneton leads to good agreement with all the measurements. Furthermore, the model yields a value for the M1 matrix-element in $\Delta^+ \rightarrow p\gamma$, which is measured in pion

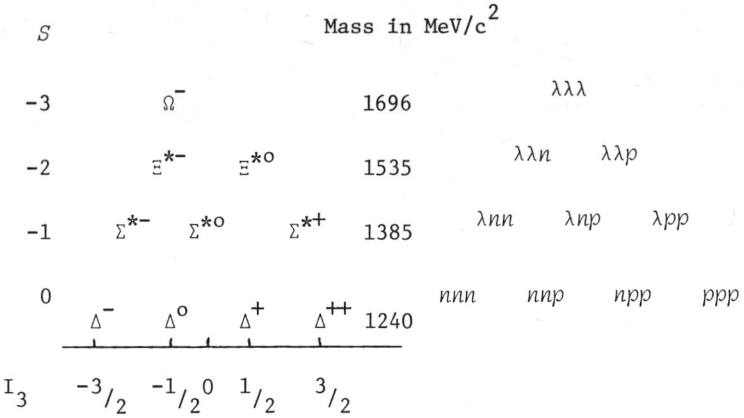

Fig. 3. The $^3/_2{}^+$ baryon decuplet and its description in the quark model.

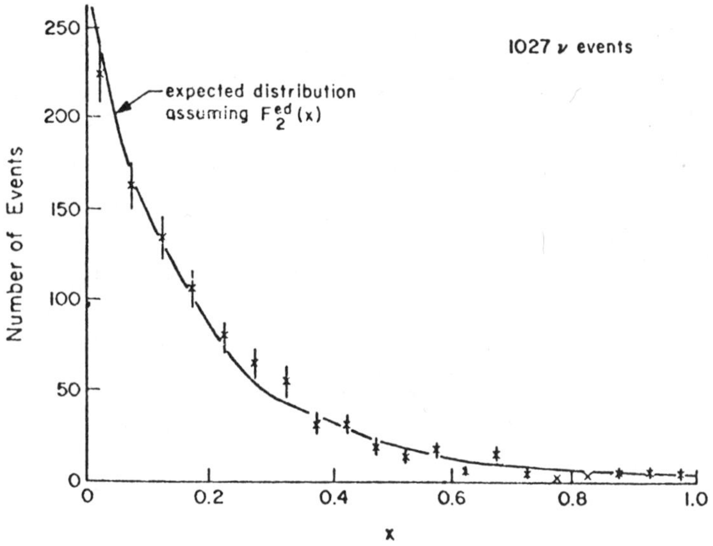

Fig. 4. The fraction of momentum carried by nonstrange quarks in the nucleon as measured by neutrinos (crosses) and by electrons (full line). From F. Sciulli, Proc. XVII Intl. Conf. on High Energy Physics, London 1974, J.R. Smith ed., Rutherford Laboratory, Chilton, 1974.

TABLE I

Baryon Magnetic Moments

Particle	Quark model (in quark units)	Experimental value (in nuclear magnetons)
p	3	+2.79
n	-2	-1.93
Λ	-1	-0.8 ± 0.08
Σ^+	+3	+3.28 ± 0.58
Ξ^-	-1	-2.2 ± 0.8
Σ^-	-1	-1.48 ± 0.37

photoproduction in the Δ^+ resonance region, which agrees with measurements within 20%. Similar predictions for mesons, although less easy to test experimentally, are consistent with the observations.

Another area in which the quark model has been strikingly successful is in the interpretation of the experiments on so-called deep inelastic scattering from protons and neutrons [8]. In deep inelastic scattering of electrons from protons, for example, high energy electrons strike a proton target and measurements of the scattered electron provide information about the internal motion of the charges within the proton. In the parton picture, proposed by Feynman, it is convenient to describe the collision in a co-ordinate system in which the target particle is moving with a very high momentum p corresponding to a velocity close to the velocity of light. Since it is known that transverse momenta are severely damped in hadron dynamics, it is not a bad approximation to neglect the transverse components of the motion in this frame. Then the state of motion of any constituent part, or "parton", is characterized in this frame by the fraction x that it carries of the target particle's total momentum p. Because of time dilatation, the state of internal motion of the partons changes very little during the collision, therefore it is permissible to consider the collision with one parton as being essentially free and independent of the influence of the others. The value of x for the parton which deflected the electron can be deduced from the energy and angle of the outgoing electron, and by studying the distribution of scattered electrons, one can infer the distribution F(x) of the partons which are effective in scattering electrons. If, taking a more general viewpoint, we think of a proton as being made up of the three kinds of quarks (and their antiquarks) then six distribution functions are required in general to represent the distribution of momentum over the different kinds of constituents. Deep inelastic electron scattering from protons then yields the weighted distribution

$$F_2^{ep}(x) = x(\frac{4}{9}[p(x)+\bar{p}(x)]+ \frac{1}{9}[n(x)+\bar{n}(x)+\lambda(x)+\bar{\lambda}(x)]) \qquad (3)$$

where p(x) is the probability that a p quark carries fractional momentum in the range (x,x+dx), etc. The factors $^4/_9$ and $^1/_9$ appear because the effectiveness of a quark in scattering an electron is proportional to the square of its charge. By charge symmetry, one obtains a corresponding distribution for neutrons

$$F_2^{en}(x) = x(\frac{4}{9}[n(x)+\bar{n}(x)]+ \frac{1}{9}[p(x)+\bar{p}(x)+\lambda(x)+\bar{\lambda}(x)]) . \qquad (4)$$

Consistency of the theoretical description adopted can be tested by using different kinematic configurations corresponding to the same value of x and seeing if the same distribution functions are obtained. Actually, the situation is slightly more complicated because there is more than one kind of interaction involved: electrons can be scattered through exchange of longitudinal or transverse photons. The angular variation of electron scattering allows one to separate the two contributions and the results show that the scattering is predominantly through the exchange of transverse photons, corresponding to the preponderance of spin-½ partons and therefore consistent with the quark hypothesis. It is then found, further, that the results do indeed scale, i.e. the observed distributions at different energies and scattering angles lead to the same values of F(x) in agreement with the parton picture.

The actual values of the parton charges could be determined from the experimental results if the numbers of n,p etc. quarks could be found in some other way. This information is obtained from the corresponding deep inelastic scattering experiments with neutrino and antineutrino beams. If we disregard for the moment strangeness-changing weak processes which are expected to occur at a relative rate of $\sin^2\theta$, the basic reactions possible with (μ-type) neutrinos and antineutrinos are

$$\nu n \to \mu^- p, \nu\bar{p} \to \mu^-\bar{n}, \bar{\nu}p \to \mu^+ n, \bar{\nu}\bar{n} \to \mu^+\bar{p}$$

If we assume that quarks (antiquarks) interact in the extreme relativistic limit in the same way as leptons (antileptons), then the scattered muon will have a different angular distribution when produced in a νn collision as compared to a $\nu\bar{p}$ collision. A consequence of this hypothesis, which is experimentally verified together with the predicted angular variation, is that neutrino cross-sections on nucleons should exceed antineutrino cross-sections by a factor of 3. Thus, by measuring the deep inelastic production of μ^- in neutrino collisions, we can deduce the distributions $n(x)$ and $\bar{p}(x)$ of n and $\bar{p}$ quarks respectively. In the same way, μ^+ production by antineutrinos will yield the distributions $p(x)$ and $\bar{n}(x)$. In this way, one can determine separately the distributions of all nonstrange quarks and antiquarks in the target. Experimental data available so far refer to targets which contain approximately equal numbers of neutrons and protons, which therefore yield these distributions for "deuteron" targets. From these results, one deduces that antiquarks are relatively rare - except at very small x, the data are consistent with zero for their probability - and that $q(x) = x[p(x)+n(x)]$ is a smoothly decreasing function of x from about $x = 0.1$ to $x = 1$. Another interesting result, from a relatively low energy CERN experiment, is [9]

$$\int [p(x)+n(x)-\bar{p}(x)-\bar{n}(x)]dx = 3.2 \pm 0.6$$

The net quark content of a nucleon, measured in this way, is consistent with the quark model expectation. Knowing the function $q(x)$, we are in a position to make a quite stringent test of the quark model. From (3) and (4), we have

$$F_2^{e(p+n)} = \frac{5}{9}\, q(x) + g(x)$$

where we have lumped together in $g(x)$ the sum of the positive-semidefinite contributions to electron scattering by antiquarks and strange quarks, which should be zero in the naive quark model and is expected to be small. When the comparison is made, see Fig. 4, the agreement between the electron and neutrino results is amazingly good, simultaneously confirming the rarity of strange quarks and antiquarks and the Gell-Mann Zweig assignment of fractional charges to the nonstrange quarks.

One further result from these measurements is that one can evaluate the fraction of the total momentum carried by quarks and antiquarks:

$$\int_0^1 x\{p(x)+\bar{p}(x) + \ldots\}dx \simeq 0.5$$

This means that about half the momentum of a nucleon, and presumably the same holds for other baryons, is carried by partons which do not participate in weak and electromagnetic interactions. These inert constituents may be identified with the "gluons" which hold the quarks together.

So far we have said very little about the problems of the quark model. The first and most obvious one is why they have not yet been seen. With their distinct fractional charges, they should be very easy to recognize and since they are the basic building blocks of hadronic matter, they should be strongly coupled to the known hadrons. One explanation of their relative rarity, that they are too massive to be produced at available energies, sets a lower limit of several GeV/c^2 on their mass, assuming a reasonable cross-section for their production above threshold. This raises more questions than it answers because quarks must then be bound very strongly in the known particles and the success of the quark model becomes very mysterious. Others have sought to make the non-observability of quarks a guiding principle, which should prove an important clue to quark dynamics, but a fully convincing scheme is yet to be presented.

The next question refers to the statistics obeyed by the quarks. If the spin and SU(3) wavefunctions are totally symmetric, as they are chosen to be in the quark model of baryons, then one would normally expect the space wavefunction to be fully antisymmetric for spin-½ particles obeying Fermi statistics. The quark model is much happier with a symmetric space wavefunction. To give just one reason, it would be difficult to avoid nodes in the electromagnetic form factors of the nucleon if the space wavefunction were totally antisymmetric. The simplest solution to this problem is to introduce a three-valued internal co-ordinate so that the overall quark wavefunction can satisfy the Pauli principle despite being totally symmetric in spin, space, and SU(3) variables, by having a wavefunction which is fully antisymmetric in the new co-ordinate. This internal variable has been named "colour" by Gell-Mann, and it is natural to postulate that there is SU(3) symmetry with respect to transformations in this internal space also. To avoid increasing the multiplicity of hadron states, it is postulated that all known hadron states are singlets of $SU(3)_{colour}$, which is just another way of saying that the colour wavefunction is required to be a fully antisymmetric combination of the three coloured units. This requirement can be related to the absence of bound states of four or more quarks. Just as in chemistry, where a valence bond is saturated when a pair of electrons with opposite spins is paired, here saturation will be reached when there is one quark of each possible colour. This explanation can be given a dynamical basis by requiring the forces between quarks to arise from the exchange of "gluons" which exchange the colours of the quarks. $SU(3)_{colour}$

symmetry would require the existence of an octet of such "colour" gluons. Non-observability of quarks could be assured if one could arrange that the gluons are unobservable, for example by making them very massive.

A further increase in the number of quarks is suggested, in unified theories of weak and electromagnetic interactions, by the absence of strangeness-changing neutral currents. These are very difficult to avoid in any theory based on the usual SU(3) triplets, but can be neatly removed if one adds [10] a fourth quark p', with charge + $^2/_3$, which forms a doublet with the λ quark analogous to the (n,p) doublet. This fourth quark must transform as a singlet under ordinary SU(3) transformations, and the quantum number which distinguishes it from the original quarks is called "charm". It is tempting to postulate an approximate overall SU(4) symmetry, which is probably even more approximate than SU(3) since charmed particles are yet to be seen. Including the colour degree of freedom, one now has 4x3 different fundamental units, and Pati and Salam have speculated [11] that a higher symmetry might correspond to adding the four observed leptons $(e^-,\nu_e,\mu^-,\nu_\mu)$ to this array. According to this view (which can also avoid fractional electric charges for the basic units), there is no fundamental distinction between baryons and leptons, except that leptons are more basic, and hadrons may, for example, decay into leptons. Fortunately, it can be arranged that this occurs very slowly indeed.

We are now clearly on very speculative ground, but the recent discovery of very sharp resonances in e^+e^- annihilation at 3.1 and 3.7 GeV definitely supports the view that hadrons possess additional internal degrees of freedom.

NUCLEAR AND ATOMIC PARITY NONCONSERVATION

Except for the "neutral current" events reported in recent high-energy neutrino reactions, the simplest description of the known weak interactions is in terms of a charged current interacting with itself:

$$H_w = \frac{G}{\sqrt{2}} J_\alpha^+ J_\alpha \qquad (5)$$

where the hadronic part of the current J_α can be written in terms of quark fields as

$$J_\alpha = \bar{p}\gamma_\alpha(1+\gamma_5)[n\cos\theta+\lambda\sin\theta] = j_\alpha^{\pi^+}\cos\theta + j_\alpha^{K^+}\sin\theta \qquad (6)$$

where the superscripts denote the transformation properties of the two pieces of the current. An immediate consequence of the current-

current interaction hypothesis is that (5) contains a part

$$H^{\Delta S=0} = \frac{G}{\sqrt{2}} \left(j_\alpha^{\pi^+} j_\alpha^{\pi^-} \cos^2\theta + j_\alpha^{K^+} j_\alpha^{K^-} \sin^2\theta \right) \tag{7}$$

which will add a small parity-nonconserving part to nuclear forces. Inclusion of neutral current interactions will add further terms to (7).

The Hamiltonian (5) describes a zero-range contact interaction, thus influence of the parity-nonconserving interaction (7) in nuclear phenomena will be rather sensitive to the nature of short-range correlations between nucleons. The relatively long-range component of the parity-violating nucleon-nucleon potential which arises from one-pion exchange as a consequence of parity-nonconserving pion-nucleon interactions induced by (7) arises only from the second term [12], and is therefore expected to be suppressed.

Table II summarizes the main experimental results and the corresponding theoretical estimates, taken from a recent review by Tadic [13]. The experimental limit for the parity-forbidden decay mode of the 2^- level of ^{16}O is consistent with the theoretical limit based on ρ-exchange. For the radiative transitions, except for the case of ^{180}Hf, where the relatively large parity-violating effect which is seen is associated with a highly forbidden transition, the reported experimental effects, which measure essentially the admixture of the "wrong" parity in nuclear eigenstates, are of the order of magnitude which one might expect from dimensional considerations: $Gm_p^2 \sim 10^{-5}$. The influence of the repulsion between nucleons tending to keep them apart (often represented by a hard core in the nucleon-nucleon potential) reduces the effectiveness of the theoretically predicted short-range parity-violating potential, consequently the detailed theoretical estimates are considerably smaller, two or perhaps even three orders of magnitude below the measurements and with the wrong sign! To explain the discrepancy, it has even been suggested that the electromagnetic interaction might itself violate parity. It is probably too early to resort to such extreme hypotheses. Neutral current effects could very well enhance the degree of parity admixture and, in particular, there is no $\sin^2\theta$ suppression of one-pion exchange if the Glashow-Iliopoulos-Maiani [10] explanation of the absence of $|\Delta S|=1$ currents is correct. In any case, the experiments should be repeated, especially the one on np capture γ rays since this is the case least subject to uncertainties arising from nuclear physics.

Shortly after the discovery of parity nonconservation in β-decay, it was noted by Zeldovich that if a similar interaction

TABLE II

Summary of Data on Nuclear Parity Nonconservation
(from D. Tadic, Ref. 13)

Experiment	Exp. Result	Change of isospin	Meson-exchange potential M_V	Theoretical predictions
O^{16}(8.82 MeV) $C^{12} + \alpha$	$\Gamma_\alpha^{PNC} \cong 10^{-10}$ eV	$\Delta I=0$	V (Vector)	0.5×10^{-10} eV $\leq \Gamma_\alpha$ $\Gamma_\alpha \leq 7\times10^{-10}$ eV
$n+p \to d+\gamma$ Circular polarization	$P_\gamma \cong (-)1.2\ 10^{-6}$	$\Delta I=0,2$	V (Vector)	$P_\gamma \approx +10^{-8};\ +10^{-9}$
$Ta^{181*} \to Ta^{181}+\gamma$	$P\gamma \cong (-)5\times10^{-6}$	$\Delta I=0,1,2$	V (Vector) V (pion)	$P_\gamma \approx +2\times10^{-7}$
$Hf^{180*} \to Hf^{180}+\gamma$	$P_\gamma = 2.5\times10^{-3}$ $A_\gamma = -1.6\times10^{-2}$	$\Delta I=0,1,2$	V (Vector) V (pion)	

existed between electrons and nuclei, small parity admixtures would be introduced into atomic states. In the zero-range approximation, the parity-violating potential between an electron and a spinless nucleus has the form

$$H' = \frac{GQ_W}{4\sqrt{2}\, m_e} [\vec{\sigma}.\vec{p}\ \delta(\vec{r}) + \delta(\vec{r})\ \vec{\sigma}.\vec{p}] \; . \tag{8}$$

Q_W is the "weak charge" of the nucleus; in Weinberg's model, for example [14],

$$Q_W = (4\sin^2\theta_W - 1)\ Z + N \; .$$

The effect of the perturbation (8) is to admix into $s_{1/2}$ states, which have a non-vanishing amplitude at the origin, a small component of $p_{1/2}$ states. The admixture of a particular $p_{1/2}$ state into a $s_{1/2}$ state will be

$$\langle p_{1/2}|H'|s_{1/2}\rangle/(E_p - E_s) \tag{9}$$

which can be estimated to be of order

$$\frac{GQ_W}{m_e} \cdot \left(\frac{Z}{a_o}\right)^4 / Z^2 Ry \sim \frac{GQ_W (Z\alpha m_e)^4}{m_e Z^2 \alpha^2 m_e} = (Gm_e{}^2) Q_W Z^2 \alpha^2 \; . \tag{10}$$

The characteristic parameter determining parity admixture is $(Gm_e{}^2)\alpha^2 \sim 10^{-16}$ and one must go to heavy atoms to have any chance of observing such parity-admixtures. The enhancement for higher Z is probably somewhat greater than indicated in (10), which was obtained using hydrogenic estimates both for the wavefunction at the origin, which enters the evaluation of the numerator, and for the energy-denominator in (9). In an actual atom of high Z, the hydrogenic approximation is a good one for the wavefunction near the origin, where the electron sees the full nuclear charge, but because of electron screening, the energy-levels for an outer valence electron correspond to a much lower value of Z. With $Q_W \sim Z$, one may therefore hope for a parity-admixture of order 10^{-8} in heavy atoms. By looking at suitable forbidden transitions, one hopes to find observable indications of parity violation. Bouchiat and Bouchiat [15] have estimated that the small E1 matrix-element which would be added as a result of such parity-admixture to the forbidden 7s → 6s M1 transition in the Cs atom would lead to a circular polarization of the emitted photon of order 10^{-4}.

Unfortunately, the sought-for line is so weak that it has yet to be seen.

It has been remarked by Moskalev and by Bernabeu, Ericson and Jarlskog [16] that the parity-admixtures may be considerably enhanced in muonic atoms. The muon density at the origin is $(m_\mu/m_e)^3$ times greater than for electrons but the energy level differences are also (m_μ/m_e) times greater, so that one apparently gains only a factor of $(m_\mu/m_e)^2$. However, one can exploit the fact that the relatively large vacuum polarization effect and the energy-shift due to finite nuclear size work in opposite directions, to find a nucleus where the $2S_{\frac{1}{2}}$-$2P_{\frac{1}{2}}$ splitting becomes as small as possible. This is expected to occur for muonic ^6Li, where relatively large effects are predicted. If polarized muons are captured in Bohr orbits around ^6Li the emitted photon in the $2P \rightarrow 1S$ transition is expected to be distributed asymmetrically with respect to the muon spin, with an asymmetry factor of several percent. The subsequent decay of the muon serves as an analyzer for its spin, thus one expects a parity-violating correlation of the photon and decay electron directions of this order of magnitude.

A related effect is the rotation of the plane of polarization of light after passing through an assembly of atoms. If atomic states are not eigenstates of parity, the forward scattering amplitudes, and consequently also the refractive indices, for right- and left-circularly polarized light will be unequal and such an effect will arise. The angle of rotation per unit distance traversed is given by

$$\Phi = \lambda N f_s \tag{11}$$

where λ is the wavelength, N the number density of atoms, and f_s the circular-polarization dependent part of the forward scattering amplitude for light. Using the estimate (10) for the parity-admixture in atomic states, we obtain

$$\Phi \sim \lambda N \frac{e^2}{m_e c^2} G m_e^2 \alpha^2 Z^2 Q_W \tag{12}$$

For heavy atoms, with $Q_W \sim Z$ and $N \sim 10^{19}$ cm^{-3}, one obtains $\Phi \sim 10^{-8}$ cm^{-1} in the optical region. To maximise the effect, one should use light of a frequency close to one of the desired forbidden transitions. Experiments to detect this effect with Bi gas are under way at several laboratories.

A related effect is expected for polarized neutrons passing through matter. The parity-violating interaction between neutrons and nuclei arising from the interaction (5) and possible similar interactions between neutrons and electrons will cause forward scattering amplitudes for neutrons to contain a small part proportional to the component of the spin along the direction of motion:

$$f_s \sim GQ_W(\vec{\sigma}.\vec{p})$$

Analogous to Eq. (11), we have a similar formula for the rotatory power for neutrons,

$$\Phi \sim \lambda N f_s \sim GQ_W N$$

For $N \sim 10^{22}$ cm^{-3}, and $Q_W \sim 10^2$, we find $\Phi \sim 10^{-8}$ cm^{-1}, which may be detectable before too long.

TIME-REVERSAL INVARIANCE?

According to the hypothesis of TCP-invariance, the observed violation of CP-invariance must be associated with a corresponding departure from T-invariance. Despite many searches which are still continuing, there is as yet no experimental evidence in support of this theoretical conclusion. At the same time, it has been shown that the observed facts of K^0-decay cannot be easily reconciled with the hypothesis of time-reversal symmetry. As long as all CP-noninvariance phenomena are restricted to the neutral kaon system, whose peculiar properties allow a super weak CP-noninvariant interaction to account for the observed effects, the only place where one may reasonably expect to observe a breakdown of T-invariance is in the neutral kaon system itself if we accept the theoretical premise that CP-noninvariance is accompanied by T-noninvariance. The effect that one may expect to see was predicted [18] some years ago but, as far as I am aware, no one has attempted to measure it. In view of the importance of the issue involved, we review the relevant argument. Since weak interactions do not conserve strangeness (hypercharge), a K^0 meson can, and does occasionally, transform itself into a $\bar{K}^0$ meson in the course of time. Similarly, a state produced initially as $\bar{K}^0$ may be found later to have changed into K^0. Time-reversal invariance, or microscopic reversibility (sometimes also called reciprocity) would require all details of the second process to be deducible from the first; in particular, it should proceed at a rate exactly equal to that of the $K^0 \rightarrow \bar{K}^0$ transformation, since time-reversal simply reverses the process in this case.

From a purely phenomenological analysis of the K^0-$\bar{K}^0$ system, where $K^o{}_S$ and $K^o{}_L$ are treated as two unstable spin zero states* which happen to have masses (energies) rather close to each other - in a situation similar to the $2S_{1/2}$ and $2P_{1/2}$ states of the hydrogen atom for example - i.e. without making any symmetry assumption whatsoever, one can show that the short lived and long lived neutral kaon states are superpositions [19]

$$|K^o{}_S\rangle = \cos(\tfrac{\pi}{4} - \delta_S)e^{i\beta}|K^o\rangle + \sin(\tfrac{\pi}{4} - \delta_S)e^{-i\beta}|\bar{K}^o\rangle \qquad (13)$$

$$|K^o{}_L\rangle = \cos(\tfrac{\pi}{4} - \delta_L)e^{-i\beta}|K^o\rangle - \sin(\tfrac{\pi}{4} - \delta_L)e^{i\beta}|\bar{K}^o\rangle \qquad (14)$$

which differ only slightly from the states of well-defined CP-symmetry

$$|K_+{}^o\rangle = (|K^o\rangle + |\bar{K}^o\rangle)/\sqrt{2} \qquad (15)$$

$$|K_-{}^o\rangle = (|K^o\rangle - |K^o\rangle)/\sqrt{2} \qquad (16)$$

which they would be if CP-invariance were universally valid. The parameters δ_S, δ_L, and β cannot exceed 10^{-2} in magnitude.

By inverting Eqs. (13), (14), we can express K^o and $\bar{K}^o$ in terms of K^o_S and K^o_L , which have a simple exponential time-dependence, and thereby follow the time development of states produced initially as K^o or $\bar{K}^o$. From such equations we can calculate the probability $P_{\bar{K}K}(\tau)$ of finding a state prepared initially as K^o to be in a $\bar{K}^o$ state at time τ, and similarly the probability $P_{K\bar{K}}(\tau)$ for the inverse transformation. The time-dependence of the two transition rates is found to be the same, so that their ratio is independent of time, and we can define a time-independent time-asymmetry factor

$$\mathcal{A}_T \equiv \frac{P_{K\bar{K}}(\tau) - P_{\bar{K}K}(\tau)}{P_{K\bar{K}}(\tau) + P_{\bar{K}K}(\tau)} = \frac{2\sin(\delta_L+\delta_S)\cos(\delta_S-\delta_L)}{1 + \sin 2\delta_S \sin 2\delta_L} \qquad (17)$$

---

*We also implicitly assume that there is no other close lying state which is significantly admixed with these two as a consequence of strong or weak interactions. This assumption is subject to experimental test, and there is some evidence in favor of it.

where the last expression is the result of the calculation we have described. To lowest order in δ_S, δ_L, β it can also be expressed as

$$\mathcal{A}_T \approx 2 \text{ Re } \langle K^o{}_L | K^o{}_S \rangle \tag{18}$$

Now the last expression can be reexpressed, using the Bell-Steinberger unitary relation [20], in terms of $K^0{}_L$ and $K^0{}_S$ decay amplitudes. Defining $\Delta = (m_L - m_S)/(\gamma_L + \gamma_S)$ we have

$$\text{Re}\langle K^o{}_L | K^o{}_S \rangle = 2(1+4\Delta^2)^{-\frac{1}{2}} \text{Re}[e^{-i\phi_w} \sum_j \eta_j \gamma_S^j / (\gamma_S + \gamma_L)] \tag{19}$$

where $\phi_w = \tan^{-1}(2\Delta)$, $\gamma_S{}^j$ is the partial rate of $K^0{}_S$ decay into the channel j and η_j is the complex amplitude ratio

$$\eta_j = \frac{A(K^0{}_L \to j)}{A(K^0{}_S \to j)}$$

Thus

$$\mathcal{A}_T = 4(1+4\Delta^2)^{-\frac{1}{2}} \text{Re}[e^{-i\phi_w} \sum_j \eta_j \gamma_S^j / (\gamma_S + \gamma_L)]$$

The $\pi^+\pi^-$ and $\pi^0\pi^0$ channels account for all but a tiny fraction of all $K^0{}_S$ decays. We should therefore expect that most of the contributions to the sum on the right-hand side of Eq. (19) would come from these channels. The relevant experimental parameters have been measured with relatively high precision, and if we neglect γ_L/γ_S in comparison with unity (an approximation which is justified by the accuracy of the data) and set $C = \gamma_S{}^{+-}/\gamma_S{}^{00}$, the 2π contributions to the second factor in the equation written above are given by

$$[C|\eta_{+-}|\cos(\phi_{+-} - \phi_w) + |\eta_{00}|\cos(\phi_{00} - \phi_w)]/(C+1) \tag{20}$$

Inserting the reported values [21] of C, η_{+-}, η_{00}, and $m_L - m_S$, γ_L, γ_S (which determine ϕ_w), the quantity (20) may be estimated as $|\eta_{+-}| \approx 2.3 \times 10^{-3}$. Contributions to the sum in (19) from other known channels are expected to be much smaller. In the absence of phase information, we can bound their contribution by using the partial decay rates $\gamma_S{}^j$ and $\gamma_L{}^j$ and Schwartz's inequality. Where even the partial decay rates are unknown, we must use experimental upper limits for those rates. A conservative limit based on available data yields 4×10^{-4} as the upper limit to the magnitude of such contributions. Thus, the contribution of other channels to the sum in (19) could not possibly cancel out the positive contribution from 2π channels given by (20). Therefore, we should definitely expect to see a positive time-asymmetry of magnitude between 5.5×10^{-3} and 7×10^{-3}, i.e. $\bar{K}^0$'s should transform into K^0's at a rate approximately 1.2% faster than K^0's transform into $\bar{K}^0$'s. Since, according to the vivid interpretation given by Landau, CP-invariance requires the mirror-image of a physical process to describe the corresponding process with particles and anti-particles interchanged, such an effect would directly demonstrate CP-noninvariance at the same time as it demonstrates T-noninvariance.

The only possible catch in the argument presented above is that measurements of the $K^0{}_L$ lifetime and the corresponding partial decay rates into various channels are still sufficiently imprecise to admit the possibility that "unknown" decay channels, viz. decay channels not explicitly identified thus far, account for as much as 10% of all $K^0{}_L$ decays. If that is the case, and the same channels account for 1% of all $K^0{}_S$ decays (this is approximately the present accuracy on measurements of the $K^0{}_S$ lifetime), then it is logically possible for those "unknown" channels to contribute to the sum in (19) an amount which cancels the contribution from known channels. Thus, if one is sufficiently attached to T-invariance one can make a last attempt to save it by invoking the possible contributions from such "unknown" channels. A very similar suggestion is that by Faissner [22] and Kenny and Sachs [23] that T-invariance may be rescued by giving up unitarity or what is equivalent, Hermiticity. This escape route could be closed by improving the measurements on $K^0{}_L$ decays sufficiently to check that the total rate of $K^0{}_L$ decay agrees with the sum of the partial rates to an accuracy of about 1%, or more directly by measuring the time asymmetry $\mathcal{A}_T$, Eq. (17) directly. A convenient way to make the measurement is to take advantage of the now well verified $\Delta S = \Delta Q$ rule. Then $\pi^- \ell^+ \nu$ and $\pi^+ \ell^- \bar{\nu}$ decays in a neutral kaon beam provide indicators of the K^0 and $\bar{K}^0$ content, respectively. The first could be used in decays of a beam produced initially as $\bar{K}^0$ to measure $\bar{K}^0 \to K^0$ conversion while the second would similarly yield the rate of $K^0 \to \bar{K}^0$ transitions. It appears that techniques for measuring electronic decay modes of neutral kaons have attained the precision to detect an $\mathcal{A}_T$ of the predicted magnitude. I hope the experiment will soon be performed.

To show that there are really no other hidden assumptions in the foregoing analysis, I shall illustrate the argument with a hydrodynamical model. Since Einstein, the greatest and most successful exponent of invariance principles, once expressed the wish that he had been a plumber, such an analogy may not be out of place in these lectures. Consider two identical systems containing water, each fitted with a drain pipe and a pump which pumps water to the other tank. Assume that the rate of loss of water from each tank is proportional to the pressure head in each case and that the rate of pumping from each tank to the other is also proportional to the level in the originating tank. This provides a model of the neutral kaon system, see Fig. 5. The drains represent decay channels while the pumps simulate the processes $K^0 \rightleftarrows \bar{K}^0$, e.g. through a 2π intermediate state. By virtue of the $\Delta S = \Delta Q$ rule, the observed leptonic charge asymmetry in $K^0{}_L$ decays shows that the K^0 and $\bar{K}^0$ levels are unequal in the state $K^0{}_L$ $(\sin(\frac{\pi}{4} -\delta_L)>\cos(\frac{\pi}{4} -\delta_L))$. We shall now show that if the ratio is to be maintained as time elapses (so that the leptonic charge asymmetry does not vary with time in a $K^0{}_L$ beam), we <u>must</u> have T or TCP violation. First suppose that the drain pipes for the two tanks are identical. Then the higher pressure in K^0 would lead to a higher rate of loss from the K^0 tank than from the $\bar{K}^0$ tank so that in the absence of the pumps, the levels would tend to equalize. With the pumps operating, the higher level in K^0 would cause more transfer to $\bar{K}^0$ than is received in return, and thus further tend to equalization of levels, if the two pumps were also identical. Therefore, the $\bar{K}^0 \rightarrow K^0$ pump must work harder than

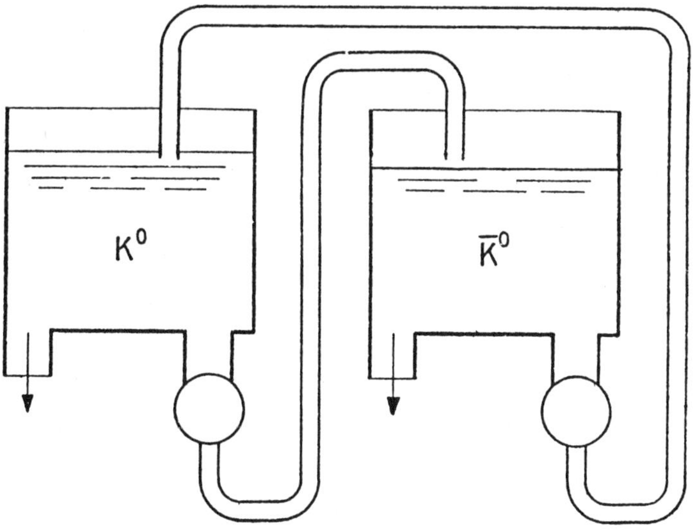

Fig. 5. Hydrodynamical model for neutral kaon decay.

the $K^o \to \bar{K}^o$ pump. Conversely, if the pumps were identical, the drain of the K^o tank must be narrower than for the other tank. The first case corresponds to T violation because it requires the rate of $K^o \to \bar{K}^o$ conversion to exceed that for $\bar{K}^o \to K^o$. The second represents TCP noninvariance because it requires K^o and $\bar{K}^o$ to decay at different rates. Of course, it is also possible to have a combination of T noninvariance and TCP noninvariance. But at least one of the symmetries must be violated. In this discussion we have assumed strict unitarity, i.e. that there is no disappearance of "kaon fluid" other than what can be accounted for by the known decay channels. Suppose, on the other hand, that there was evaporation, and that the rate was greater from the K^o tank than from the $\bar{K}^o$ tank. Then the constant unequal ratio of the levels in the two tanks could be maintained, solely as a consequence of the differential rate of evaporation, even if the two pumps <u>and</u> drain pipes were identical. This corresponds to a failure of unitarity, in a TCP violating manner, corresponding to the suggestion of Faissner and Kenny and Sachs.

REFERENCES

[1] For a slightly more expanded discussion of these arguments, see e.g. P.K. Kabir in "Symmetries in Elementary Particle Physics", A. Zichichi, ed. (Academic Press, New York & London, 1965).

[2] E.P. Wigner, Nobel Lectures (Elsevier, Amsterdam & New York, 1964).

[3] W. Heisenberg, Introduction to the Unified Field Theory of Elementary Particles (Interscience, London, 1966).

[4] See e.g. footnote 20 of R. Houtappel, H. Van Dam, and E.P. Wigner, Rev. Mod. Phys. 37 (1965) 595.

[5] J.J. Sakurai, Phys. Rev. Lett. 10 (1963) 4461.

[6] M. Gell-Mann and Y. Ne'eman, The Eightfold Way (Benjamin, New York, 1965).

[7] For a review, see G. Morpurgo, Ann. Rev. Nucl. Sc. 20 (1970) 105.

[8] R.P. Feynman, Proc. Intl. Conf. on Neutrinos, Philadelphia 1974, (American Inst. of Physics, New York 1974).

[9] Denen et al. Nucl. Phys. B85 (1975) 269.

[10] S. Glasgow, J. Iliopoulos, and L. Maiani, Phys. Rev. D2 (1970) 675.

[11] J.C. Pati and A. Salam, Phys. Rev. Lett. 34 (1975) 617.

[12] G. Barton, Nuovo Cim. 19 (1961) 512.

[13] D. Tadic, Proc. GIFT Seminar 1974, Zaragoza, 2, 168.

[14] S. Weinberg, Phys. Rev. D5 (1972) 1412.

[15] M. Bouchiat and C. Bouchiat, Phys. Lett. 48B (1974) 111.

[16] J. Bernabeu, T. Ericson, and C. Jarlskog, Phys. Lett. 50B (1974) 467.

[17] F.C. Michel, Phys. Rev. B329 (1964) 133.
[18] P.K. Kabir, Phys. Rev. D2 (1970) 540.
[19] F.S. Crawford, Phys. Rev. Lett. 15 (1965) 1045.
P. Eberhard, Phys. Rev. Lett. 16 (1966) 150.
[20] J.S. Bell and J. Steinberger, Proc. Oxford Intl. Conf. on Elementary Particles 1965 (Rutherford Laboratory, Chilton) (1966).
[21] K. Kleinknecht, Proc. XVII Intl. High Energy Physics Conf. London 1974 (Rutherford Laboratory, Chilton) (1974).
[22] H. Faissner, Proc. XV Intl. Conf. on High Energy Physics, Kiev 1970 (Naukova Dumka, Kiev 1972).
[23] B.G. Kenny and R.G. Sachs, Phys. Rev. D8 (1973) 1605.

MUONIC AND HADRONIC ATOMS

H. Koch

Universität Karlsruhe, Germany

and

CERN, Geneva, Switzerland

1. INTRODUCTION

In these five lectures I will try to give a more or less self-consistent picture of the majority of experiments done so far in the field of muonic and hadronic atoms. The purpose of these lectures is not to confront you with all available data, but rather to explain the ideas and data of selected experiments. A survey of the data existing so far can be found, for example, in review articles and conference contributions[1-9]. In contrast with many of these reviews here I have not tried to treat all atoms separately, but to search for similar features and treat them together. Theoretical considerations will only be taken into account if it is necessary for the understanding of the topics. A detailed discussion of the theory can be found elsewhere[9-13]. I have tried also to cover some very recent ideas and experimental results (presented at the Santa Fe Conference) and I apologize in advance if I have forgotten important contributions. Because of lack of time I was unable to treat the topics of very light exotic atoms ($\mu^- p$, $\pi^- p$, $K^- p$, $\bar{p}p$) where interesting effects occur and of the observation of γ-rays after π^- and K^- absorption. References to these topics are given elsewhere[14-20].

2. PROPERTIES OF THE MUONIC AND HADRONIC ATOMS

Until now X-ray transitions in muonic, pionic, kaonic, antiprotonic and Σ^--hyperonic atoms have been observed. The properties of these particles are listed in Table 1. To show which important contribution, in particle physics also, comes from X-ray measurements in exotic atoms, the quantities derived from such measurements are underlined.

Table 1

	Mass [MeV/c^2]	Mean life-time [sec]	Spin [$\hbar$]	Magnetic moment
μ^-	106	2.2×10^{-6}	½	$1.001\ 166\ 16(31)\mu_\mu$ *)
π^-	<u>140</u>	2.6×10^{-8}	0	0
K^-	<u>493</u>	1.2×10^{-8}	0	0
$\bar{p}$	<u>938</u>	∞	½	$-2.8\mu_N$ **)
Σ^-	<u>1197</u>	1.5×10^{-10}	½	$(-1.5$ or $+0.6)\mu_N$

*) Muonic magneton

**) Nuclear magneton

To understand how an exotic atom is formed we investigate what happens if a negatively-charged particle is stopped in matter. The history of a muon and a hadron is basically different (the muon interacts only weakly and electromagnetically with a nucleus, whereas the hadron feels in addition the strong interaction) and the discussion is therefore started with the muon alone.

A muon of, let us say, 100 MeV is produced in an accelerator and enters a moderator. Via electromagnetic interactions with the electrons of the absorber material it is slowed down in 10^{-11}-10^{-9} sec to an energy of 2 keV (step 1). It has then the same velocity as the electrons of normal electronic atoms. The following process (step 2) consists of the electromagnetic interaction of the muon with the electrons in the neighbourhood of that nucleus, whose Coulomb field finally attracts the muon. The time for this interaction from 2 keV to about 0 keV is 10^{-15}-10^{-14} sec. In the step 3 the muons cascade down to hydrogen-like bound states and reach finally the 1s-ground state, where they stay until they decay (light atoms) or where they are captured via the weak interaction by the nucleus (heavy atoms). The transition of the muon between two hydrogen-like bound states is accompanied by the emission of Auger electrons (transitions in the upper part of the cascade) and X-rays, the emission of which dominates in the lower part of the cascade. The time which is needed for step 3 is strongly dependent on the atomic surrounding of the system under investigation. In normal material (e.g. metal) at medium Z it is about 10^{-15}-10^{-14} sec. An energy level scheme of a muonic atom is given in Fig. 1. As a result of step 2 of the capture process a distribution of muons over the different ℓ values at some high n values occurs. In many cases it has been found out that this distribution is essentially statistical, that means it is proportional to $(2\ell + 1)$. Exceptions to this rule occur and are discussed in Section 5.1. As a result of this distribution the transitions at the edge of the level scheme, that is the transitions between circular orbits $(n, \ell = n - 1) \rightarrow (n - 1, \ell = n - 2)$, are the most intense ones, a fact which could be experimentally verified in many cases. A general word of warning should be said at this stage: the capture process is very badly understood for the moment. It is dependent on a lot of chemical and solid-state effects and cannot be properly calculated (for some recent developments see Section 5.1). Only the X-rays of the lower part of the cascades have been observed until now (no proper Auger-electron spectra are available) and everything which is deduced about initial distribution of high n-levels and the capture process is indirectly obtained and therefore open to doubts. The only thing which can be stated is that the assumption of a statistical (or a little bit modified statistical distribution) is not inconsistent with the data in many cases.

The most important features of the muonic atoms can already be seen in the simple Bohr model. The binding energy of the levels is given by

$$E_B = -\mu c^2 \frac{(Z\alpha)^2}{2n^2} \tag{1}$$

(μ = muon-nucleus reduced mass) and the corresponding Bohr radius is

$$r_n = \frac{\hbar^2}{\mu c^2} \frac{n^2}{Z} . \tag{2}$$

Compared with electronic atoms, the binding energies are about three orders of magnitude bigger. The energies of the 2p-1s transitions,

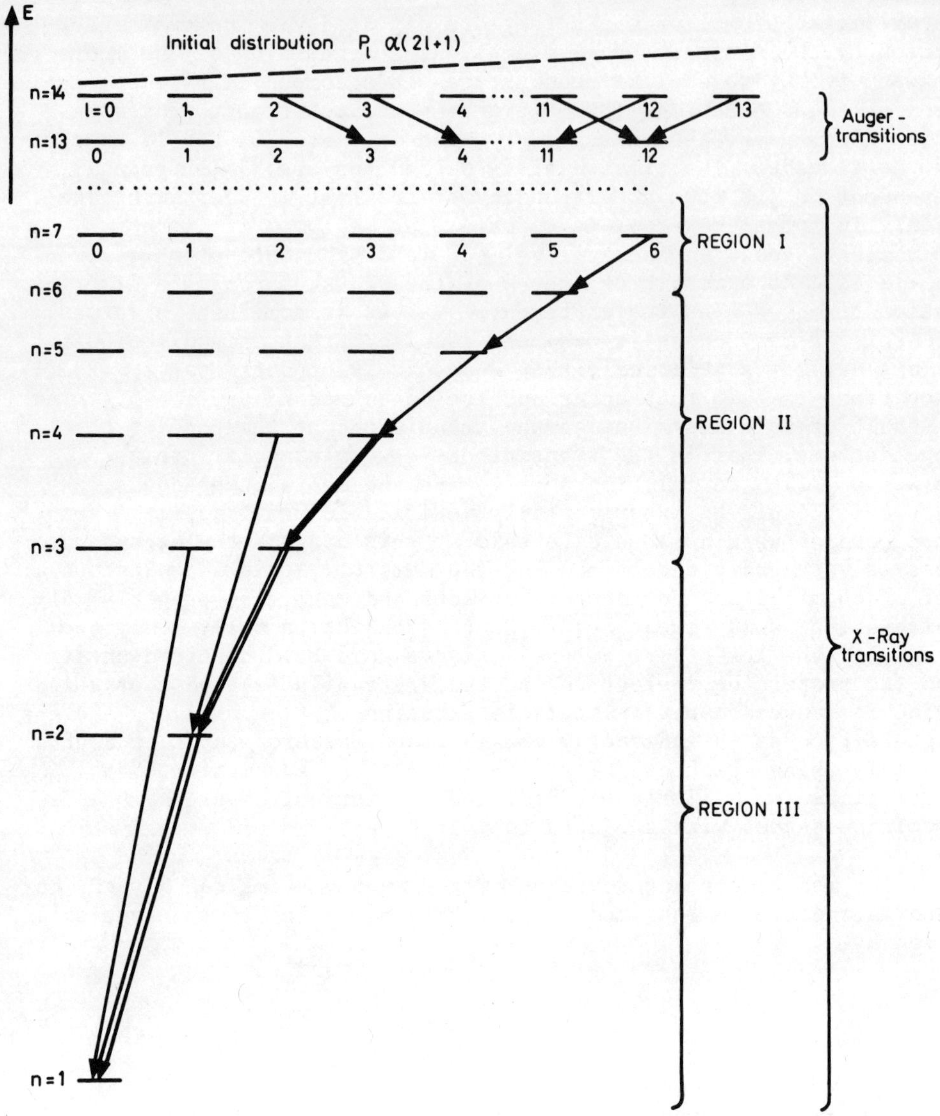

Fig. 1 Muonic atom level scheme

in muonic atoms, for example, range in dependence on Z between some keV and 10 MeV. The radii of the Bohr orbits are three orders of magnitudes smaller than in electronic atoms. In muonic Pb, for instance, the 1s radius is 4 fm, i.e. smaller than the nuclear radius which is 7 fm. This explains the fact that the observation of muonic X-rays yields very precise results about the proton distributions and magnetic moments of the nuclei.

For the strong interacting particles (π^-, K^-, $\bar{p}$, Σ^-) the slowing-down process and the capture process (steps 1 and 2) are assumed to be quite similar to the muonic case, because only electromagnetic processes play a role. The same is true for the upper part of the X-ray cascade, but essential differences occur, when the particle's wave function overlaps a little bit with the nucleus and the short-range strong interaction becomes important. What actually happens is best demonstrated in Fig. 2, where the energy level scheme for the pionic oxygen atom is sketched. When the pion reaches the 2p level the strong interaction becomes important and gives rise to two effects.

i) The elastic π^--nucleus scattering leads to a shift of the 2p level, compared to its purely electromagnetically determined value. This shift (ε_{up}) is very small and cannot be detected with the accuracies reached nowadays.

ii) The pion reacts inelastically with the nucleus and is absorbed, for example by the reaction $\pi^- pn \rightarrow nn$. The pion disappears and no 2p-1s X-ray transition is observed. This process leads to a weakening of the 2p-1s X-ray intensity and gives rise to a broadening of the 2p level width (Γ_{up}) which is comparable to the 2p-1s electromagnetic transition width.

Both processes also happen when the pion has reached the 1s level, but now the effects are about a factor of 1000 stronger, because the overlap between pionic wave function and nucleus has increased by this factor. The width and the shift (Γ_{low}, $\varepsilon_{low} = E^{2p-1s}_{meas.} - E^{2p-1s}_{em}$) of the 1s level become as large as several keV, an effect which can be immediately seen by looking at the energy and line shape of the 2p-1s X-ray. At which levels the strong interaction effects can be observed depends on the charge number of the nucleus Z, the mass (the mass determines the Bohr radius) of the captured particle, and the strength of the strong interaction processes. For example, in the case of $20 \leq Z \leq 30$ for pionic atoms, the strong interactions occur in the 3d and 2p level; the 2p-1s transition is no longer observable. In the same Z range, for kaonic atoms the effects happen already in the 5 g and 4 f levels, all succeeding transitions disappear.

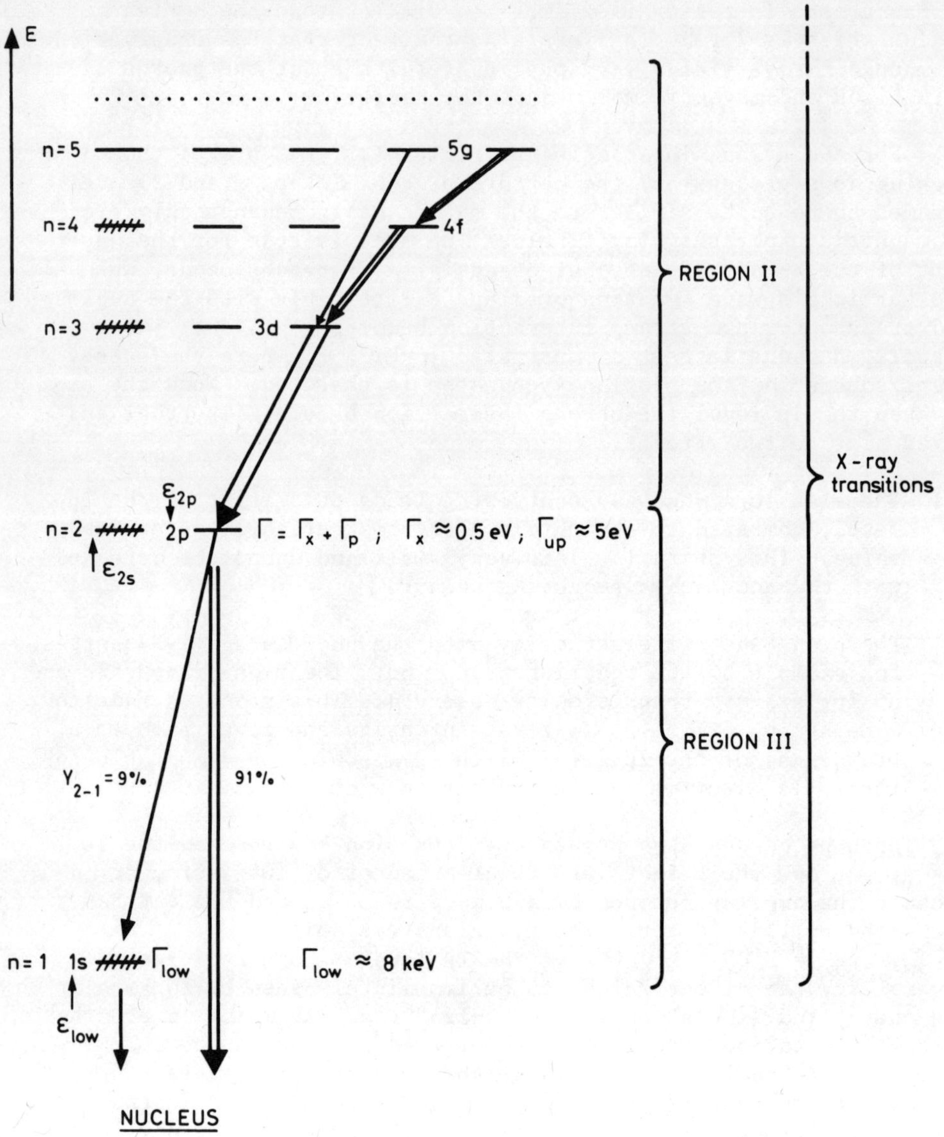

Fig. 2 Hadronic atom level scheme ($\pi^{-}\ {}^{16}O$)

Thus the main difference between hadronic and muonic atoms is the following: In hadronic atoms, there is always a *last observable transition*. It can be the 2p-1s transition (light pionic atoms), but it can also be the 6h-5g transition (heavy antiprotonic atoms). This transition is shifted in energy (compared to the purely electromagnetic value), it is broadened and reduced in intensity.

3. WHAT CAN BE LEARNT FROM MEASUREMENTS OF ENERGIES, INTENSITIES AND LINE SHAPES OF X-RAY TRANSITIONS?

The quantities measured so far in exotic X-ray research are (a) absolute intensities/stopped particle, (b) relative intensities, (c) energies, (d) energy level splittings (only when captured particle has a non-zero spin), and (e) line shapes (Lorentzian widths). Depending on the chosen transition in a cascade one can derive information of a quite different kind from such measurements and in order to illustrate this the X-ray cascade is artificially divided into three regions (see Figs. 1 and 2).

Region I: Influence of the electrons not negligible

Region II: Small electron influence, negligible finite size effects, and strong interaction effects.

Region III: Strong finite nuclear size (muons) and strong interaction effects (π^-, K^-, $\bar{p}$, Σ^-).

3.1 Absolute and relative intensities in regions I, II and III (muons)

The intensity of a selected transition in the cascade depends on the population of the upper level and thus on the interactions which govern the capture and the cascading down of the particle. These electromagnetic interactions depend very sensitively on the electron configuration around the investigated atom and strong *chemical and solid-state effects* are observed in the measurements. For hadronic atoms, these effects can only be observed in regions I and II. Measurements of this type are discussed in Section 5.1.

3.2 Precise energy measurements in region II

The energy of the levels in region II is mainly determined by Eq. (1) -- that means it is determined by the particle mass -- and by smaller corrections, the largest one of which is the vacuum polarization, an effect calculable by QED. Thus, precise energy measurements in region II allow one to determine the *particle masses* and *test QED predictions* with high accuracy (see Sections 5.2 and 5.3). The *electrical polarizability* of the particles yields also a contribution to the level energies, but the effects are for present accuracies still too small. Only upper limits can be deduced.

3.3 Precise measurements of fine structure splittings in region II

The fine structure splitting of a level is proportional to the *magnetic moment* of the captured particle; whereas the magnetic moment of the muon can be very accurately measured by other methods (g-2), such measurements do not exist for antiprotons and Σ^- particles and the most precise results have been obtained from X-ray measurements (Section 5.3).

3.4 Measurements of intensities, energies, and line shapes in region III

3.4.1 Muons

The energies of the s levels are extremely sensitive to the *charge monopole distribution in the nucleus* and very precise *root mean square radii* can be deduced from the data. In deformed nuclei with non-zero spin the *nuclear magnetic moment* and the charge *quadrupole moment* give rise to HFS of the levels. In the low-lying levels the *finite spatial distribution* of the moments yields considerable effects which can be used to test predictions of nuclear models. In the higher levels the HFS effects are smaller and the spatial distribution of the moments becomes negligible. Here the *spectroscopic values of the moments* can be measured, with an accuracy which is often much higher than the accuracy of other (optical) methods. In cases of deformed nuclei, very often low-lying nuclear energy levels can be excited by the muon on its way down to the ground state. This leads to *dynamic effects* in the observed muonic X-ray pattern which are dependent on the *nuclear properties of the excited state*. Thus, for example, also statements about *root mean square radii* and *quadrupole moments of excited nuclear states* can be deduced. Examples for these effects are discussed in Section 5.4.1.

3.4.2 Hadrons

The strong interaction effects (energy shifts, level broadenings, intensity reductions of X-ray lines) are only measurable on the last observable transition. They depend on the following parameters:

a) Elementary elastic and inelastic interaction between the hadron and one unbound nucleon at low energies (in pionic atoms the pion needs at least *two* nucleons to be absorbed; therefore in this case the inelastic elementary process is at least a *two*-nucleon interaction). In principle, the amplitudes describing these processes can be deduced from low-energy hadron-nuclear scattering experiments. The amplitudes are better known in some cases than in others, according to the different particles.

b) The distribution of protons and neutrons in the interaction region. It can be easily shown[21] that for K^-, Σ^- and $\bar{p}$ atoms the interaction region is confined to the nuclear tail ($20\% \leq \rho/\rho_{centre} \leq 0.1\%$), whereas for pions the nuclear interior contributes more.

c) Binding of the nucleons in the nucleus and short-range correlations between the nucleons.

All effects yield considerable contributions to the observed strong interaction data. The accuracy of the data is partly very high. In some cases it is in the region of several per cent, which is quite uncommon for observation of strong interaction effects. The points (a)-(c) contribute always simultaneously to the observed effects, but their relative contributions depend on the selected hadron and the measured nucleus. Thus at least, a part-separation of the effects is possible.

Depending on the chosen hadron, different topics are emphasized. For light pionic atoms, for example, the free elastic π^--N interaction and the proton and neutron distributions are sufficiently well-known from other experiments and one can concentrate on point (c) and the π^--absorption process. For light kaonic atoms point (b) is known, whereas different phase analyses still give different answers for (a). So, the emphasis here lies at point (a) [given that (c) will be made more transparent by pionic atom results or other experiments], which is in this case of special interest, because at the K^--N threshold a resonance [Y_0^* (1405 MeV)] is located. The explanation of the observed effects might be that it is not the kaon that interacts with the nucleus, but that the Y_0^* resonance interacts with the rest of the nucleons. Thus, the *nucleus would be used as a laboratory for the investigation of a resonance-nucleon interaction at low energies*, a problem, which is very often discussed nowadays. If similar resonances exist in the $\bar{p}p$ system, as predicted recently[22], they would also influence the interpretation of the data. Measurements with kaonic and antiprotonic atoms on heavy nuclei could help to solve the old question of a possible *neutron halo*, but to answer this question the points (a) and (c) must be understood better.

A lot of data -- some of them already accurate enough -- already exists and is discussed in Sections 5.4.2.1 and 5.4.2.3. The discussion about the interpretation of the data is still in progress. The conclusions which can already today be drawn by comparison between calculations and data are briefly discussed in Sections 5.4.2.2 and 5.4.2.3.

4. EXPERIMENTAL SET-UP

The experimental set-up consists generally of three different parts: beam, counter telescope, X-ray detector.

4.1 Beam

The beam elements consist of quadrupoles and bending magnets which produce an image of the production target at the place where the material under investigation is placed. They must be as short as possible (exception: $\bar{p}$ beam) so that not too many particles decay. In order to get a high μ-flux, usually a so-called muon-channel is used. It consists of a series of quadrupoles, or a solenoid, which have the task to collect as many muons from the pion decay as possible and transport them to the beam telescope. The separation of muons and pions can be done using time-of-flight techniques or the different range of pions and muons of the same momentum is an absorber material. The part-separation of kaons and antiprotons from the pions is done by an electrostatic separator. The momenta chosen for the beam lines lie for pions and muons around 200 MeV/c and lower, for kaons and antiprotons around 800 MeV/c. The production of low-energy Σ^- beams is not possible because of the short Σ^- lifetime. The observation of Σ^- atoms is only possible via K^--N reactions, for example, $K^-p \rightarrow \Sigma^-\pi^+$. Reactions of this kind happen whenever the strong interaction between a K^- and a nucleon finishes the K^--cascading process in a kaonic atom. Σ^-'s occur in about 8% of all K^--captive processes. Their energies lie between 20-30 MeV. This energy is so high that a part of the particles leave the nucleus where they were produced and run a short distance (mm) through the target material until they are absorbed by another atom. Σ^- X-rays are always observed together with K^- X-rays (the time for the process described above is short compared to electronic resolution times); their intensity is only about 8% of the K^- X-ray intensities. This is the reason why experiments with Σ^- X-rays are the most difficult of all the measurements discussed here.

4.2 Counter telescope

A typical set-up for a K^- beam counter telescope is given in Fig. 3. It consists of scintillation and Čerenkov counters and the moderator which diminishes the energy of the K^-'s so that the largest part of them stops in the target. In spite of the electrostatic separator the ratio between slow kaons and fast pions in the beam is about 1%, so that the counter telescope must have a good rejection power for pions to give a clean trigger signal for a stopping kaon. The telescope for an antiproton beam works very similarly, whereas a telescope for a pion or muon beam consists only of four counters ($123\bar{4}$), because of the smaller contamination problems. Typical stopping rates/sec in a 8 g/cm^2 thick target are:

Stopping rates/sec:	Old machines	Meson factories
Pions :	10^5	10^8
Muons :	5×10^4	10^6
Kaons :	$10^3 - 10^4$	-
Antiprotons:	$5 \times 10^2 - 5 \times 10^3$	-

4.3 X-ray detectors

Most of the X-ray spectra have been obtained with Si(Li)-, Ge(Li)- or intrinsic Ge-detectors; only in the very low energy region proportional chambers have been used. In the meson factories which are just now coming into operation crystal spectrometers can also be used for lines of high yield. In Table 2 a list of essential properties of the X-ray detectors in use is given. In a very limited number of cases the energy resolution can be considerably increased by using the technique of critical absorption edges. The detectors are usually triggered with a stopped particle signal produced by the counter telescope to decrease the background. Time resolutions of this coincidence depend on the energy range and can be as good as several nsec.

The problems in using the detectors for very accurate energy and intensity determination are connected with the different behaviour of the detectors in the laboratory and under beam conditions. All beams are quite short (to minimize the particle decay) and the detectors have to be used in an area where a lot of neutrons and charged particles are present. The high pulses induced by the bombardment of these particles cause considerable

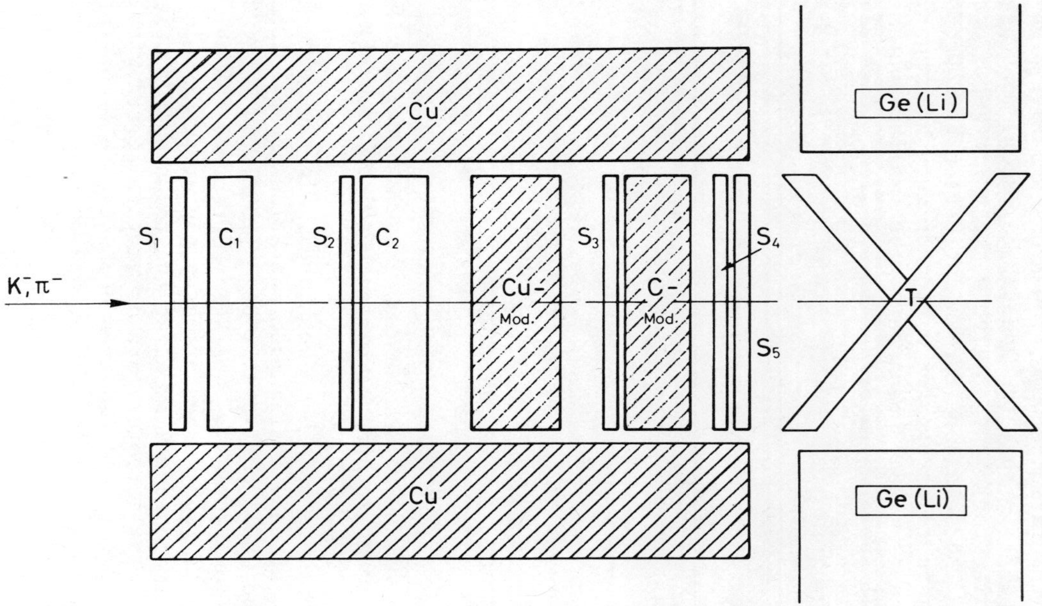

Fig. 3 Counter telescope (K^-, $\bar{p}$ beam)

Table 2

Type of detector	Energy region, where detector is best applicable	Solid angle	Energy resolution: Absolute values	Energy resolution: Relative values
Proportional chamber	Low energies (several keV)	Large		≥ 20%
Si(Li)-detectors	0-40 keV	Medium	6 keV: 180 eV	3%
Ge(Li)-detectors	0-several MeV (upper limit dependent on size)	Medium	400 keV: 800 eV 10 MeV: 10 keV	2‰ 1‰
Na(I)-detectors	100 keV-100 MeV (upper limit dependent on size)	Large		7%
Crystal spectrometer	0-80 keV	Small	50 keV: 5 eV	0.1‰

dead-times and energy shifts. To achieve good results it is therefore absolutely necessary to monitor the X-ray efficiency and the position of calibration lines *during the run under the same counting rate conditions* as the X-rays under investigation. How this is done in practice is discussed at a later stage (Section 5.2), when the high resolution experiments are treated.

5. MEASURABLE EFFECTS

5.1 Effects of atomic physics, chemistry and solid state-physics

From absolute and relative intensity measurements of X-rays of exotic atoms in different physical and chemical surroundings, it has been found that changes of the electronic configuration around the investigated atom give rise to high effects. These effects have not yet been understood, and therefore here the emphasis is put on a more or less systematic summary of some examples of the observed effects. Attempts to explain the data are mentioned, but not discussed in detail.

a) Capture on a single independent atom

This case is realized only in dilute gases and is the basic process for the understanding of the more complicated cases. The first ideas of how the capture process proceeds via the interaction between the slowed down particle and the atomic electrons go back to 1947 [23]; more recent calculations have been performed in 1974 and 1975 [24,25]. They all use a classical description for the interaction and how the classical orbit of the particle during the capture process might look is shown in Fig. 4. The result of the capture process might be a statistical population of the ℓ sublevels of one or several levels with a high n-quantum number. This calculated result seems not to be inconsistent with the few existing data on capture processes in pure dilute gases.

b) Atoms are not independent of each other

This case is realized, for instance, in every solid where the binding between the individual atoms occurs via the electrons. Going from one atom (Z) to the neighbouring atom (Z + 1) (all metals) big effects have been observed already a long time ago with muonic atoms[26] and pionic atoms[27] and very recently also with kaonic atoms[20]. A schematic picture of the results is given in Figs. 5a, b, c. The expected curve (same capture process for all atoms) would be a straight line in all cases. The loss of X-ray intensity of the π^- 4-3 and the K^- 6-5 transition and the increase of the μ^- 4-2 transition in the Z-region under investigation means that the initial ℓ distribution -- produced by the capture process --

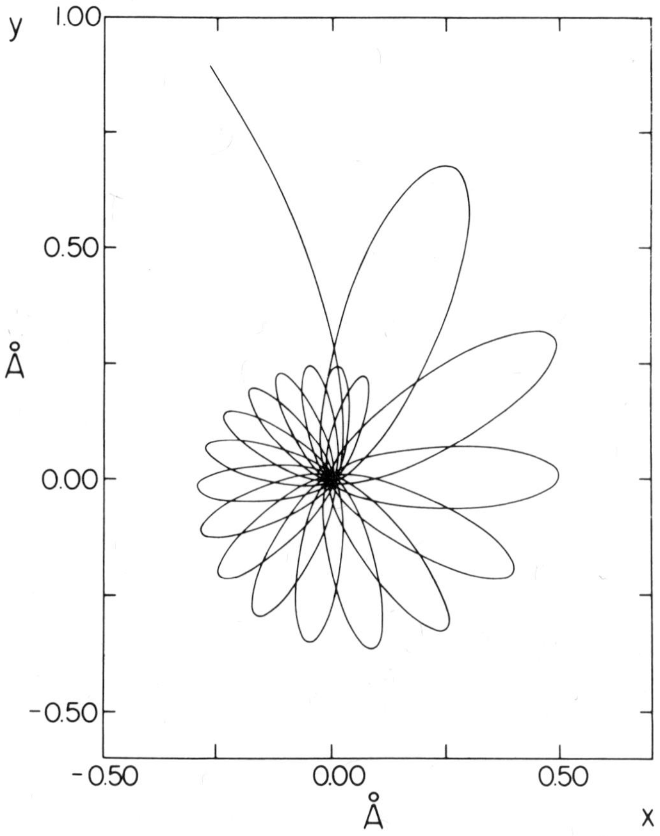

Fig. 4 Trajectory of a muon approaching an argon atom [from P. Vogel et al.[25]]

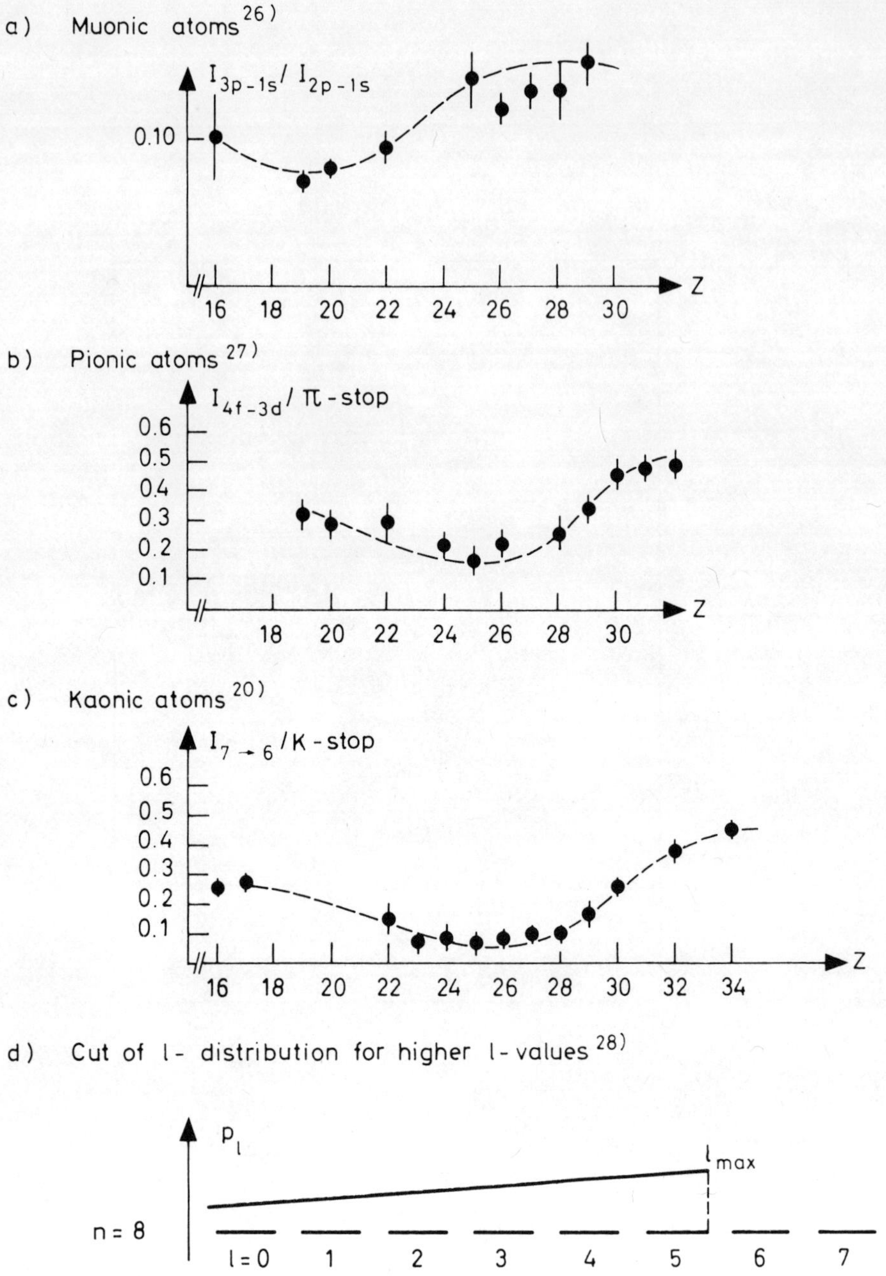

Fig. 5 Solid-state physics effects on transition intensities [from D. Quitmann et al.[26], A.R. Kunselman[27], C.W. Wiegand et al.[20] and G.T. Condo[28]]

changes considerably in this region. Capture models working in case (a) seem not to be able to explain this effect. A possible explanation might lie in the following observation[28]: The distances between the atoms in their metallic lattices follow the same trend as the data presented here. The finite distance between two atoms determines the maximal angular momentum which can be involved in the capture process and thus the maximal possible ℓ value (ℓ_{max}) of the initial distribution. With ℓ_{max} as a function of atomic distances the change of the initial distribution would be explained. The cut of the distribution suggested by this idea is illustrated in Fig. 5d.

c) Atoms have different atoms as neighbours (chemical effects)

The comparison of the X-ray intensities of one exotic atom in different chemical compounds shows very big differences. They are best expressed in terms of double ratios. Some examples:

μ^- atoms[29,30]: $$\mathrm{Ti}\left(\frac{7p-1s}{2p-1s}\right)_{\mathrm{TiO_2}} \Big/ \mathrm{Ti}\left(\frac{7p-1s}{2p-1s}\right)_{\mathrm{Ti\text{-}metal}} = 0.6 \pm 0.06$$

$$\mathrm{C}\left(\frac{5p-1s}{2p-1s}\right)_{\mathrm{CH_2}} \Big/ \mathrm{C}\left(\frac{5p-1s}{2p-1s}\right)_{\mathrm{graphite}} = 1.99 \pm 0.15$$

$\bar{p}$ atoms[31] : $$\mathrm{O}\left(\frac{7g-4f}{5g-4f}\right)_{\mathrm{D_2O}} \Big/ \mathrm{O}\left(\frac{7g-4f}{5g-4f}\right)_{\mathrm{CO_2}} = 4.0 \pm 1.0 .$$

Without the influence of chemical surroundings, these ratios should be equal to one. Another way of checking the predictions of capture models is to investigate how many particles are captured in the different atoms of a chemical compound. Fermi and Teller predict for a molecule consisting of n atoms with the charge Z_1 and m atoms with charge Z_2 for the relative capture ratio

$$\frac{W(nZ_1)}{W(mZ_2)} = \frac{n}{m}\frac{Z_1}{Z_2} , \qquad (3)$$

while Vogel et al. obtain

$$\frac{W(nZ_1)}{W(mZ_2)} = \frac{n}{m}\left(\frac{Z_1}{Z_2}\right)^{1.15} . \qquad (4)$$

These predictions have been checked for a variety of compounds[8] but no clear picture for their validity exists yet. Particularly well investigated with pions in this respect are compounds containing hydrogen[8]. The π^- capture process on hydrogen is clearly detectable by the reactions $\pi^- p \rightarrow n\pi^0$ and $\pi^- p \rightarrow n\gamma$ and it was found that Eq. (3) is not valid at all. For example, the experimental number for the ratio

$$W_H(N_2H_4)/W_H(N_2 + 2H_2)$$

is $^1/_{30}$, while the predicted number is $^2/_7$. These observations led to the hypothesis of molecular orbits of the particle around the chemical compound which was able to explain the experiements[32]. However, no direct transitions between molecular and atomic orbits have been observed yet[33].

d) Transfer reactions in gases

In a gas mixture consisting of two components, transitions of a muon captured in the lighter atoms to the heavier atoms are observed. Particularly well investigated is the special process

$$\mu^- p + X \rightarrow p + \mu^- X , \quad (5)$$

where X represents an atom or molecule heavier than hydrogen. The transfer process leads to a population of the ℓ sublevels quite different from the usual (statistical) one which is demonstrated, for example, in the following two experiments:

i) The intensity ratio between all lines of the K series except the 2p-1s transition and the 2p-1s line was observed for pure argon gas and a mixture of argon and hydrogen gas[34]. For pure argon gas the ratio was found to be 7%; for the mixture, where the muon is transfered from the μ^-p system to the μ^--A atom, it was measured to be 50%. This result can be interpreted in such a way that the initial population resulting from the transfer is peaked much more towards low ℓ values than in the usual capture case.

ii) The capture process in pure SF_6 gas was compared with the transfer process between hydrogen and SF_6 [35]. The intensity ratios of many transitions of the fluor-K-series were separately determined and it turned out again that the transfer process tends to populate lower ℓ levels.

At present, the interpretation of the data is not quite clear. Everybody agrees that the μ^-p system is in its ground state when it collides with the heavier atom. Either a highly excited molecular state of both atoms is formed which consequently de-excites to states of the heavier atom with high binding energy, or the muon is directly (no intermediate molecular state) transferred to a bound state of the heavier atom with an energy similar to the 1s μ^-p state. Then no transitions from levels of higher n-values should be observable. The population of sublevels with low ℓ values is then explained by the lack of big angular momenta in the process, which of course must conserve the total angular momentum. Quantitative calculations have been tried for different systems[25,34,36].

Concluding the discussion of the capture processes in different physical and chemical surroundings one should say that very large effects are observable. With the improved intensity of the meson factories also finer effects will be open to accurate measurements. The problem lies clearly in the interpretation of the data. My feeling is that a more systematic research is required and close contact should be kept with chemists and solid-state physicists. From a thorough discussion of the possibilities of these methods and a detailed comparison with other methods of chemistry or physics it may turn out that unique ways for the investigation of physical and chemical properties of materials can be found.

5.2 Test of fundamental theories: QED

Quantum electrodynamics (QED) is a theory which allows most detailed predictions in electrodynamic processes. It has to be taken into account in many fields of physics, but it is in the field of exotic atoms where it yields relatively large contributions and it is there where its validity can be checked best experimentally. How big contributions QED gives in muonic atoms is shown in Fig. 6, which shows schematically the different contributions which affect the energy of the $5g_{9/2}-4f_{7/2}$ transition in μ^--Pb. The largest contribution comes, of course, from the solution of the Dirac equation for a point nucleus; minor contributions of different signs come from the finite size of the nucleus and from nuclear polarization. The QED effects are dominated by the vacuum polarization term of order $\alpha Z\alpha$, smaller contributions come from higher order terms. The smallest term is the Lamb shift, quite in contrast to electronic atoms, where the self-energy term ($\sim$ Lamb shift) is the dominant one. A non-negligible term comes from the shielding of the nuclear Coulomb field by the electrons, present around the muonic atom. If one is able to calculate the electron screening effect accurately enough and to measure the transition (431 keV energy) with an accuracy of some eV, the QED vacuum polarization predictions (as given in Fig. 6) can be checked. There are a lot of transitions in various μ^- atoms which can serve as test transitions for QED effects, but the most appropriate ones have to fulfil the following conditions:

i) All corrections, particularly the ones due to strong interaction effects and electron screening, must be small or exactly calculable.

ii) Energy of the transition must lie in an easily measurable region. These conditions are met in 4-3 and 5-4 transitions in heavy muonic atoms and it is there where the measurements presented up to now were performed.

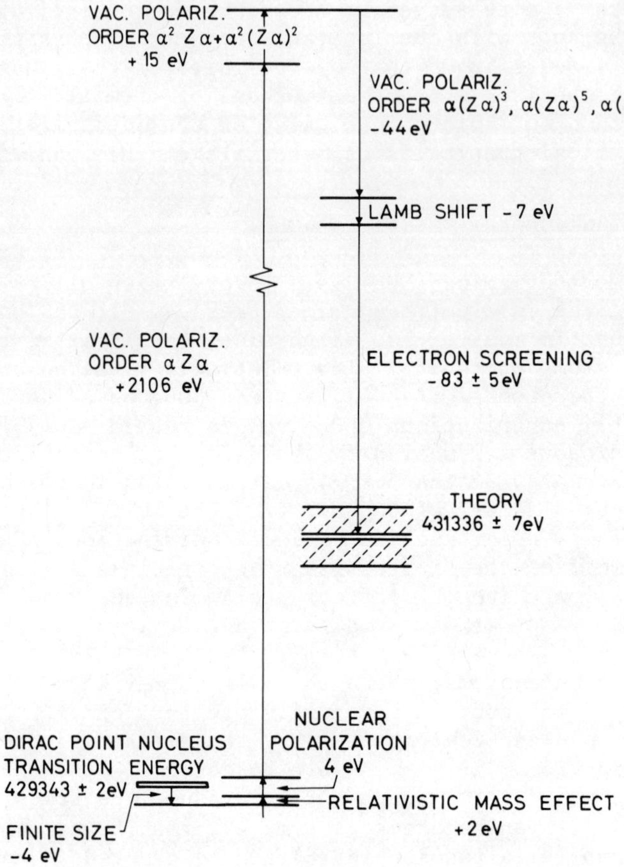

Fig. 6 $5g_{9/2}-4f_{7/2}$ transition in muonic Pb [M.S. Dixit, private communication]

In all cases two problems must be solved:

a) Precise (eV order of magnitude) calculation of corrections: Most corrections (including strong interaction effects) are so small that there no problem can arise. The only exception is the electron screening. The biggest effect is produced by the electrons in the K- and L-shell. The calculation of the effect is easy if one knows to which extent the electron shells are filled when the muonic transition under study occurs. The population of the shells is usually calculated in a cascade program which allows for refilling of the electron shells after an Auger process has happened. These calculations are open to criticism, but fortunately they can be checked in some cases: In heavy muonic atoms the upper part of the cascade is dominated by Auger processes. But there the energy difference between two levels with $\Delta n = 1$ is too small to allow the emission of a K-shell Auger electron. In the lower part of the cascade X-ray transitions dominate; no K electrons will be expelled. In these cases the K-electron shell is populated with big probability to nearly 100%. The cascade calculations are checked in these cases and can be believed with some confidence also for lighter nuclei. An experimental check of cascade program predictions was performed in the case of muonic niobium[37], where the screening shifts of 6 transitions to the $n = 5$ level were measured. It was found that during these transitions both K electrons were present. The screening contribution caused by the L-shell electrons is calculable with less accuracy and it is this term which gives the largest error in the calculation of the corrections.

b) Precise determination of the experimental energy value: A measurement of a line of 500 keV with an error of 10 eV means a relative accuracy of 2×10^{-5}. The resolution of Ge-detectors around 500 keV is about 1.5 keV, that means the centre of the line must be determined to better than 1% of the resolution. Small effects can spoil the measurements, and the precise calibration of the spectra under beam conditions has turned out to be the most difficult point. The change of counting rate under beam on/out conditions can shift lines considerably and methods had to be invented to overcome this problem. Three of them are sketched in Fig. 7.

In method I the memory of a multichannel analyser or computer is split into two equal parts. A calibration source, which emits γ-quanta of well-known energy in the energy region under investigation, is present during the recording of the X-ray spectra. The left part of the memory is triggered with a coincidence of the stopped muon and an X-ray event in the Ge-detector. Thus it records the X-rays under study and to a small extent events from the γ-source which are fed in by accidental coincidences (feed-through line). The other part of the memory is triggered with a

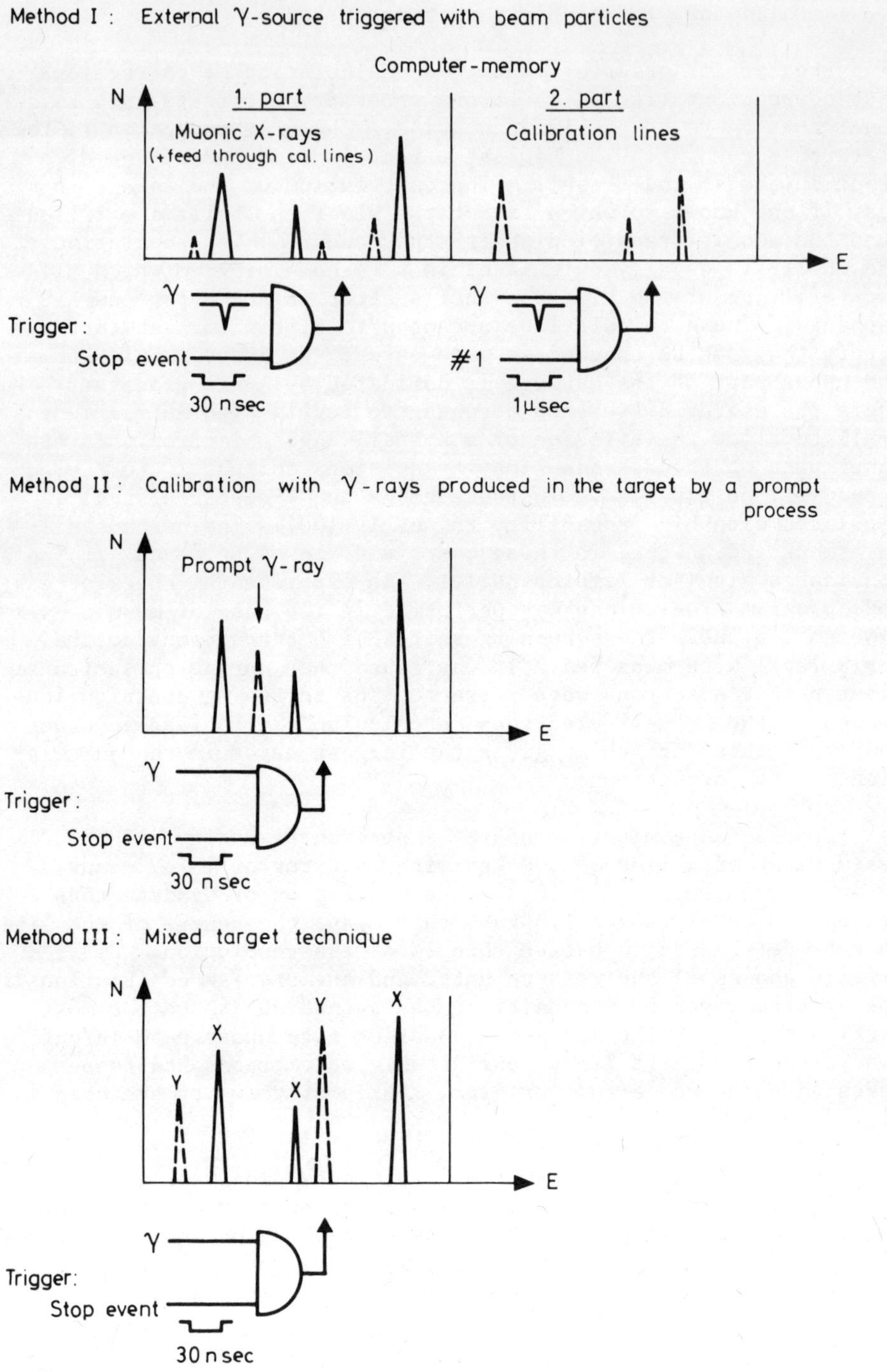

Fig. 7 Scheme of usual energy calibration methods

broad coincidence (1 μsec) of passing particles and Ge-events. Only events from the γ source are found here. The reason for the trigger is that the data in both parts of the memory are recorded under more or less the same counting rate conditions. The comparison between the feed-through line (bad statistics) and the γ lines shows, if the calibration in both parts of the analyser took place under identical conditions. With this technique the first QED measurements (1970/1971) were done.

Subsequent measurements tried to use method II. In special cases the beam traversing the target excites the target nuclei which de-excite via a more or less prompt (10-100 nsec) γ-ray transition. If the energy of this γ-ray is well known and if it de-excites quickly enough it is produced under practically identical counting rate conditions as the line in question and can be taken as an ideal calibration standard. In the most recent QED experiment on μ^--Ba, for example, a ^{137}Cs line was activated and was used as reference.

The most certain way (method III) for on-line energy calibration is the use of a mixed target, consisting of two materials. The first material (X) is the substance under study, the second one (Y) is selected such that it produces muonic X-rays near the line of interest. In many cases the energies of the transitions in Y can be calculated with a precision of some eV -- it can be done, if all corrections are known or very small -- and serve as calibration standards taken under exactly the same conditions as the lines to be investigated. The relative energies of transitions in muonic Ba and Pb could be determined with very high accuracy using this method.

Since 1970 five different experiments have been performed to test the QED predictions. One of them[39)], when it came out, seemed to disprove the validity of QED, showing in one case a difference between theory and experiment of six standard deviations. Much of this discrepancy disappeared when errors in the calculations were found[40)], and the two most recent experiments[38,41)] find -- averaged over four transitions in muonic Ba and Pb -- less than one standard deviation difference between theory and experiment. QED seems to work also in muonic atoms -- i.e. in big electric fields -- with its usual accuracy.

5.3 Particle properties: Masses, magnetic moments

5.3.1 Masses

The most accurate values for the masses of negative pions, kaons and antiprotons nowadays available come from exotic X-ray measurements. The principle behind these measurements and the experimental procedure is quite similar to the QED test experiments. The measurements must be performed at transitions lying at region II of the cascade. The energies of selected transitions are measured and calculated as accurately as possible. The transitions are chosen such that the corrections to the Dirac energy are as small as possible, particularly the QED and electron screening corrections. The energies then depend only on the mass of the particle and it is varied such that calculated and measured values coincide. To get higher precision, not only one but all clearly visible transitions of that element are used for the analysis. The tricky points are again the correct energy calibration under beam conditions and the calculation of the electron screening corrections, and it turns out that the final error in the masses is determined by the errors of these effects.

The evaluation of the measured spectra must be done very carefully. For example, one has to correct for transitions parallel to the one under investigation of the type $(n, \ell = n - 2) \rightarrow (n - 1, \ell = n - 3)$ and for possible excitations of the nucleus by the cascading particle (see, Section 5.4.1.2, Dynamic effects). The results of the mass determinations done so far are listed in Table 3a.

The π^- mass was measured by two different X-ray-detectors in different energy regions. The lines observed with the crystal spectrometer[42] were around 80 keV. The error in the mass resulting from this measurement is nearly completely due to the bad statistics of the lines. With Ge-detectors lines of higher energy can be observed best and the experiment was performed on transitions of about 300 keV. In this case the error on the resulting mass is not only due to experimental uncertainties, but also to problems in the calculation of the transitions (screening effects). By improving the statistics of the crystal spectrometer measurements, which should be no problem at the meson factories, or by improving the accuracy in the calculation of the disturbing effects, still considerably better mass values can be expected for the future.

From the measurement of the muon momentum of the π-μ decay at rest $(\pi \rightarrow \mu + \nu)$[46] and the accurately-measured muon and pion rest masses, an upper limit for the muon-neutrino mass can be deduced:

$$m_\nu^2 = (-0.29 \pm 0.90)\ (\mathrm{MeV}/c^2)^2 .$$

It should be noted that the recent measurements[45] of the $\bar{p}$ mass give a value which is more than one standard deviation different from the p-mass value:

$$m_p - m_{\bar{p}} = (100 \pm 58)\ \mathrm{keV}/c^2 .$$

Whether this result is statistically significant will be cleared up in future experiments.

Table 3

a) Masses

Particle		Mean of masses [MeV/c^2]	Observed transitions and target nucleus	Remarks	References
π^-	139.566 ± 0.010	139.568 ± 0.005	4-3 in Ca and Ti	Crystal spectrometer	42
	139.569 ± 0.006		5-4 in I, 6-5 in Tℓ	Ge-detectors	43
K^-	493.688 ± 0.030	493.667 ± 0.017	11-10, ..., 7-6 in Ba 12-11, ..., 8-7 in Au	Ge-detectors	44
	493.657 ± 0.020		13-12, ..., 8-7 in Pb	"	45
$\bar{p}$	938.179 ± 0.058		16-15, ..., 11-10 in Pb	"	45
Σ^-	1197.24 ± 0.14		15-14, ..., 11-10 in Pb	"	45

b) Magnetic moments

Particle	Magnetic moment [n.m.]	Observed transitions	References
$\bar{p}$	-2.819 ± 0.056	11-10 in Pb and U	47
	-2.790 ± 0.021	"	45
Σ^-	-1.48 ± 0.37	12-11 in Pb, ^{208}Pb, pt	48
	$-1.40^{+0.41}_{-0.28}$	12-11 in Pb	45
	$+0.65^{+0.28}_{-0.41}$		

5.3.2 Magnetic moments

Although there are different methods available to determine the magnetic moments of elementary particles, no way is known to measure the moments of antiprotons and Σ^- hyperons except from exotic X-ray measurements. The magnetic moment of a spin-½ hadron (H) consists of a Dirac- (g_0) and an anomalous- (g_1) part:

$$\mu_H = (g_0 + g_1)\mu_N \tag{6}$$

with $\mu_H = eh/2m_Hc$ being the Bohr magneton for the corresponding particle mass m_H. In the Pauli approximation this magnetic moment yields a dublet splitting of a level with the quantum numbers n,ℓ

$$\Delta E_{n,\ell} = (g_0 + 2g_1)\frac{(Z\alpha)^4}{2n^3}\frac{m}{\ell(\ell+1)} \tag{7}$$

(where m = reduced mass of hadron and nucleus). This fine structure gives rise to a splitting of a transition of the type (n + 1, ℓ + 1) → (n,ℓ) in three components (see Fig. 8). This splitting is comparable to the instrumental resolution in all cases where strong interaction effects can be neglected, and it is there (region II) where the magnetic moments can be deduced from the data. The magnitude of the moments can be obtained from the observed energy splitting alone; for the determination of the sign the relative intensities of the transitions must be known. This is illustrated in Fig. 8 for the case of a negative moment and a circular transition between the n = 12 and n = 11 levels. Assuming a statistical population of the 2j + 1 sublevels of each j state one obtains for the intensity ratio of the transitions a, b, c:

$$I_a : I_b : I_c = 252 : 1 : 230 . \tag{8}$$

The spin-flip transition b can be practically neglected and the ratio of the low-energy component (a) to the high-energy component (c) is 11/10. For positive magnetic moment this ratio would just be the inverse.

Table 3b lists the results obtained recently by two different groups. In heavy $\bar{p}$ atoms the splitting of lines (a) and (c) is bigger than 1 keV and can be completely resolved in the 11-10 U transition. Both experimental results agree within the errors. The value obtained is in magnitude the same as the magnetic moment value for the proton, but opposite in sign, in agreement with the TPC prediction. The splitting in Σ^- atoms is always smaller than 300 eV and cannot be resolved with present experimental accuracies. The result of the most recent measurement[45] is shown in Fig. 9, where the Σ^- 12-11 transition in Pb is sitting on a high background. The poor peak/background ratio occurs in all measured

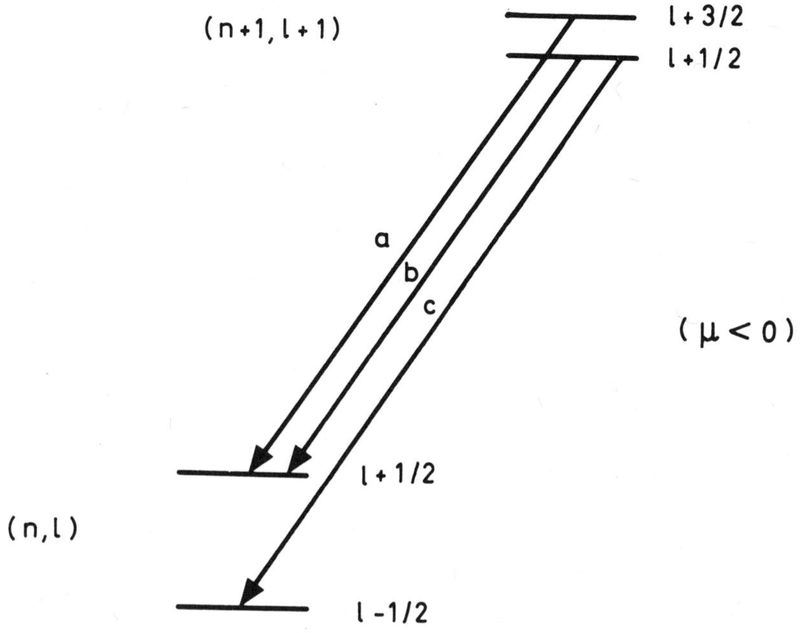

Statistical population: $n = 11$

$a : b : c = 252 : 1 : 230$

↳ $I_a / I_c \approx 11/10$

Fig. 8 Transition intensities of fine structure components

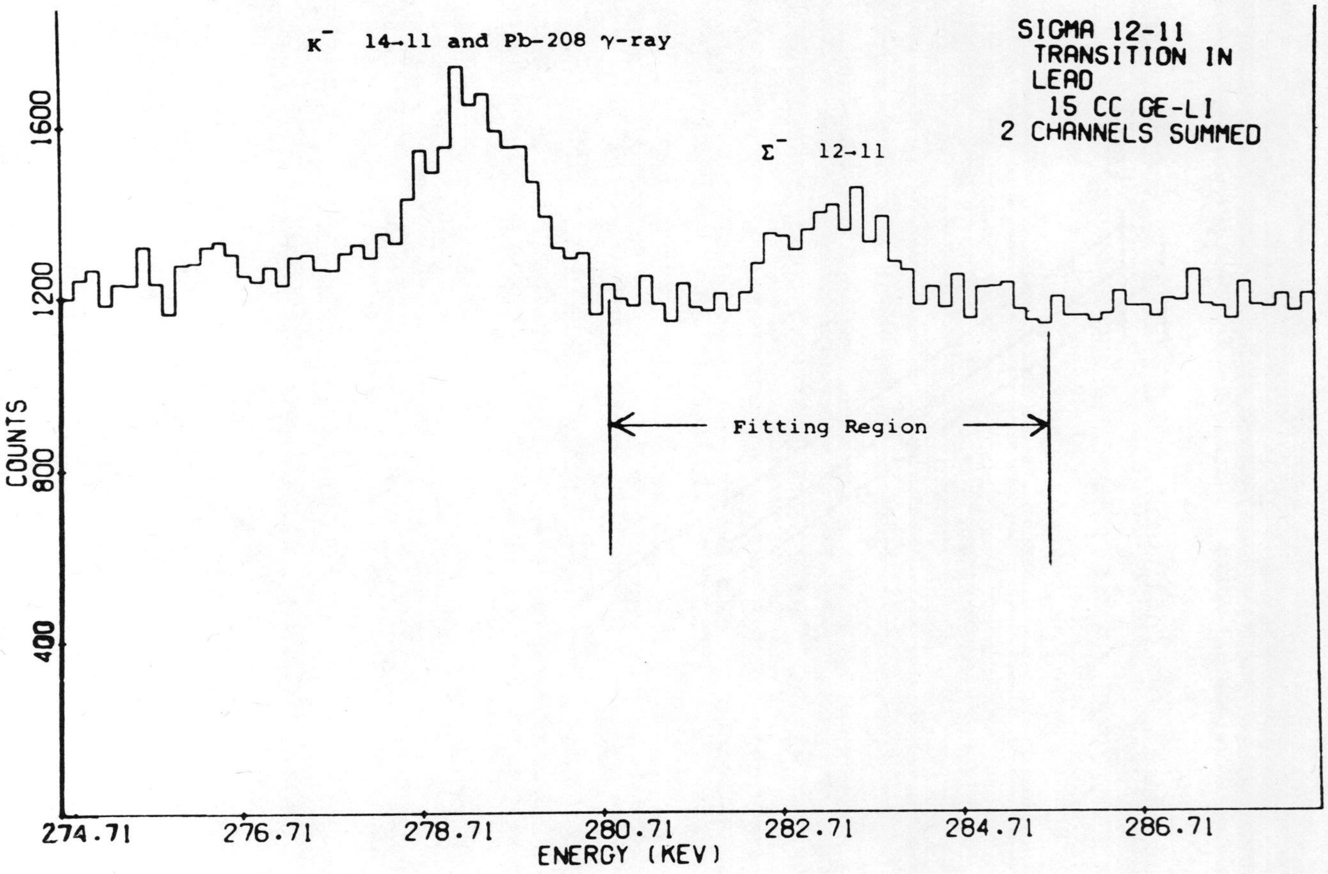

Fig. 9 The 12-11 transition in Σ^--Pb [from G. Dugan et al.[45]]

Σ^- spectra and is due to the fact that K^-- and Σ^--atomic lines appear simultaneously in the spectrum, but the Σ^- lines only with 8% probability compared to the kaonic lines (see Section 4). The analysis of the line shape and hitherto the determination of the splitting can only be done with statistical (χ^2) methods. An assumed pattern is folded in the observed spectrum and an observed minimum in the χ^2 distribution indicates the best splitting assumption. The biggest problem is the determination of the sign of the magnetic moment which determines the relative intensities of both components of the pattern. Whereas one group[48] seems to see an indication for a negative sign of the moment (ratio $I_a/I_c > 1$) -- their χ^2 distribution is shown in Fig. 10 -- the other group does not yet see a clear indication and gives two values as a result of their analysis. Further measurements are in progress and will certainly clear up this point. The determination of μ_{Σ^-} is of special importance for elementary particle physics theory, because an SU(3) prediction, based on the quark model, exists, which predicts the value

$$\mu_{\Sigma^-} = -0.88 \text{ nucl. magn.}$$

5.4 Investigation of nuclear properties

5.4.1 Muons: monopole-, quadrupole-, hexadecupole-charge distribution moments, magnetic moments and their spatial distribution

In the first part of this section a survey is given of nuclear effects which can be seen in muonic X-ray spectra. The transitions in non-deformed nuclei and the higher transitions in deformed nuclei are simpler to interpret than the transitions between low n-values in deformed nuclei and therefore the former ones are discussed first. In the latter case the muon on its way down to the ground state is able to excite the nucleus, which gives rise to interesting effects but needs a more complex treatment.

5.4.1.1 Transitions in non-deformed nuclei and higher transitions in deformed nuclei

Common to these cases is that the nucleus is in the ground state when the muonic transitions occur and only nuclear properties of the ground states play a role. The energies of the levels (centres of gravity of the HFS multiplets) are determined by the following effects:

a) Point nucleus Coulomb field

The Dirac equation yields in lower order of (αZ) for the muonic energy levels

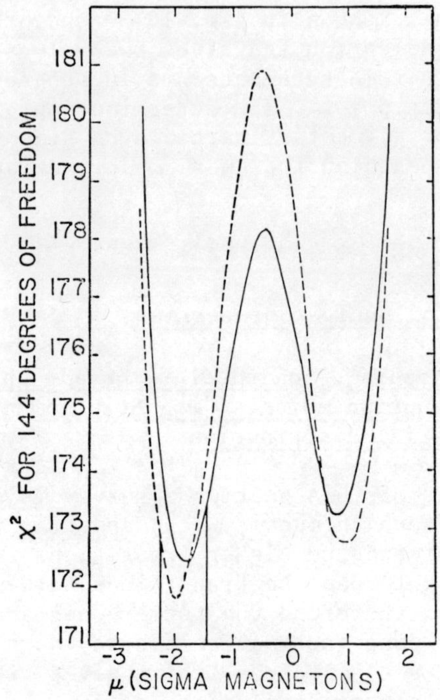

Fig. 10 Results of χ^2 fits for different values of the Σ^--magnetic moment [from B.L. Roberts et al.[48]]

$$E^0_{n,j} = -\mu c^2 \frac{(\alpha Z)^2}{2n^2} \left\{ 1 + \frac{(\alpha Z)^2}{n^2} \left(\frac{n}{j + \frac{1}{2}} - \frac{3}{4} \right) + \cdots \right\} . \qquad (9)$$

The binding energies of the 1s levels range from a few keV in hydrogen to 21 MeV in Pb. The fine structure splitting can be quite large, it is 550 keV in the 2p level of Pb.

b) Nuclear finite size

The finite nuclear size (monopole part of the charge distribution) reduces the binding energies of the muon considerably. In the 1s state of μ-Pb it is as big as 10 MeV (half of the point nucleus value), and in the 2p state of μ-Pb it still amounts to 180 keV. The effect is of short range and goes with the overlap between the muonic ware functions and the nuclear monopole proton distribution. The radial dependence of the wave functions $|\psi|^2$ and of the overlap $\rho|\psi|^2$ for a typical case (μ^--Nb) is shown in Fig. 11. The overlap varies quite drastically with the charge number Z of the nuclei and with the muonic angular momentum ℓ, so that the effects become more and more important when heavier nuclei and smaller ℓ values are observed.

c) Vacuum polarization

An important contribution to the binding energies of the levels is given by the QED effects, particularly the vacuum polarization. The binding energies are always increased by this effect. In the 1s-state of μ^--Pb it amounts, for example, to 66 keV. As we have seen in Section 5.2, these effects seem to be understood very well and can be calculated with high precision. It is a long-range effect and influences practically all transitions seen in X-ray spectra.

d) Nuclear polarization

A muon, the wave function of which has some overlap with the nucleus, can be elastically scattered on the nucleus with the formation of intermediate excited nuclear states. The calculation of this effect is difficult, because it depends on a lot of nuclear degrees of freedom. It gives a small, but not negligible contribution to the binding energies which is in the case of μ^--Pb (1s state) about 6.8 keV[49].

An HFS pattern of the muonic levels is observed when the nuclei are deformed or have a non-zero magnetic moment. In contrast to electronic atoms, the effect of the quadrupole deformation is much bigger than the effect of the nuclear magnetic moment. This is due to the μ- magnetic moment which is about 200 times

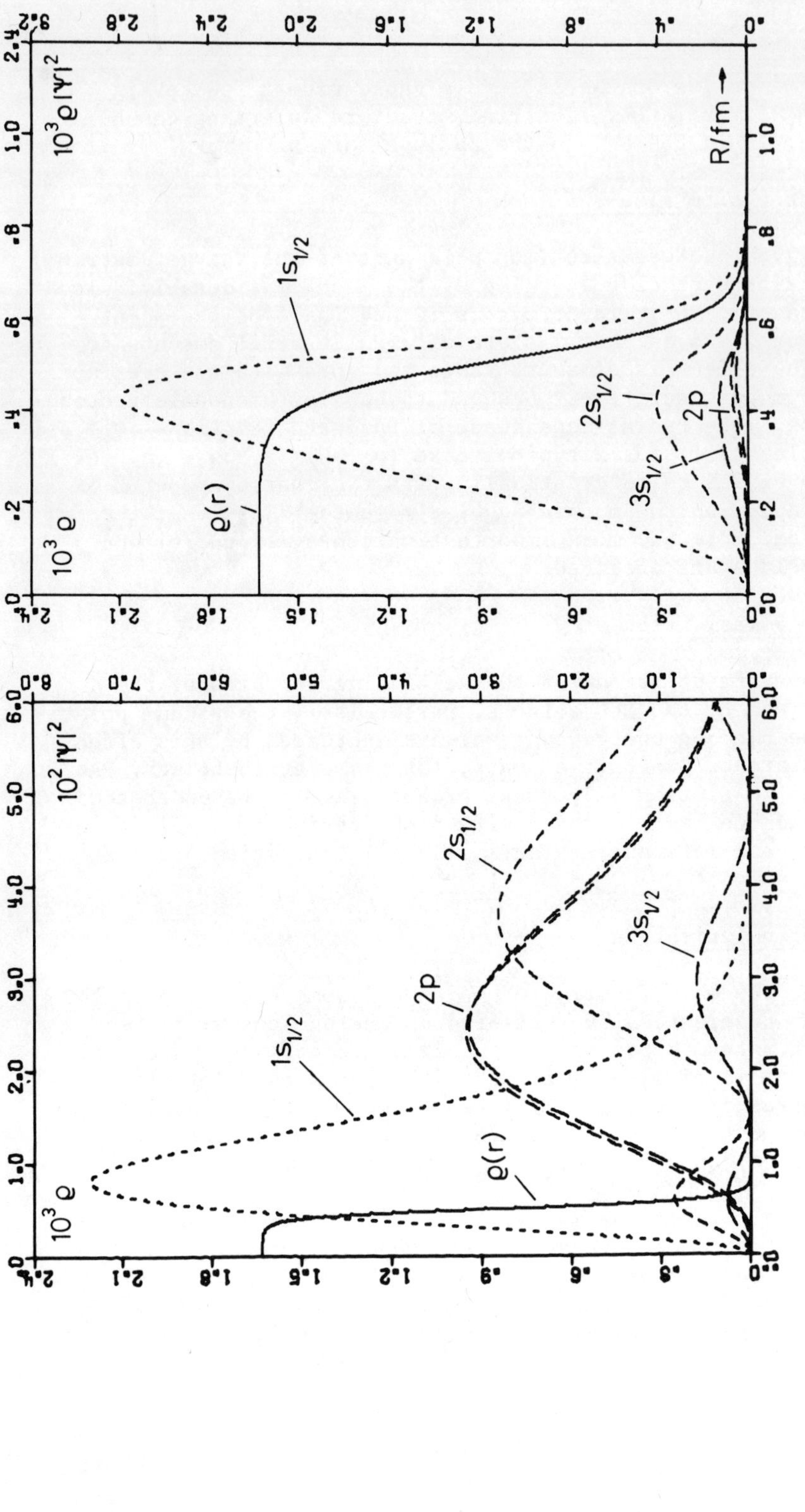

Fig. 11 Muonic atom wave functions, nuclear charge distribution and overlaps between wave functions and charge distributions for μ⁻-niobium [from H.P. Povel[37])]

smaller than the magnetic moment of the electron. In one case[50] there is even some indication for a higher electric moment, due to a hexadecupole part of the charge distribution. The different effects giving rise to an HFS pattern of muonic levels are discussed in the following separately:

e) Quadrupole moment

A charge distribution of general shape is usually expanded in a series of multipoles:

$$\rho(\vec{r}) = \rho_0(r) + \rho_2(r)Y_{20}(\theta,\phi) + \rho_4(r)Y_{40}(\theta,\phi) + \ldots \quad (10)$$

The intrinsic quadrupole moment Q_0 is connected with $\rho(\vec{r})$ via the relation

$$Q_0 = 2\sqrt{\frac{4\pi}{5}} \int \rho(\vec{r})Y_{20}(\theta,\phi)r^2\, d\tau \ . \quad (11)$$

The quadrupole (E2) interaction between a muon and the deformed nucleus has the following form[51]:

$$H(E2) = -\frac{e^2}{2}\sqrt{\frac{4\pi}{5}}\, Q_0 f(r)Y_{20}(\theta,\phi) \ . \quad (12)$$

When electronic HFS effects are discussed, $f(r)$ is put equal to r^{-3}. This corresponds to the case that the charge distribution producing the quadrupole moment is so far away from the probing particle that only a point quadrupole is seen which produces a r^{-3} field. The same replacement can be done in muonic atoms when the splitting of a level with high n and ℓ values is measured, for example, 5g, 4f, 3d levels. In these cases the muon is so far away from the nucleus that finer details of the $\rho_2(r)$ distribution producing the quadrupole moment are lost. Measurements of this kind determine directly the so-called spectroscopic quadrupole moment which is simply related to Q_0. The splitting in these cases is given by the well-known formula

$$\Delta E(E2) = A_2 \frac{\frac{3}{2}K(K+1) - 2I(I+1)j(j+1)}{4I(2I-1)j(2j-1)} \quad (13)$$

[where $K = F(F+1) - I(I+1) - j(j+1)$; $|F| \leq \mathrm{Min}(I,j)$]. The splitting is only observable for nuclear spins $I \geq 1$ and muonic angular momenta $j \geq {}^3{}_2$. The quadrupole HFS constant A_2 can be easily worked out in these cases and depends only on Q_0:

$$A_2^0 = \frac{2_j + 1}{2_j + 2}\, e^2 Q_0 \left\langle \frac{1}{r^3} \right\rangle_{n,j} \ . \quad (14)$$

Though the splittings of higher levels in muonic atoms are small, recently some measurements came up which give very accurate values for Q_0. The errors are much smaller than in the corresponding electronic atom HFS measurements. A discussion of these results is found in Section 5.4.1.3.

In general, the expression f(r) is unequal to r^{-3}. It is connected with the charge distribution $\rho(\vec{r})$ in the following way[51]:

$$Q_0 f(r) = 2\sqrt{\frac{4\pi}{5}}\left\{\frac{1}{r^3}\int_0^r \rho(\vec{r}\,')Y_{20}(\theta',\phi')\,d\tau' + \right.$$

$$\left. + r^2\int_r^\infty \rho(\vec{r}\,')\,\frac{1}{r'^3}\,Y_{20}(\theta',\phi')\,d\tau'\right\}. \tag{15}$$

and depends on finer details of the quadrupole distribution. Different models for the ρ distribution of the deformed nucleus yield different functions f(r) and different A_2 values, which can be measured. These effects are only visible in muonic atoms, and there only in the low-lying levels. As such levels practically never can be observed without excitation of the nucleus, the discussion of the function f(r) or the discussion of different ρ distributions is delayed until the dynamical effects are treated.

f) Magnetic moment

Whenever a nucleus has a magnetic moment unequal to zero, an HFS magnetic splitting of *all* muonic levels via the M1 magnetic interaction is obtained. Because of the small muon magnetic moment it is a relatively small effect, which does not exceed some keV also in the 1s levels of heavy muonic atoms. The energy splitting is given by the expression

$$\Delta E(M1) = A_1\,\frac{1}{2}\left\{F(F+1) - I(I+1) - j(j+1)\right\}. \tag{16}$$

The magnetic HFS constant A_1 is generally dependent on the nuclear magnetic moment, its spatial distribution in the nucleus, and the muonic wave function. In cases where only a point-like magnetic distribution can be seen -- this is realized in all atomic HFS-A_1 measurements and in muonic measurements of the M1 splitting of higher levels -- A_1 is completely determined by the nuclear magnetic moment $g_I\mu_N$:

$$A_1^0 = \mu_\mu g_I \mu_N 2\,\frac{\ell(\ell+1)}{j(j+1)}\left\langle\frac{1}{r^3}\right\rangle_{n,j}. \tag{17}$$

In muonic atoms the splitting can be properly observed only in the 1s state. There Eq. (17) is no longer valid. An A_1 different from A_1^0 is observed due to the finite distribution of the magnetic moment in the nucleus. That such an effect exists was predicted by Bohr and Weisskopf[52] and has been observed (as a very small effect of a few per cent) already in M1-HFS patterns in electronic atoms. In muonic atoms it gives rise to a dramatic (factor 2) decrease of A_1 (observed) compared to A_1^0. Thus, M1 pattern determination in muonic atoms is an ideal method to determine the spatial distributions of magnetic moments of nuclei. First results of such measurements are discussed in Section 5.4.1.3.

g) Electric hexadecupole moments

According to Eq. (10), the third term in the nuclear charge distribution expansion is a hexadecupole moment. First indications of the existence of this contribution exist and a recent measurement on muonic ^{165}Ho claims to have seen it, too. It can be observed only, if $j \geq {}^5\!/_2$, that means that it is always a small effect. The observation of such small effects is intimately connected with the statistics of the lines, and it is expected that similar measurements on the meson factories will help to determine such moments also in other nuclei.

5.4.1.2 Muonic transitions in deformed nuclei

Which big differences exist between muonic spectra of a non-deformed and a deformed nucleus is best seen when the 2p-1s line patterns are compared. In the first case the 2p-1s transition consists of a doublet ($2p_{3/2}-1s_{1/2}$; $2p_{1/2}-1s_{1/2}$), whereas in the deformed nucleus a pattern of sometimes more than ten lines is observed which are spread in a range of 100-200 keV. The differences between the observed patterns become smaller and smaller the higher transitions are compared. The explanation of this phenomenon is illustrated in Fig. 12a, where on the left side a typical excitation spectrum of a deformed nucleus is sketched, whereas on the right side the level scheme of the muonic atoms is given. It happens just that in the region of deformed nuclei the difference between two nuclear levels (rotation band) has the same order of magnitude (ε) as the fine structure splitting of the muonic 2p or 3d states ($\approx$ 100 keV). Then with a relatively high probability via the E2 interaction the process sketched in Fig. 12b occurs: The muon arrives at the $2p_{3/2}$ level, for example, while the nucleus is in its ground state. Instead of cascading down to the muonic 1s state the muon falls into the $2p_{1/2}$ state. The energy gained in this transition is used to excite the nucleus to the 2^+ state. The $2p_{1/2}-1s_{1/2}$ muonic transition occurs then in the presence of the excited nucleus. If such a situation exists the E2 interaction between the muon and the nucleus can no longer be treated as a small perturbation -- as was done until now -- and higher orders

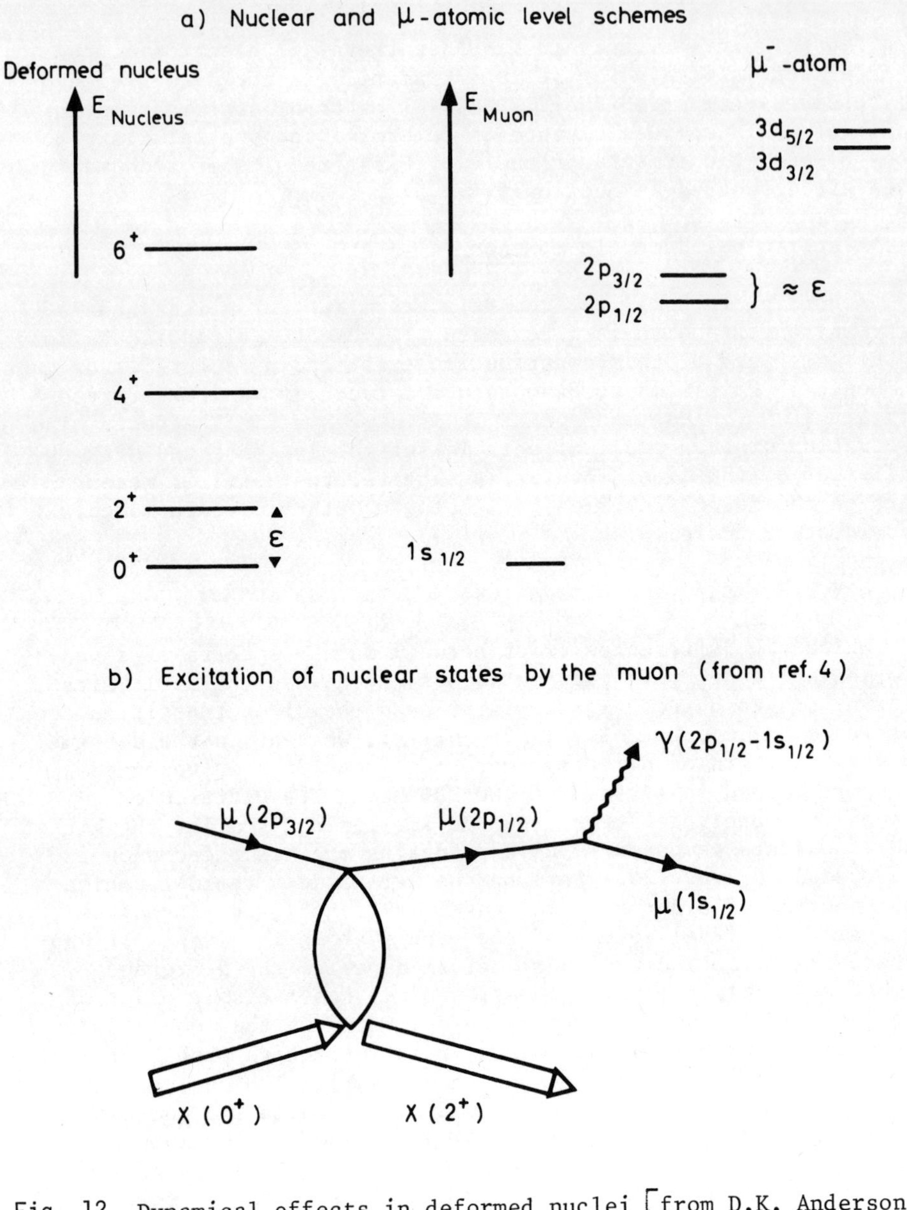

Fig. 12 Dynamical effects in deformed nuclei [from D.K. Anderson et al.[4]]

which take into account as intermediate states excitations of the nucleus must be considered. The states of the system can no longer be characterized by n and j of the muon alone, but the nuclear spins I are equally important. In Fig. 13 it is indicated that without E2 interaction no coupling of $\vec{I}$ and $\vec{j}$ takes place, but that the E2 interaction couples states with the same F-quantum number to new states which have energies different from the uncoupled ones. This effect gives rise to a 2p-1s pattern, quite different from the usual doublet. What it looks like is shown in Fig. 13.

The energies of the new coupled levels and the relative intensities of the transitions depend not only on properties of the ground state, but also on properties of the excited states, for example, their quadrupole moments, and on transition properties, for example, the transition quadrupole moment $\langle 2^+|H(E2)|1^+\rangle$. It is here, where data on excited states can be determined, which otherwise are not at all available, or only with great uncertainties. It is this situation which usually occurs in the cases where the spatial distributions of the quadrupole moments cause big effects, and it is obvious that a lot of information is contained in such spectra. Similar, but smaller, mixing effects are observed in higher levels, and it must be generally checked in every investigated spectrum, if not such dynamical effects are present, which cause energy shifts and intensity variations. Until now, it has not yet been possible to analyse the observed patterns showing strong dynamical effects without the use of a model. Examples for the analysis of such spectra in terms of the Bohr-Mottelson model are given in Section 5.4.1.3.

It should be clear now that the energies and energy splittings of muonic levels are very sensitive against the monopole-, quadrupole- (hexadecupole-) nuclear charge distribution, against the nuclear magnetic moment and its spatial distribution and in the case of deformed nuclei also sensitive to similar properties of the excited nuclear levels. Some examples for the precision which is reached today in the determination of such properties are given in the following two sections.

5.4.1.3 Examples for the determination of nuclear properties

i) Monopole charge distributions

Spherical nuclei with no magnetic moment can be analysed most easily in terms of the monopole-charge distribution $\rho_0(r)$. From detailed discussion in recent years it was found that a model-independent determination of the shape of $\rho_0(r)$ seems very difficult from muonic X-ray data alone. Crudely speaking, only the root-mean-square radii can be determined, but with a very high accuracy. It allows one to obtain relative errors smaller than 1%; for relative measurements of isotopes the accuracy is as high as

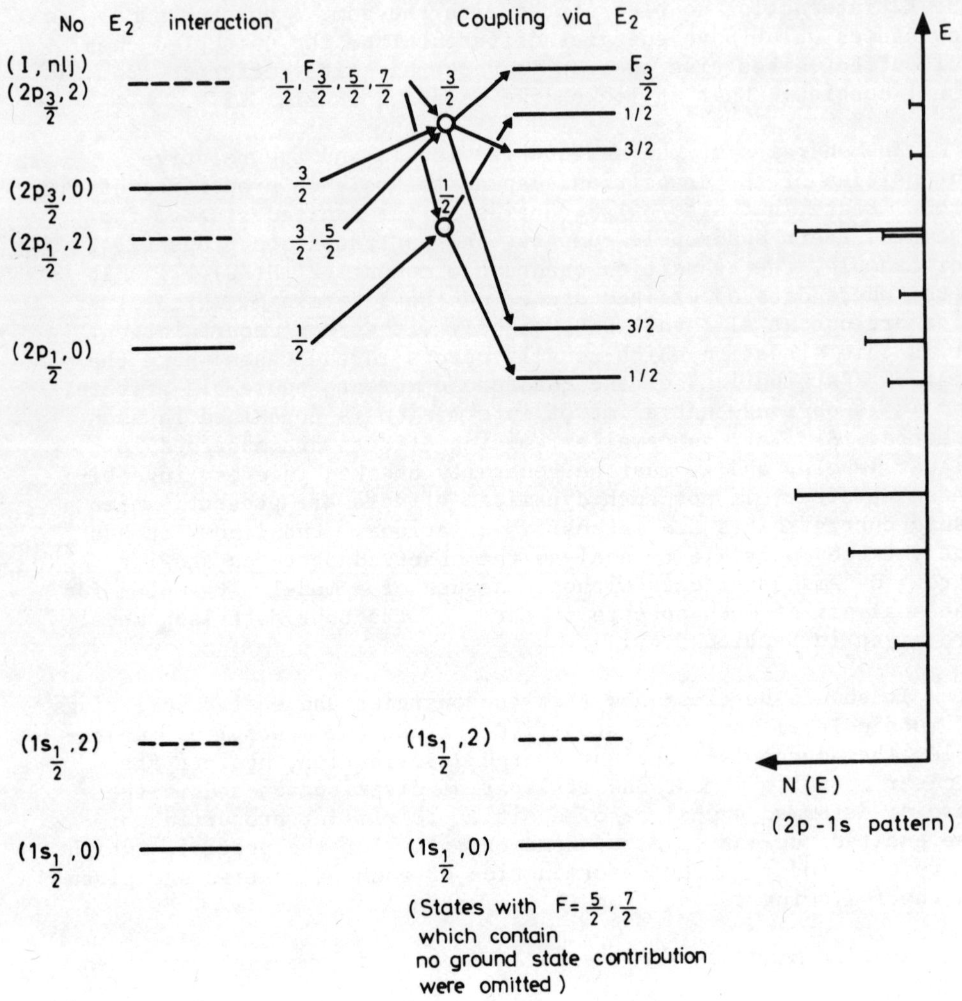

Fig. 13 Coupling of nuclear and muonic states via the E2 interaction in deformed nuclei [from S. Devons et al.[1]]

some hundreds of a fermi[53]. Why the accuracy achieved is so high is best demonstrated in terms of a model for $\rho_0(r)$. A widely-used model uses a Fermi-type distribution with the parameters c_0 and t_0:

$$\rho_0(r) \propto (1 + \exp\{4 \ln 3(r - c_0)/t_0\}^{-1} \ . \tag{18}$$

changing the c_0 parameter of $\rho_0(r)$ by 1 fm changes the energy of the 2p-1s transition of a medium Z nucleus by about 200 keV; the energy of the 3d-2p transition is changed by about 3 keV. The relative change in the energy of the lines is as high as about 10% (2p-1s) and about 0.5% (3d-2p). As the position of a line can be measured with a precision in the order of magnitude of 100 eV (for lines of high energy) it becomes clear that the high precision stated above can be obtained. One problem in the analyses of the spectra is the contribution from the nuclear polarization which cannot yet be calculated accurately enough and is therefore often treated as an additional free parameter in the fit. Recent analyses work not only with the 2p-1s and the 3d-2p lines, but use also the weaker transitions 2s-2p, 3p-2s, and so on, which show equally strong effects as the main transitions 2p-1s and 3d-2p.

The largest problem in the analysis of the data is the model dependence. Until several years ago, every analysis used a model distribution, mostly of the type of Eq. (18), and fitted the parameters to the data. A detailed discussion of these analyses is found elsewhere[54]. The actual shape of the distribution thus found is open to doubt, but the r.m.s. radii obtained from the distributions are fairly model independent. In 1969 Ford and Wills published a method[55] which allows a more model-independent analysis of the data. According to them every muonic transition defines a specific moment of the ρ_0 distribution

$$\left\langle r^k \right\rangle = \int \rho(\vec{r}) r^k \, d\tau \ . \tag{19}$$

k is generally no integer and is different for each transition and each Z. It can be obtained from a simultaneous discussion of all measured muonic lines[55]. Typical values range between 0.5 and 2.0. Thus, each transition determines an equivalent radius, given by

$$R_k = \left[\frac{1}{3}(k + 3)\left\langle r^k \right\rangle\right]^{1/k} \ . \tag{20}$$

For k = 2 the well-known expression for the root mean square radius is obtained. From a measurement of several transitions in a nucleus a set of R_k values is determined. These values are the only (nearly) model-independent values which can be deduced from muonic X-ray measurements and limit the possible charge distributions. A survey of charge distribution parameters obtained by using for the analysis of the muonic spectra both the methods discussed above is found elsewhere[56]. A detailed discussion of the practical application of the Ford and Wills method can be found in the literature[37,57].

The most recent development in the analysis of muonic spectra in terms of ρ_0 distributions is discussed elsewhere[58,59]. The method developed there divides the nucleus into charge shells with different charge density. The energy of a specific muonic transition is determined by the charge contained in one or a number of these charge shells. Measurements of elastic electron-nucleus scattering for different momentum transfers allow already quite detailed statements about the charges contained in the different shells. Adding to this information the very precise muonic data a very accurate and practically model-independent ρ_0 distribution can be obtained[59].

ii) Magnetic moments and their spatial distributions

The magnetic nuclear moment gives rise to an HFS splitting of all levels. It is generally quite small and is only clearly observable in the 2p-1s transition. The HFS was firstly observed in ^{209}Bi and the 2p-1s pattern is shown in Fig. 14 [1]. More recently[37,60] the effect could also be observed on a lighter nucleus (Nb) which is more difficult because the splitting becomes quite small (3.5 keV in the 1s state). Both measurements give a large Bohr-Weisskopf effect, which means that not a point-like magnetic moment but an extended one is observed. This statement means that A_1 (observed) is for the 1s level about 20-40% smaller than A_1^0 from Eq. (17), which has been obtained by using the value of $g_I\mu_N$ from other (optical or NMR) measurements. By comparing A_1 (observed) with predictions of different nuclear models, statements about the spatial magnetic moment distribution can be derived.

The models discussed here are the single-particle model and the configuration-mixing model. In Fig. 14 the predictions of the different models for the 2p-1s pattern are indicated and it is seen which different shapes are obtained. The single-particle model assumes that the magnetic moment is produced only by the angular momentum and the spins of the nucleons outside the nuclear core. A detailed discussion of this is found, for instance, in another paper[61]. The configuration-mixing model allows in addition to this for an interaction between core and outer nucleons leading to a configuration mixing of the states. For the Bi nucleus it is found that the configuration-mixing model, which already gives the absolute magnetic moment with good accuracy, also seems to give a good description for the spatial distribution of the moment. For Niobium the best predictions for the observed 1s splitting are given by the pairing plus quadrupole model[60].

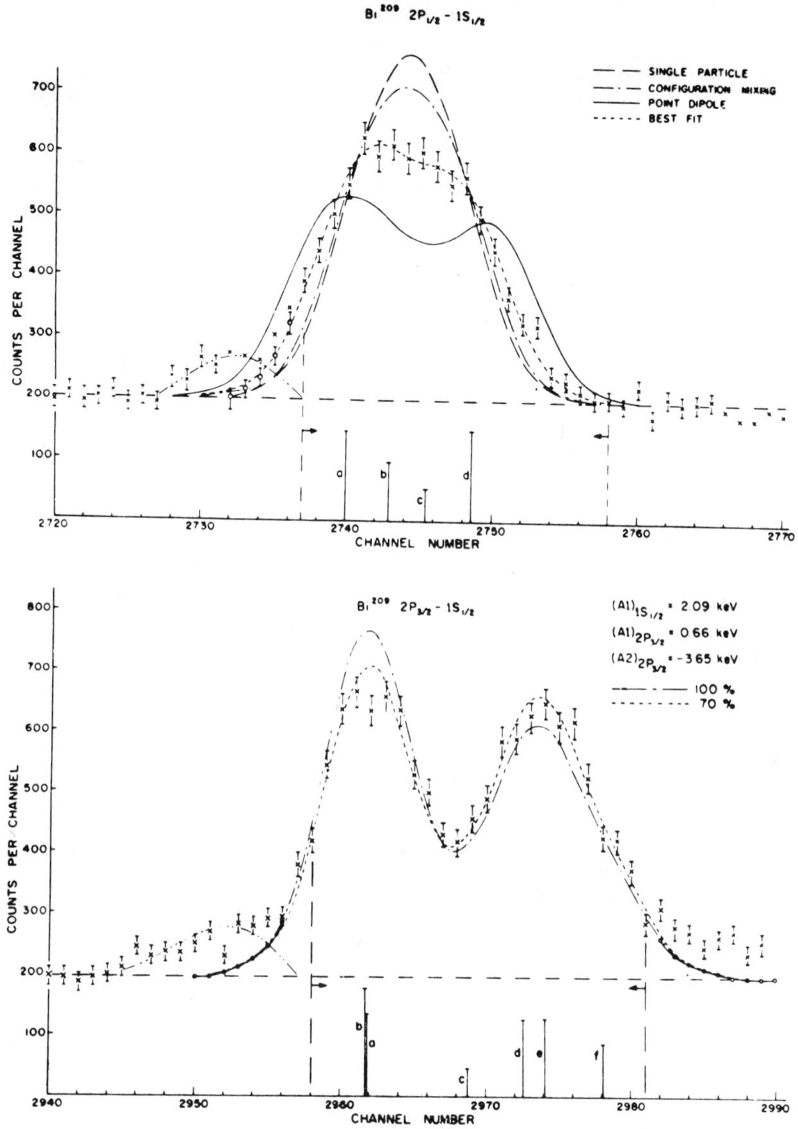

Fig. 14 Muonic transitions $2p \rightarrow 1s$ in Bismuth (0.855 keV/channel). In the $2p\frac{1}{2}$–$1s\frac{1}{2}$ transition, the hyperfine structure is purely magnetic dipole and the predictions of different nuclear models are shown. In the $2p\frac{3}{2}$–$1s\frac{1}{2}$ transition, there is both magnetic-dipole and electric-quadrupole hyperfine structure, and the hyperfine-interaction constants for the best fit are indicated. The agreement is improved if the intensity of the $F = 6$ to $F = 5$ component (designated b) is reduced to 70% of the statistical value, consistent with a small mixing of the muonic and nuclear states. [from S. Devons et al.[1]]

iii) Spectroscopic quadrupole moments

The spectroscopic quadrupole moments deduced from other than muonic X-ray measurements -- optical methods, Coulomb excitation, giant resonance measurements, and so on -- often yield different results because of the smallness of the effects or because of problems in the interpretation of the data. All these difficulties are not present if the HFS splitting of higher muonic levels can be measured accurately enough. The splittings in, for instance, 5g-4f and 4f-3d transitions are small, but can be determined with a high precision if a careful analysis of a spectrum with good statistics is performed. One of the most recent measurements[62)] on muonic ^{175}Lu 5g-4f and 4f-3d transitions gave for the intrinsic quadrupole moment Q_0 = 3.49 b with the remarkably small error of ± 0.02 b.

An experiment of similar accuracy[50)] was performed on the muonic ^{165}Ho nucleus, which in addition to a quadrupole moment with a small error gave also a first indication for a hexadecupole moment, derived from the splitting of the $3d_{5/2}$ level.

iv) Quadrupole moments of nuclear ground and excited states and their spatial distributions derived from dynamical effects.

The HFS pattern of transitions between low-lying muonic levels in deformed nuclei with low-lying nuclear rotational levels is simultaneously determined by the quadrupole moments of the ground states, by the transition E2-moment between ground and excited states and their spatial distributions, respectively. As too many unknown parameters are involved in the process, practical analyses make use of the collective model[51)]. It relates the quadrupole moments of the excited states and the transition quadrupole moments to the intrinsic quadrupole moment Q_0 of the ground state. The spatial distributions of all quadrupole moments involved are taken to be equal and are tested by using two extreme models for $\rho(\vec{r})$ which yield two different spatial distributions and thus give two different functions f(r) defined in Eq. (12). Both distributions are based on the spherical distribution of Eq. (18), but are modified to allow for the deformation of the nucleus

$$\rho(r) \propto \{1 + \exp(4X \ln 3)\}^{-1} \qquad (21)$$

with

$$X = \left[r(1 - \beta Y_{20}(\theta)) - c\right]/t \quad \text{deformed model (I)} \qquad (22)$$

and

$$X = \left[r - c(1 + \beta Y_{20}(\theta))\right]/\left[t + 2\beta c Y_{20}(\theta)\right] \quad \text{hard core model (II)} \qquad (23)$$

The difference between both models is illustrated in Fig. 15, where isodensity lines, the $\rho_0(r)$ and $\rho_2(r)$ distributions [see Eq. (10)],

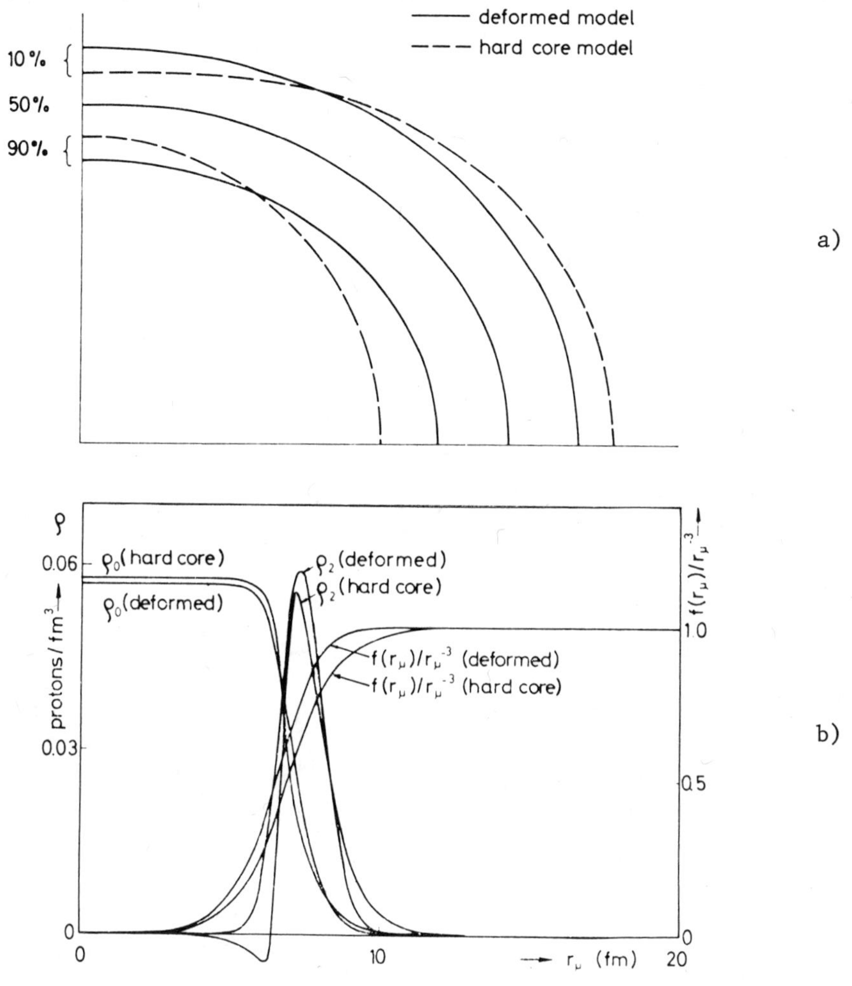

Fig. 15 Isodensity lines of a deformed model and hard core model charge distribution (a). Monopole and quadrupole distributions, and quadrupole form factors for both models [from S.A. de Wit et al.[51]]

and the functions f(r) divided by r^{-3} (the quadrupole moment form factors) are given. In model I all the nuclear matter is homogeneously deformed; in model II the nuclear core stays spherical and only the outer regions are deformed. The deformation parameter β of Eqs. (22) and (23) is related to the intrinsic quadrupole moment Q_0. Thus, the observed pattern must be fitted with three parameters $[c, t, \beta(Q_0)]$ for each assumed model for the quadrupole moment distribution. Many examples of this type of analysis are discussed in Ref. 51 and also in these days the method is still applied in more precise measurements[50,57]. The results of the analysis of the spectra are the following:

i) The collective model description is in agreement with the data.

ii) Model II (hard-core model) seems to be ruled out. The spatial quadrupole moment distribution is more similar to the one predicted from model I.

A survey of the data taken until now can be found in Ref. 56. One of the great advantages of this method is that, also, in cases where the spin of the nuclear ground state is too small to yield an observable quadrupole moment, the deformation of the ground state can be measured. Then only the transition and excited state E2 moments give rise to the effect via the dynamical interaction and the ground-state quadrupole moment can be deduced from the collective model.

Apart from the E2 mixing in muonic atoms still other dynamical effects exist, which yield important information about the nuclear structure. All these effects depend on the fact that nuclear levels are excited during the muonic cascade and de-excite while the muon still stays in the 1s level. The observation of the nuclear transitions in the presence of the muon allows the measurement of small differences in the ρ_0 distribution between the excited nuclear states and the ground state (isomeric shifts) and to determine via the interaction between the nuclear magnetic moment and the muon magnetic moment the spatial distribution of the nuclear magnetic moment (dynamical M1 interaction). References to these effects are also found in Ref. 56.

5.4.2 Hadrons: strong interaction effects between the hadrons and the nucleus

As already discussed earlier (Section 3) the strong interaction effects consist of shifts of the energy of the levels (relative to the purely electromagnetically determined values) and of natural level widths caused by the strong hadron-nucleus interaction. The shifts observed until now range between several eV (1s level of π^--deuterium) and some tens of keV $[\varepsilon_{1s}(\pi^- \mathrm{Na}) \approx 50$ keV$]$, the widths between 10^{-3} eV (Γ_{2p} in π^{-4}He) and 17 keV (Γ_{2p} in π^{-}Zn). The natural level widths which are not more than 5-10 times smaller than the experimental resolution -- that is widths larger than about 50 eV -- can be determined directly from the Lorentz-broadened shape of the transitions; the smaller widths have to be deduced from intensity measurements of the lines. A very elegant way to determine the small natural widths from relative intensity measurements was first discussed elsewhere[63] and is explained in the following on the example of the pionic 2p-level width. The pionic level scheme with the interesting transitions is given in Fig. 2. The competition between radiative transitions and strong absorption effects in the 2p level leads to a total 2p-level decay width Γ which is the sum of Γ_x = 0.45 eV (radiation width) and of Γ_{up} which is to be determined. The yield of the pionic 2p-1s transition accompanied by X-ray emission is

$$Y_{2-1} = P_2 \frac{\Gamma_x}{\Gamma_x + \Gamma_{up}} \tag{24}$$

(a small correction due to Auger processes is here neglected). P_2 is the population of the 2p level which is quite difficult to determine. Solving Eq. (24) for the unknown quantity Γ_{up} yields

$$\Gamma_{up} = \Gamma_x \left(\frac{P_2}{Y_{2-1}} - 1 \right) . \tag{25}$$

The calculation of Γ_x is simple, but the Fried-Martin factor[64] should not be forgotten. The problem is the determination of P_2/Y_{2-1}. An independent determination of Y_{2-1} and P_2 needs absolute yield measurements and absolute cascade program predictions. The simpler way is to take the population P_2 from the same spectrum where the 2p-1s transition appears. P_2 is equal to the sum of pionic transitions feeding the 2p level. The most important transitions responsible for the filling of the 2p level are indicated in Fig. 2. Thus

$$P_2 = \sum_{i=3}^{\infty} Y_{i-2} \sim Y_{3-2} + Y_{4-2} + Y_{5-2} + Y_{6-2} + Y_{7-2} + \ldots \tag{26}$$

In practice the series can be cut after 4-6 transitions, because the intensities of the transitions decrease with increasing i. With Eq. (26) one obtains

$$\Gamma_{up} = \Gamma_x \left[\sum_{i=3}^{\infty} Y_{i\to 2}/Y_{2-1} - 1 \right] . \tag{27}$$

This expression contains only *relative* intensities of transitions occuring in the same spectrum. The still necessary corrections (energy dependence of *relative* detector efficiencies and of *relative* X-ray absorption in the target) can be performed quite easily and the results obtained with this method yield quite small errors.

5.4.2.1 Pionic atoms

A summary of the greatest part of strong interaction data of pionic atoms is found in Ref. 3. The interest in the last years concentrated on very precise measurements of light pionic atoms and a list of some new data is given in Table 4. The most spectacular and most fruitful measurements were these on pionic deuterium and ^{4}He. The deuterium experiment was done using the critical absorber edge technique together with a proportional gas counter, the ^{4}He experiment was done with a high resolution Si-detector. The pionic ^{4}He spectrum is shown in Fig. 16. The shifts and widths of higher levels and of heavier nuclei are much less well measured. The errors for the shifts are equal to or larger than 200 eV, the errors for Γ_{low} range between 200 and 1000 eV, and the accuracy for Γ_{up} lies at 25-40%. Several isotope measurements have been performed (^{6}Li/^{7}Li; ^{10}B/^{11}B; ^{16}O/^{18}O; ^{58}Ni/^{60}Ni) and particularly the light isotopes show very pronounced differences. A general result of all measurements is that for N = Z nuclei the shift of the 1s level is always negative (strong interaction gives a repulsive effect), whereas all higher levels (2p, 3d, ...) show a positive shift (attractive strong interaction).

5.4.2.2 Comparison between pionic data and calculations

The most desirable theoretical description of the measured effects would be a microscopic treatment of the pion-nucleus interaction which uses the properties of the free pion-nucleon interaction and contains all important nuclear properties, such as nucleon distributions, binding effects, long-range Pauli correlations, short-range nucleon-nucleon correlations and so on. The comparison between theoretical predictions and the data would then allow statements about unknown nuclear effects which play a role in the pion-nucleus interaction. Such a description is not yet available, because of the complexity of the phenomena involved. In light nuclei the chances for a progress in this field seem to be greatest and it is here that already some rough microscopic predictions exist. Some predictions for the pionic ^{4}He data are summed up in the following:

Table 4

Recent results (after 1970) on strong interaction effects in pionic atoms [from Tauscher (Ref. 7)]

Level	Element	ε [eV]	Γ [eV]	Remark
1s	Deuterium	$-4.6 \begin{array}{l} +1.6 \\ -2.0 \end{array}$	-	Critical absorption edge
1s	^{3}He	54 ± 20	110 ± 120	a)
1s	^{4}He	-76 ± 2	45 ± 3	a)
1s	^{6}Li	-324 ± 3	195 ± 12	a)
1s	^{7}Li	-570 ± 4	195 ± 13	a)
1s	^{9}Be	-1595 ± 9	591 ± 14	a)
1s	^{20}Ne	-33340 ± 500	14500 ± 3000	a)
2p	^{4}He	-	$(7.2 \pm 3.3) \times 10^{-4}$	b)
2p	^{16}O	4.1 ± 2.3	11 ± 6	b)
2p	Al	212 ± 23	-	
2p	P	366 ± 31	-	
2p	A	825 ± 100	1170 ± 170	a)
3d	Ba	5440 ± 270	4300 ± 900	a)
3d	^{140}Ce	7030 ± 290	5600 ± 1000	a)
3d	^{142}Ce	7210 ± 330	6500 ± 900	a)
4f	Ho	350 ± 80	210 ± 40	a)
4f	Lu	670 ± 70	230 ± 70	a)

a) Width determined from line shape

b) Width determined from intensities

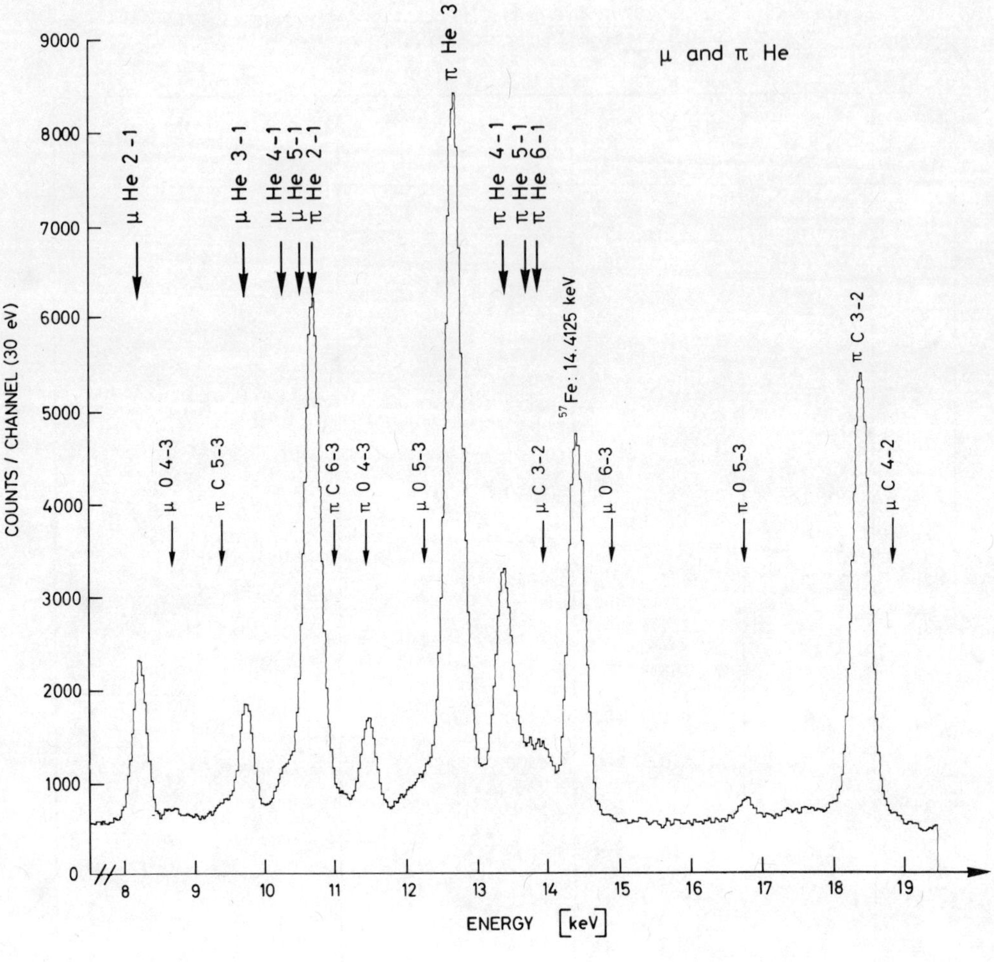

Fig. 16 Pionic ^{4}He spectrum [from G. Backenstoss et al.[66]]

		ε_{1s} [eV]	Γ_{1s} [eV]	Ref.
Experiment		-75.7 ± 2.0	45.2 ± 3.0	66
Predictions:	a)	-56 ± 4	40	67
	b)		400 ± 240 or 141 ± 85	68
	c)		64	69

The prediction (a) for the shift uses a simple perturbation approach; for the determination of the widths always the π absorption on a deuteron pair was assumed as the dominating absorption process. The comparison between the data and calculations shows that the data nowadays are quite accurate enough to distinguish between different predictions. At the moment, this is only true in some selected cases, for instance in ^{4}He and, in the future, experiments of similar precision should be performed at the meson factories. A candidate, particularly well suited is the ^{3}He nucleus, the description of which, in terms of the elementary pion nucleon interaction and in terms of nuclear correlations, seems feasible[70]. Another example for microscopic calculations is the analysis of the observed isotopic effects. A first analysis of the isotopic effects in $^{16}O/^{18}O$ has already shown special features of the nuclear wave functions which could be tested on the experimental results[71].

Most of the information known today about the π^--nucleus interaction , however, has been extracted from calculations of the effects in terms of an optical potential. The merits and limitations of this potential have already been discussed in the lectures about pion-nucleus interaction and thus only some of the most important points will be stressed here. The strong pion-nucleus interaction is described[10,7] by

$$-2\mu V_{\pi^- \text{-nucleus}}(r) = 4\pi\{[p_1(a_p\rho_p + a_n\rho_n) + p_2 B_0(\rho_n + \rho_p)^2$$

$$+ \text{ short-range correlation corrections}$$

$$+ \text{ Pauli correlation corrections (long-range)}$$

$$+ \text{ terms of order } 1/A \text{ (only important at light nuclei)}$$

$$+ \vec{\nabla}\frac{1}{f}\alpha(r)\vec{\nabla}\} \ . \qquad (28)$$

$$\alpha(r) = q_1(c_p\rho_p + c_n\rho_n) + q_2 C_0(\rho_n + \rho_p)^2 \ . \qquad (29)$$

$$f = 1 + \xi\frac{4\pi}{3}\alpha(r) \ . \qquad (30)$$

ρ_p, ρ_n are the proton and neutron densities; p_1, p_2, q_1, q_2 are kinematical factors. a_p, a_n and c_p, c_n are the free pion-nucleon scattering amplitudes at kinetic energy zero (= scattering lengths) of s-wave and p-wave interactions, respectively. They are real and can be deduced from low-energy scattering experiments[72]. The constants B_0 and C_0 are generally complex numbers and are connected with the elastic and inelastic π^--2 nucleon interaction. ξ determines the strength of the Lorentz-Lorenz effect. $\xi = 1$ would mean that the effect is completely present. A lot of the effects determining the effective potential (28) are not known and cannot be inferred from the results of other experiments. That is particularly true for the short-range nucleon-nucleon correlation -- in other words, what is the nucleon-nucleon interaction at small distances -- for the real part of B_0, which describes the π^--2 nucleon elastic interaction (probably again connected with the short-range nucleon-nucleon interaction) and for the importance of the Lorentz-Lorenz effect. Without the possibility of disentangling the effects, an analysis of the data in terms of the unknown, but on the other hand very interesting, processes would be hopeless, but fortunately the observed shifts and widths of the different ℓ levels depend only on specific parts of the potential. This is illustrated in the following: The usual way to obtain the complex energy values and wave functions of the levels (and thus the shifts and widths) is to solve the Klein-Gordon equation with the optical potential of Eq. (28):

$$\left[h^2\nabla^2 + (E_{n,\ell} - V_{\text{Coulomb}})^2 - m_\pi^2 c^4\right]\psi_{n,\ell} = 2\mu V_{\pi^- \text{-nucleus}}\,\psi_{n,\ell} \ . \qquad (31)$$

The wave function thus obtained for the 1s state is practically constant in the nuclear interior where $V_{\pi^- \text{-nucleus}}$ is different from zero. That means that terms of the kind $a_p\rho_p\psi_{1s}$ and $\text{Re } B_0(\rho_n + \rho_p)^2\psi_{1s}$ are unequal to zero, but terms belonging to the p-wave π-nucleon interaction, for example, $\vec{\nabla} c_p\rho_p\vec{\nabla}\psi_{1s}$ and $\vec{\nabla} \text{ Re } C_0(\rho_n + \rho_p)^2\vec{\nabla}\psi_{1s}$, are practically equal to zero, because ψ_{1s} has no gradient in the nuclear interior. Thus, for the s-level effects only s-wave π-nucleon interaction terms matter. Similarly, it can be shown that the effects on levels with $\ell = 1, 2, 3, \ldots$ depend only on the gradient term of (28) and thus only on C_0 and ξ. The shifts ε are essentially determined by the real part, the widths of the levels by the imaginary part of the potential, in other words, the dispersion effects are small. Concluding, one arrives at the following results for the relation between the measured shifts and widths and the unknown parameters of the potential:

ε_{1s}: The 1s-level shift depends on Re B_0 and the short-range correlations. The Pauli correlations and the 1/A terms can be independently determined with sufficient accuracy. The 1s shift is of particular interest, because $a_p \approx -a_n$. That means that the elastic pion-nucleon interaction plays practically no role and all of the measured effect is dependent on the nucleon-nucleon interaction. Thus, the observation of 1s shifts is an ideal tool to get information about still unknown properties of the nuclear matter. Microscopic calculations of these processes are in progress and can easily be checked by comparison with the data.

Γ_{1s}: The 1s-widths are mainly dependent on the π^--absorption process. It is here where the assumption of the dominant π^--2 nucleon-absorption mechanism can be checked.

$\varepsilon_{2p,3d,\ldots}$: These effects depend on c_p, c_n, Re C_0 and ξ. Re C_0 is probably considerably smaller than c_n [10] and c_p turns out to be about a factor 10 smaller than c_n. Thus, the shifts of the higher levels are dominated by the π^--neutron p-wave interactions and the Lorentz-Lorenz effect strength ξ. It is also here where predicted neutron distributions $\rho_n(r)$ can be tested.

$\Gamma_{2p,3d,\ldots}$: The widths of the higher levels are determined by the p-wave π^--2 nucleon absorption process and the Lorentz-Lorenz parameter ξ. As a first step in the data analysis a fit of the unknown parameters of the potential (28) to all existing data was done using a computer code for the solution of Eq. (31) [73]. Because of the dependence of the data on specific parameters small errors (sometimes less than 1%) for the obtained parameters can be given. The results of more recent fits -- including 1/A terms -- can be found elsewhere [74,75]. The influence of the short range correlation and Pauli correlation corrections and the influence of Re C_0 was treated in most cases as an addition to the terms $p_1(a_p\rho_p + a_n\rho_n)$ and $q_1(c_p\rho_p + c_n\rho_n)$, leading thus to *effective* $a_{p,n}$ and $c_{p,n}$ values. The general agreement between the data and the values calculated with the potential (28) using the fitted parameters is quite good. That means that the potential is a reasonable way of describing the data.

In the following, some of the first results obtained from a comparison between the fitted and the calculated parameters of Eq. (28) are listed:

i) The inclusion of the Lorentz-Lorenz effect ($\xi = 1$) gives a considerably better fit to the data[74]. $\xi = 0$ also fits the data, but then the fitted values for c_0 and Im C_0 are considerably different from the predicted ones. That would mean that a pion sees a nucleus in the same way as a light quantum sees condensed matter and that therefore also the nucleus has a granular structure[13].

ii) The deduction of a value for Im B_0 from π-production experiments yields a much smaller value than is needed for fitting the data. If this is due to the poor π-production data or to another π-absorption mechanism is one of the problems which hopefully will be solved soon.

iii) The consideration of 1/A terms and the deduction of the effect of short-range correlation effects from data on heavier nuclei made it possible to get information on the elastic π-2N process. The result for light nuclei was that Re B_0 = $-$Im B_0 [75], a result, which gave rise to detailed theoretical considerations.

iv) A fit for the parameters of the potential was made which allowed, in principle, for different proton and neutron distributions[76]. The result was that the difference between the r.m.s. radii of neutron and proton distributions is equal to (-0.01 ± 0.16) fm and thus consistent with zero. This result seems not to rule out the existence of big differences between proton and neutron distributions, because only an effect present for all investigated nuclei would show up.

v) A recent experiment on deformed nuclei[77] showed an HFS pattern in the 5g-4f transition in pionic Lutetium. It is due to the strong interaction between the pion and the deformed nucleus and can be evaluated according to a method developed by Scheck[78]. The most interesting quantity is the ratio between the strong interaction shifts of the individual members of the HFS pattern characterized by the parameter ε_2 and the strong interaction shift of the centre of gravity of the pattern (ε_0). $\varepsilon_2/\varepsilon_0$ is more or less independent of the parameters of the optical potential and essentially determined by the quadrupole part ρ_2 of the *neutron* distribution ($c_n \gg c_p$). A preliminary analysis of the ^{175}Lu data shows no big difference between ρ_2 (proton) and ρ_2 (neutron).

Further experiments in the field of pionic atoms will be concentrated on the determination of still more precise results in spherical and deformed nuclei (see point v) and on accurate measurements of isotope effects. Such experiments can easily be done with the high intensity pion beams available in the meson factories. In addition to the X-ray observation, the detection of the particles emitted after a π^- absorption will be started, in light nuclei, if possible, in coincidence with the pionic 2p-1s X-ray. For example, the detection of neutral and charged particles in the π^- ^3He absorption process allows a kinematically complete experiment, where initial *and* final state are exactly known. It is hoped that the analysis of such experiments on light nuclei will help to shed light on the question of the π-absorption mechanism -- experiments not mentioned here indicate π^- absorption on an α-cluster in the nucleus[17-20] -- and the elastic π-2N interaction which should allow relevant statements about the N-N short-range interactions. Such information will also help us to understand still better the strong interaction in pionic atoms which gives rise to very large and easily observable effects.

5.4.2.3 Kaonic and antiprotonic atoms

In recent years, spectra of kaonic and antiprotonic atoms showing strong interaction effects have become available. Because of the low-beam intensities and the high-particle background at the outlet of the short beams, only a few elements have been measured and the statistics of the lines showing strong interaction effects is smaller than in the case of pionic atoms. One of the most accurate measurements was done on kaonic sulphur[79] where the spectrum shown in Fig. 17 was measured. One of the most recent antiprotonic spectra[80] ($\bar{p}{}^4He$) is shown in Fig. 18. It was measured on liquid helium using high resolution Si-detectors. The windows in the target were so thin that lines with energies as low as 2-3 keV could be clearly observed.

Only in a few cases a complete determination of all measurable strong interaction effects (ε_{low}, Γ_{low}, Γ_{up}) was possible. A list with all measured effects in kaonic and antiprotonic atoms was given by Tauscher[7] and by Koch[6]. Also in Σ^--hyperonic atoms strong interaction effects could be observed, but only in a few cases and with small statistical accuracy. A survey of these results has just been published[81], but because of lack of time the data are not discussed here.

One of the main motivations to study X-ray spectra of very heavy hadrons is that the strong interaction effects are very sensitive against the proton *and* neutron distribution in the nuclear *tail*. This is so, because only in cases where there is a *part* overlap between the hadronic wave function and the nuclear matter

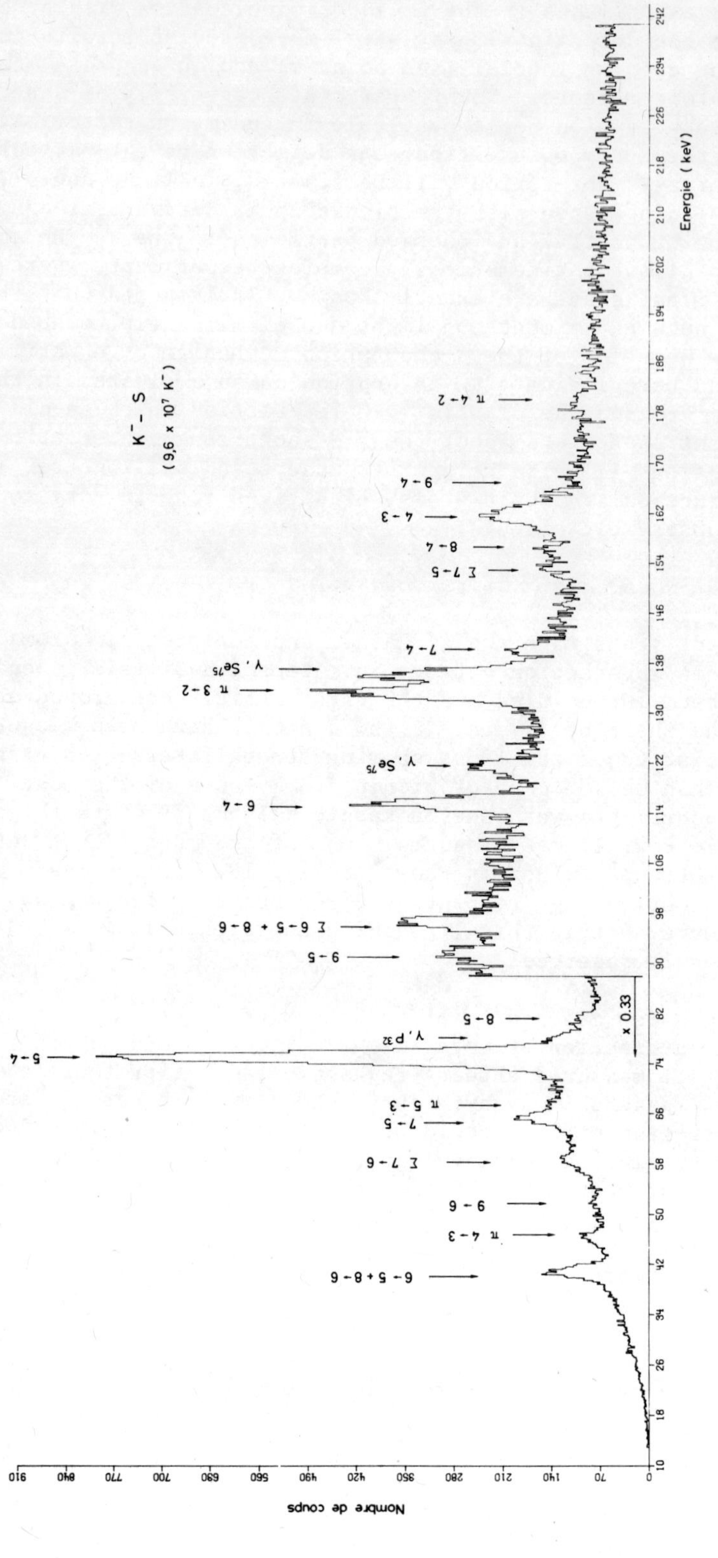

Fig. 17 Kaonic sulphur spectrum [from G. Backenstoss et al. [82]]

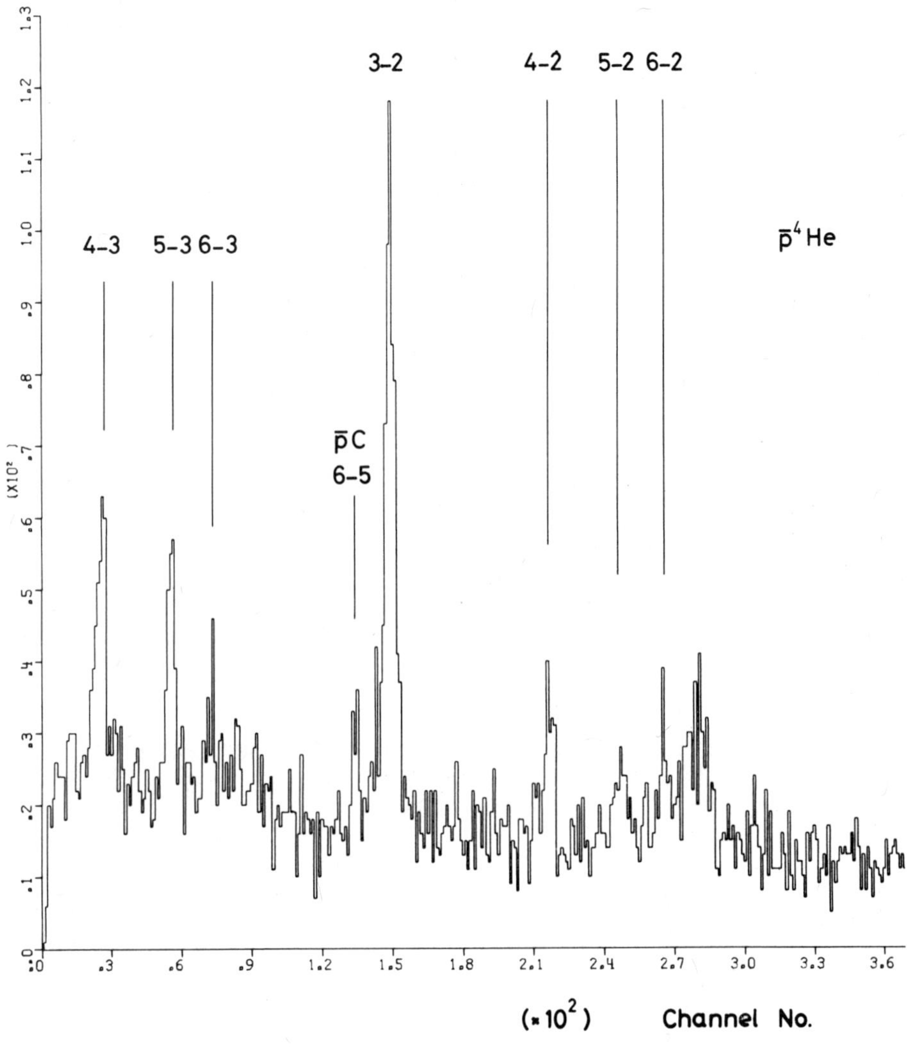

Fig. 18 Antiprotonic ^{4}He spectrum, measured at CERN (to be published).

distribution the transitions turn out to be observable. In pionic atoms the absorptive interaction is an order of magnitude smaller than in kaonic, antiprotonic and Σ^--hyperonic atoms and therefore also the inner part of the nuclear matter distribution contributes to the strong interaction effects. A further advantage of the heavy particles is that the s-wave interaction dominates at zero energy and the main absorption channels occur on *one* nucleon only. This fact could make the theoretical description of the processes much easier than in the pionic case.

Before the first data came out, it was hoped that they could be interpreted in terms of an optical potential of the following form:

$$-2\mu V_{\text{hadron-nucleus}}(r) = 4\pi\, p_1{}'(a_p\rho_p + a_n\rho_n) \tag{32}$$

$p_1{}'$ is a kinematical factor, a_p and a_n are the free hadron-proton and hadron-neutron scattering lengths, respectively, ρ_p and ρ_n are the proton and neutron distributions. Introducing this potential into Eq. (31) and for spin-½ particles into the Dirac equation yields values for the strong interaction effects, as a function of a_p, a_n, $\rho_p(r)$, and $\rho_n(r)$. With a_p, a_n, and $\rho_p(r)$ being known from other experiments, statements about $\rho_n(r)$ can be made, particularly in the region which is defined by $20\% \lesssim \rho_n/\rho_n$ (centre) $\lesssim 0.1\%$, because only there the overlap between the wave function and $\rho_n(r)$ is considerably different from zero. The potential (31) has exactly the same form as in the pionic case [see Eq. (28)], but is much simpler because p-wave, 2-nucleon processes and the corrections were not taken into account.

A comparison between the kaonic and antiprotonic data and predictions using Eq. (32) and the simple assumption $\rho_n(r)/\rho_p(r) = (A-Z)/Z$ (A = total number of nucleons, Z = proton number) yields the result[6,82] that the ansatz (32) does not work at all with a_p and a_n being the *free* hadron-nucleon scattering lengths. However, it could be shown that Eq. (32) allows a reasonably good explanation of the data, if a_p and a_n are replaced by a_p^{eff} and a_n^{eff}. The imaginary parts of a_p and a_n^{eff} coincide, but for the real parts even the signs are opposite to each other[5,6,79]. Different reasons could be responsible for that discrepancy:

a) The hadron-nucleon scattering lengths used in Eq. (32) could be wrong. This is easily possible in the $\bar{p}$-nucleon case, where only rough data from scattering experiments are available. The recently found resonances near the $\bar{N}$-N threshold can easily change the extrapolation from higher energies to the threshold. In the K^--N case the data seem to be more firm, but it should be kept in mind, that the K-N system near threshold is very complicated and can only be analysed via the K-matrix formalism which uses a lot of not extremely well known reaction channels. Thus, different analyses are not yet in agreement with each other and yield different results.

b) The ansatz (32) assumes tacitly that the hadron-nucleon interaction takes place at a relative energy exactly equal to zero. If the scattering amplitude near threshold is more or less constant as a function of the relative energy, this assumption is fulfilled. However, in the K^--N case the scattering amplitude shows a resonance structure near threshold, which means that it is varying very quickly as a function of the energy. This behaviour is shown in Fig. 19. The existence of resonances near threshold in the $\bar{p}p$ system as predicted by Shapiro[22] would give rise to a similar structure in the amplitude and it is quite obvious that these irregularities could completely spoil the validity of the ansatz (32).

c) Similar to the pionic atoms there could be strong contributions from corrections, for instance, hadron-2N processes and N-N correlations. Such processes give important contributions only, if the hadrons do not interact with a single, quasi-free nucleon, but with a cluster of nucleons.

d) The ansatz (32) could not be adequate for the problem. In contrast to pionic atoms the real and imaginary parts of the strong interaction potential became quite strong and the approximations leading to (32) could no longer be valid. The potential introduced into the Klein-Gordon or Dirac equation gives rise to large dispersive effects[83]. The widths of the lower level (Γ_{low}), for instance, are about equally strongly determined by the imaginary and the real part of the potential, which means that a proper disentangling of the effects of Re V and Im V on the shifts and widths is no longer possible, in contrast to the pionic atoms[5].

Many theoretical studies concerning the K^--atoms which attack the problems (a)-(d) have already been published, but a definite answer is not yet available. Very little theoretical work has been done on $\bar{p}$ atoms. A quite successful and simple approach is due to Bardeen and Torigoe[84] in the case of kaonic atoms. They believe in the ansatz (32), but take into account the proper relative K^--nucleon energy and the effect of the Y_0^* resonance at 1405 MeV, quite close to the K^--N threshold. The effect of the resonance can easily be seen in Fig. 19 (zero crossing of Re f_{K^-p} and the bump of Im f_{K^-p} at 1405 MeV). Averaging over the K^--N energies which occur in the nucleus they result in values of a_p^{eff} and a_n^{eff}, quite close to the fitted ones. More complete approaches[85,86] taking into account the K^--N resonances and some nuclear properties, for instance, off-shell effects and N-N correlations, yield results similar to Bardeen et al., but do not improve the agreement between data and calculations. The same is true for an approach by Deloff[87), which uses separable K^--N potentials. The range of the K^--N force is obtained from K^--^{4}He scattering, the potential depth from the free K^--N scattering lengths.

If Bardeen and other authors are right, the existence of the resonances play a dominant role in the explanation of the hadronic atom data. To find out if this is true, more accurate measurements are needed. They should be done in the low Z-region, because there ρ_p and ρ_n are quite well known, also in the very dilute region, which is only of importance here[59]. Measurements in this Z-region can help to find out how a resonance at low energies reacts with nuclear matter. This is of special interest in the case of antiprotonic atoms, where it is not clear if resonances exist or not. However, their existence has been theoretically predicted and recently indications have been found in experiments at low energies. Particularly, the existence of resonances *below* threshold can be studied. The fact that Re a^{eff} has a sign opposite to that of Re a in antiprotonic atoms could already be a first hint for the existence of the resonances, if a mechanism similar to the one proposed by Bardeen et al., is responsible for the $\bar{p}$-nucleus interaction. Thus, the first part of the future program is quite clear: Measurement of precise strong interaction data in the low Z-region, if possible down to hydrogen. Comparison of the data with improved calculations then will show a proper way of describing the effects, and will allow an interpretation of the data in terms of the behaviour of a resonance in nuclear matter and other effects. A first step towards this aim was already made at CERN, where measurements on $\bar{p}$-^{4}He [7] and $\bar{p}$ ^{16}O/^{18}O [88] were performed. Particularly, the study of isotope effects might show the way to a proper description of the effects. If this first step is finished, an interpretation of data in the higher Z-region in terms of neutron distributions seems hopeful. It is there, where with other techniques more or less big differences between proton and neutron distributions have been found, and here the study of heavy hadronic atoms may yield valuable contributions. Of particular interest, of course, is the investigation of the K^-p and $\bar{p}p$ systems, which allows the determination of the K^-p and $\bar{p}p$ scattering length in a very direct way and gives information on the existence of resonances in the systems near threshold, which, in the $\bar{p}p$ case, is especially of very great interest in these days.

6. FURTHER EXPERIMENTS: TEST OF PARITY VIOLATING CHARACTER OF NEUTRAL CURRENTS

At the end of each of the foregoing sections a summary of proposals for future experiments was given. However, all these proposals made use of existing techniques, and therefore here an example for an experiment is discussed which is quite different from the topics discussed above.

Recently, the existence of neutral currents has been experimentally proved. In spite of this, additional confirmation is highly desirable and the properties of the neutral currents have to be studied. One of the most important problems is the question of whether neutral currents are parity-violating or not, and two proposals came up recently which show that this question may be solved by looking at asymmetries of muonic X-rays relative to the muon spin. The first proposals of this kind came from Bernabeu et al.[89] and then from Feinberg et al.[90]. They predict an asymmetry of the 2s-1s M1 X-ray transition in light elements (e.g. Li, Be) relative to the muon spin. The effect occurs as follows:

The unified theories, for instance that of Weinberg[91], couple electromagnetic and weak interactions together. Without this coupling, the energy level scheme of μ^--Li for instance, looks as indicated in Fig. 20a. No coupling exists between the 2s and 2p levels; the angular momentum ℓ is a good quantum number. A very weak (only allowed by relativistic effects) M1 transition occurs between the 2s and 1s levels. If the unified theories are right, a coupling exists between the 2s and the 2p level (see Fig. 19b), which has its origin in the parity violation (weak) part of the unified interaction. The coupling can be described in a non-relativistic approach by the potential

$$V^{PV} = \frac{G_F}{\sqrt{2}} \frac{Q}{4m_\mu c} \left[\vec{\sigma}\vec{p}\rho(r) + \rho(r)\vec{\sigma}\vec{p} \right] \tag{33}$$

with G_F being the Fermi coupling constant, $\vec{\sigma}$, $\vec{p}$ the spin and momentum operators of the muon, and $\rho(r)$ the nucleon density. The parameter Q is dependent on the model. In the Weinberg model it is given by

$$Q = - \left[(4 \sin^2 \Theta_W - 1)Z + N \right] \tag{34}$$

with Θ_W = Weinberg angle ($\sin^2 \Theta_W \approx 0.4$ according to neutrino experiments) and Z and N being the proton and neutron number of the nucleus, respectively. The state wave functions $|2s\rangle$ and $|2p\rangle$ are mixed by V^{PV} and the new 2s wave function is given by

$$|2\rangle = |2s\rangle + \eta|2p\rangle \tag{35}$$

with

$$\eta = \frac{\langle 2p|V^{PV}|2s\rangle}{E_{2s} - E_{2p} - \frac{i}{2}(\Gamma_{2s} - \Gamma_{2p})} , \tag{36}$$

where E_{2s}, E_{2p} and Γ_{2s}, Γ_{2p} are the unperturbed energies and widths of the 2s and 2p levels. The admixture of the 2p-level wave

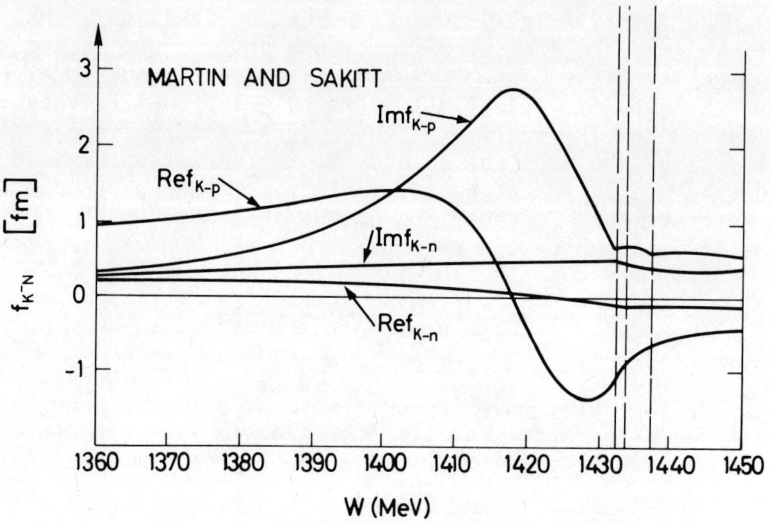

Fig. 19 $\bar{K}$-nucleon scattering amplitude as a function of energy [from W.A. Bardeen et al.[84]]

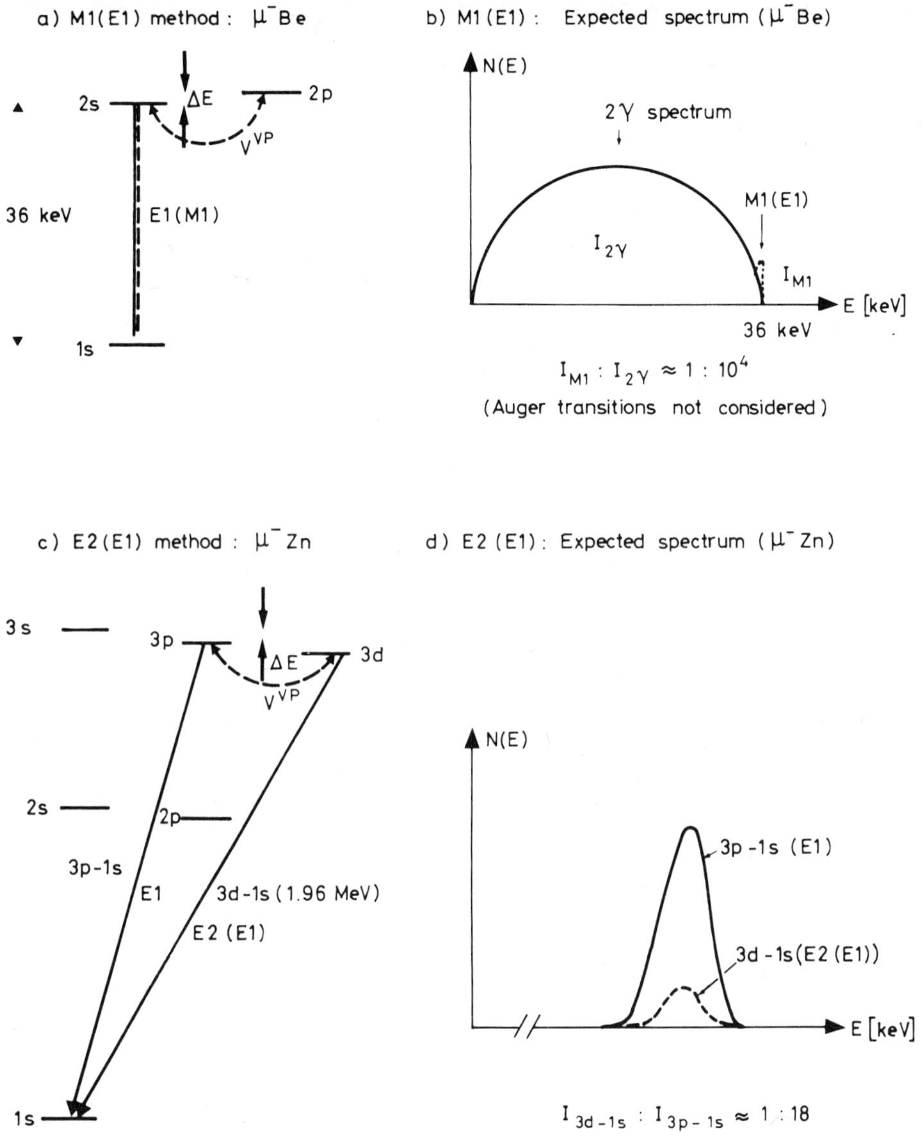

Fig. 20 Parity violating effects in muonic atoms

function to the 2s-wave function leads to an asymmetry of the 2s-1s X-rays relative to the muon spin direction, because the former M1 transition has now a small E1 contribution. The asymmetry A of the M1(E1) X-ray radiation relative to the muon spin is given by[92]

$$A = 2CB_1\eta \frac{\langle E1 \rangle}{\langle M1 \rangle}\ , \qquad (37)$$

where C is an angular correlation factor of the order of magnitude 1, B_1 is the orientation parameter which is proportional to the muon polarization, $\langle E1 \rangle$ and $\langle M1 \rangle$ are the reduced matrix elements for the transition amplitudes. Though the admixture parameter η is as small as 10^{-7} in light muonic elements, the asymmetry A can become as big as a few per cent due to the big ratio $\langle E1 \rangle / \langle M1 \rangle$ (enhancement factor). This effect is very large compared to parity violating effects in nuclei which are as small as 10^{-7}-10^{-6}.

However, a critical discussion of the effect from the experimental point of view shows that the measuring times would be very long, because of the weak population of the 2s level and because of other processes which depopulate the 2s level more quickly than the M1(E1) transition. One of the main processes is the 2γ 2s-1s transition, which gives rise to a continuous background. The line in question would lie at the high-energy end of the 2γ bump and a detection of the line and its asymmetry would be very difficult because of the high background, as indicated schematically in Fig. 20b. Under these circumstances, a measurement of the line at higher Z seems more hopeful, though the energy denominator of η is much larger than in the low Z-region and diminishes the effect to about 10^{-5}.

Another method which seems to be the most promising one has been proposed by Simons[92]. It uses the mixing between higher muonic levels, for instance, the $3p_{3/2}$ and $3d_{3/2}$ levels. The observed line is the 3d-1s E2 transition which has via the parity-violating 3p-3d interaction an E1 admixture of opposite parity and shows therefore again an asymmetry relative to the muon spin. This situation is sketched in Fig. 20c. The asymmetry observed is, in this case, determined by

$$A = 2CB_1\eta \frac{\langle E1 \rangle}{\langle E2 \rangle} \qquad (38)$$

and can be as big as 10^{-5} in selected cases. For instance, in the region around Z = 30 finite size effect and vacuum polarization effect, which determine the difference $E_{3p}-E_{3d}$ are such that the energy difference is as small as some eV. The E2 transition in question has a much higher yield (a few per cent) than the M1 transition in light muonic atoms. It has practically the same energy as the 3p-1s transition which shows no asymmetry effect. Therefore, the observed transition pattern has the composition shown in Fig. 20d. The effect of 10^{-5} refers already to the total observed line, not only to the smaller E2(E1) transition. Although the asymmetry is quite small, it is hoped that an experiment of this kind is feasible at one of the very powerful muon channels of the meson factories. Pre-experiments have already started. Similar experiments are possible in electronic atoms[93]. However, there the effects are several orders of magnitude smaller (smaller overlap between the electron wave function and the nucleus) and the corrections are more difficult. It is here again, that the simplicity of the system of an exotic atom allows very clear statements, a fact which has shown up in many cases considered in the previous sections.

7. SUMMARY

Looking back to the different topics discussed here three points seem to me the most essential ones. The first point is concerned with the variety of phenomena which are observable in spectra of muonic and hadronic X-rays. They range from observations of chemical and solid-state effects over clear very detailed nuclear physics information to the accurate determination of properties of elementary particles and to the investigation of the strong interaction between the captured hadrons and the nuclei. The second point of interest is the accuracy with which all this information can be gathered. For instance, the masses of the particles can be determined with a relative error approaching 10^{-5} and also the accuracy in the determination of the strong interaction effects is sometimes as good as a few per cent, which is rather uncommon in the investigation of such processes. The third point which should be stressed is the simplicity of the systems. One deals always with a hydrogen-like system, which is an enormous advantage in the interpretation of the data. This gives hope that even in the field of chemistry and solid-state physics, where a lot of other exploration methods are available, valuable contributions can arise from the study of mesic X-ray data.

In all fields mentioned above, the observation of mesic X-ray data gives rise to very clear and pronounced effects. The main difference between the fields lies in the degree to which the interpretation of the data has already proceeded. The most clear answers are obtained in cases where only the electromagnetic interaction is important, as is the case in muonic atoms and in the determination of masses and magnetic moments of the captured particles. Thus, the monopole, quadrupole and very recently the hexadecupole moments of nuclear charge distributions and the magnetic moments of nuclei could be determined with high accuracy. The results are in many cases superior to the data obtained from other techniques. For instance, the spectroscopic quadrupole moments could be measured with an accuracy which is at least one order of magnitude better than the precision obtained from optical methods. Furthermore, the spatial distributions of these moments give rise to big effects seen in the data, which allow a clear test of proposed nuclear models. Making use of dynamical effects also very detailed information about the properties of nuclei in excited states can be obtained.

Not quite as clear as these phenomena is the interpretation of effects where the strong interaction dominates. However, it can be shown that the data very clearly contain information about nuclear properties which cannot be seen with the same accuracy in other experiments, for instance, nucleon-nucleon interactions at small distances, the interaction between resonances and nuclear matter and nuclear matter distributions. The information in this field is nowadays limited to a phenomenological description of the data, the parameters of which are not yet fully understood in terms of the elementary interactions. First results, however, have already come up. The existence of the Lorentz-Lorenz effect in nuclear matter, the neutron distribution in deformed nuclei and some indications of the existence of $\bar{p}p$ resonances near threshold could be deduced. It is here where more accurate experimental information and more detailed calculations are needed to get a clearer picture. The experiments should, however, not only be done on mesic X-rays, but scattering experiments at higher energies and the observation of the particles emitted after the hadron capture are needed to get a satisfying answer to the problems. Such experiments form a substantial part of the programmes of the meson factories now coming into operation.

The observation of chemical and solid-state effects has shown very pronounced effects. The problem there lies in the interpretation of the data which according to my way of thought can only be solved by a close co-operation between specialists in these fields. It may well be that in some cases measurements on mesic atoms are the only possibility to get information about interesting effects which cannot be obtained by other methods.

From all this it should have become clear that measurements on mesonic X-rays are not at all confined to a systematic investigation of all available elements, but that they are a powerful tool for attacking interesting physical problems of all kinds. The experiments have to be performed adequately for a given problem and thus the experimental set-up has to be modified each time. The basic conception is very similar in all cases, but the energy range of X-rays, the very precise energy and efficiency calibrations, and the different background conditions enforce a different strategy each time. Also, in the future, very important problems can probably be attacked by measurements of the type discussed here. This is best demonstrated in the case of the test of unified field theories, where not energies, intensities and line shapes are measured, but angular distributions of X-rays contain the desired information.

REFERENCES

1) S. Devons and I. Duerdoth, Advances Nuclear Phys. 2, 295 (1968).

2) C.S. Wu and L. Wilets, Annu. Rev. Nuclear Sci. 19, 527 (1969).

3) G. Backenstoss, Annu. Rev. Nuclear Sci. 20, 467 (1970).

4) D.K. Anderson and D.A. Jenkins, Nuclear spectroscopy and reactions, Part B, (Ed. J. Cerny), (Academic Press, New York and London, 1974).

5) H. Koch, Externer Bericht der Gesellschaft für Kernforschung, Karlsruhe, Gfk-Ext. 3/73-3 (1973).

6) H. Koch, Proc. 5th Int. Conf. on High-Energy Physics and Nuclear Structure, Uppsala, 1973 (Almgvist and Wiksell, Stockholm, 1974), p. 225.

7) L. Tauscher, Invited talk at 6th Int. Conf. on High-Energy Physics and Nuclear Structure, Santa Fe, New Mexico, 1975, (To be published in the Proceedings).

8) L.I. Ponomarev, Annu. Rev. Nuclear Sci. 23, 395 (1973).

9) J. Hüfner, Pions interact with nuclei, MPI preprint MPIH-1975-V1, Heidelberg, 1975, to be published in Phys. Reports.

10) M. Ericson and T.E.O. Ericson, Ann. Phys. (NY) 36, 323 (1966).

11) M. Krell and T.E.O. Ericson, Nuclear Phys. B11, 521 (1969).

12) T.E.O. Ericson and F. Scheck, Nuclear Phys. B19, 450 (1970).

13) T.E.O. Ericson, Pion nucleus interactions, Lectures given at the International School of Nuclear Physics, Erice, 23-30 September 1974, (to be published in the Proceedings).

14) A. Bertin, A. Vitale and A. Placci, Atomic and molecular processes involving hydrogen and deuterium muonic systems in matter (to be published in Rivista del Nuovo Cimento).

15) H.S.W. Massey, E.H.S. Burhop and H.B. Gilbody, Electronic and ionic impact phenomena, (Clarendon Press, Oxford 1974), vol. 5.

16) M. Leon and H.A. Bethe, Phys. Rev. 127, 636 (1962).

17) W.J. Kossler, H.O. Funsten, B.A. MacDonald and W.F. Lankford, Phys. Rev. C 4, 1551 (1971).

18) H. Ullrich, E.T. Boschitz, H.D. Engelhardt and C.W. Lewis, Phys. Rev. Letters 33, 433 (1974).

19) P.D. Barnes, R.A. Eisenstein, W.C. Lam, J. Miller, R.B. Sutton, M. Eckhouse, J. Kane, R.E. Welsh, D.A. Jenkins, R.J. Powers, R. Kunselman, R.P. Redwine, R.E. Segel and J.P. Schiffer, Phys. Rev. Letters 29, 230 (1972).

20) C.E. Wiegand and G.L. Godfrey, Phys. Rev. A 9, 2282 (1974).

21) D.H. Wilkinson, Phil. Mag. 4, 215 (1959).

22) I.S. Shapiro, Uspekhi Fiz. Nauk 109, 431 (1973).

23) E. Fermi and E. Teller, Phys. Rev. 72, 399 (1947).

24) M. Leon and R. Seki, Phys. Rev. Letters 32, 132 (1974).

25) P. Vogel, P.K. Haff, V. Akylas and A. Winther, Nuclear Physics, A254, 445 (1975).

26) D. Quitmann, R. Engfer, U. Hegel, P. Brix, G. Backenstoss, K. Goebel and B. Stadler, Nuclear Phys. 51, 609 (1964).

27) A.R. Kunselman, Thesis, UCRL-18654 (1969).

28) G.T. Condo, Phys. Rev. Letters 33, 126 (1974).

29) D. Kessler, H.L. Anderson, M.S. Dixit, H.J. Evans, R.M. McKee, C.K. Hargrove, R.D. Barton, E.P. Hincks and J.M. McAndrew, Phys. Rev. Letters 18, 1179 (1967).

30) H. Daniel, H. Koch, G. Poelz, H. Schmitt, L. Tauscher, G. Backenstoss and S. Charalambus, Phys. Letters 26B, 281 (1968).

31) H. Poth, private communication.

32) S.S. Gershtein, V.I. Petrukhin, L.I. Ponomarev, Yu.D. Prokoshkin, Uspekhi, Fiz. Nauk 97, 3 (1969).

33) L.I. Ponomarev, Invited talk at 6th Int. Conf. on High-Energy Physics and Nuclear Structure, Santa Fe, New Mexico, 1975 (to be published in the Proceedings).

34) Yu.G. Budyashov, P.F. Ermolov, V.G. Zinov, A.D. Konin and A.I. Mukhin, Sov. J. Nuclear Phys. 5, 426 (1967).

35) H. Daniel, H.-J. Pfeiffer and K. Springer, Phys. Letters 44A, 447 (1973).

36) G. Holzwarth and H.J. Pfeiffer, Z. Phys. A272, 311 (1975).

37) H.P. Povel, Thesis, Gesellschaft für Kernforschung, Karlsruhe, KFK-Bericht 1602 (1972).

38) L. Tauscher, G. Backenstoss, K. Fransson, H. Koch, A. Nilsson and J. de Raedt, Test of QED by muonic atoms: an experimental contribution (to be published in Phys. Rev. Letters).

39) M.S. Dixit, H.L. Anderson, C.K. Hargrove, R.J. McKee, D. Kessler, H. Mes and A.C. Thompson, Phys. Rev. Letters 27, 878 (1971).

40) J. Blomquist, Nuclear Phys. B48, 95 (1972).

41) C.K. Hargrove, E.P. Hincks, H. Mes, R.J. McKee, M.S. Dixit, A.L. Carter, D. Kessler, J.S. Wadden and H.L. Anderson, Contributed paper 6th Int. Conf. on High-Energy Physics and Nuclear Structure, Santa Fe, New Mexico, 1975.

42) R.E. Shafer, Phys. Rev. D 8, 2313 (1973).

43) G. Backenstoss, H. Daniel, H. Koch, Ch. von der Malsburg, G. Poelz, H.P. Povel, H. Schmitt and L. Tauscher, Phys. Letters 43B, 539 (1973).

44) G. Backenstoss, A. Bamberger, I. Bergström, T. Bunaciu, J. Egger, R. Hagelberg, S. Hultberg, H. Koch, U. Lynen, H.G. Ritter, A. Schwitter and L. Tauscher, Phys. Letters 43B, 431 (1973).

45) G. Dugan, Y. Asano, M.Y. Chen, S. Cheng, E. Hu, L. Lidofsky, W. Patton, C.S. Wu, V. Hughes and D. Lu, Contributed paper to 6th Int. Conf. on High-Energy Physics and Nuclear Structure, Santa Fe, New Mexico, 1975.

46) E.V. Shrum and K.O.H. Ziock, Phys. Letters 37B, 115 (1971).

47) B.L. Roberts, C.R. Cox, M. Eckhouse, J.R. Kane, R.E. Webb, D.A. Jenkins, W.C. Lam, P.D. Barnes, R.A. Eisenstein, J. Miller, R.B. Sutton, A.R. Kunselmann, R.J. Powers and J.D. Fox, Phys. Rev. Letters 33, 1181 (1974).

48) B.L. Roberts, C.R. Fox, M. Eckhouse, J.R. Kane, R.E. Welsh, D.A. Jenkins, W.C. Lam, P.D. Barnes, R.A. Eisenstein, J. Miller, R.B. Sutton, A.R. Kunselman and R.J. Powers, Phys. Rev. Letters 32, 1265 (1974).

49) M.Y. Chen, Phys. Rev. C 1, 1167 (1970).

50) R.J. Powers, F. Boehm, P. Vogel, A. Zehnder, T. King, A.R. Kunselman, P. Roberson, P. Martin, G.H. Miller, R.E. Welsh and D.A. Jenkins, Phys. Rev. Letters 34, 492 (1975).

51) S.A. de Wit, G. Backenstoss, C. Daum, J.C. Sens and H.L. Acker, Nuclear Phys. 87, 657 (1967).

52) A. Bohr and V.F. Weisskopf, Phys. Rev. 77, 94 (1950).

53) E.B. Shera, E.T. Ritter, A.B. Perkins, L.K. Wagner, H.D. Wohlfahrt, G. Fricke and R.M. Steffen, Contributed paper to 6th Int. Conf. on High-Energy Physics and Nuclear Structure, Santa Fe, New Mexico, 1975.

54) H.L. Acker, G. Backenstoss, C. Daum, J.C. Sens and S.A. de Wit, Nuclear Phys. 87, 1 (1966).

55) K.W. Ford and J.G. Wills, Phys. Rev. 185, 1429 (1969).

56) R. Engfer, H. Schneuwly, J.L. Vuilleumier, H.K. Walter and A. Zehnder, Atomic data and nuclear data tables 14, Nos. 5-6 (1974).

57) A. Zehnder, Thesis, ETH, Zürich, No. 5280 (1974).

58) J. Friedrich and F. Lenz, Nuclear Phys. A183, 523 (1972). R.C. Barrett, Rep. Progr. Phys. 37, 1 (1974).

59) I. Sick, Invited talk at 5th Int. Conf. on High-Energy Physics and Nuclear Structure, Santa Fe, New Mexico, 1975 (to be published in the Proceedings).

60) H.P. Povel, Nuclear Phys. A217, 573 (1973).

61) M. Le Bellac, Nuclear Phys. 40, 645 (1963).

62) W. Dey, P. Ebersold, B. Aas, R. Eichler, H.J. Leisi, W.W. Sapp and F. Scheck, Contributed paper to the 5th Int. Conf. on High-Energy Physics and Nuclear Structure, Santa Fe, New Mexico, 1975.

63) H. Koch, Thesis, University of Karlsruhe, Germany,

64) Z. Fried and A.D. Martin, Nuovo Cimento 29, 3684 (1963).

65) J. Bailey, D.V. Bugg, U. Gastaldi, P. Hattersley, D.R. Jeremiah, E. Klempt, K. Nenbecker, E. Polacco and J. Warren, Phys. Letters 50B, 403 (1974).

66) G. Backenstoss, J. Egger, T. von Egidy, R. Hagelberg, C.J. Herlander, H. Koch, H.P. Povel, A. Schwitter and L. Tauscher, Nuclear Phys. A232, 519 (1974).

67) S. Deser, W. Baumann, W. Thirring, Phys. Rev. 96, 774 (1954). K.A. Brueckner, Phys. Rev. 98, 769 (1955).

68) S.G. Eckstein, Phys. Rev. 129, 413 (1962).

69) D.S. Koltun and A. Reitan, Nuclear Phys. B4, 629 (1968).

70) A.C. Phillips and F. Roig, Nuclear Phys. B60, 93 (1973).

71) K. Chung, M. Danos and M.G. Huber, Phys. Letters 29B, 265 (1969).

72) D.V. Bugg, A.A. Carter and J.R. Carter, Phys. Letters 44B, 278 (1973).

73) M. Krell and L. Tauscher, Program PIATOM. M. Krell and T.E.O. Ericson, Nuclear Phys. B11, 521 (1969).

74) L. Tauscher, Proc. Int. Seminar on π-Meson Nucleus Interactions, Strasbourg, 1971 (CNRS, Strasbourg, 1971), p. 45.

75) L. Tauscher and W. Schneider, Z. Phys. 271, 409 (1974).

76) D.K. Anderson, D.A. Jenkins and R.J. Powers, Phys. Rev. Letters 24, 71 (1970).

77) P. Ebersold, Thesis, ETH, Zürich (1975).

78) F. Scheck, Nuclear Phys. B42, 573 (1972).

79) J. Egger, Thesis, ETH, Zürich (1974).

80) Measurement of $\bar{p}$-^{4}He at CERN. Results not yet published (see Ref. 7).

81) G. Backenstoss, T. Bunaciu, J. Egger, H. Koch, A. Schwitter and L. Tauscher, Z. Phys. A273, 137 (1975).

82) G. Backenstoss, J. Egger, H. Koch, H.P. Pavel, A. Schwitter and L. Tauscher, Nuclear Phys. B73, 189 (1974).

83) M. Krell, Phys. Rev. Letters 26, 584 (1971).

84) W.A. Bardeen and E.W. Torigoe, Phys. Letters 38B, 135 (1972).

85) S. Wycech, Nuclear Phys. B28, 541 (1971).

86) M. Alberg, E.M. Henley and L. Wilets, Phys. Rev. Letters 30, 255 (1973).

87) A. Deloff and J. Law, Phys. Rev. C 10, 1688 (1974).

88) H. Koch, G. Backenstoss, P. Blum, W. Fetscher, R. Hagelberg, A. Nilsson, P. Parlopoulos, H. Poth, I. Sick, L. Simons and L. Tauscher, Contributed paper to the 5th Int. Conf. on High-Energy Physics and Nuclear Structure, Santa Fe, New Mexico, 1975.

89) J. Bernabeu, T.E.O. Ericson and C. Jarlskog, Phys. Letters 50B, 467 (1974).

90) G. Feinberg and M.Y. Chen, Phys. Rev. D10, 190 (1974).

91) S. Weinberg, Phys. Rev. Letters 19, 1264 (1967) and 27, 1688 (1971).

92) L. Simons, Helv. Phys. Acta 48, 141 (1975).

93) M.A. Bouchiat and C.C. Bouchiat, Phys. Letters 48B, 111 (1973).

ON THE APPLICATION OF POLARIZED POSITIVE MUONS IN SOLID STATE PHYSICS

Alexander Schenck

Laboratorium für Hochenergiephysik

E.T.H., Zürich, Switzerland

I. INTRODUCTION

It is only another point in favor of the new generation of medium high energy accelerators - the meson factories - that in a summer school on the physics at meson factories not only nuclear and particle physics will be covered, but also topics belonging to solid state physics and chemistry [1]. It is thus due to the wide-ranging research power of these new facilities that such separate fields of physics will come together once again. I hope that this encounter will not only stimulate mutual interest, but will also lead to a mutual inspiration and exchange of ideas and methods. In the course of these lectures I will present one example where such an encounter on the theoretical level has resulted in a very interesting treatment of the screening of an impurity charge in metals using ideas pertaining more to non-relativistic potential scattering as applied to nucleon-nucleon scattering.

As for the topics discussed in these lectures, the unifying link is set up by the availability of high intensity muon beams from meson factories and by the properties of muons which make them ideal probes for studying a wide variety of solid state phenomena. To be more specific, only the applications of positive muons will be treated in these lectures, although recently there have also been some very encouraging experiments involving negative muons [2]. The applicability of positive muons to solid state physics rests in the fact that magnetic interactions of the muon (which carries a magnetic moment and spin = ½) can easily be made visible by exploiting the distribution of muon decay positrons which is anisotropic with respect to the muon spin direction. This anisotropy is a consequence of parity violation in the muon decay, and in fact the implicit idea of using muons as probes is as old as the detection of parity violation in the muon decay [3]. By studying the positron distribution in a time differential fashion, it is possible to observe the dynamics of the muon's spin under the action of magnetic fields which have their origin and special character in the very nature of the piece of matter in which the muons have been stopped. And it is this very nature that one hopes to probe and investigate in this manner.

Another prerequisite for such measurements, namely the availability of spin polarized muon beams, is also given since muon beams naturally show a high degree of polarization due to the way by which these beams are formed from pions decaying in flight. As is well known, muons are 100% spin polarized with respect to their momentum in the rest frame of the pion. This is another consequence of the parity violating weak interaction that also governs pion decay. Muon beams with polarization of 80% or more are quite common.

In order to investigate certain properties of a sample of matter with the help of muons, it is necessary to stop or implant the muons in that sample. It is common experience that during the slowing down process of the muons no polarization is lost, so that the thermalized muons preserve their full spin polarization [4]. This is a very important fact.

The slowing down process is in another respect very important. Towards their final thermalization, muons will capture an electron to form the hydrogen-like atom, muonium (μ^+e^-). If no other effects were present, this would be the usual state in which muons would be thermalized. Indeed, muons are in the muonium state in many insulating crystals while in metals the electron is very quickly lost to the conduction band due to Coulomb screening by conduction electrons, which prevents the existence of bound states. More on this later.

A third possibility is observed in insulating crystals: while muonium still possesses epithermal energies (1-20 eV) it is capable of entering into so-called hot atom reactions. These are reactions that are usually forbidden because of lack of energy. As a result of such reactions, the muon may replace, abstract or add to one of the constituents of the target material, thus becoming part of a chemical compound. The study of hot atom reactions of muonium is in itself a very interesting subject. However, in the context of the present discussions we will only refer to it occasionally, with regards to its bearing on other effects and their interpretations.

The existence of thermalized muonium - particularly in solutions and in gases - opens up another field of study, namely that of fast chemical reactions of the hydrogen-like muonium atom [5] in analogy to reactions of atomic hydrogen. This very interesting subject, that has already led to a number of beautiful results, will also not be treated in these lectures, as it has found detailed coverage elsewhere [4,1]. Instead I will concentrate exclusively on the role the muon may play as a tool in solid state physics.

Why is it that we think that the muon is an ideal tool in solid state physics as compared with other implanted ions or probes such as Mössbauer nuclei or other techniques like NMR, $\gamma\gamma$ PAC, etc., and even neutron diffraction?

The reasons can be listed as follows:

1. Only about 10^5 - 10^7 muon decays have to be sampled (corresponding to about 10^6 - 10^8 muon stops in the target). This may be compared with about 10^{18} nuclear spins needed to produce a good NMR-signal.

2. Only one muon is present in the target at a time. This presents the case of infinite dilution. No interaction between the probes themselves is present. The disturbance of the host lattice is minimal.

3. Virtual absence of radiation damage. This is due first to the exceedingly small stopping rate and second - probably - due to the fact that in the last phase of the slowing down the muon exists in the charge neutral state of muonium, thus avoiding any ionization in the neighborhood of the final stopping position.

4. The muon has no complicated electron core or no core at all. Thus many effects are absent that usually complicate the analysis of Mössbauer, NMR, etc. data. The magnetic interaction of the ion core with the nucleus is generally much stronger than the interaction of the nucleus with the magnetic bulk properties of the crystal under investigation.

5. The muon has no electric quadrupole moment. Thus, all electric quadrupole interactions that usually mask and interfere with the magnetic interactions are absent.

6. The muons will be stopped almost homogeneously over the target volume, and in particular, all interstitial sites will probably be sampled by the muons. In contrast, the common probes as Mössbauer nuclei, NMR-nuclei, etc., yield information about the regular lattice sites. In many instances, however, the magnetic properties of interstitial sites are much more interesting, as they provide clues as to how magnetic ordering comes about for example. In these instances the muon method appears to be superior even to neutron diffraction, which measures in $\vec{k}$-space, while the muons measure in $\vec{r}$-space.

Of course, there are also many problems related to the muon applications. It will be one of the purposes of these lectures to discuss some of them in somewhat more detail. Most of these problems are already quite interesting in themselves, for example, muon diffusion and muon charge screening in metals.

The selection of topics presented here is somewhat arbitrary and reflects mostly the main directions of our own research interests at SIN. Therefore, this treatment will not comprehensively cover what has already been done and what is presently being discussed at the various laboratories active in this field.

Nevertheless, it is hoped that these lectures will provide a collection of useful conceptions, ideas and formulas. Most of these concepts and ideas are quite old and can be found in well-known books like the one of A. Abragam, and in a widely scattered number of original papers, and in different contexts. Therefore, another purpose of these lectures is to collect, present, and discuss these conceptions in the context of muon applications.

The method of muon application has been named 'μSR', which stands for 'muon spin rotation, relaxation, research etc'.

This mnemonic acronym was chosen to indicate the close analogy of this method with NMR (nuclear magnetic resonance) and ESR (electron spin resonance). This analogy as well as relations to other methods like PAC (perturbed angular correlations), PAD (perturbed angular distributions) and oriented nuclei will become clearer when we discuss the principles of the μSR-method.

In the following we will discuss the principle and various types of μSR-experiments, the types of magnetic interactions of the muon, and related phenomena like spin-spin relaxation and motional narrowing. Then we will consider muon diffusion, charge screening of positive muons in metals and, in some depth, the possibilities of μSR for studying magnetic problems. Finally, the behaviour of muonium (μ^+e^-) in solids will be discussed.

II. THE PRINCIPLES OF THE μSR-METHOD

a) Muon properties

In the following we list some of the basic properties of muons [4]

Spin: $\frac{1}{2}$

Mass: $m_\mu = 105.6595(3)\ \text{MeV} = 206.7684(6)\ m_e$
$= 0.1126123(6)\ m_p$

Magnetic Moment: $\mu_\mu = |g_\mu\ S_z|\ \dfrac{e\hbar}{2m_\mu c}$

$= 3.183346(g)\ \mu_p{}^*$
$= 28.0272(2)\quad 10^{-18}\ \text{MeV/gauss}$

Compton Wavelength: $\lambda\!\!\!^{-}_\mu = \dfrac{\hbar}{m_\mu c} = 1.86758\ \text{fm}$

* μ_p = magnetic moment of the proton

The positive muon will not be captured by or interact strongly with nuclei. Although it belongs to the class of leptons, the implanted muon will behave much more like a proton, the only important difference in this context being its lighter mass, which, by comparison with the proton behaviour, will allow the study of isotope effects. On the other hand, it compares much less with the positron, which, by its very much lighter mass, behaves in many respects completely differently.

b) The decay of the positive muon

The muon decays in the following way:

$$\mu^+ \rightarrow e^+ + \nu_e + \bar{\nu}_\mu$$

with an average lifetime of:

$$\tau_\mu = 2.1994(6)\cdot 10^{-6}\ \text{sec}$$

The decay is governed by the weak interaction which leads to a violation of parity.

The positron spectrum is given by the following expression that can be derived rigorously from the 4 fermion current-current interaction:

$$\frac{dN(w,\theta)}{dw d\Omega} = \frac{w^2}{2\pi} \quad [(3-2w)-P\cdot(1-2w)\cos\theta]$$

$$= \frac{c}{2\pi} \quad [1+D\cos\theta] \tag{2.1}$$

where $w = E/E_{max}$ is the positron energy measured in units of the maximal possible energy $E_{max} = \frac{m_\mu}{2} = 52.8$ MeV. The expression in the brackets shows the spatial anisotropic distribution of the positrons, where θ is the angle between the spin of the decaying muon and the positron trajectory. The asymmetry parameter D is a function of the positron energy

$$D = P\cdot \frac{2w-1}{3-2w} \tag{2.2}$$

with P = degree of the spin polarization of the decaying muons. The energy spectrum and the asymmetry parameter (P=1) are plotted in Fig. 1.

In practice, the positrons are detected with an efficiency $\varepsilon(w)$ which will not be constant over the entire energy range due to absorption and scattering in the target and the counters as well as to the effect of an external magnetic field on the positron trajectories.

The observed distribution probability is then (integrated over energy)

$$\frac{dN(\theta)}{d\Omega} = \int_0^1 \left[\frac{dN(\theta,w)}{dw d\Omega}\right] \varepsilon(w)\,dw$$

$$= \frac{1}{4\pi}\,\tilde{\varepsilon}(1 + \tilde{A}\cos\theta) \tag{2.3}$$

If all positrons were detected with the same efficiency, the observed average asymmetry $\tilde{A}$ would be 1/3 P. Usually the low energy positrons have only a reduced detection efficiency which results in an $\tilde{A}$ greater that 1/3 P. In practice, however, this effect is counterbalanced by a reduction of the average asymmetry due to the finite detection angle. The resulting effective asymmetry $\tilde{A}$ varies in the different experiments from about 0.22 to 0.3. In Fig. 2, typical shapes of the $(1 + \tilde{A}\cos\theta)$ - law are shown. The average energy of the positrons is about 35 MeV. This

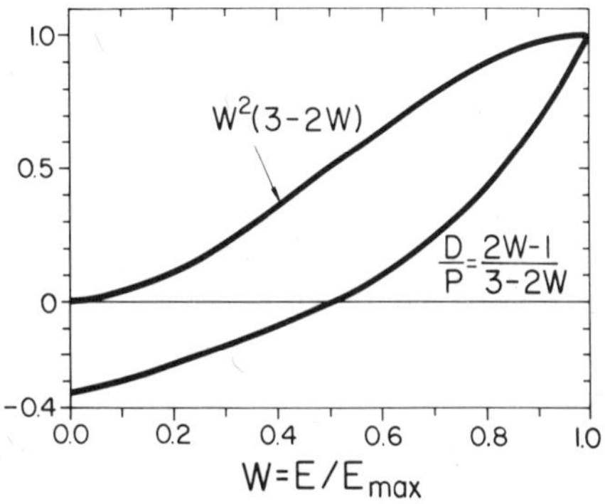

Fig. 1. Energy spectrum of positrons from the muon decay and energy dependence of the asymmetry parameter. The energy is given in units of the maximum possible positron energy E_{max} = 52.8 MeV.

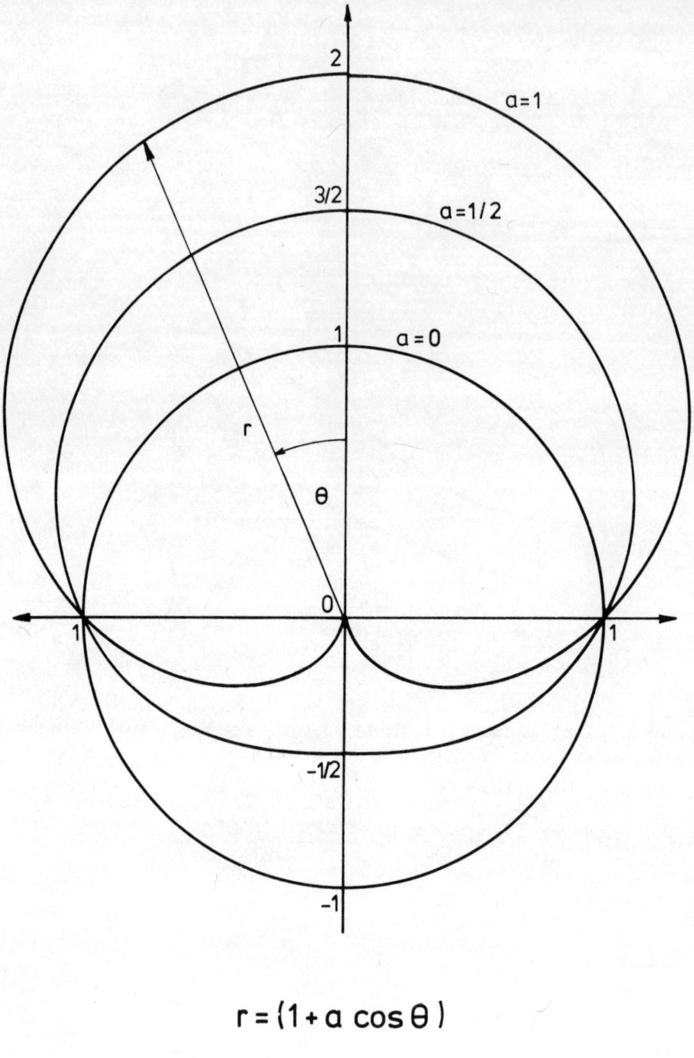

Fig. 2. Plot of the $(1 + a \cos \theta)$ distribution for various a values.

corresponds to some typical radiation length of ∿15 g. It is therefore no problem to observe most positrons, even if the decaying muons were placed deep inside some extended target. The detectors for both muons and positrons are simple plastic scintillators that can be arranged in any convenient geometry.

For future application we write eq. (2.3) in a different way by splitting the effective asymmetry into the polarization, P, of the observed muon ensemble and a remaining effective asymmetry a:

$$dN(\theta) \simeq \frac{1}{4\pi} (1 + P \cdot a \cdot \cos\theta) d\Omega \tag{2.4}$$

We note that the distribution of positrons is thus proportional to the projection of the polarization onto the detection axis

$$dN(\theta) \simeq \frac{1}{4\pi} (1 + a \cdot P_{\perp}) d\Omega \tag{2.5}$$

with

$$P_{\perp} = P \cdot \cos\theta = |\vec{P}| \cos\theta \tag{2.6}$$

c) <u>The μSR-method</u>

The measurement of the positron distribution will determine the direction and, except for the factor a, the value of the polarization of the observed muon ensemble. This is the basic principle of the μSR-method. As we are interested in the magnetic interactions of the muon ensemble with the target material and their time dependent effects on the polarization, the measurement of the positron distribution has to be performed as a function of elapsed muon life time; that is, in a time differential fashion. The time dependent distribution reads:

$$dN(\theta, t) = \frac{1}{4\pi\tau_\mu} e^{-t/\tau_\mu} (1 + a|\vec{P}(t)| \cos\theta) \, d\Omega dt \tag{2.7}$$

where the exponential factor takes the decay into account.

This implies that we have to measure for each implanted muon its individual lifetime and the direction of the emitted positron. In an experiment the instant at which a muon is stopped is recorded as well as the instant when the positron appears. The thus measured time interval determines the individual lifetime. Of course this is done for one muon after the other. We may equally well visualize the muon ensemble thus collected in time as being implanted as a whole at one time.

In principle one would now have to plot the spatial positron distribution as a function of time in order to see the time development of the polarization. We know, however, that at the instant of time zero the polarization of the muon ensemble is given by the polarization of the muon beam, that is, the polarization is either directed parallel or antiparallel to the beam momentum. It then suffices for most purposes to measure only the positron rate in that direction as a function of elapsed muon life time. For more complicated situations it would however be desirable to measure the evolution of the polarization in the three directions of a 3-dimensional orthogonal coordinate system as will be discussed in a moment. Thus far the discussion has been fairly general. We will now discuss three specific experimental arrangements, the first two are the ones that have generally been applied so far.

1. μSR in external zero or longitudinal magnetic fields

By longitudinal we mean that the magnetic field is parallel or antiparallel to the initial polarization, the beam polarization. The experimental arrangement is shown in Fig. 3. Two detectors (M1, M2) in front of the target (T) will track in coincidence the incoming muons. If no signal was obtained from the counters (E1, E2) behind the target, the muons must have been stopped in the target. Likewise a positron is identified by requiring coincident signals from E1 and E2, but no signal from M1 or M2. Positrons emitted in the opposite direction will be identified in a similar way as stopped muons with the additional requirement that no muon was entering the target at the same time (using additional counters further upstream in the beam). With respect to the initial polarization, the angles of positron detection are fixed at $\theta = 0^o$ and $\theta = 180^o$.

The distributions thus obtained are

$$dN(0°,t) \simeq \exp(^{-t/\tau_\mu})(1+\tilde{a}P(t)) \, dt$$

$$dN(180°,t) \simeq \exp(^{-t/\tau_\mu})(1-\tilde{a}'P(t)) \, dt \tag{2.8}$$

where we have integrated over the solid angle covered by the counters. If the forward and backward emitted positrons are collected with the same efficiency ($\tilde{a} = \tilde{a}'$) we can form the ratio

$$\frac{dN(0°,t) - dN(180^o,t)}{dN(0°,t) + dN(180°,t)} = \tilde{a} \, P(t) \tag{2.9}$$

which displays directly the time dependent polarization P (t). P(t) is always understood to be the projection of the polarization vector on the beam axis.

If P(t) decreases irreversibly in time, depolarization of the muon spin ensemble must be taking place. This may for instance come about by spin flip processes that lead to an equal population of spin parallel and spin antiparallel states with respect to the initial direction of polarization. If an external magnetic field were applied, the two spin states differ by their Zeeman energy. Any spin flip event must be therefore accompanied by the absorption or emission of an energy quantum. This energy quantum must be supplied or absorbed by the host lattice in which the muon has been implanted. In this instance one speaks of a spin-lattice relaxation of the muon spins. P(t) may assume an exponential decay function:

$$P(t) = \exp(-t/T_1) \tag{2.10}$$

where T_1 is called the longitudinal or spin lattice relaxation time [6]. In solids T_1 is generally too long to be measurable over the muon life time. However, in paramagnetic aqueous solutions, T_1 may be in the range of a few μsec and should be quite visible, although no such measurements have become known so far.

2. μSR in a transverse magnetic field

By transverse we mean that the magnetic field is oriented perpendicularly to the initial muon polarization. This implies that the stopped muons will start to precess in a plane perpendicular to the direction of the magnetic field. The precession or Larmor frequency is given by

$$\nu = \left| \frac{2\mu_\mu B}{\hbar} \right| = \left| \frac{g_\mu eB}{4\pi m_\mu c} \right| = 13.55 \frac{\text{kHz}}{\text{gauss}} \times B \tag{2.11}$$

and $\omega = 2\pi\nu$

For the positive muon the sense of precession is left-handed when viewing the precession in the direction of the magnetic field. The Larmor frequency is equal to the splitting frequency of the two Zeeman levels in a longitudinal field.

The experimental arrangements for making the precession visible is shown in Fig. 4. It is the same as in Fig. 3, except that now the spin polarization vector and the asymmetric positron

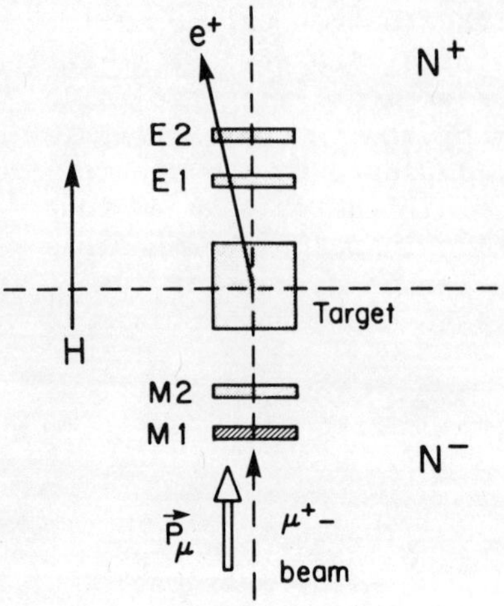

Fig. 3. Schematic experimental arrangement in a longitudinal field M_1, M_2, E_1, E_2 are counters.

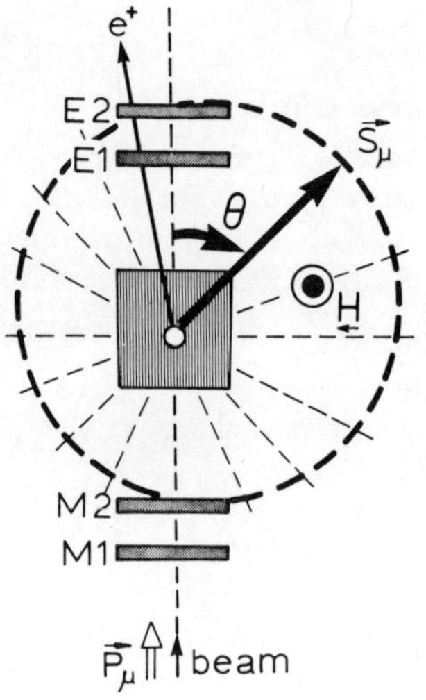

Fig. 4. Schematic experimental arrangement in a transverse field. The asymmetric decay pattern is rotating past the counters.

distribution rotate in the paper plane. When observing the positrons with the detector telescope (E1 · E2) the angle between the spin polarization vector and the positron trajectories will change according to

$$\theta = \omega t \tag{2.12}$$

and the distribution of positrons in time is recorded like

$$dN(t) \cong \frac{1}{\tau_\mu} \exp(-t/\tau_\mu)(1+\tilde{a}\cdot P\cdot \cos(\omega t+\phi))dt$$

or

$$dN(t) \cong \frac{1}{\tau_\mu} \exp(-t/\tau_\mu)(1+\tilde{a}P_x(t))\, dt \tag{2.13}$$

with $P_x(t) = P\cdot\cos(\omega t+\phi)$, the projection of the polarization on the axis of positron observation which is supposed to be the x-direction ($\vec{B}||$Z-axis). ϕ is a phase that indicates the angle between the initial muon polarization and the axis of the positron detector telescope. Eq. (2.13) shows that the precession manifests itself by a cosine modulation of the time dependent positron rate. The time dependence of $P_x(t)$ may not only be introduced by the rotation of the polarization vector, but also by time dependent changes of the value of the polarization vector

$$P_x(t) = P(t)\cos(\omega t+\phi) \tag{2.14}$$

An irreversible decrease of P(t) in time is again a consequence of some spin depolarizing processes. The nature of these may, however, be quite different from those in a longitudinal field. Depolarization in a transverse field is a consequence of the loss of phase coherence between the precessing spins. No energy exchange need be involved. The loss of phase coherence always takes place when the magnetic field is not infinitely sharp but instead shows a certain distribution about some average value. Some muons will thus precess a little bit faster than others, etc., which, after some time, leads to the presence of a more or less random distribution of spin directions in the muon ensemble.

If the probability density distribution of fields or precession frequencies is given by $f(\omega)$, eq. (2.13) should be replaced by

$$dN(t) \cong \frac{1}{\tau_\mu} \exp(-t/\tau_\mu)(1+P\cdot\tilde{a}\int f(\omega)\cos(\omega t+\phi)d\omega) \tag{2.15}$$

Very often the integral can be replaced by

$$\int f(\omega)\cos(\omega t+\phi)d\omega = F(t)\cos(\tilde{\omega}t+\phi) \qquad (2.16)$$

where $\tilde{\omega}$ is the average frequency and F(t) a time-dependent function that describes the depolarization of the muon ensemble. Often F(t) is given by a Gaussian ($F(t) = e^{-\sigma^2 t^2}$) or an exponential ($F(t) = e^{-t/T_2}$) decay law. In this instance T_2 is the so-called transverse or spin-spin relaxation time [6].

We are now in the position to relate the μSR-method to other methods. The way spin precession is measured by a time differential technique is the same as in γγ perturbed angular correlation (PAC) or perturbed angular distribution experiments (PAD) [7]. In the first instance the precession of a nucleus in an intermediate excited state is observed. The intermediate state is prepared in a known spin aligned configuration by measuring the direction of the γ-photon that was emitted when the nucleus decayed from some higher excited level to the level under consideration. Subsequently the direction of the second γ-photon is measured with respect to the first one when the intermediate state decays by γ emission to e.g. the ground state. In the presence of a transverse magnetic field the intermediate spin will precess in just the same way as the muon spin. Because the distribution of the emitted γ's is that of electromagnetic multipole radiation, the recorded angles between the first and the second γ's will show a distribution analogous to that of eq. (2.13) but with a more complex frequency spectrum. Similar techniques apply to a γ-electron correlation experiment.

More closely related to the μSR-technique is the PAD method. Here from the beginning the nuclear spin has a well defined polarization or orientation created, for instance, by the absorption of polarized neutrons or by specific nuclear reactions. It then suffices to record the distribution of the decay photons or electrons with respect to the initial direction of polarization or the initial axis of orientation.

We already mentioned that the Larmor frequency is equal to the Zeeman-splitting frequency in a longitudinal field. If the Zeeman states are populated differently (e.g. by the Boltzmann distribution) an equal population can be achieved by inducing transitions between the Zeeman states with help of a radio frequency field with a frequency equal to the splitting frequency. This is the principle of nuclear magnetic resonance (NMR) experiments [6]. Without going into the details, we merely note that the shape of a NMR signal as a function of the frequency of the rf-field is given by the function f(ω) introduced in eq. (2.16). The precession signal of the spin ensemble is thus the Fourier transform of the corresponding NMR-signal and vice versa.

A Gaussian shaped NMR signal corresponds to a Gaussian damping function of the free precession signal and a Lorentzian shaped NMR signal corresponds to an exponential damping function of the free precession signal. The FWHM of the Lorentzian signal is related to T_2 as follows

$$\frac{1}{T_2} = \pi\Delta\nu \tag{2.17}$$

This relation can also be derived from the energy time uncertainty principle: $\Delta E\Delta t = \hbar$. The observation of the free precession signal (μSR) and the method of NMR are completely complementary and lead to the same amount of information.

In Fig. 5 we show an example of the positron distribution from precessing muons in copper nicely displaying the cosine modulation. The external applied magnetic field had a strength of ~2000 gauss. These data were obtained with the SIN μSR facility.

3. μSR - in an arbitrarily oriented magnetic field

If the initial muon spin polarization is not perpendicular to the magnetic field, the polarization vector will start to precess on the surface of a cone (see Fig. 6). The precession frequency is independent of the angle between the polarization vector and the magnetic field and is still given by eq. (2.11); that is

$$\nu = \left| 2 \frac{\mu_\mu H}{\hbar} \right|$$

The initial polarization defines the z-direction.

The direction of the magnetic field is described by the angle ν between H and the z-axis (this is half the aperture of the cone) and the angle ϕ between the projection of H into the x-y plane and the x-axis.

The x-, y- and z components of the polarization vector exhibit the following time dependences:

$$\begin{aligned} P_x(t) &= P\{\tfrac{1}{2}\sin 2\nu \cos\phi\ (1-\cos\omega t) + \sin\nu \sin\phi \sin\omega t\} \\ P_y(t) &= P\{\tfrac{1}{2}\sin 2\nu \sin\phi\ (1-\cos\omega t) - \sin\nu \cos\phi \sin\omega t\} \\ P_z(t) &= P\{\cos^2\nu + \sin^2\nu \cos\omega t\} \end{aligned} \tag{2.18}$$

By measuring the positron rate in the x-, y- and z-directions, it is possible to determine the direction and the value of the magnetic field unambiguously. This is of interest if an unknown

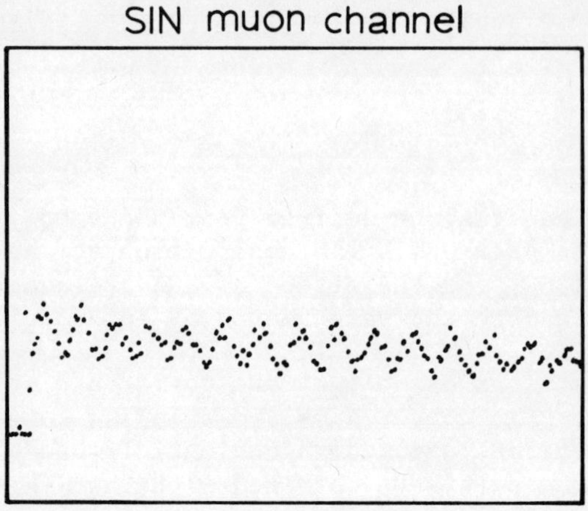

Fig. 5. Precession pattern of positive muons in Cu.

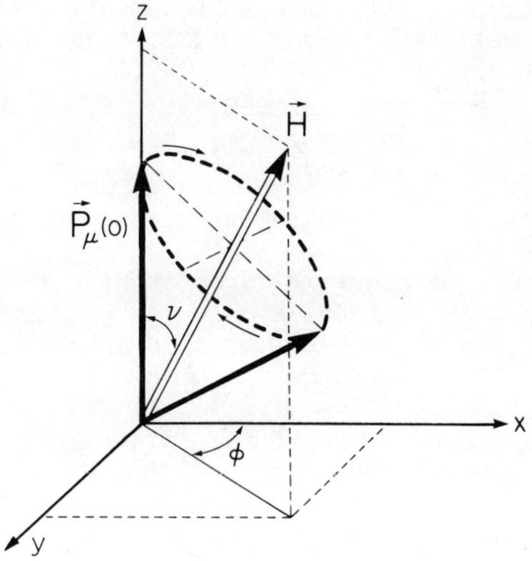

Fig. 6. Precession of the polarization vector for an arbitrary direction of the magnetic field. The polarization vector rotates on the surface of a cone.

magnetic structure is to be investigated by muons. We will demonstrate this on three examples:

1. The magnetic structure is antiferromagnetic ($\phi = 0, 180°$; $\nu = 90°$). The muon will occupy with equal probability places where the internal field is parallel and antiparallel to the z-axis. The total polarization has the components:

$$P_x(t) = 0, \qquad P_y(t) = 0, \qquad P_z(t) = P\cdot\cos\omega t$$

2. The magnetic structure is canted antiferromagnetic ($\phi=0°, 180°$, $\nu < 90°$). Again the muon is assumed to sample the complete structure. The total polarization has the components

$$P_x(t) = P\tfrac{1}{2} \sin 2\nu \; (1-\cos \omega t)$$

$$P_y(t) = 0$$

$$P_z(t) = P\{\cos^2\nu + \sin^2\nu \cos \omega t\}$$

3. The magnetic structure is helical, ($0\leq\phi\leq 2\pi, \nu<90°$). We get

$$P_x(t) = 0, \quad P_y(t) = 0, \quad P_z(t) = \cos^2\nu+\sin^2\nu\cos\omega t$$

More complicated structure may be investigated by adding an external magnetic field and changing the relative orientations of the initial muon polarization, the extended field and the target.

c) Limitations of the Time Differential μSR-method

Several limitations of the time differential μSR-method are apparent.

1. The method requires the observation of individual muon decays. This limits the muon stopping rate to about several 10^4/sec. Otherwise the identification of individual decay events is increasingly impossible. In view of the high stopping density at the meson factories ($10^6\ \mu^+$/sec·g at SIN) this is a very unwanted property of the time differential μSR-method. For the case of muon precession an alternative integral technique is discussed in the next section.

2. The best time resolution that can be achieved at present is of the order of 0.5 nsec. This will limit the highest resolvable muon Larmor frequency to about 500 MHz.

3. Relaxation times shorter than several 10^{-8} sec will render the observation of the time dependent polarization impossible.

This is primarily due to dead time effects connected with anticoincidence requirements. Compared with NMR, however, this is an improvement of about two orders of magnitude.

On the other hand, the lifetime of the muon of about 2.2 μsec will prevent the observation of relaxation times much greater than 200 μsec.

d) The Stroboscopic Method

This method can be applied if the intensity of the muon beam can be modulated periodically. The principle of the method consists in letting the muons precess with a frequency close to an integral multiple of the frequency with which the intensity of the muon beam is modulated. Muons entering the target thus have almost the same spin phase as the ones that have already spent some time in the target. In effect, most muons, independent of their arrival time, will precess more or less coherently. The coherence is strongest at

$$\omega_\mu = n\ \omega_{beam} \qquad (n = 1,2,...)$$

and vanishes for

$$|\omega_\mu - n\ \omega_{beam}| > \frac{1}{\tau_\mu}$$

The degree of coherence is detected by measuring the positron rate in a time window of suitable length and suitable position with respect to the phase of the beam intensity modulation.

Such a method has been first applied by Christiansen et al. in γγPAD experiment. [8].

I would like to illustrate this method by referring to the special condition of the muon beam at SIN (see Fig. 7). The muon beam at SIN still reflects the microscopic time structure of the primary proton beam which consists of small bursts with a repetition rate of 50 MHz. The 50 MHz-structure is related to the accelerator rf, which is derived from a very stable quartz oscillator. This signal is also available to the experimenters. The time structure of the muon beam can be determined with respect to this reference signal. The bursts in the muon beam have a typical width of about 5 nsec. The average burst shape is expressed as $F(t) = F(t+nT)$, $n = 1,2...$

$$\int_0^T F(t) = 1 \qquad ; \qquad T = 20\ \text{nsec}$$

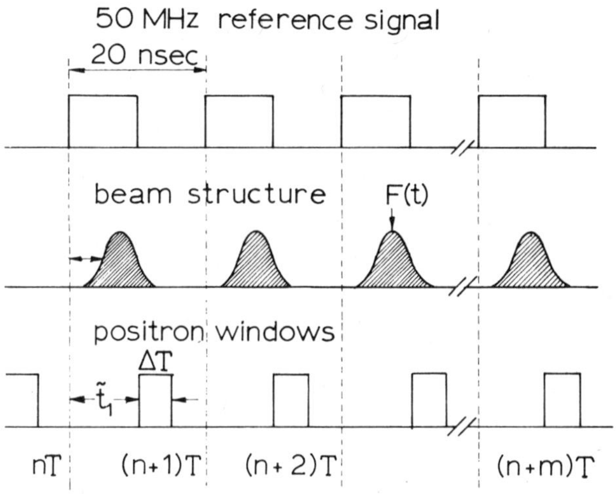

Fig. 7. Schematic arrangement of the reference signal the muon beam bursts and the time windows of observation for the stroboscopic method.

Let us assume that one has begun to stop muons in the target at time $t = 0$. What is the positron rate at some time $t_o \gg \tau_\mu$ in a positron telescope whose axis has an angle ϕ to the initial muon spin polarization?

The positrons that are counted at time t_o in an interval $d\tilde{t}_o$ originate from all the muons that have been stopped in the interval $t_o - 0 = t_o$. This leads to

$$\frac{dN(t_o)}{d\tilde{t}_o} \sim \frac{1}{\tau_\mu} \int_0^{t_o = n_o T + \tilde{t}_o} \exp\left(-(t_o - t)/\tau_\mu\right) F(t)\left(1 + a \cdot P \cdot \cos(\omega_\mu (t_o - t) + \phi)\right) dt \tag{2.19}$$

Taking the periodicity of F(t) into account, we arrive after various calculations at

$$\frac{dN(\tilde{t}_o)}{d\tilde{t}_o} \sim \frac{1}{T}\left\{ C + \frac{a \cdot P}{1 + (\Delta\omega\tau_\mu)^2}\left[(A(\omega_\mu) + B(\omega_\mu)\Delta\omega\tau_\mu) \cos(\omega_\mu \tilde{t}_o + \phi + \Delta\omega T) + (B(\omega_\mu) - A(\omega_\mu)\Delta\omega\tau_\mu)\sin \omega_\mu \tilde{t}_o + \phi + \Delta\omega T)\right]\right\} \tag{2.20}$$

with

$$A(\omega_\mu) = \int_0^T e^{t'/\tau_\mu} F(t') \cos \omega t' dt'$$

$$B(\omega_\mu) = \int_0^T e^{t'/\tau_\mu} F(t') \sin \omega t' \; t'$$

$$C = \int_0^T e^{t'/\tau_\mu} F(t') dt' \qquad \Delta\omega = \omega_\mu - n\omega_{beam} \tag{2.21}$$

The time is measured in units of the period T = 20 nsec. Only the fraction $\tilde{t}_o$ is still present in eq. (2.20).

To obtain the actual counting rate, eq. (2.20) has to be integrated over the time window, in which the positrons are counted. The time window(s) or data gate(s) can be derived synchronously from the 50 MHz reference signal. The position of the data gate with respect to the reference signal and its width are indicated by times $\tilde{t}_1$ and ΔT.

$$\Delta N \simeq \int_{\tilde{t}_1}^{\tilde{t}_1+\Delta T} \frac{dN(\tilde{t}_o)}{d\tilde{t}_o}\, d\tilde{t}_o = 1 + \frac{A^*\left(\dfrac{\sin \frac{\Delta T}{2}\omega_\mu}{\frac{\Delta T}{2}\omega_\mu}\right)}{1+(\Delta\omega\tau_\mu)^2}$$

$$[\cos\,(\Delta\omega\tau^*+\phi^*) + \tau_\mu\Delta\omega\,\sin\,(\Delta\omega\tau^*+\phi^*)] \qquad (2.22)$$

with the effective asymmetry

$A^* = a\cdot P\cdot D \; (D \lesssim 1)$

The parameters τ^*, ϕ^*, D contain all the information on the beam structure, the relative position of the beam bursts with respect to the reference signal and the relative position and width of the data gate. This information is actually not needed, rather the parameters can be used as fitting parameters. Eq. (2.22) consists of a superposition of a Lorentzian and dispersive parts. The width of the signal is solely determined by the muon lifetime τ_μ and not by the beam structure or the data gate arrangement. The relative fraction of the Lorentzian and dispersive parts, however, depends on the experimental details. This is shown in Fig. 8, which displays simulated data for different gate positions.

The method "simply" consists of recording the positron rate in some counter telescope in a chosen time window of suitable position and length. No limitations concerning an upper value for the muon stopping rate are present.

The muon precession frequency, or the corresponding magnetic field can be measured with a high statistical power and largely free of systematic errors that are connected with time measurements.

This method will be applied at SIN for a new high precession determination of the magnetic moment of the positive muon [9]*. Later it is planned to apply this method also to solid state experiments where a high precision in the frequency determination is required (as e.g. in Knight shift measurements). The disadvantage of this method for general application in μSR is related to the fact that only a particular magnetic field can be applied (e.g. one that leads to a precession frequency of 50 MHz).

Finally, it should also be mentioned that transverse relaxation times can be measured with the stroboscopic method. As in NMR,

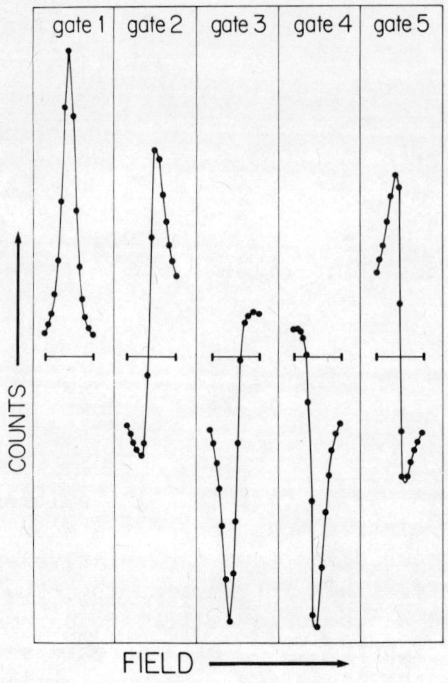

Fig. 8. Simulated data showing different superpositions of absorptive and dispersive curves. The 20 nsec interval is covered by five time windows. Absorptive stroboscopic signal with SIN muon beam (p_μ = 115 MeV/c, ω_μ = 50 MHz).

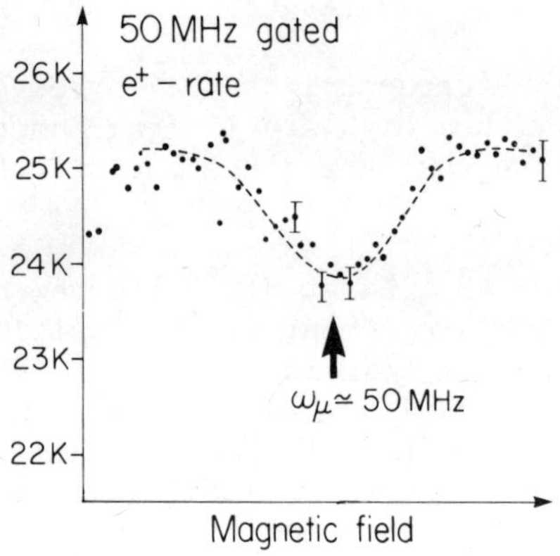

Fig. 8b. Experimental straboscopic signal obtained on the SIN muon beam (p_μ = 115 MeV/c, ω_μ = 50 MHz). Time window chosen so as to produce a pure absorption signal.

relaxation will lead to a larger line width. The line width (FWHM) in the absence of relaxation is given by eq. (2.22)

$$\delta\omega = \frac{2}{\tau_\mu}$$

and in the presence of relaxation (T_2)

$$\delta\omega = 2\left(\frac{1}{\tau_\mu} + \frac{1}{T_2}\right)$$

*Addendum

Since the end of the summer school, at SIN we have been able to obtain for the first time the stroboscopic signal with a high energy meson beam, i.e. the SIN muon beam. (p_μ = 115 MeV/c, ω_μ = 50 MHz). The time window was chosen such as to produce an almost pure absorption signal. The signal is shown in Fig. 8 b.

The width of the signal is larger than the natural width of $\Delta\omega = \frac{2}{\tau_\mu}$ and is compatible with the transverse relaxation time observed directly via the damping of the precession signal.

III. MAGNETIC HYPERFINE AND DIPOLE-DIPOLE INTERACTIONS AND THE TIME EVOLUTION OF THE MUONS' POLARIZATION.

We will now consider in somewhat more detail the various types of magnetic interactions that the muon will be subject to when implanted in a solid. Sources of internal fields are electrons and nuclei. In addition an externally applied magnetic field may be present. The general form of the interaction Hamiltonian will consist of the following contributions:

$$H_\mu = H_{\mu e} + H_{\mu N} + Z_\mu + Z_e + Z_N \tag{3.1}$$

with $H_{\mu e}$, $H_{\mu N}$ describing the magnetic interactions with electrons and nuclei, and Z_μ, Z_e and Z_N are the Zeeman terms of muons, electrons and nuclei. The Zeeman terms are of the form

$$Z = g\,\mu_B \cdot \vec{S} \cdot \vec{H}_{ext} \tag{3.2}$$

and need no further explanation. The general form of $H_{\mu e}$ can be obtained as follows:

The muon's magnetic dipole $\vec{\mu}_\mu$ will create a magnetic field $\vec{H}(\vec{r})$ at an electron situated in a distance $\vec{r}$ from the muon. The magnetic field is related to the vector potential

$$\vec{A} = \frac{\vec{\mu} \times \vec{r}}{r^3} = \vec{\nabla} \times \frac{\vec{\mu}}{r} = \mathrm{curl}\left(\frac{\vec{\mu}}{r}\right) \tag{3.3}$$

with $\vec{\nabla} \cdot \vec{A} = \text{div}\,\vec{A} = 0$

and $\vec{H}(\vec{r}) = \vec{\nabla} \times \vec{A} = \text{curl}\,\vec{A}$ (3.4)

For $r \to 0$, $\vec{A}$ becomes singular. This point needs special attention. Using the nonrelativistic Pauli approximation for electrons we can write down the Hamilton operator for the electron:

$$H = \frac{1}{2m}(\vec{p} + \frac{e}{c}\vec{A})^2 + 2\,\mu_B\,\vec{S}_e \cdot \text{curl}\,\vec{A}$$

$$= \frac{p^2}{2m} + H_1 + \frac{e^2}{2mc^2}\vec{A}^2 \tag{3.5}$$

with

$$H_1 = H_{\mu e} = \frac{e}{2mc}(\vec{p}\cdot\vec{A} + \vec{A}\cdot\vec{p}) + 2\mu_B\,\vec{S}_e \cdot \text{curl}\,\vec{A} \tag{3.6}$$

The term quadratic in $\vec{A}$ is small and can be neglected. The operator H_1 describes the mutual magnetic interaction between the muon and electron. The first term of $H_{\mu e}$ describes the interaction originating from an orbital movement of the electron. This term will not be considered further as it has not been seen to play a role in μSR. The spin dependent term, $2\mu_B\vec{S}_e \cdot \text{curl}\,\vec{A}$, can be rewritten as follows:

$$H^s_{\mu e} = 2\mu_B\vec{S}_e \cdot [\vec{\nabla} \times (\vec{\nabla} \times \frac{\vec{\mu}}{r})] = 2\,\mu_B\,[(\vec{S}_e \cdot \vec{\nabla})(\vec{\mu} \cdot \vec{\nabla}) - (\vec{S}_e \cdot \vec{\mu})\nabla^2]\frac{1}{r}$$

$$= 2\,\mu_B[(\vec{S}_e \cdot \vec{\nabla})(\vec{\mu} \cdot \vec{\nabla}) - \frac{1}{3}(\vec{S}_e \cdot \vec{\mu})\nabla^2]\frac{1}{r} - \frac{4}{3}\mu_B(\vec{S}_e \cdot \vec{\mu})\nabla^2\frac{1}{r} \tag{3.7}$$

We note, that $\vec{\mu}\,\vec{\nabla}\,\frac{1}{r} = -\frac{\vec{\mu}\cdot\vec{r}}{r^3}$

and $(\vec{S}_e \cdot \vec{\nabla})(\vec{\mu}\cdot\vec{\nabla})\,\frac{1}{r} = -\frac{\vec{S}_e \cdot \vec{\mu}}{r^3} + 3\,\frac{(\vec{\mu}\cdot\vec{r})(\vec{S}_e \cdot \vec{r})}{r^5}$

and $\nabla^2\,\frac{1}{r} = -\,4\pi\,\delta(r)$

For $r \neq 0$ we get $(\vec{\mu} = \hbar\gamma_\mu\,\vec{S}_\mu)$

$$H_D = 2\,\mu_B\,\hbar\,\gamma_\mu\,(-\frac{\vec{S}_\mu \cdot \vec{S}_e}{r^3} + 3\,\frac{(\vec{S}_\mu \cdot \vec{r})(\vec{S}_e \cdot \vec{r})}{r^5}) \tag{3.8}$$

which corresponds to the classical dipole-dipole interaction.

$H^{S}_{\mu e}$ becomes singular for r = 0. Since we have excluded any orbital motion of the electron from our considerations, the only electron state which may involve r = 0 is an s-state. Exploiting the rotational symmetry properties of the Hamiltonian eq. (3.7) one can show [6] that in this case only the second term of eq. (3.7) (last line) will contribute to the magnetic interaction energy $<\psi_s|H^S_{\mu e}|\psi_s> = <\psi_s|H_c|\psi_s>$ with

$$H_c = \frac{16\pi}{3} \mu_B \gamma_\mu \hbar \vec{S}_\mu \cdot \vec{S}_e \, \delta(r) \qquad (3.9)$$

This is the famous Fermi Contact interaction Hamiltonian, which has no classical counterpart. H_c can also be expressed as

$$H_c = \frac{8\pi}{3} g_\mu \mu^\mu_B g_e \mu^e_B \, |\psi(0)|^2 \vec{S}_\mu \cdot \vec{S}_e \qquad (3.10)$$

$|\psi(0)|^2$ is the electron density at the muon site.

The contact interaction is responsible for the hyperfine structure splitting of the 1s-ground state of muonium. It is also present in other instances, in particular in metals, where it is responsible for the magnetic interaction of positive muons with the conduction electrons.

The magnetic interaction of muons with nuclei is a dipole-dipole interaction. The Hamiltonian is given as in eq. 3.8 and reads

$$H_{\mu N} = \frac{\hbar^2 \gamma_\mu \gamma_N}{r^3} \left(\vec{S}_\mu \cdot \vec{I} - \frac{3(\vec{S}_\mu \cdot \vec{r})(\vec{I} \cdot \vec{r})}{r^2}\right) \qquad (3.11)$$

where r is the distance between the two dipoles.

We will now discuss the evolution of the muon polarization under the action of a contact interaction in the muonium 1s ground state and subsequently the effects of a dipole-dipole interaction.

1) Evolution of the Muon Polarization in the Muonium ($\mu^+ e^-$)-state

In the presence of an external field $\vec{H}$ the total spin Hamiltonian will read:

$$H^S = A \vec{S}_\mu \cdot \vec{S}_e + g_\mu \mu^\mu_B \vec{S}_\mu \cdot \vec{H} + g_e \mu^e_B \vec{S}_e \cdot \vec{H} \qquad (3.12)$$

$$\text{with} \quad A = \frac{8\pi}{3} g_\mu \mu^\mu_B g_e \mu^e_B |\psi(0)|^2 \qquad (3.13)$$

The electron is in the 1s-state:

$$\psi_{1s}(r) = \frac{1}{\sqrt{\pi}} \frac{1}{a_o^{3/2}} e^{-r/a_o}, \quad a_o = \frac{\hbar^2}{e^2}\left(\frac{1}{m_e} + \frac{1}{m_\mu}\right) \tag{3.14}$$

we then have

$$A = \frac{8}{3} g_\mu \mu_B^\mu g_e \mu_B^e \frac{1}{a_o^3} = \hbar \omega_o \tag{3.15}$$

with $\omega_o = 2\pi \cdot 4.46 \cdot 10^9$ rad/sec.

The hyperfine coupling will lead to a splitting of the 1s ground state. We have to solve the equation

$$H^s \psi = E \psi \tag{3.16}$$

It is easily seen that the manifold of solutions is fourfold (four combinations of muon and electron spin).

The eigenvalues are given by the well known Breit-Rabi-formula

$$\begin{aligned} E_1 &= \frac{A}{4}(1 + 2\,d\,x) \\ E_2 &= \frac{A}{4}(-1 + 2\sqrt{1+x^2}) \\ E_3 &= \frac{A}{4}(1 - 2\,d\,x) \\ E_4 &= \frac{A}{4}(-1 - 2\sqrt{1+x^2} \end{aligned} \tag{3.17}$$

with

$$d = \frac{1 + g_\mu \mu_B^\mu / g_e \mu_B^e}{1 - g_\mu \mu_B^\mu / g_e \mu_B^e} \cong 0.99$$

$$x = \frac{\mu_B^e g_e - \mu_B^\mu g_\mu}{A} \cdot H = \frac{H}{H_0} \tag{3.18}$$

$$H_o \cong 1585 \text{ Gauss}$$

H is measured in units of the magnetic field that the muon produces at the electron site. The eigenvalues are plotted in Fig. 9 as a function of magnetic field (Breit-Rabi-diagram).

The four eigenfunctions are given by:

$$
\begin{aligned}
& & & x \rightarrow 0 \\
\psi_1 &= |+\,+> & & \rightarrow |\, F = 1, m_F = 1 > \\
\psi_2 &= \sin\rho\, |+\,-> + \cos\rho\, |-\,+> & & \rightarrow |\, F = 1, m_F = 0 > \\
\psi_3 &= |-\,-> & & \rightarrow |\, F = 1, m_F = -1 > \\
\psi_4 &= \cos\rho\, |+\,-> - \sin\rho\, |-\,+> & & \rightarrow |\, F = 0, m_F = 0 >
\end{aligned}
\qquad (3.19)
$$

$$
\begin{aligned}
\text{with } \cos\rho &= \frac{1}{\sqrt{2}} \sqrt{1 + \frac{x}{\sqrt{1+x^2}}} \\
\sin\rho &= \frac{1}{\sqrt{2}} \sqrt{1 - \frac{x}{\sqrt{1+x^2}}}
\end{aligned}
\qquad (3.20)
$$

For low magnetic field the eigenfunctions can also be labelled in terms of the eigenvalues of the total angular momentum.

We introduce the following notation

$$\chi_1 = |+\,+> , \quad \chi_2 = |+\,-> , \quad \chi_3 = |-\,->$$

$$\chi_4 = |-\,+>$$

Using these spin functions as basis vectors, the density operator expressing the initial spin ensemble is given by

$$\rho = \sum_{i=1}^{4} p_i |\chi_i >< \chi_i| \qquad (3.21)$$

Any other suitable basis may be used instead. p_i is the probability of finding the system in the state $|\chi_i>$ and $\sum_i p_i = 1$. In the Heisenberg picture the time dependence of the spin operator $\vec{S}_\mu = \frac{1}{2}\vec{\sigma}_\mu$ can be expressed in the usual way.

$$\vec{\sigma}_\mu(t) = e^{iH^S t/\hbar} \vec{\sigma}_\mu \, e^{-iH^S t/\hbar} \qquad (3.22)$$

The expectation value is then given by

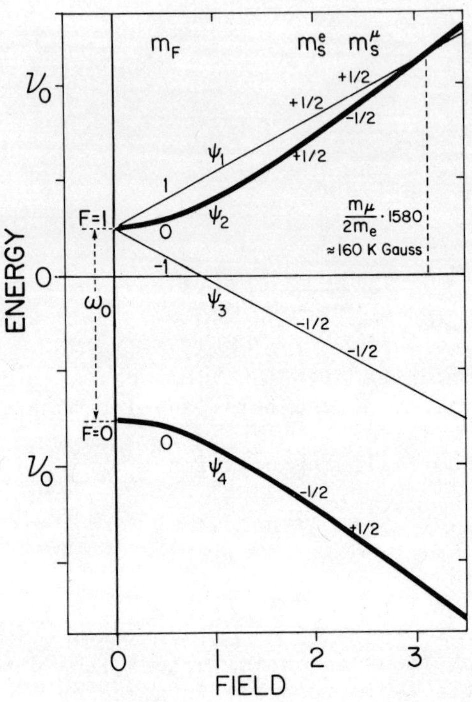

Fig. 9. Schematic Breit-Rabi diagram for muonium

$$< \vec{\sigma}_\mu (t) > = \mathrm{Tr}\, \rho\, \vec{\sigma}_\mu(t) = \mathrm{Tr} \left(\sum_{i=1}^{4} p_i |\chi_i\rangle\langle\chi_i| \vec{\sigma}_\mu(t) \right)$$

$$= \sum_{i=1}^{4} p_i \langle\chi_i|\vec{\sigma}_\mu(t)|\chi_i\rangle \tag{3.23}$$

$$= \sum_{i=1}^{4} p_i \langle\chi_i| e^{iH^s t/\hbar}\, \vec{\sigma}_\mu\, e^{-iH^s t/h} |\chi_i\rangle$$

It is now convenient to go by a unitary transformation from the basis χ_i to the basis ϕ_i where the ϕ_i are eigenvectors of H^s:

$$|\chi_i\rangle = U_{ij}\, |\phi_j\rangle, \quad \langle\chi_k| = \langle\phi_l| U^{-1}_{lk} \tag{3.24}$$

and

$$\langle\vec{\sigma}_\mu(t)\rangle = \sum_{ijk} p_i\, U^{-1}_{ki}\, U_{ij} < \phi_k| e^{\frac{i}{\hbar}H^s t}\, \vec{\sigma}_\mu\, e^{-\frac{i}{\hbar}H^s t} |\phi_j\rangle$$

$$= \sum_{ijk} p_i\, U_{ik}\, U_{ij}\, e^{\frac{i}{\hbar}(E_k - E_j)t} \langle\phi_k|\vec{\sigma}_\mu|\phi_j\rangle \tag{3.25}$$

The calculation of the matrix elements is most simply done in the basis χ_i.

$$|\phi_i\rangle = U^{-1}_{jl}|\chi_l\rangle, \quad \langle\phi_k| = \langle\chi_m|U_{mk} \tag{3.26}$$

Hence

$$\langle\vec{\sigma}_\mu(t)\rangle = \sum_{ijklm} p_i\, U_{ik}\, U_{ij}\, U_{mk}\, U_{lj}\, e^{\frac{i}{\hbar}(E_k - E_j)t} \langle\chi_m|\vec{\sigma}_\mu|\chi_l\rangle \tag{3.27}$$

The transformation matrix U follows directly from eqs. 3.19.

$$U = \begin{pmatrix} 1 & 0 & 0 & 0 \\ 0 & \sin\rho & 0 & \cos\rho \\ 0 & 0 & 1 & 0 \\ 0 & \cos\rho & 0 & -\sin\rho \end{pmatrix} = U_{!!} \tag{3.28}$$

Eq. (3.27) will now be evaluated for two cases: a) with the initial muon polarization parallel to the applied field b) with the initial muon polarization transverse to the applied field.

a) polarization vector and external field are in z-direction. Two spin configurations are possible at the instant of muonium formation:

$$\underset{\mu}{\uparrow}\,\underset{e}{\uparrow} \equiv \chi_1 \qquad \underset{\mu}{\uparrow}\,\underset{e}{\downarrow} \equiv \chi_2$$

We therefore have $p_1 = \frac{1}{2} \quad p_2 = \frac{1}{2} \quad p_3 = p_4 = 0$.

We wish to know $\langle\sigma_\mu^z(t)\rangle = P_\mu^z(t)$.

After a straightforward calculation eq. (3.27) reduces to

$$\langle\sigma_\mu^z(t)\rangle = 1 - \frac{1}{2}\frac{1}{1+x^2}(1 - \cos\omega_o\sqrt{1+x^2}\,t) \tag{3.29}$$

The evolution of the muon polarization in the z-direction is thus characterized by a time independent but field dependent term (called the residual polarization) and a time and field dependent term with periodicity $\frac{\omega_o}{2\pi}\sqrt{1+x^2} \gtrsim 4.46$ GHz.

This is too fast to be resolved experimentally and the term will be averaged to zero in an experiment. This effect is often referred to as depolarization. The "residual polarization" is given by

$$P_\mu^z = \frac{1+2x^2}{2+2x^2} \tag{3.30}$$

For zero field P_μ^z is 50% and it approaches 100% when the magnetic field is increased to large values (Fig. 10). This is a consequence of the decoupling of electron and muon spin in strong magnetic fields (Paschen-Back effect) and is usually referred to as "quenching of depolarization". As an example Fig. 11 shows a quenching curve obtained in quartz [10]. The data are fitted nicely by eq. (3.30) assuming that a small fraction of muons are not thermalized in the muonium state.

b) The external field is in the z-direction and the initial polarization is assumed to be along the x-direction. The initial muon spin state is given by

$$\psi_\mu^o = \frac{1}{\sqrt{2}}(|+\rangle_\mu + |-\rangle_\mu) \tag{3.31}$$

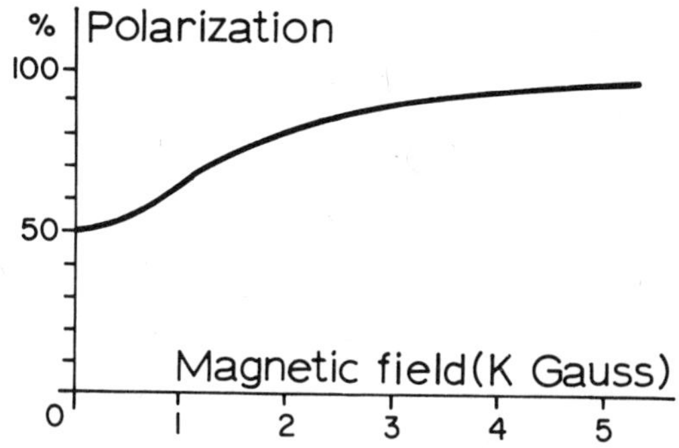

Fig. 10. Field dependence of residual polarization (quenching curve) in a longitudinal field.

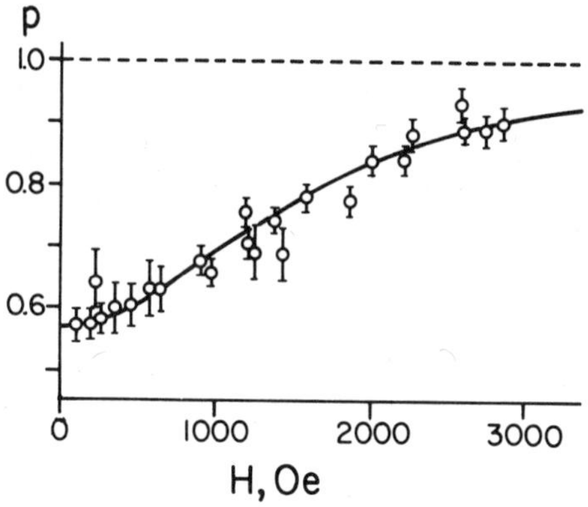

Fig. 11. Residual polarization of μ^+ in quartz [10] as a function of longitudinal field strength.

and $\langle\psi_\mu^o|\sigma_\mu^x|\psi_\mu^o\rangle = 1$

The initial muonium spin states are

$$\psi_1 = \psi_\mu^o|+\rangle_e = \frac{1}{\sqrt{2}}(\chi_1 + \chi_4) \qquad P_1 = 1/2$$

$$\psi_2 = \psi_\mu^o|-\rangle_e = \frac{1}{\sqrt{2}}(\chi_2 + \chi_3) \qquad P_2 = 1/2$$

with 50% population each.

The two other possible states $\psi_3 = \frac{1}{\sqrt{2}}(\chi_1 - \chi_4)$ and $\psi_4 = \frac{1}{\sqrt{2}}(\chi_2 - \chi_3)$ have weights of $P_3 = 0$ and $P_4 = 0$. In a condensed form we can write $\psi = U^1\chi$.

We are now interested e.g. in the evolution of the x-component of the polarization. ψ_i, i = 1,2,3,4 are used as basis vectors in this case and replace the corresponding χ_i in eq. (3.27). To evaluate eq. (3.27) we have to find the transformation matrix that relates ψ_i to the eigenvectors of H^s. This matrix is obtained from eq. (3.24) as follows:

Using
$$\psi_i = U^1_{ij}\chi_j = U^1_{ij}U^{\parallel}_{jk}\phi_k = U^{\perp}_{ik}\phi_k$$

From which follows:

$$U^{\perp} = \begin{pmatrix} \frac{1}{\sqrt{2}} & \frac{1}{\sqrt{2}}\cos\rho & 0 & -\frac{1}{\sqrt{2}}\sin\rho \\ 0 & \frac{1}{\sqrt{2}}\sin\rho & \frac{1}{\sqrt{2}} & \frac{1}{\sqrt{2}}\cos\rho \\ \frac{1}{\sqrt{2}} & -\frac{1}{\sqrt{2}}\cos\rho & 0 & \frac{1}{\sqrt{2}}\sin\rho \\ 0 & \frac{1}{\sqrt{2}}\sin\rho & -\frac{1}{\sqrt{2}} & \frac{1}{\sqrt{2}}\cos\rho \end{pmatrix} \qquad (3.32)$$

A lengthy but straightforward calculation leads to the result:

$$\langle\sigma_\mu^x(t)\rangle = P_x(t)$$

$$= \frac{1}{4}\{(1+\delta)\cos\omega_{12}t + (1-\delta)\cos\omega_{14}t + (1+\delta)\cos\omega_{34}t + (1-\delta)\cos\omega_{23}t\} \quad (3.33)$$

with $\delta = \frac{x}{\sqrt{1+x^2}}$

Likewise we can calculate

$$P_y(t) = \sigma_\mu^y(t)$$

$$= \frac{1}{4}\{-(1+\delta)\sin\omega_{12}t + (\delta-1)\sin\omega_{14}t + (\delta-1)\sin\omega_{23}t + (1+\delta)\sin\omega_{34}t\} \quad (3.34)$$

The frequencies ω_{14} and ω_{34} are larger than $\omega_o = 2\pi \cdot 4.46 \cdot 10^9$ rad/sec and the corresponding terms will again be averaged to zero in the experiment. For very low applied fields ($\delta \simeq 0$) ω_{12} and ω_{23} are equal (linear Zeeman region) and the observable evolution of the polarization in e.g. the x-direction is simply

$$P^x_{obs}(t) = \frac{1}{2}\cos\omega_{12}t \quad (3.35)$$

This muon spin rotation is due to the precession of the triplet F = 1 state in the external field. The precession frequency is half the precession frequency of a free electron.

At somewhat larger fields (100 gauss) the deviation from the linear Zeeman effect leads to $\omega_{12} \neq \omega_{23}$. The observable part of the polarization in x-direction is then

$$P^x_{obs}(t) = \frac{1}{2}\left[\cos\Omega t\cos\bar{\omega}t + \frac{x}{\sqrt{1+x^2}}\sin\Omega t\sin\bar{\omega}t\right] \quad (3.36)$$

with

$$\bar{\omega} = \frac{1}{2}(\omega_{12}+\omega_{23}), \quad \Omega = \frac{1}{2}(\omega_{12}-\omega_{23}) = \frac{\omega_o}{2}(\sqrt{1+x^2}-1) = \frac{(\bar{\omega})^2}{\omega_o} \quad (3.37)$$

The muon precession is again determined by the Larmor frequency of the electron in the $F = 1$ state, however this time it is modulated with a second smaller frequency Ω which is related to the hyperfine frequency ω_o. This beating behavior in the muonium precession was first predicted and experimentally verified by Gurevich et al [11]. Fig. 12 shows the data that were obtained using a quartz target. It is now generally referred to as the two frequency precession of muonium.

At higher fields ($\delta \to 1$) only the terms with ω_{12} and ω_{34} survive. The term with ω_{12} has a curious behavior as the frequency $\omega_{12} = \frac{1}{\hbar}(E_1 - E_2)$ drops to zero at an external field of ca. 160 kGauss. This field corresponds to the crossing of the two upper energy levels which occurs when the external field is equal to the field generated by the electron at the muon position.

The density matrix formalism is especially suited when more complicated effects have to be included, like relaxation of the muonium electron itself.

The density matrix of muonium is, as we have already noted, a 4 x 4 matrix and $\mathrm{Tr}\rho = 1$. The four linear independent spinors $\chi_i (i = 1,4)$ may serve as the basis vectors of the space in which the density operator is represented by the 4 x 4 matrix.

It is well known that a 4 x 4 matrix can most generally be expressed in terms of 16 linear independent matrices formed by a tensor product of Pauli spin matrices and the unit matrix: In the case of muonium we can specify the Pauli matrices to be those of the muon and the electron. The 16 matrices are

$$\left.\begin{array}{llll} \sigma_\mu^i \otimes \sigma_e^j & i,j = 1,2,3 & (9) \\ \sigma_\mu^i \otimes \mathbb{1} & i = 1,2,3 & (3) \\ \mathbb{1} \otimes \sigma_e^j & j = 1,2,3 & (3) \end{array}\right\} \mathrm{Tr} = 0$$

$$\begin{array}{lll} \mathbb{1} \otimes \mathbb{1} & & \underline{(1)} \\ & & (16) \end{array}$$

It is convenient to also express ρ in terms of these matrices, since we know the commutation relations of the Pauli matrices and since we can easily calculate all traces.

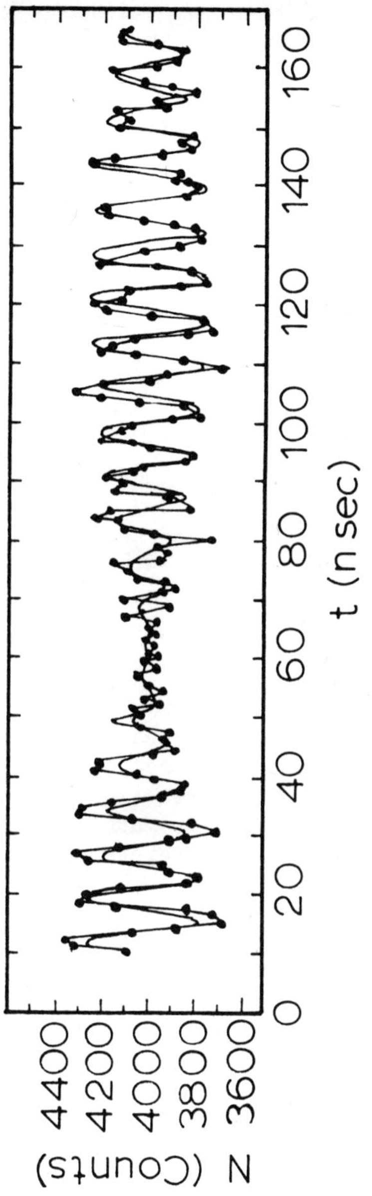

Fig. 12. Two frequency precession in quartz [11].

The density matrix then reads

$$\rho = \frac{1}{4}(\mathbb{1} \otimes \mathbb{1} + \vec{\pi}_\mu \vec{\sigma}_\mu \otimes \mathbb{1} + \vec{\pi}_e \mathbb{1} \otimes \vec{\sigma}_e + \sum_{i,j=1}^{3} \pi_{ij} \sigma_\mu^{\ i} \otimes \sigma_e^{\ j}) \tag{3.38}$$

with

$$\vec{\pi}_\mu = \langle\vec{\sigma}_\mu\rangle = \mathrm{Tr}\,\rho\,\vec{\sigma}_\mu = \text{polarization of muon}$$

$$\vec{\pi}_e = \langle\vec{\sigma}_e\rangle = \mathrm{Tr}\,\rho\,\vec{\sigma}_e = \text{polarization of electron}$$

$$\pi_{ij} = \langle\sigma_\mu^{\ i}\,\sigma_e^{\ j}\rangle = \mathrm{Tr}\,\rho\,\sigma_\mu^{\ i}\,\sigma_e^{\ j}$$

The time evolution of the density matrix obeys the differential equation:

$$i\hbar \frac{d}{dt}\rho(t) = [H, \rho(t)] \tag{3.39}$$

which has to be solved in order to find the evolution of the muon polarization.

$$\vec{P}_\mu = \mathrm{Tr}(\vec{\sigma}_\mu\,\rho(t)) \tag{3.40}$$

The Hamiltonian, H, contains the muonium Hamiltonian and any other Hamiltonian of relevance. Eq. (3.39) lead to the Ivanter and Smilga formalism and the phenomenelogical theory of muon depolarization in the presence of chemical reactions of muonium [11]. A detailed account of this theory and its applications and extensions, as well as a complete list of references, can be found in [4].

2) Effects of the Dipole-Dipole Interaction

In treating these effects we will mainly have the interaction of the muon with nuclei in mind. The results can, however, be applied to any other case where dipole-dipole interactions are present, as e.g. in ferromagnetic substances with localized electronic magnetic moments.

The Hamiltonian reads (eq. 3.11)

$$H_{\mu N} = H_D = \frac{\hbar^2\,\gamma_\mu\,\gamma_N}{r^3}(\vec{S}_\mu \cdot \vec{I} - \frac{3(\vec{S}_\mu \cdot \vec{r})(\vec{I} \cdot \vec{r})}{r^2}) \tag{3.41}$$

By introducing polar coordinates H_D can be rewritten as follows: [6]

$$H_D = \frac{\hbar^2 \gamma_\mu \gamma_N}{r^3} (A + B + C + D + E + F) \tag{3.42}$$

with

$$A = S^z I^z (1 - 3 \cos^2 \theta)$$

$$B = -\frac{1}{4} (S^+ I^- + S^- I^+) (1 - 3 \cos^2 \theta)$$

$$C = -\frac{3}{2} (S^+ I^z + S^z I^+) \sin\theta \cos\theta \, e^{-i\phi}$$

$$E = -\frac{3}{4} S^+ I^+ \sin\theta \, e^{-2i\phi}$$

$$D = C^\dagger \quad \text{and} \quad F = E^\dagger$$

θ is the angle between the radius vector $\vec{r}$ and the z-axis, which is chosen as the axis of quantization.

a) Isolated dipole pair in an external field. Usually, as in all NMR experiments, the dipole interaction energy is only a small fraction of dominating Zeeman energy in strong external magnetic fields. The total Hamiltonian is

$$H = H_D + \gamma_\mu \vec{S} \cdot \vec{H} + \gamma_N \vec{I} \cdot \vec{H} \tag{3.43}$$

For the case of two unlike spins - as for a system of a muon spin and a nuclear spin - the terms B, C, D, E, F do not contribute, since they involve spin flip processes which are forbidden for energy conservation reasons. This is of course not true for zero external magnetic field. The terms B - F are also very important if time dependent perturbing fields are present that may induce transitions between different spin states.

We are thus left with the term A, that has the same form as two classical interacting dipoles. We also assume the external field to coincide with the z-direction.

The total Hamiltonian is simply given by:

$$\begin{aligned} H &= \frac{\hbar^2 \gamma_\mu \gamma_N}{r^3} (1 - 3 \cos^2 \theta) S^z I^z + \gamma_\mu S^z H + \gamma_N I^z H \\ &= \hbar d \, S^z I^z - \hbar\omega_o S^z - \hbar\omega_1 I^z \end{aligned} \tag{3.44}$$

As can be seen, the dipole term and the Zeeman terms commute. For simplicity we will assume the nucleus to have spin $I = \frac{1}{2}$ (e.g. a proton). The eigenfunctions are

$$\chi_1 = |+ + >, \quad \chi_2 = |+ - >, \quad \chi_3 = |- - >, \quad \chi_4 = |- + > \tag{3.45}$$

and the energy eigenvalues are obtained from the set of equations:

$$\begin{aligned}
H \chi_1 &= \hbar(- \frac{\omega_o}{2} - \frac{\omega_1}{2} + \frac{d}{4}) \chi_1 = E_1 \chi_1 \\
H \chi_3 &= \hbar(\frac{\omega_o}{2} + \frac{\omega_1}{2} + \frac{d}{4}) \chi_3 = E_3 \chi_3 \\
H \chi_2 &= \hbar(- \frac{\omega_o}{2} + \frac{\omega_1}{2} - \frac{d}{4}) \chi_2 = E_2 \chi_2 \\
H \chi_4 &= \hbar(\frac{\omega_o}{2} - \frac{\omega_1}{2} - \frac{d}{4}) \chi_4 = E_4 \chi_4
\end{aligned} \tag{3.46}$$

We now assume the initial muon polarization to fall into the x-direction (transverse to the external field) and ask for the evolution of the polarization.

As defined previously, the initial muon spin state is described by

$$|\phi> = \frac{1}{\sqrt{2}} \{|+>_\mu + |->_\mu\}$$

The two nuclear states are $|+ >_N$, $|- >_N$ and the total initial spin states are

$$\begin{aligned}
\phi_1 &= \frac{1}{\sqrt{2}} (\chi_1 + \chi_4) \\
\phi_2 &= \frac{1}{\sqrt{2}} (\chi_2 + \chi_3)
\end{aligned} \tag{3.47}$$

with weights $P_1 = \frac{1}{2}$ and $P_2 = \frac{1}{2}$.

We now proceed as in the case of muonium in a transverse field. The transformation matrix U is given

$$U = \begin{pmatrix} \frac{1}{\sqrt{2}} & 0 & 0 & \frac{1}{\sqrt{2}} \\ 0 & \frac{1}{\sqrt{2}} & \frac{1}{\sqrt{2}} & 0 \\ \frac{1}{\sqrt{2}} & 0 & 0 & -\frac{1}{\sqrt{2}} \\ 0 & \frac{1}{\sqrt{2}} & -\frac{1}{\sqrt{2}} & 0 \end{pmatrix} \qquad (3.48)$$

Evaluation of eq. (3.27) leads to

$$\begin{aligned} P_\mu^x(t) &= \langle \sigma_\mu^x(t) \rangle = \frac{1}{2} \cos\,(\omega_o - \frac{d}{2})t + \frac{1}{2}\cos(\omega_o + \frac{d}{2})t \\ &= \cos\frac{d}{2}\,t\,\cos\,\omega_o t \end{aligned} \qquad (3.49)$$

This result has a very simple interpretation. The muons precess in the external field H (corresponding to the Larmor frequency ω_o) to which the magnetic field of the nuclear magnetic dipole has to be added or subtracted depending on the direction of the nuclear spin (up or down with respect to the external magnetic field). The total precession exhibits a beating behavior with a beat frequency $\omega^* = \frac{d}{2}$ and an average precession frequency ω_o.

This effect has been experimentally verified in an experiment where positive muons have been stopped in a single crystal of $Ca\,SO_4 \cdot 2H_2O$ (gypsum) [12]. About 50% of the stopped muons occupied the site of a proton, possibly placed there by a substitutional hot atom reaction of muonium. The proton muon dipole pairs are sufficiently separated to be treated as fairly isolated systems. The unit cell of gypsum contains two H_2O molecules with different orientations with respect to the crystal axes. One therefore expects up to four different precession frequencies (Fig. 13) whose values depend on the crystal orientation with respect to the external field.

In addition there will be weaker interactions with protons farther away. These interactions will lead to a quasi-continuous distribution of frequencies around each of the four main frequencies. As will be discussed later the weight distribution is very close to a Gaussian shape. Applying eq. (3.49) we can write the evolution of the muons' polarization:

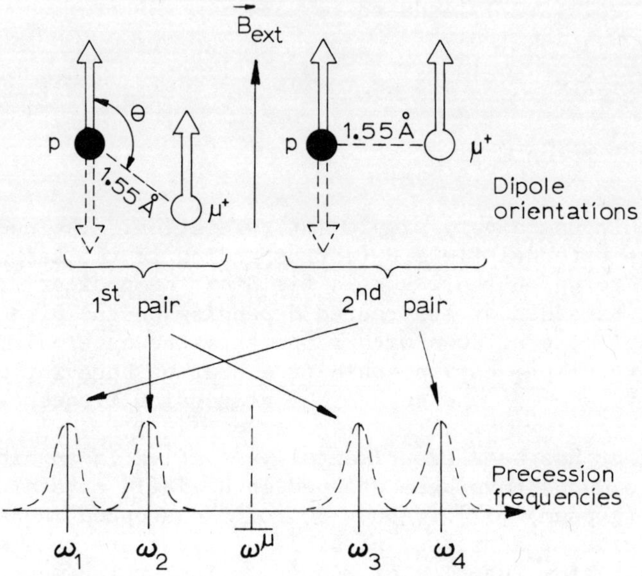

Fig. 13. Schematic arrangement of the two possible dipole pairs in the unit cell of Ca $SO_4 \cdot 2H_2O$ and the corresponding frequency spectrum.

$$
\begin{aligned}
P^x_\mu(t) &= \frac{1}{4} e^{-\sigma^2 t^2} \{\cos(\omega_0 - \frac{d_1}{2})t + \cos(\omega_0 + \frac{d_1}{2})t \\
&\quad + \cos(\omega_0 - \frac{d_2}{2})t + \cos(\omega_0 + \frac{d_2}{2})t\} \\
&= e^{-\sigma^2 t^2} \cos\left(\frac{d_1 + d_2}{4} t\right) \cos\left(\frac{d_1 - d_2}{4} t\right) \cos \omega_0 t \qquad (3.50) \\
&= F(t) \cos \omega_0 t
\end{aligned}
$$

d_1, d_2 refer to the first and second dipole pair, respectively. Fig. 14 shows the experimentally obtained distribution of F(t) values versus time for two different crystal orientations.

The solid lines represent calculations for F(t) (eq. 3.50) in which only the information on the crystal and the muon-proton dipole-dipole orientation (r = 1.5 Å) and the known magnetic moments of proton and muon enter. The agreement between data and calculation is considered very good in view of the poor knowledge of the crystal orientation available in that experiment. Fourier transformation of F(t) into the ω-space yields the frequency spectrum that can be directly compared with the spectrum obtained from proton NMR measurements in gypsum [13]. The agreement between data and calculations and the correspondence to the NMR-data proves the basic assumption that indeed the observed muons are to be found at the sites of protons. The missing 50% were probably thermalized as muonium and were not visible in the strong external magnetic field used (4.5 kGauss).

b) Isolated dipole pair in zero external field. Although not yet tried, the application of the muons will in principle allow one to study the effects of a coherent pure dipolar interaction which, to our knowledge, has so far not been considered. The application of conventional methods usually requires strong magnetic fields, so that all dipoles are decoupled from each other (Paschen-Back effect).

In view of possible future experiments and in order to deepen the understanding for the kinds of phenomena involved in μSR we will also discuss here how the evolution of the muon polarization behaves in the presence of a coherent dipolar interaction. These effects can in principle be studied in gypsum in zero external magnetic field.

Now the Hamiltonian to be used is given by eq. (3.42).

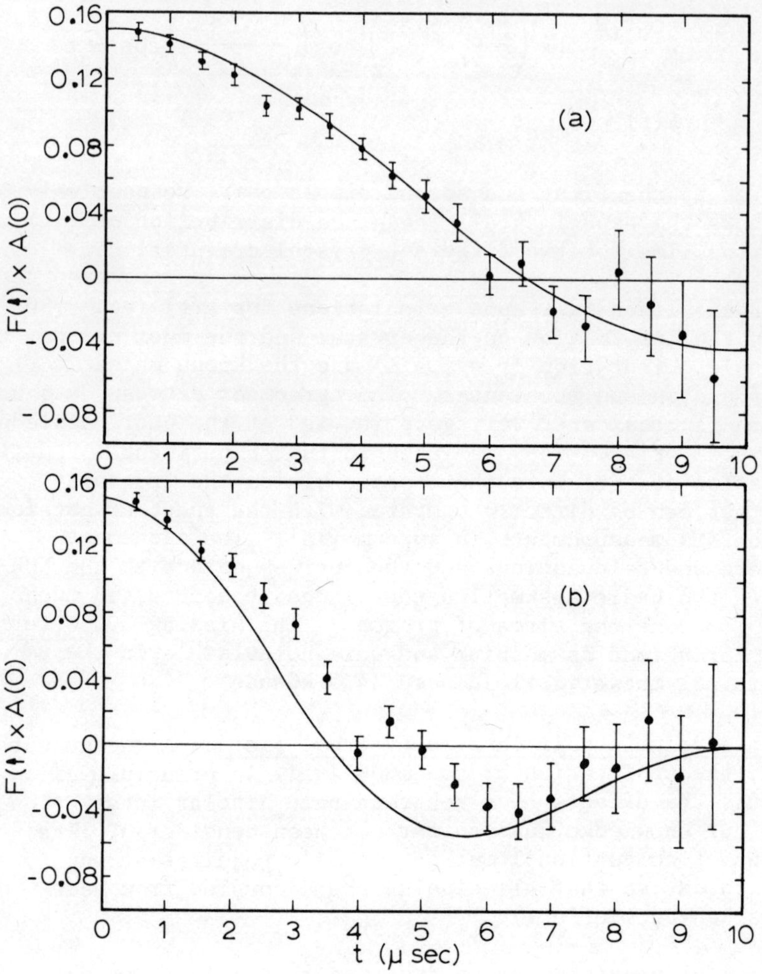

Fig. 14. Envelope of the precession amplitude in a single crystal of Ca $SO_4 \cdot 2H_2O$ for the two different crystal orientations. The solid line was calculated according to eq. (3.50) [12].

The manifold of eigenfunctions is again four. The eigenfunctions will be expressed in terms of the basis spin states χ_i.

$$\phi_j = \sum_{i=1}^{4} a_{ij}\, \chi_i \tag{3.51}$$

and

$$H_D\, \phi_j = E_j\, \phi_j \tag{3.52}$$

The energy eigenvalues are obtained from the secular equation:

$$\begin{vmatrix} a-E & b & b & c \\ b^\dagger & -a-E & -a & -b \\ b^\dagger & -a & -a-E & -b \\ c^\dagger & -b^\dagger & -b^\dagger & a-E \end{vmatrix} = 0 \tag{3.53}$$

with

$$\begin{aligned} a &= \frac{1}{4}(1 - 3\cos\theta) \\ b &= -\frac{3}{4}\sin\theta\cos\theta\, e^{-i\phi} \\ c &= -\frac{3}{4}\sin^2\theta\, e^{-2i\phi} \end{aligned} \tag{3.54}$$

The Hamiltonian H_D consists of scalar products of the spins and the radius vector, $\vec{r}$. The Hamiltonian is therefore invariant under a rotation of the coordinate system. The energy eigenvalues can thus be determined by selecting a coordinate system of most convenient orientation e.g. $\phi = 0$ and $\theta = 0$.

The secular equation reduces then to

$$E\{(a - E)(a - E)(2a + E)\} = 0 \tag{3.55}$$

from which we obtain the energy eigenvalues (in units of $\frac{\gamma_\mu \gamma_N}{r^2}\hbar^2$)

$$E_1 = 1, \quad E_2 = 0, \quad E_3 = -\frac{1}{2}, \quad E_4 = -\frac{1}{2} \tag{3.56}$$

Next we calculate the corresponding eigenfunctions for two special cases

$$1)\quad \theta = 0 \qquad\qquad 2)\quad \theta = 90^{\circ}\ .$$

The azimuthal angle ϕ will only enter through a phase factor and without loss of generality we can set $\phi = 0$.

1) $\theta = 0$

$$\begin{aligned}
E_1 &= 1 & \psi_1 &= \frac{1}{\sqrt{2}}(|+\,-> + |-\,+>) = |F = 1,\ m_F = 0> \\
E_2 &= 0 & \psi_2 &= \frac{1}{\sqrt{2}}(|+\,-> - |-\,+>) = |F = 0,\ m_F = 0> \\
E_3 &= -1/2 & \psi_3 &= |+\,+> \qquad\qquad\qquad = |F = 1,\ m_F = 1> \\
E_4 &= -1/2 & \psi_4 &= |-\,-> \qquad\qquad\qquad = |F = 1,\ m_F = -1>
\end{aligned} \tag{3.57}$$

2) $\theta = 90^{\circ}$

$$\begin{aligned}
E_1 &= 1 & \psi_1 &= \frac{1}{\sqrt{2}}(|+\,+> - |-\,->) = \frac{1}{\sqrt{2}}\{|F = 1,\ m_F = +1> \\
& & & \qquad\qquad - |F = 1,\ m_F = -1>\} \\
E_2 &= 0 & \psi_2 &= \frac{1}{\sqrt{2}}(|+\,-> - |-\,+>) = |F = 0,\ m_F = 0> \\
E_3 &= -1/2 & \psi_3 &= \frac{1}{\sqrt{2}}(|+\,-> + |-\,+>) = |F = 1,\ m_F = 0> \\
E_4 &= -1/2 & \psi_4 &= \frac{1}{\sqrt{2}}(|+\,+> + |-\,->) = \frac{1}{\sqrt{2}}\{|F = 1,\ m_F = +1> \\
& & & \qquad\qquad + |F = 1,\ m_F = -1>\}
\end{aligned} \tag{3.58}$$

In contrast to the muonium case in zero external field we have three instead of two energy levels. For $\theta = 0$ the eigenfunctions are identical to those of muonium. We recognise the curious fact that the triplet state (F = 1) is split into two levels. This is of course a consequence of the properties of a dipole field which leads to different interaction energies depending on whether the spins are lined up along the radius vector $\vec{r}$ or are sitting parallel side by side and perpendicular to the radius vector $\vec{r}$. The time evolution of the muon's polarization can now be obtained again from eq. (3.27). The initial muon polarization is assumed to fall in the z-direction. The initial spin states are χ_1 and χ_3 with weight factors $P_1 = \frac{1}{2}$ and $P_3 = \frac{1}{2}$.

The matrix U for the two cases $\theta = 0$ and $\theta = 90^o$ follows directly from eqs. (3.57) and eqs. (3.58).

Evaluation of eq. (3.27) leads to

$$\theta = 0 \qquad P_\mu^z(t) = \langle\sigma_\mu^z(t)\rangle = \frac{1}{2}(1 + \cos\omega_{12}t) \tag{3.59}$$

$$\theta = 90^o \qquad P_\mu^z(t) = \langle\sigma_\mu^z(t)\rangle = \frac{1}{2}(\cos\omega_{14}t + \cos\omega_{23}t) \tag{3.60}$$

The energy scheme and the relevant transitions are shown in Fig. 15. For arbitrary θ, $P^z(t)$ will depend on all three transition frequencies ω_{12}, ω_{14}, ω_{23}.

In contrast to the muonium case, where the time dependence of the polarization in zero external magnetic field is characterized by the occurrence of one frequency, the dipole-dipole interaction will involve up to three frequencies. For a dipole field of about 7 gauss ($Ca\ SO_4 \cdot 2H_2O$) the three frequencies will be of the order of 1 MHz and should be easily visible on the scale of the muon lifetime.

c) Interactions with many magnetic dipoles. A muon implanted e.g. in a metal like copper where all nuclei possess a magnetic moment will in fact interact with a large number of magnetic dipoles. We already mentioned that in a case such as ($Ca\ SO_4 \cdot 2H_2O$), interactions with remote magnetic dipoles have to be taken into account. Let us assume that the muon interacts with N nuclei with spin I and that a strong external field perpendicular to the initial muon polarization will be present (in z-direction). It is to be expected that the precession will be mainly governed by the external field.

The total Hamiltonian reads

$$H = \sum_{j=1}^{N} \alpha_j S_\mu^z I_j^z - \gamma_\mu \hbar S_\mu^z H = H_D + H_o \tag{3.61}$$

$$\alpha_j = \frac{\hbar^2}{r_j^3} \gamma_N \gamma_\mu (1 - 3\cos^2\theta_j)$$

We have neglected the Zeeman energies of the nuclei, the dipole-dipole interactions among nuclei and, as before, the dipole terms B - F.

Following the general treatment outlined above we introduce the density operator representing the initial spin ensemble with

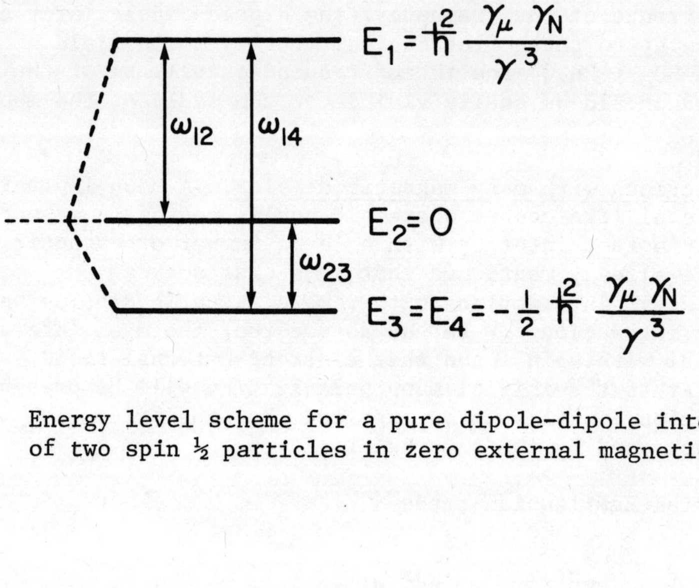

Fig. 15. Energy level scheme for a pure dipole-dipole interaction of two spin ½ particles in zero external magnetic field.

$|n\rangle = |\psi^i_\mu\rangle\, m|_1\rangle m_2\rangle \ldots |m_N\rangle$ which span a $2(2I+1)^N$ dimensional space

$$\rho = \sum_n c_n\, |n\rangle\langle n| \tag{3.62}$$

and

$$|\psi^1_\mu\rangle = \frac{1}{\sqrt{2}}(|+\rangle_\mu + |-\rangle_\mu), \quad \langle\psi^1_\mu|\sigma^x_\mu|\psi^1_\mu\rangle = 1$$

$$|\psi^2_\mu\rangle = \frac{1}{\sqrt{2}}(|+\rangle_\mu - |-\rangle_\mu)$$

$|m_i\rangle$ describes the spin state of the ith-nucleus with $\langle I^z_i\rangle = m_i$.

The c_n are determined by the initial occupation of the spin states. Therefore

$$c_n = c_{\mu_i m_1 m_2 m_3 \ldots m_N} = \begin{cases} \dfrac{1}{(2I+1)^N} & \text{if } \mu_i = 1 \\ 0 \quad \text{if } \mu_i = 2 \end{cases} \tag{3.63}$$

Eq. (3.23) now reads

$$P^x_\mu(t) = \sum_n c_n \langle n| e^{\frac{iH}{\hbar}t}\, \sigma^x_\mu\, e^{\frac{-iH}{\hbar}t} |n\rangle \tag{3.64}$$

Since $[H_o, H_D] = 0$ we can write

$$e^{\frac{i\,Ht}{\hbar}} = e^{\frac{i\,H_D t}{\hbar}}\, e^{\frac{i\,H_o t}{\hbar}} = e^{\frac{i\,H_D t}{\hbar}} \cdot e^{\frac{-i\omega_o \sigma^z_\mu t}{2}}$$

The operator exp. $(-\frac{i\omega_o}{2} t\, \sigma^z_\mu)$ represents a rotation of angle $\omega_o t$ about the z-axis. We can therefore write

$$e^{\frac{-i\omega_o}{2} t\, \sigma^z_\mu}\, \sigma^x_\mu\, e^{\frac{i\omega_o}{2} t\, \sigma^z_\mu} = \sigma^x_\mu \cos \omega_o t + \sigma^y_\mu \sin \omega_o t$$

and

$$P^x_\mu(t) = Q(t) \cos \omega_o t + Q'(t) \sin \omega_o t \tag{3.65}$$

The second term vanishes since H is invariant under time reversal and consequently $P^x_\mu(t)$ has to be invariant under time reversal also.

$$Q(t) = \sum_n c_n \langle n| e^{-iH_D t/\hbar} \sigma_\mu^x e^{iH_D t/\hbar} |n\rangle \tag{3.66}$$

In order to evaluate Q(t) we involve the concept of moments. By a Fourier transformation of Q(t) we obtain a frequency distribution f(ω) in ω-space

$$f(\omega) = \int_0^\infty Q(t) \cos \omega t \, dt \tag{3.67}$$

or

$$Q(t) = \frac{2}{\pi} \int f(\omega) \cos \omega t \, d\omega \tag{3.68}$$

The nth-moment is now defined as

$$M_n = \int_{-\infty}^{+\infty} f(\omega) \omega^n d\omega \neq 0 \quad \text{only if } n = \text{even} \tag{3.69}$$

from which we derive

$$M_{2n} = (-)^n \left(\frac{d^{2n} Q(t)}{dt^{2n}}\right)_{t=0} \tag{3.70}$$

The moment M_2 is the most important [6] and we will restrict ourselves to its consideration. Inserting Q(t) from eq. (3.66) into eq. (3.70) one obtains:

$$M_2 = \frac{-1}{\hbar^2} \sum_n c_n \langle n| [H_D, [H_D, \sigma_\mu^x]] |n\rangle \tag{3.71}$$

M_2 can be evaluated in a straightforward calculation using the spin commutator relations and noting that

$$Tr(I_j^z)^2 = Tr(I_j^x)^2 = Tr(I_j^y)^2 = \frac{1}{3} Tr \vec{I}^2 = \frac{1}{3} Tr[I(I-1)] =$$

$$= \frac{I(I+1)}{3} (2I + 1):$$

$$M_2 = \frac{I(I + 1)}{3} \sum_j \alpha_j^2 \tag{3.72}$$

With the ansatz

$$Q(t) = e^{-\Omega^2 t^2} \tag{3.73}$$

one obtains

$$M_2 = -2\Omega^2 \tag{3.74}$$

Identification yields

$$\Omega^2 = \frac{I(I+1)}{6\hbar^2} \sum_j \frac{(\hbar^2 \gamma_\mu \gamma_I)^2 (1 - 3\cos^2\theta)^2}{r_j^6} \tag{3.75}$$

The precession of muons in the presence of many magnetic dipoles is thus characterized by a time dependent damping of the precession amplitude represented by a Gaussian law. This damping has been seen in many substances like LiH, LiF, $Ca(OH)_2$ [4] and copper (see next section).

IV. DIFFUSION OF POSITIVE MUONS

Although the diffusion of positive muons has only been studied in detail in copper so far [14], this field will be very likely to attain a place of much importance in μSR studies. Muon diffusion is a part of the highly popular and much studied field of hydrogen diffusion or - more generally - of the field of "hydrogen in metals" [15]. At least three reasons are apparent that motivate the study of muon diffusion:

1. Study of isotope effects.

2. Study of diffusion in the absence of any other "hydrogen".

3. Study of diffusion in metals in which hydrogen cannot be dissolved in sufficient quantities to allow the application of conventional methods such as neutron diffraction, Gorsky effect, NMR etc. In particular, the μSR-technique may allow to study diffusion at very low temperatures where quantum effects may become dominant (tunneling).

In the following we will first discuss the principle of diffusion measurements via the "motional narrowing" effect; secondly, the experiment by Gurevich et al. [14] will be discussed, and finally a short review of existing diffusion theories with special emphasis on the problem of muon diffusion will be presented.

1) Motional Narrowing

According to the principle of the μSR method, diffusion of the muons will only manifest itself if it leads to a modification of the magnetic interactions that determine the evolution of the muons' polarization. A necessary prerequisite is thus the presence of internal magnetic fields in the target, whose actions on the muons are modulated in time by the latter's motion through the crystal. Therefore, diffusion studies will be restricted to substances, in particular metals, that possess nuclear magnetic dipoles or that are paramagnetic or ferromagnetic. In all instances the time dependence of the magnetic interactions introduced by spin flip processes of the nuclear or electronic spins must be "slow" in comparison with the time dependence introduced by the diffusional motion of the muon. This is usually fulfilled for nuclear spins and ferromagnetically etc. ordered electronic spins, but not so, however, for paramagnetic

In the following we discuss the modification of the evolution of the muon's polarization if the interaction with many neighboring magnetic dipoles becomes time dependent [6]. The Hamiltonian eq. (3.61) reads now

$$\mathcal{H}(t) = \mathcal{H}_z + \mathcal{H}_D(t) \tag{4.1}$$

and

$$\mathcal{H}_D(t) = \sum_{j=1}^{n} \alpha_j(t)\, S_\mu^z I_j^z , \qquad \mathcal{H}_z = \gamma_\mu \hbar H S_\mu^z$$

$$\alpha_j(t) = (\hbar^2 \gamma_\mu \gamma_I)/r_j^{\,3}(t)(1 - 3\cos^2\theta_j(t))$$

Following the treatment in Chapt. III exactly, eqs. (3.62–3.66) (with the initial muon polarization in the x-direction) we obtain

$$P_\mu^x(t) = Q(t)\cos\omega t \tag{4.2}$$

with

$$Q(t) = \sum_n C_n \langle n|\exp\!\left(\frac{i}{\hbar}\int_0^t dt'\, \mathcal{H}_D(t')\right)\cdot\sigma_\mu^x\cdot\exp\!\left(-\frac{i}{\hbar}\int dt'\, \mathcal{H}_D(t')\right)|n\rangle \tag{4.3}$$

The dipole Hamiltonian is now a stochastic function of time. For a time independent $\mathcal{H}_D$, $P_\mu^x(t)$ could be expressed as (see eq. 3.68)

$$P_\mu^x(t)_0 = \cos\omega_0 t \int_0^\infty \cos(\omega t)\, f(\omega)\, d\omega = \cos\omega_0 t \langle\cos\omega t\rangle \tag{4.4}$$

where the symbol <> means average taken over the distribution $f(\omega)$. When we pass over to the time dependent Hamiltonian $\mathcal{H}_D(t)$, we will write instead

$$P^x_\mu(t) = \cos \omega_o t \ \langle \cos \chi(t) \rangle \tag{4.5}$$

$$\chi(t) = \int_0^t \omega(t')\, dt' \tag{4.6}$$

The latter expression follows from eq. (4.3) where we have to integrate over $\mathcal{H}_D(t)$ in the exponent. The next step consists in finding the distribution function $P(\chi,t)$ defined by

$$P^x_\mu(t) = \cos \omega_o t \int_0^t P(\chi,t) \cos \chi(t)\, d\chi(t) \tag{4.7}$$

In the absence of any time dependence we have shown that $f(\omega)$ can be approximated by a Gaussian law.

We now assume - mainly for mathematical simplicity - that $\omega(t)$ is also distributed according to a Gaussian law, that it is stationary, i.e. $\overline{\omega(t')\ \omega(t'')} = \overline{\omega(t''-t')\omega(o)}$, and that $\langle\omega^2\rangle$ has the same value as in the absence of motion. The physical meaning of these assumptions is that at each instant the microscopic distribution of local fields is the same as in the absence of motion, viewing the field distribution from the probe, e.g. the muon. However, the local field fluctuates at each point randomly and the rate of change is expressed by a correlation function

$$\begin{aligned} G_\omega(\tau) &= \langle \omega(t)\ \omega(t-\tau) \rangle \\ &= \iint p(\omega_1,t;\ \omega_2,t-\tau)\ \omega_1\ \omega_2\ d\omega_1\ d\omega_2 \end{aligned} \tag{4.8}$$

Here $p(\omega_1,\ t;\ \omega_2,\ t-\tau)$ is the probability that ω takes the value ω_1 at the instant t and ω_2 at the instant $t-\tau$. Because $\omega(t)$ is assumed to be stationary, the probability function will only depend on the difference τ and not on t. $G_\omega(\tau)$ is thus a measure of how fast $\omega(t+\tau)$ will change with time τ.

In the absence of motion:

$$G^o_\omega(\tau) = \int f(\omega)\ \omega^2\ d\omega = \langle\omega^2\rangle = M_2 \tag{4.9}$$

which is the second moment introduced previously (eq. 3.69). It is then convenient to write

$$G_\omega(\tau) = \langle\omega^2\rangle\ g_\omega(\tau)$$

$g_\omega(\tau)$ is called a reduced correlation function with $g_\omega(0) = 1$. $\omega(t)$ is assumed to be Gaussian distributed. It follows then that $\chi(t)$ must be Gaussian distributed, too:

$$P(\chi,t) = [2\pi <\chi^2(t)>]^{-\frac{1}{2}} \exp\left\{- \frac{1}{2} \frac{\chi^2}{<\chi^2(t)>}\right\} \tag{4.10}$$

and

$$\begin{aligned} <e^{-i\,\chi(t)}> &= [2\pi<\chi(t)^2>]^{-\frac{1}{2}} \int_{-\infty}^{+\infty} \exp\left\{- \frac{\chi^2}{<\chi^2(t)>}\right\} e^{-i\chi}\, d\chi \\ &= \exp <- \frac{1}{2}\,\chi^2(t)> \end{aligned} \tag{4.11}$$

and

$$\begin{aligned} <\chi^2> &= <[\int \omega(t')\, dt']^2> \\ &= <\int_0^t dt' \int_0^t dt''\ \omega(t')\ \omega(t'')> = 2\int_0^t (t-\tau)\ g_\omega(\tau)\ d\tau \end{aligned} \tag{4.12}$$

Putting everything together, we arrive finally at

$$P_\mu^x(t) = \cos\omega_o t\ \exp\{- <\omega^2> \int_0^t (t-\tau)\ g_\omega(\tau)\ d\tau\} \tag{4.13}$$

The simplest expression for the correlation function that seems to be a good approximation in the case of random diffusion by jumping from one crystal site to the other, is

$$g_\omega(\tau) = \exp(- {}^{\tau}/\tau_c) \tag{4.14}$$

τ_c is the so-called correlation time, which is a measure of the duration of a coherent interaction with the neighbor magnetic dipoles. τ_c may be considered the average residence time in a crystal site. Inserting $g_\omega(\tau)$ from eq. (4.14) into eq. (4.13), we obtain

$$P_\mu^x(t) = \cos\omega_o t \cdot \exp\{-<\omega^2>\tau_c^2[\exp(-{}^t/\tau_c) - 1 + {}^t/\tau_c]\} \tag{4.15}$$

If $<\omega^2>^{\frac{1}{2}}\,\tau_c >> 1$, which implies a very slow diffusion, eq. (4.15) reduces to

<u>slow</u>
$$P_\mu^x(t) = \cos\omega_o t\ \exp\left(- \frac{<\omega^2>}{2}\ t^2\right) \tag{4.16}$$

This is identical to eq. (3.73) in the absence of motion.

For very rapid motion, i.e. $<\omega^2>^{\frac{1}{2}} \tau_c << 1$ eq. (4.15) reduces to

$$\underline{\text{rapid}} \qquad P_\mu^x(t) = \cos \omega_0 t \exp(-<\omega^2>\tau_c t) \qquad (4.17)$$

This represents a pure exponential damping. The Fourier transform has a Lorentzian shape with a FWHM of $<\omega^2>\tau_c$. This shape would also be the shape of the corresponding NMR-signal. The faster the diffusion, the smaller τ_c, the smaller would be the NMR line width, from which the expression motional narrowing originates.

The reduced correlation function $g_\omega(\tau)$ should be calculated from a detailed theory of diffusion [6, 16]. In all instances, one is confronted with a correlation time τ_c, that is a measure of the duration of the coherent interaction with the neighbor magnetic dipoles, and which may thus be considered the average residence time in a crystal site.

The study of hydrogen diffusion by neutron diffraction involves quite different correlation functions than the one occurring in motional narrowing. With respect to the diffusional process the relevant correlation function is given by [21]

$$G_s^d(\vec{r},t) \cong < \sum_{i=1}^{N} \sum_{j=1}^{N} \delta(\vec{r} + \vec{r}_i(0) - \vec{r}_j(t))> \qquad (4.18)$$

which describes the mutual spatial correlations among the hydrogen atoms as a function of time. As may be expected, this correlation function involves a correlation time that has the same interpretation as the one in "motional narrowing", and it is therefore possible to directly compare results of these quite independent methods without much ambiguity.

All microscopic diffusion theories are mainly directed at calculating this correlation time or, equivalently, the diffusion rate (see Chapt. IV.3).

Finally, it may be in order to evaluate the range of correlation times τ_c measurable by the μSR technique.

Table 1 contains the range of τ_c-values accessible to μSR for various $<\omega^2>$, derived under the assumption that the longest damping time that can be measured is of the order of 100 μsec, and that the error is of the order of 10%.

Table 1

$\langle\omega^2\rangle^{\frac{1}{2}}$ μsec^{-1}	τ_c, min [sec]	τ_c, max [sec]
0.1	10^{-6}	10^{-4}
0.2	$2.5\cdot10^{-7}$	$5\cdot10^{-5}$
0.5	$4\cdot10^{-8}$	$2\cdot10^{-5}$
1.0	10^{-8}	10^{-5}
10.0	10^{-10}	10^{-6}

Finally, the correlation or jump time τ_c can be related to the classical diffusion coefficient D that is usually measured by macroscopic methods

$$\frac{1}{\tau_c} = \frac{6D}{d^2} \qquad (4.19)$$

where d is the jump distance to the next neighbor site.

2) Muon Diffusion in Copper

Copper has a fcc crystal structure. Natural copper consists of the two isotopes ^{63}Cu and ^{65}Cu with abundances of 69.1% and 30.9%, respectively. Each has spin I = 3/2; the magnetic moments are +2.23 n.m and 2.38 n.m, respectively. The implanted muons will be most likely situated in the interstitial sites with octahedral symmetry (Fig. 16) and the diffusion will proceed by jumps between neighboring interstitial sites.

This experiment was performed by Gurevich et al. [14] at Dubna. They measured the damping of the muon precession signal between 77°K and room temperature. The damping time was defined by the value τ, at which the precession amplitude had decreased to 1/e. The thus obtained damping rates $\lambda = 1/\tau$ are shown in Fig. 17a. A narrowing effect is clearly visible. The authors report that for T = 77°K the damping curve approaches the Gaussian law eq. (4.16) and that the calculated $\langle\omega^2\rangle_{av}$ averaged over all crystal orientations gives a good agreement with the

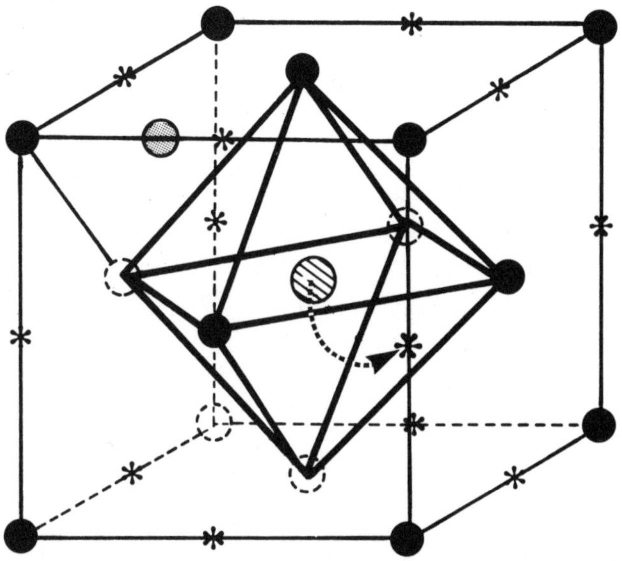

Fig. 16. Schematic arrangements of atoms in a face centered cubic (fcc) crystal. The center position in the cube is the octahedral position. Other octahedral positions are marked by a star. One of the tetrahedral positions is also shown ◍. The dashed arrow indicates the diffusion path from one interstitial site to the next.

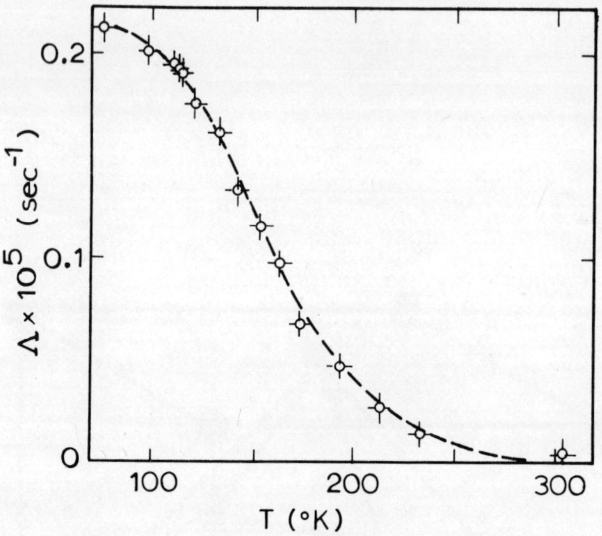

Fig. 17a. Temperature dependence of the μ^+ depolarization rate in Cu.

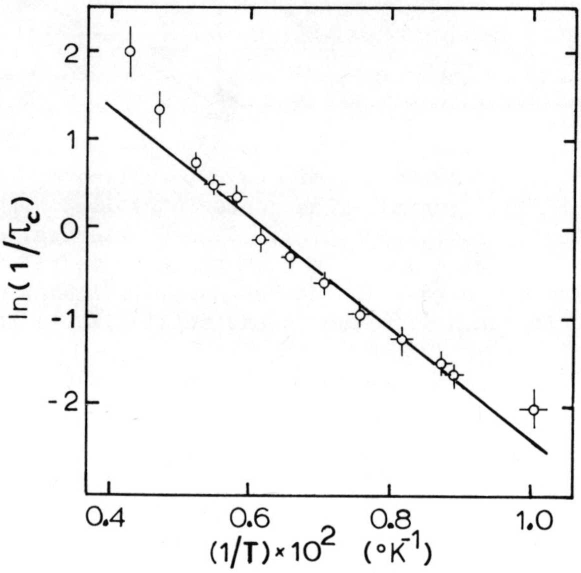

Fig. 17b. Dependence $\ln(1/\tau_c) = f(1/\tau)$ obtained from the relaxation rates in Cu.

measured value

$$\frac{1}{2} \langle\omega^2\rangle^{1/2}_{exp} = (0.219 \pm 0.008)\ \mu sec^{-1}$$

They conclude that at 77^oK the diffusion of muons through the Cu-lattice is not important (e.g. slow as compared with the muon lifetime) and that the measured $\langle\omega^2\rangle_{exp}$ represents the expected "dipolar line width" very well, averaged properly over all crystal orientations. (The Cu-sample used was a polycrystal.)

The correlation times for higher temperatures were now extracted by setting equal

$$-\langle\omega^2\rangle\ \tau_c^2[\exp(-{}^{\tau}/\tau_c) - 1 + {}^{\tau}/\tau_c] = -1$$

The logarithms of the thus obtained τ_c values are plotted in Fig. 17b versus temperature.

In this semi-logarithmic plot the data seem to fall on a straight line, which implies a law of the form

$$\frac{1}{\tau_c} \sim \exp(-{}^{E}/kT)$$

This law is of the Arrhenius type and is typical for barrier jump diffusion. E is an activation energy which may be identified with the potential energy barrier height. The diffusion coefficient is usually expressed as

$$D = D_o\ e^{-E/kT} \qquad (4.20)$$

Hence, (eq. 4.19)

$$\frac{1}{\tau_c} = \frac{6\ D_o}{d^2}\ e^{-E/kT} \qquad (4.21)$$

As remarked before, the muon diffusion will probably proceed via jumping from one octahedral interstitial site to the next. The jump distance in units of the lattice constant $a = 3.61$ Å is

$$d = \frac{1}{\sqrt{2}}\ a$$

and

$$\frac{1}{\tau_c} = \frac{12\ D_o}{a^2}\ e^{-E/kT} = \nu\ e^{-E/kT} \qquad (4.22)$$

Table 2

	investigated temperature range	$E\ (\frac{Kcal}{Mole})$	$\nu\ (sec^{-1})$	
μ^+	77^oK - 300^oK	1.072	$3.12\ 10^7$	[14]
H	430° - 635^oC	9.75	$1.06\ 10^{14}$	[17]
	450° - 915^oC	9.286	$1.04\ 10^{14}$	[18]

Table 2 contains the fitted E and ν values for muons in copper together with the E and ν values obtained for hydrogen in copper [17, 18].

As can be seen, the difference in the pre-exponential factors is huge (almost 7 orders of magnitude). The ratio of the activation energies is

$$\frac{E_H}{E_\mu} = \frac{9.5}{1.07} \simeq 8.66 \simeq \frac{m_p}{m_\mu}$$

The near equality to the ratio of the proton and muon mass is a curiosity that is not explained as yet. Classical rate theories would predict that the activation energies should not show an isotope effect, while the pre-exponential factors should behave as

$$\frac{\nu_H}{\nu_\mu} = \sqrt{\frac{m_\mu}{m_p}} \simeq 2.98$$

By comparing the muon and hydrogen diffusion data, one has of course to consider that the data have been obtained in quite different temperature ranges. It is evident that the muon behaviour at low temperatures follows a different pattern. Preliminary data obtained at SIN on muon diffusion in Palladium [19] (frozen in muons should exhibit a damping time of about 34 μsec, while even at 4^oK the damping time was larger than 50 μsec) and measurements by Kossler et al. [20] in Nb also indicate that muon diffusion in these metals proceeds much faster than would be expected from the hydrogen diffusion rates at higher temperatures. In the next section we will consider these phenomena against the background of existing diffusion theories.

3) Diffusion Models

a) Classical and semi-classical models. The most developed classical model for rate processes in solids is the model of Vineyard [22]. Consider a crystal containing a vacant lattice site and let Γ be the average rate at which a specific atom adjacent to the vacancy jumps into the vacant site. The crystal is supposed to contain $^N/_3$ atoms and the configuration space is made of $x_1, x_2 \ldots x_N$ coordinates. Before the jump the total potential energy $\phi(x_1 \ldots x_N)$ will possess a minimum for a given point A in configuration space. In order that an atom will jump to the vacant site, which corresponds to another minimum of the potential energy at the point B in configuration space, it has to overcome a saddle point P. Following Vineyard, the jump rate can be expressed as the ratio of two configurational partition functions

$$\Gamma = \sqrt{\frac{kT}{2\pi}} \frac{\int_S e^{-\phi/kT} dS}{\int_A e^{-\phi'/kT} dv} \tag{4.23}$$

The integral in the numerator is taken over the hypersurface S, which contains the saddle point P, and which is perpendicular to the contours of constant ϕ. S thus separates the region A from the region B. The integral in the denominator is taken over the configuration space of the A region. Further it is assumed that each atom is bound harmonically to its equilibrium site. The vibrations are labelled $\nu_1, \nu_2 \ldots \nu_N$ and the potential energy can be written as

$$\phi(x) \cong \phi(A) + \sum_{j=1}^{N} \frac{1}{2} (2\pi\, \nu_j)^2\, q_j^{\,2} \tag{4.24}$$

q_j are the normal coordinates $(x_j(A) - x_j(X))$ and likewise

$$\phi'(x') = \phi(P) + \sum_{j=1}^{N-1} \frac{1}{2} (2\pi\, \nu_j')^2\, q_j'^{\,2} \quad \text{constrained to S.} \tag{4.25}$$

Then

$$\Gamma = \left(\prod_{j=1}^{N} \nu_j\right) \Big/ \left(\prod_{j=1}^{N-1} \nu_j'\right) \cdot \exp[-(\phi(p) - \phi(A))/kT] \tag{4.26}$$

The normal frequencies ν_j are proportional to $\sqrt{m_j}^{-1}$ (harmonic oscillator). After some calculations one obtains

$$\Gamma = \frac{c}{\sqrt{m^*}} e^{-[\phi(P)-\phi(A)]/kT}$$

$$= \frac{c}{\sqrt{m^*}} e^{-E_a/kT} \qquad (4.27)$$

with m* = mass of diffusing atom.

The pre-exponential factor has a $\frac{1}{\sqrt{m}}$ dependence on the mass, while the activation energy E_a is independent of the mass of the diffusing particle. The ratio of the diffusion rates of μ^+ and hydrogen in this classical picture should be

$$\frac{\Gamma_\mu}{\Gamma_H} = \sqrt{\frac{m_H}{m_\mu}} \qquad (4.28)$$

Another way of writing Γ is

$$\Gamma = N \cdot \nu \exp\left(-\frac{E_a}{kT}\right) \exp\left(\frac{\Delta S}{k}\right) \qquad (4.29)$$

where N is the number of nearest neighbor positions open to the jumping particle and ν is the vibration frequency of the particle in its momentary equilibrium position and S is the entropy of activation.

A number of authors have tried to apply quantum corrections to the classical Vineyard theory [18, 23, 24]. The classical partition functions were replaced by quantum partition functions, the vibration frequency ν was corrected, and the possibility of tunneling was taken into account. A number of new parameters were thus introduced that had to be fitted by the data. With regard to the diffusion of muons neither the pure classical description nor the quantum corrected versions seem to be applicable, not to mention the conceptual problems that are inherent in the classical and semi-classical approaches [25].

b) Quantum mechanical models. A muon implanted e.g. in a metal will see the periodic electrostatic potential of the lattice and so from the very beginning one is inclined to describe the muon more in terms of a band model than to view it in a well localized state. This has indeed been tried by a number of authors [26, 27, 28, 29, 97] with regard to the diffusion of hydrogen. As a first approach one may represent the periodic potential by a one-dimensional sinusoidal potential with a periodicty of ℓ. The muon wave function then has to obey the Schrodinger equation

$$\left[\frac{\hbar^2}{2m_\mu}\frac{d^2}{dx^2} + E - V\cos(2\pi\frac{x}{\ell})\right]\psi_\mu = 0 \qquad (4.30)$$

This equation can be reduced to the Mathieu equation

$$\frac{d^2f}{dy^2} + (a - 2q\cos 2y)f = 0 \qquad (4.31)$$

with the parameters

$$a = \frac{8\, m_\mu\, \ell^2\, E}{h^2} \qquad\qquad q = \frac{4\, m_\mu\, \ell^2\, V}{h^2}$$

Fig. 18 shows that the range of values of a(q), for which one gets bound solutions of this equation (taken from ref. [28, 30]), the allowed energy bands. The width of the energy band for a given q is indicated by $\delta_i(q)$.

The eigenfunctions are Bloch functions and it is evident that the muon is not localized but rather propagates with a typical velocity of [32]

$$\langle|\upsilon|\rangle = \frac{1}{\hbar}\langle|\vec{\nabla}\, E_i(k)|\rangle$$

$$\simeq \frac{\delta_i}{\hbar\, k_{max}} \simeq \frac{\delta_o}{\hbar}\,\frac{\ell}{\pi} \qquad (4.32)$$

where k_{max} is taken at the zone boundary, $k_{max} = \frac{\pi}{\ell}$.

For a potential energy of typically 2V = 0.4 eV (fcc - metals) and site distance ℓ = 2.5 Å we get q = 3.5 and obtain from Fig. 18 a band width of

$\delta_o \approx$ 0.5 meV (ground state)

$\delta_1 \approx$ 20 meV (1st excited state)

Rather than speaking of a velocity of propagation we can introduce the term tunneling frequency:

$$\nu = \frac{\langle\upsilon\rangle}{\ell} = \frac{2\,\delta_i}{h}$$

This is the frequency with which the propagating particle crosses or penetrates the potential barrier. With the given number we get

$$\nu_o = 2.4 \cdot 10^{11}\ \mathrm{sec}^{-1}$$

$$\nu_1 = 9.6 \cdot 10^{12}\ \mathrm{sec}^{-1}$$

The jump time or the earlier introduced correlation time τ_c is just the inverse of this frequency. As can be seen, the diffusion rate thus obtained is much too high to explain the rates obtained in copper, while it could explain the muon behaviour in niobium and palladium.

With regard to the ground state of muons in a metal, a pure band model is probably not a very realistic assumption. Random strains in the crystal or local strains, due to the presence of a muon, will distort the periodicity, which in effect leads to a localization of the muon.

Of particular importance will be the displacement of the neighbor lattice atoms of the muon (polaron theory) which, if the muon lattice interaction is sufficiently strong, will lead to what one calls self-trapping. A variety of dynamical effects result [31].

One effect, for weak coupling at least, is that the effective mass of the particle will be increased. The lattice distortion will follow the particle motion. The effective mass will presumably be a weighted average of the particle mass and those of the lattice atoms. Consequently, any isotope effect will be reduced. Also, the character of a positive motion of the self-trapped system will be altered. Instead of propagating, the self-trapped particle moves by uncorrelated random jumps. At high temperatures the classical picture may again become applicable.

That a localized particle, e.g. the muon, starts to move at all is a consequence of the coupling to phonons, that is, the lattice vibrations. One can distinguish several processes:

1) Phonon induced tunneling. Transitions occur directly from one localized state to the other. The initial and the final states lie below the top of the barrier and therefore this process is called "phonon induced tunneling". The particle lattice coupling must be weak. One has to consider one phonon, [25] two phonon [31, 33] and many phonon transitions [25, 32]. The different temperature behaviour is listed in Table 3. Ref. [32] also predicts the isotope dependence for different simple models (Table 3). The single phonon transition probability is proportional to the transparency of the potential barrier given grossly by the WKB formula

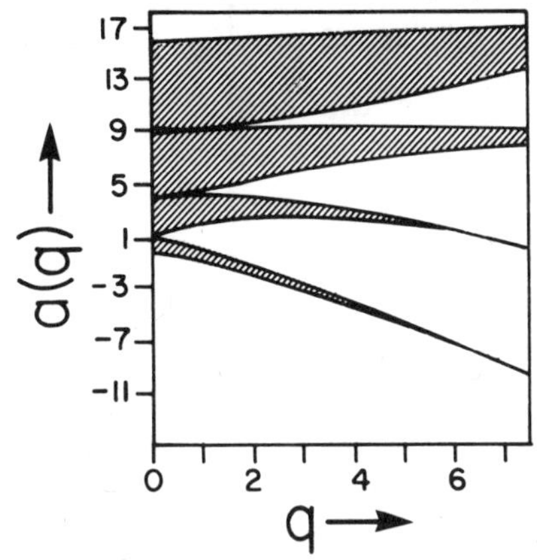

Fig. 18. Allowed energy bands for solutions of the Mathieu equation.

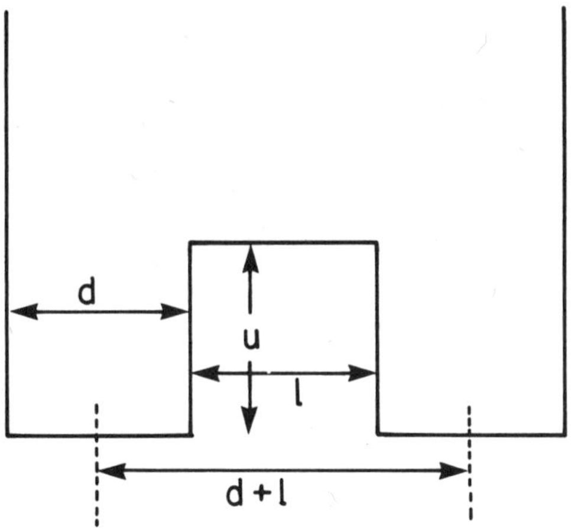

Fig. 19. Schematic arrangement of the square well potentials in the simple tunneling model for Cu.

Table 3: Predictions of Temperature and Isotope Dependence of Various Diffusion Models

Model	Temperature Dependence	Isotope Dependence
Vineyard [22]	$\nu \simeq \nu_o \exp(-E_{act}/kT)$	$\nu_o \simeq (m)^{-\frac{1}{2}}$
Phonon induced tunneling	$\nu \simeq f(\delta,U,f)\cdot T$ $kT >> U$	
a) one phonon [25]	$\nu \simeq f'(\delta,U,f) \exp(-U/kT)$ $kT < U$ U = lattice accommodating energy	
b) two phonon [31]	$\nu \sim T^R$ $5 \leq R \leq 9$ for $\delta << kT << k\theta_D$ $R = 2$ " $\delta << k\theta_D << kT$	
c) multi-phonon [32]	$\nu \approx \lvert J_{pp'}\rvert^2 T^7$ for $T << \frac{1}{2}\theta_D$ * $\nu \approx \left(\frac{1}{E_a T}\right)^{\frac{1}{2}} \lvert J_{pp'}\rvert^2 \exp(-E_a/kT)$ direct process * $\nu \approx \frac{1}{(E_a E_s)^{\frac{1}{2}}} \lvert J_{lim}\rvert^2 \exp(-(E_a+E_s)/kT)$ lattice activated process * for $T >> \frac{1}{2}\theta_D$ E_a = lattice accomodation energy E_s = deformation energy	oscillator: $E_a \simeq m^{-1}$; $E_s = m^o$; $\frac{\lvert J_{pp'}\rvert^2}{E_a^{\frac{1}{2}}} \approx m^{\frac{1}{2}} e^{-\xi\sqrt{m}}$; $\frac{\lvert J_{lim}\rvert^2}{E_a^{\frac{1}{2}}} \approx m^{-\frac{1}{2}}$ well: $E_a \simeq m^{-2}$; $E_s = m^o$; $\approx me^{-\xi\sqrt{m}}$; $\approx m^o$

Table 3 continued

Model	Temperature Dependence	Isotope Dependence
Orbach-type [25]	a) excitation accomplished in one step: $\nu \cong \nu_0\, e^{-V/kT}$ for $T \ll \theta_D$ $\nu \cong \nu_0 \left(\frac{T}{\theta_D}\right)^{V/k\theta_D} e^{-V/2kT}$ for $\theta_D \ll T$ $V = E_i - E_o$ E_o = energy of ground state E_i = energy of excited state	
	b) excitation accomplished in a large no. of steps: $\nu \simeq \exp(-E_n/kT)$ E_n = energy of excited state	
tunneling bottle neck [25]	a) $\nu \simeq \frac{\delta}{\hbar} \exp(-E_n/kT)$ b) $\nu \simeq \frac{\delta}{\hbar} \exp(-E_n/kT)$ $T < \theta_D$ $\nu \simeq \frac{\delta}{\hbar} \frac{1}{T} \left(\frac{T}{\theta_D}\right)^{E_n/k\theta_D} \exp -\left(\frac{E_n}{2kT}\right)$ $T > \theta_D$ θ_D = Debeye Temperature	

$$\Gamma = \exp\left[-\frac{2}{\hbar}\int\sqrt{2m(U(x)-E)}\,dx\right] \qquad (4.33)$$

2) Orbach-type process [25]. In this case the particle is excited to some intermediate state by the absorption of one or more phonons of total energy Δ. The particle then moves and decays to the final state in a different lattice site by the re-emission of one or more phonons of similar total energy. The particle-lattice coupling is assumed to be weak and the transition rates are dominated by the excitation probability. See Table 3 for the type of temperature dependence.

3) Tunneling bottle neck process [25]. This process may happen when the lattice-particle coupling is strong. It also involves intermediate excited states (band states). The strong coupling case is characterized by $\Gamma > \delta$ where δ is the width of the band to which the particle has been excited and Γ is the decay constant of the excited band state. Consequently the excitation process will lead to the coherent excitation of many states forming a wave packet. The motion or spreading of the wave packet, i.e. tunneling, will be the rate limiting process, forming a bottle neck. The diffusion rate can be expressed as

$$\nu_D = \frac{\delta_n}{\hbar}\, f\!\left(\frac{\delta_n}{\sqrt{n}}\right) \exp\left(\frac{-E_n}{kT}\right) \qquad (4.34)$$

where E_n is the energy difference between ground state and the center of the band to which the particle has been excited. Γ is of course also temperature dependent.

Finally, we will apply a very simple model to the diffusion of muons in copper. We describe the potential energy distribution by a simple one-dimensional square well potential with a barrier height V_o, a barrier width of ℓ and well width d. The distance from the center of one well to the other is $d + \ell$ (see Fig. 19). The barrier height V_o is assumed to be given roughly by the activation energy for hydrogen diffusion at high temperatures; V = 0.4 eV. The zero point motion of the muon in the well can be obtained from the uncertainty relation

$$\Delta p \cdot \Delta x = \Delta p \ell = \hbar$$

and the zero point vibration frequency is

$$\nu_o = \frac{E}{h} = \frac{\hbar}{4\pi\, m_\mu\, \ell^2}$$

For simplicity we assume $d = \ell = \frac{1}{2} \frac{a}{\sqrt{2}} \cong 1.3$ Å.
Hence,

$$\nu_o = \frac{2\hbar}{4\pi m_\mu a^2} \simeq 6.7 \cdot 10^{11} \text{ sec}^{-1}$$

In order to calculate the diffusion rate we use the WKB expression for the transition probability through the potential barrier. The diffusion rate is then given by

$$\nu_D = \nu_o \exp \left[- \frac{2}{\hbar} \sqrt{2m_\mu U}\, d\right]$$

from which we obtain

$$\nu_D = \nu_o\, 4.10^{-6} \simeq 2.7 \cdot 10^6 \text{ sec}^{-1}$$

Multiplying this by the number of next neighbor interstitial sites (n = 12) (eq. 4.27), the total tunneling rate should be $\nu_t = 3.2 \cdot 10^7$ sec^{-1}. This is very close to the actually obtained pre-exponential factor for muon diffusion in copper. The measured temperature dependence and the activation energy can of course not be explained by this simple model. It is however tempting to assume that the observed activation energy is the one occuring in the Sussmann formula for the one phonon induced tunneling model. The measured activation energy could then be identified with the lattice accommodation energy, i.e. the energy change due to the rearrangement of the lattice around the muon.

Only future measurements will show how much can be learned about diffusion behaviour of muons as a function of temperature. Of paramount importance is of course more information on the potential well in which the muon is situated. The binding energy depends not only on the arrangement of positive ion cores, but in addition on the screening of the positive charge of the muon by the conduction electrons [34]. This very important effect will be the subject of the next chapter.

V. KNIGHTSHIFT AND THE SCREENING OF THE POSITIVE MUON CHARGE IN METALS

We already mentioned in chapter III that the magnetic interaction of a positive muon with the conduction electrons is of the Fermi contact type. Possible dipolar contributions are averaged to zero because of the isotropic distribution of electrons around a muon. We write

$$\mathcal{H}_c = \frac{8\pi}{3} \gamma_\mu \gamma_e \hbar^2 \vec{S}_\mu \sum_i \vec{S}_{e,i} \, \delta(\vec{r}_i) \tag{5.1}$$

where the sum is taken over all conduction electrons.

Before dealing further with this Hamiltonian, we have to mention some of the properties of conduction electrons. They are described by Bloch waves

$$\psi_k(\vec{r}) = \mathcal{U}_k(\vec{r}) \exp(i\vec{k}\cdot\vec{r}) \tag{5.2}$$

where $\mathcal{U}_k(\vec{r})$ reflects the periodicity of the lattice. Including the spin we have

$$\psi_k(\vec{r},\sigma) = \psi_k(\vec{r}) \, \chi_\sigma \tag{5.3}$$

where χ_σ is a spinor.

The electrons, which are fermions, have to obey Fermi statistics. The occupation number as a function of energy $E_{K\sigma}$ is given by the Fermi-Dirac distribution law

$$n_{k\sigma} = [\exp((E_{k\sigma} - \mu)\beta) + 1]^{-1} \tag{5.4}$$

with $\beta = {}^1/kT$ and μ the chemical potential which for $T \to 0^o$ is the Fermi energy E_F. The density of states for a certain energy $E_{k\sigma}$ is indicated by $N(E_{k\sigma})$. The total number of electrons is then given by

$$N_\sigma = \frac{1}{2} \int_0^\infty dE_{k\sigma} \, N(E_{K\sigma}) \, n_{K\sigma} \tag{5.5}$$

For simplicity we will assume that the conduction electrons can be described as a non-interacting electron gas. From this model we obtain [35].

$$\text{Fermi energy:} \quad E_F = \frac{\hbar^2 k_F^2}{2m} = \frac{\hbar^2}{2m} (3\pi^2 n)^{2/3} \tag{5.6}$$

Fermi momentum: $$K_F = \sqrt[3]{\frac{9\pi}{4}} \; \frac{1}{r_s \, a_B} \tag{5.7}$$

a_B = Bohr radius, n = density of electrons
and

$$\frac{1}{n} = \frac{4\pi}{3} \, r_s^{\,3} \, a_B^{\,3} \tag{5.8}$$

r_s is the radius in units of the Bohr radius of a volume that contains just one electron.

Density of state: $$N(E_k) = \frac{(2m)^{3/2}}{2\pi^2\hbar^3} \sqrt{E_k} \tag{5.9}$$

with $$E_k = \frac{\hbar^2 k^2}{2m} \tag{5.10}$$

Next we consider the effect of a magnetic field on the electron gas. The magnetic field is supposed to be in the Z-direction. According to the two spin states of the electrons, we have energies of

$$\begin{aligned} E_{k,+1} &= E(k) - \mu_B \cdot B \\ E_{k,-1} &= E(k) + \mu_B \cdot B \end{aligned} \tag{5.11}$$

Fig. 20 shows the energy bands for the spin up and spin down states. The spin up band is lowered by the energy μ_B B while the spin down band is raised by the same energy. Both bands are filled up to the Fermi energy, E_F. Fig. 20 shows immediately that the external field leads to a redistribution of populations of the two spin states, favoring the spin up configuration. The difference of the total number of electrons with spin up and spin down is:

$$\begin{aligned} N_+ - N_- &= \frac{1}{2} \int_0^\infty dE_k \; N(E_k) \; (n_{k,+1} - n_{k,-1}) \\ &= \mu_B \;\; B \cdot N(\mu \simeq E_F) \end{aligned} \tag{5.12}$$

One obtains a magnetization of

$$M = \mu_B^{\,2} \; B \; N(E_F) \tag{5.13}$$

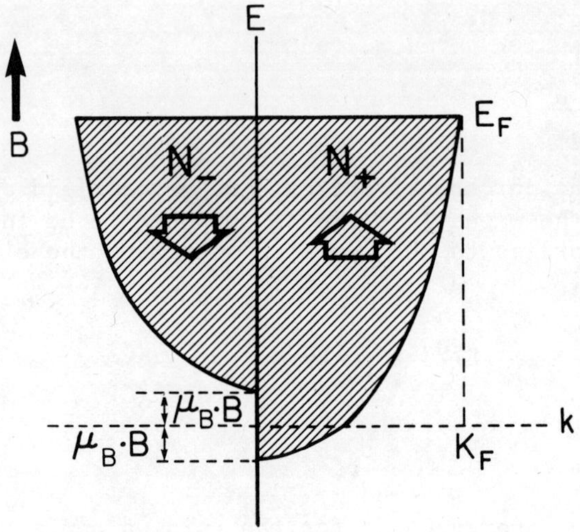

Fig. 20. Schematic plot of the relative shift of the spin up and spin down bands due to an external magnetic field in the free electron gas model.

From this equation follows the expression for the Pauli electron spin susceptibility

$$\chi = \frac{M}{B} = \mu_B^{\;2}\, N(E_F) = \mu_B^{\;2}\, \frac{mk_F}{\pi^2\hbar^2}\, V \qquad (5.14)$$

with V the total volume of the sample. This phenomenon is also known as Pauli paramagnetism.

We now return to eq. (5.1). Including an external field B_Z, we have to consider the Hamiltonian

$$\begin{aligned} H &= H_c + \gamma_\mu\, \hbar\, \vec{S}_\mu . \vec{B} \\ &= \gamma_\mu\, \hbar\, \vec{S}_\mu\, \Big(\vec{B} + \frac{8\pi}{3}\gamma_e\, \hbar \sum_i \vec{S}_{e,i}\, \delta(\vec{r}_i)\Big) \end{aligned} \qquad (5.15)$$

The second term acts like an additional field which will lead to a shift of the muon precession frequency. This shift, which is of course also manifested in NMR-experiments, is called the Knight-shift [36].

Neglecting relaxation processes, we are only interested in the static part of the additional field which is given by the average

$$\begin{aligned} \Delta B_z &= \frac{8\pi}{3}\, \gamma_e\, \hbar \,\Big\langle \sum_i S^z_{e,i}\, \delta(\vec{r}_i)\Big\rangle \\ &= \frac{8\pi}{3}\, \mu_B \Big\langle \sum_{k\sigma} |U\;(0)|^2 \langle \chi_\sigma | S^z | \chi_\sigma\rangle \Big\rangle \\ &= \frac{8\pi}{3}\, \frac{\mu_B}{2} \sum_{k=} U_k^2(0)\; (n_{k,+1} - n_{k,-1}) \\ &= \frac{8\pi}{3}\, \frac{\mu_B}{2} \int_0^\infty dE\; |U\;(0)|^2 N(E)\,(n_{+1}(E) - n_{-1}(E)) \end{aligned} \qquad (5.16)$$

$n_{+1}(E) - n_{-1}(E)$ will only be different from zero in the vicinity of the Fermi energy. Hence,

$$\Delta B_z = \frac{8\pi}{3} \langle |U_k(0)|^2\rangle_F\;\; \chi\; B_z \qquad (5.17)$$

where χ is the earlier defined Pauli spin susceptibility, and

$\langle\ \rangle_F$ denotes the average taken over all states at the Fermi level.

The Knight shift constant is defined as

$$K = \frac{\Delta B}{B} = \frac{8\pi}{3} \langle |U_k(0)|^2 \rangle_F \, \chi \qquad (5.18)$$

In the free electron model we have $|U_k(0)|^2 = 1$.

If the conduction electrons are spin polarized, as in ferromagnetic metals, they likewise produce a magnetic field at the site of the muon, which we call a hyperfine field

$$\begin{aligned} B_{hf} &= \frac{8\pi}{3} \mu_B \left(N_{+1}(0) - N_{-1}(0)\right) \\ &= \frac{8\pi}{3} \mu_B \sum_k \{ |\psi_{k,+1}(0)|^2 n_{k,+1} - |\psi_{k,-1}|^2 n_{k,-1} \} \end{aligned} \qquad (5.19)$$

The very important question now arises; how much is the electron density at the muon site changed by the presence of the positive "impurity" charge as compared to the undisturbed case? As one may well imagine, the positive impurity charge will attract electrons and a screening cloud is formed around the muon. The knowledge of the enhancement of electron density at the muon site is a prerequisite for a meaningful analysis of internal fields measured by the muon in ferromagnetic metals.

The number in question is the ratio:

$$E(k) = \frac{|\psi_k(0)|^2}{|\psi_k{}^o(0)|^2} \qquad (5.20)$$

where $|\psi_k{}^o(0)|^2$ is the undisturbed electron density at the muon site. We will assume that only s-electrons are contributing to the screening cloud.

The problem of the electron distribution around a point charge is a rather complicated one and is essentially a many-body problem. To begin with, the ordinary linear response theory, eg. the Thomas-Fermi model, does not apply to the present problem as the large enhancement of electron density at the impurity site is in contradiction to the presuppositions of linear response theory. An adequate procedure would have to solve the following set of equations consistently [37]

1) The electron density is given by

$$n(r) = \sum_{i=n}^{N} |\psi_i(r)|^2 \tag{5.21}$$

2) An effective potential is defined as

$$V_{eff}(r) = -\frac{Ze}{r} + e\int d^3r' \frac{n(r')}{|r-r'|} + V_{xc}[n(r)] \tag{5.22}$$

The first term is the Coulomb potential introduced by the impurity (for the muon Z = 1), the second term is the potential originating from the electron charge distribution around the impurity, the third term accounts for exchange and correlation effects among the electrons and is a function of the density n(r), too. No simple expressions exist for this term, and the various calculations use more or less approximate expressions for it [38].

3) The single particle wave functions $\psi_i(r)$ are finally obtained from the Schrödinger equation using the effective potential $V_{eff}(r)$

$$\left(-\frac{\hbar^2}{2m}\Delta + V_{eff}(r)\right)\psi_i(r) = \varepsilon_i\,\psi_i(r) \tag{5.23}$$

We do not want to discuss here the various approaches to a non-linear treatment of the screening problem, but would instead like to present a quite recent calculation by P.F. Meier, who refers to scattering theory for solving eq. (5.23), adopting an effective potential that has been used in nucleon-nucleon scattering - namely the Hulthen potential [39,98].

$$V_{eff}(r) = -Ze^2 \frac{\lambda}{e^{\lambda r}-1} \tag{5.24}$$

His results agree quite well with more elaborate calculations and have the advantage of being simple and instructive at the same time [40].

The potential introduced by the impurity, e.g. the muon, acts as a scattering center on the conduction electrons that are approaching the potential with momentum k. On the average, electron charge will be displaced towards the impurity charge by an amount equal to the impurity charge, thus screening the impurity charge completely for distances larger than a certain distance $1/\lambda$, the screening length.

The total displaced charge is now connected with the phase shifts of scattered particle waves by the Friedel sum rule [41].

$$Z = \frac{2}{\pi} \sum_{\ell=0}^{\infty} (2\ell + 1)\, \eta_\ell(k_F) \tag{5.25}$$

This fundamental sum rule can be obtained as follows [42].

For simplicity, we again consider a free electron gas. The scattering potential is assumed to be spherically symmetric: V(r).

From standard collision theory we know that a particle scattered by the potential V(r) can be described by the wave function

$$U(r,\theta) = \sum_{\ell=0}^{\infty} \frac{\phi_\ell(r)}{r} P_\ell (\cos\theta) \tag{5.26}$$

where P_ℓ is a Legendre polynomial and $\phi_\ell(r)$ is a solution of the differential equation

$$\frac{d^2\phi_\ell(k,r)}{dr^2} + \left[k^2 - U(r) - \frac{\ell(\ell+1)}{r^2}\right] \phi_\ell(k,r) = 0 \tag{5.27}$$

where U(r) = 2 m V(r) and m the mass of the scattered particle, e.g. the electron mass. The asymptotic behaviour of $\phi_\ell(k,r)$ is

$$r \to 0 \qquad \phi_\ell(k,r) \to 0$$

$$r \to \infty \qquad \phi_\ell(k,r) \to \left(\frac{1}{2\pi R}\right)^{\frac{1}{2}} \sin\left(kr + \eta_\ell(k) - \tfrac{1}{2}\ell\pi\right) \tag{5.28}$$

where $\eta_\ell(u)$ is the phase shift produced by the scattering potential. The factor $\left(\frac{1}{2\pi R}\right)^{\frac{1}{2}}$ is chosen to normalize $\frac{\phi_\ell}{r}$ in a sphere of a large radius R.

In the absence of the scattering potential we would have the asymptotic wave function

$$r \to \infty \qquad \phi_\ell^{o}(k,r) \to \left(\frac{1}{2\pi R}\right)^{\frac{1}{2}} \sin\left(kr - \tfrac{1}{2}\ell\pi\right) \tag{5.29}$$

The change in the number of particles in the state (k,ℓ) inside the sphere of radius R is

$$\Delta n(R) = 4\pi \int_0^R dr\, \left[\phi_\ell^{2}(k,r) - \phi_\ell^{o2}(k,r)\right] \tag{5.30}$$

The integral $\int \phi_\ell^2 \, dr$ can be expressed, using eq. (5.27), as [42]

$$\int \phi_\ell^2 \, dr = \frac{1}{2k} \left[\frac{d\phi_\ell}{dk} \frac{d\phi_\ell}{dr} - \phi_\ell \frac{d^2\phi_\ell}{dkdr} \right] \tag{5.31}$$

Using the asymptotic expressions of ϕ_ℓ and ϕ_ℓ^{o} for large R we calculate

$$\Delta n(R) = \frac{1}{R} \left[\frac{d\eta_\ell}{dk} - \frac{1}{k} \sin \eta_\ell \cos(2kR + \eta_\ell - \ell\pi) \right] \tag{5.32}$$

To obtain the total number of particles displaced by the potential into the sphere of radius R, we have to sum over all states described by k and ℓ up to the Fermi momentum. The density of states k,ℓ per unit wave number inside the large sphere with radius R is $2(2\ell+1)R/\pi$. The first factor 2 stems from counting both spin directions, the second factor describes the number of states with angular momentum ℓ and the third factor is the number of states for a given ℓ and m_L per unit wave number. The last one arises from the quantization condition for k, which implies that the allowed values of k differ asymptotically by $k = \pi/R$.

We get

$$\Delta N(R) = \frac{2}{\pi} \sum_\ell (2\ell+1) \int_0^{k_F} dk \left[\frac{d\eta_\ell}{dk} - \frac{1}{k} \sin \eta_\ell \cos(2kR+\eta_\ell-\ell\pi) \right]$$

$$= \frac{2}{\pi} \sum_\ell (2\ell+1)\eta_\ell(k_F) + \text{oscillatory terms} \tag{5.33}$$

If the impurity has a charge of +Ze relative to that of the lattice nuclei, one expects that negative charge will be piled up at the impurity until the impurity is completely screened. This leads to the effective potential $V(r) = V_{eff}$ which in turn determines the screening mechanism. Neglecting the oscillatory terms, whose average is zero, we thus obtain the self consistency condition

$$Z = \frac{2}{\pi} \sum_\ell (2\ell+1)\eta_\ell(k_F) \tag{5.34}$$

i.e. the Friedel sum rule.

The sum rule eq. (5.25) puts rather stringent conditions on the behaviour of V_{eff} near $r = 0$, the region which we are mainly interested in. It will be used to determine the value of λ occurring in the expression of the Hulthen potential.

The important quantity to be calculated is, as already stated, the ratio

$$E(k) = \frac{|\psi_k(0)|^2}{|\psi_k^o(0)|^2}$$

i.e. the enhancement factor of the charge density at the muon site, which involves only s-electrons.

As is well known from potential scattering theory, this enhancement is related to the Jost function $f_0(k)$ for $\ell = 0$ and $r = 0$ as follows [43,44]

$$E(k) = [f_0(k)]^{-2} \tag{5.35}$$

The Jost function is a particular solution of eq. (5.27) defined by the boundary condition

$$\lim_{r\to\infty} e^{ikr} f(r,k) = 1 \tag{5.36}$$

For the derivation of eq. (5.35) see ref. [44].

Another important property of f(k) is the occurrence of bound states for zeros of f(k) on the imaginary k-axis.

Using the Hulthen potential, the Jost function can be calculated analytically for s-waves [44]

$$f_0(k) = \prod_{n=1}^{\infty} \frac{k + i\,(\frac{1}{an} - \frac{n\lambda}{2})}{k - i\,\frac{n\lambda}{\mu^2}} \tag{5.37}$$

with $a = Z\,\frac{e^2}{\hbar^2\mu}$, μ is the reduced mass of the electron and the impurity. It is immediately seen from eq. (5.37) that the Hulthen potential has no bound states for $\lambda > 2/a$.

The enhancement factor is now obtained in closed form [40]

$$E(k) = \frac{2\pi}{ka} \frac{\mathrm{Sinh}(\frac{2\pi k}{\lambda})}{\mathrm{Cosh}(\frac{2\pi k}{\lambda}) - \cos\left(\frac{2\pi k}{\lambda}\sqrt{\frac{2\lambda}{ak^2} - 1}\right)} \tag{5.38}$$

In order to determine the parameter λ from the Friedel sum rule one needs to calculate the higher order phase shifts. This has been done numerically, using the variable phase function approach [45]. Table 4 lists a number of λ- values for $Z = 1$, $a = 1.058\ a_B$ (Bohr-radius of muonium) and various electron densities (represented by r_s see eq. (5.8)).

Finally, using eqs. (5.38), the Knight shift constant for muons was calculated as a function of average electron density shown in Fig. 21. The upper solid curve was calculated by using the free electron Pauli paramagnetic susceptibility χ_p alone (eq. (5.18)), whereas the lower one was calculated by including the diamagnetic (Landau) contribution $\chi_L = -1/3\chi_p$. Also shown in Fig. 21 are the measured shifts $\frac{\Delta B}{B}$ for muon precession in some metals as obtained by Hutchinson et al. [46]. The Meier model gives the correct order of magnitude with the exception of Li.

On the other hand, the experimental numbers have to be taken with caution, as these experiments were not specifically designed to measure Knight shifts.

In spite of the crudeness of the model used (free electron gas) the qualitative agreement nevertheless should be considered a success.

An improved theory has to take into account the actual band structure of the conduction electron states, as well as the ionic-structure of the host lattice. Very precise Knight shift measurements are most likely to produce highly valuable information that can be used to check on more advanced many body calculations and the intricate effects involved therein. Compared with Knight shift measurements by NMR, it should be stressed again that the absence of any electron core makes the "bare" muon an ideal probe to study the properties of a many electron system in a solid.

For $\lambda < \lambda_c = 2/a$ and $n = 1$, the Jost function has a zero on the negative imaginary axis and the potential has a bound state with a

Table 4: Values of λ ($10^8\ cm^{-1}$) for various electron densities

r_s	2	3	4	5	6
λ	2.42	2.07	1.90	1.80	1.75

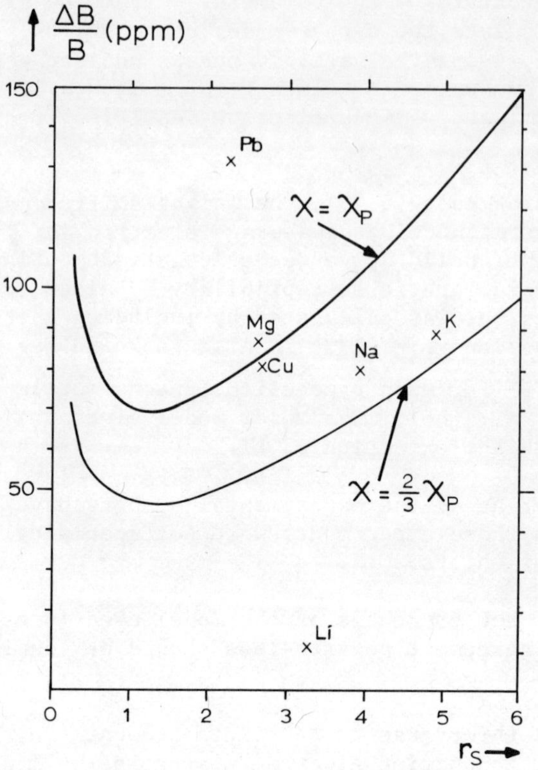

Fig. 21. Knight shift of the μ^+ calculated from eq. (5.38) with $\chi = \chi p$ and $\chi = 2/3\ \chi p$. The crosses are measurements of Hutchinson et al. [46].

binding energy $E_1 = -\frac{\hbar^2}{4\mu}(\lambda_c - \lambda)^2$. Table 4 shows that this can occur for $T_s > 2$ for a muon impurity. The question is whether one may really observe bound states in a metal. Since the binding energy is in any case much smaller than for muonium in vacuum, one expects to have a system with a large radius. The bound electron would extend over many lattice sites and the ionic cores would play an important role probably making a bound state impossible. This possibility in any case needs a careful check, particularly in metals with a very low conduction electron density.

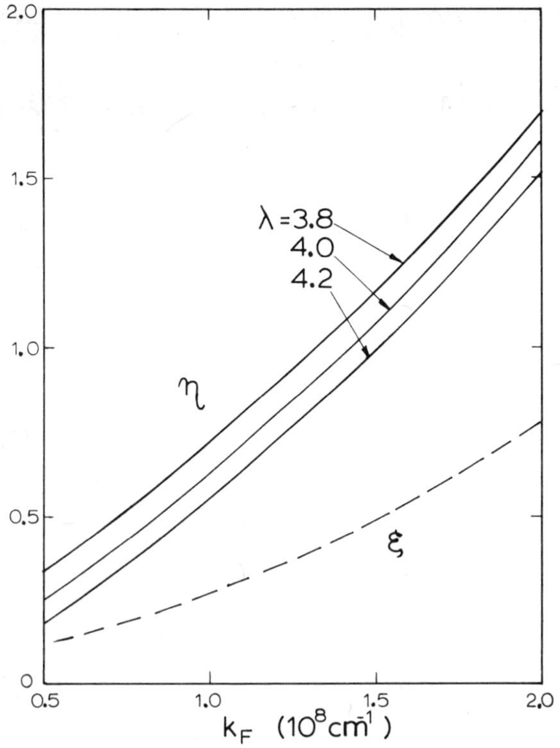

Fig. 22. Full curves: contact density $\eta = \frac{n(0)}{\eta_{Mu}}$ versus k_F; dashed curves $\xi = E(k_F) \cdot \frac{\eta_o}{\eta_{Mu}}$ versus k_F.

For later use, we present in Fig. 22 the total contact density expressed in units of the contact density in free muonium $\eta = \frac{n(0)}{n_{Mu}}$ and the contact density of the electrons at the Fermi level, also in units of the contact density of free muonium $\xi = \frac{n_0}{n_{Mu}} \cdot E(k_F)$ versus the Fermi wave vector $k_F = (3\pi^2 n_0)^{1/3}$. The total density n(0) is calculated from

$$\frac{n(0)}{n_0} = \frac{3 \int_0^{k_F} dk\, k^2\, E(k)}{k_F{}^3}$$

VI. MUON PRECESSION STUDIES IN FERROMAGNETIC METALS

We are now going to discuss that area of μSR-application that has attracted most of the attention and will perhaps be the most interesting for some time to come. It is also the area where the advantages and disadvantages of the μSR-method are most visibly rivaling each other. The disadvantages concern the question of location or dislocation (i.e. diffusion behaviour) and the screening properties of the implanted muons. We have discussed these two topics in the preceeding chapters in some detail, and it may not be too unrealistic to expect that a sufficient practical and theoretical understanding of these will be developed in the near future.

The advantages rest again in the property of the "bareness" of the muon and the absence of any electron core effects that change, mask or complicate the original ferromagnetic structure which one wishes to investigate.

The nature of ferromagnetism or any other magnetic order in a solid is quite an old subject the more it is surprising when one becomes aware that we are still far away from a consistent, quantitative and complete understanding of the involved mechanisms. This is particularly true for the 3d-transition metals like iron, cobalt and nickel; and also the 4f-magnetic metals like gadolinium etc. which are still investigated in many laboratories, not to speak of alloys and amorphous systems.

In any case, the magnetic order is a consequence of the particular electronic structure of the metal in question.

The approach to the study of magnetism on an atomic scale proceeds from many directions: study of the electronic (band) structures, including the polarization of electrons as a function of energy (Photoelectron spectroscopy), studies of the distribution of magnetization by neutron diffraction, study of hyperfine interactions with test probes (Mossbauer effect, NMR, γγ-PAC etc. and μSR).

Parallel to the wealth of experimental methods, an equal wealth of theoretical ideas is aimed at a thorough understanding of magnetic order. The situation now is that on one hand the experimental information from the various methods, despite all the work that has been invested so far, is still insufficient, incomplete, contradictory, and ambiguous, and on the other hand agreement with theory may be good in one instance while failing completely in another.

Of prime interest, of course, are the mechanisms that establish a magnetic order. In principle the responsible interactions are all of the nature of a spin dependent exchange interaction, and thus lead to a splitting of the energy bands and a redistribution in the population of spin down and spin up states in just the same manner as described for the Knight shift. The role of the external field is now taken by the exchange interaction. What makes a detailed understanding so difficult is that we are dealing with a many body problem.

Of particular interest is the role that the conduction electrons may play in the formation of magnetic order. As is well known, a local magnetic moment in a metal will polarize the spins of the conduction electrons by the famous RKKY (Rudermann, Kittel, Kasuya, Yosida) interaction [47] that will lead to spin density oscillations as one goes away from the local magnetic moment. A second local moment will likewise interact with the conduction electrons, which in effect will lead to an indirect coupling between the two local magnetic moments or spins. By this, magnetic order can be established. The details of this indirect interaction will determine e.g. the type of magnetic order, ferromagnetic or antiferromagnetic.

Information, concerning the spin polarization of conduction electrons, is now particularly ambiguous - not only with regard to the average magnitude but also with regard to the overall sign, not to mention the details of the spatial distribution. The conduction electron polarization at normal lattice sites has been investigated by the Mössbauer effect, NMR and $\gamma\gamma$ PAC. The hyperfine field due to the conduction electrons is however only a very small fraction of the hyperfine field acting on the nuclear probes. The main contributions stem from the spin polarized electron core (core polarization CP) and from possible unquenched orbital moments. As these contributions cannot be calculated reliably, it is very doubtful how much confidence one can have in quoted conduction electron polarization values extracted from measured hyperfine fields. [48]

From neutron diffraction measurements one can in principle obtain information on the magnetization in the interstitial region, which may be ascribed to the conduction electrons. The spatial distribution of magnetization is thereby obtained from a Fourier transformation of the measured magnetic scattering amplitude or form factor [49]. This procedure is not model independent. Also, the diffusive scattering from the interstitial magnetization involves only small momentum transfers, i.e. small scattering angles, which are difficult to measure.

The positive muon on the other hand - if we assume for the moment that it will not diffuse - is expected to reside in the available interstitial sites, which guarantees a minimum in potential energy. Trapping in vacant regular lattice sites or other dislocations may also be possible, but it has not yet been found of relevance. For the time being we will exclude this possibility from our considerations. Although it may in fact be a very important one.

Via the contact interaction the muon will monitor the conduction electron polarization at the interstitial site. There is of course the screening problem and we will come back to this when we discuss some results. Except for possible dipole contributions from the localized magnetic moments, which can however be calculated, the contact interaction with the conduction electrons is the only other internal field acting on the muon. It is therefore hoped that the muon will provide some unambiguous information on the conduction electron polarization at least in the interstitial region. In connection with the discussion of the data it will also become clear that fast diffusion of the muon may not necessarily pose a serious problem, provided that all temporary sites are of the same type.

The total local field that the muon will interact with at its site consists of the following contributions:

$$\vec{B}_{loc} = \vec{B}_{ext} - \vec{B}_{DM} + \vec{B}_{L} + \vec{B}_{dip} + \vec{B}_{hf} \tag{6.1}$$

with

$\vec{B}_{ext}$ = external magnetic field

$\vec{B}_{DM}$ = demagnetization field = $(N)\vec{M}_s(T)$ depends on the geometry of the target sample

$\vec{B}_L$ = Lorentz field = $\frac{4\pi}{3}\vec{M}_s(T)$, with $\vec{M}_s(T)$ = saturation magnetization of the sample metal

$\vec{B}_{dip} = \sum \vec{b}_i$ = sum of dipole fields $\vec{b}_i$ from sources in the neighbourhood of the muon (inside the Lorentz sphere)

$\vec{B}_{hf}$ = hyperfine field or contact field due to conduction electrons. If bound states would exist there could be further contributions to $\vec{B}_{hf}$.

By writing the total field this way, one has taken into account the microscopic origin of the fields. Sources far away from the muon, i.e. outside a fictive sphere of suitable radius (the Lorentz sphere) which is centered at the muon site, are contributing some average field $\vec{B}_{ext} - \vec{B}_{DM} + \vec{B}_L$. The Lorentz field $\vec{B}_L$ originates from the magnetization of the surface of the Lorentz sphere. Sources inside the Lorentz sphere are considered individually and lead to the contribution $\vec{B}_{dip} + \vec{B}_{hf}$.

We now turn to the discussion of the μSR-measurements performed in ferromagnetic metals so far. Since the pioneering work of Kossler and collaborators in Ni and Fe [50], practically all other groups active in μSR have taken up this subject. The Crowe group at LBL-Berkeley has reported measurements with a single crystal of Ni [51] and some measurements in poly and single crystal Fe [52], the Gurevich group at Dubna has performed measurements in Ni, Fe and Co [53], later Kossler and a group from Bell Labs reported on measurements in Dy [54] and recently the Russians came up with results obtained in Gd [55]. Preliminary results on Fe Pd alloys from the Yamazaki group [57] and spin glasses from the Kossler group are also available, but will not be discussed here. [56]

The measurements generally yield three types of information, according to the formula

$$N_{e^+}(t) \sim \frac{1}{\tau_\mu} \exp\left(-\frac{t}{\tau_\mu}\right) \left[1 + P \cdot a \cdot e^{-t/\tau} \cos(\omega_\mu t + \phi)\right] \qquad (6.2)$$

1) the average polarization of the precessing muon P usually normalized to the polarization P_c obtained in a carbon target.

2) the damping time τ (slow depolarization)

3) the precession frequency $\omega_\mu = \gamma_\mu |\vec{B}|$ and from the phase the sense of the precession, i.e. the direction of $\vec{B}$

To give you an impression how the data look, Fig. 23 and 24 show the rate histograms as obtained by Kossler et al [74], and Gurevich et al. [53] in ferromagnetic Ni.

We will now present the results on the local fields as obtained directly from the measured precession frequencies:

1.) Ni Fig. 25 shows the local field B plotted versus temperature. The distribution follows the known temperature dependence of the magnetization in Ni, only properly scaled (Brillouin curve). For T = 0° K one obtains a saturation field of about B(T = 0° K) ≃ 1480 gauss. Fig. 25 contains all data from refs. [50,51]. The Russian data point at room temperature fits into this description [53].

Very peculiar at first sight is the observation that $\vec{B}_{loc}$ is essentially independent of the external field below a certain value [50,51,53]. According to Patterson [51] this is due to the high premeability of Ni. Below saturation, magnetic shielding prevents the external field from penetrating the sample. This is due to the movement of the domain walls until they reach their limiting positions, where the saturation is complete and no further shielding

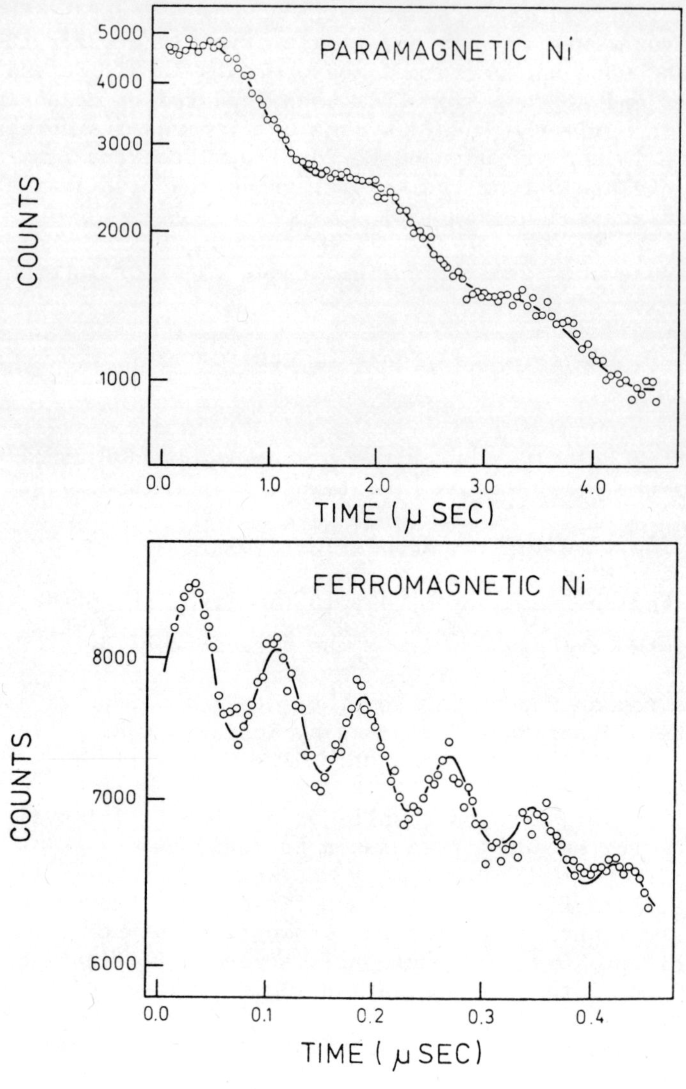

Fig. 23. Muon precession signal in paramagnetic and ferromagnetic Ni [74].

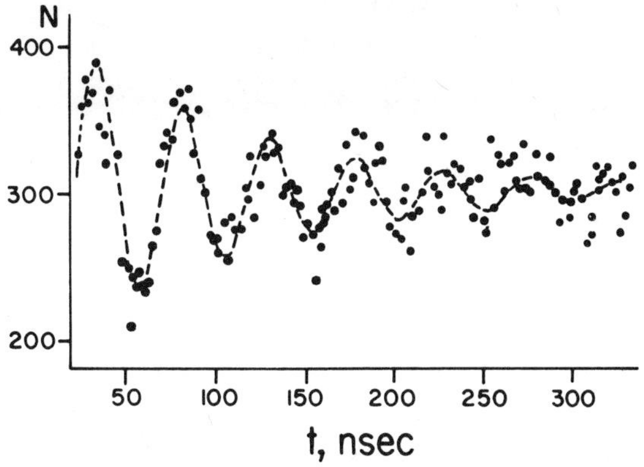

Fig. 24. Muon precession signal in ferromagnetic Ni. The exponential decay due to the muon lifetime has been divided out [53].

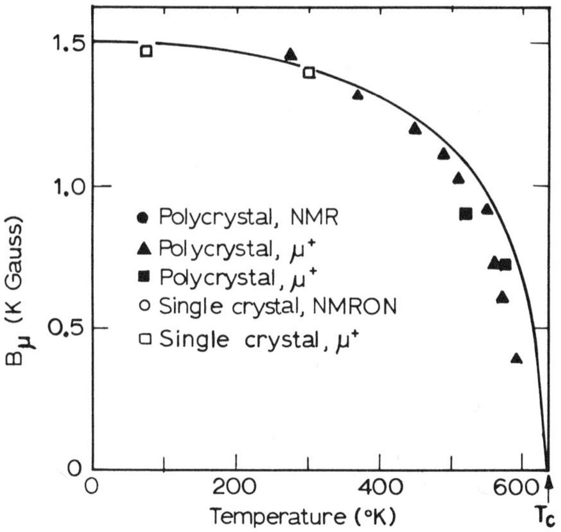

Fig. 25. Temperature dependence of the local magnetic field at the muon in nickel. The data are well represented by a Brillouin function with a saturation field at T = 0°K of 1480 gauss [50,51].

is possible. Patterson has investigated this effect in some more detail, using as a target an ellipsoid (see Fig. 26). As long as the external field is smaller than $(N)\cdot M_s(T)$ (M_s = saturation magnetization at temperature T), the internal field stays independent of the external field. However, as soon as the external field is larger than $(N)\cdot M_s(T)$, the internal field rises linearly with the external field. The data shown in Fig. 26 are obtained from two different crystal orientations with respect to the external field, involving two different demagnetization factors (N). From the observed sense of the spin rotation it was deduced that the local field $\vec{B}_{loc}$ had the same direction as the external field $\vec{B}_{ext}$. This is in agreement with the linear rise of $\vec{B}_{loc}$ with $\vec{B}_{ext}$ above $(N)\cdot M_s(T)$, which shows that the internal hyperfine field and the penetrating external field are pointing in the same direction.

2.) Fe Data on muon precession in ferromagnetic Fe are not as complete and rich as in Ni. Kossler et al. [50] measured the temperature dependence of $|\vec{B}_{loc}|$ in the temperature range room-temperature to 675°K from which they estimated a saturation field of B_{loc} (0°K) = 4.2 kGauss. This is only in marginal agreement with the results of Gurevich et al. [53], that gives a value of $\vec{B}_{loc} \cong 3500$ gauss at room temperature. Patterson [52] measured the internal field at room temperature in a polycrystal sample (B_{loc} = 3.46 kG) and in a single crystal (B_{loc} = 3.39 kGauss). This is in agreement with the measurements of Gurevich et al. Like in Ni, Gurevich et al. found that the internal field in Fe is independent of the external field up to 2 kGauss. There is some contradiction in the determination of the sense of the spin rotation. From the measurement of the phase of the spin precession with a positron telescope perpendicular to the beam axis it was inferred that $\vec{B}_{loc}$ has the same direction as $\vec{B}_{ext}$ [53]. A later measurement of the same authors [58], where they measured the dependence of the internal field on the external field (Fig. 27) at higher external fields, suggests on the contrary that $\vec{B}_{loc}$ is directed antiparallel to $\vec{B}_{ext}$. As can be seen from Fig. 27, $|\vec{B}_{loc}|$ decreases when $|\vec{B}_{ext}|$ is increased beyond about 2 kGauss.

3.) Co Only one measurement at room temperature and in zero external field has been reported [53]. The result is

$$|\vec{B}_{loc}\ (300°K)| = 858\ (\pm\ 60)\ \text{gauss}$$

No information on the sense of the spin rotation is available.

4.) Gd Gurevich et al. [55] have measured $|\vec{B}_{loc}|$ for various temperatures. The results are shown in Fig. 28. The data fell roughly on a Brillouin curve with a saturation field of B_{loc} (0°K) = 1760 gauss. The quite large scattering of the data suggests, however, that some more refined description is probably necessary. The

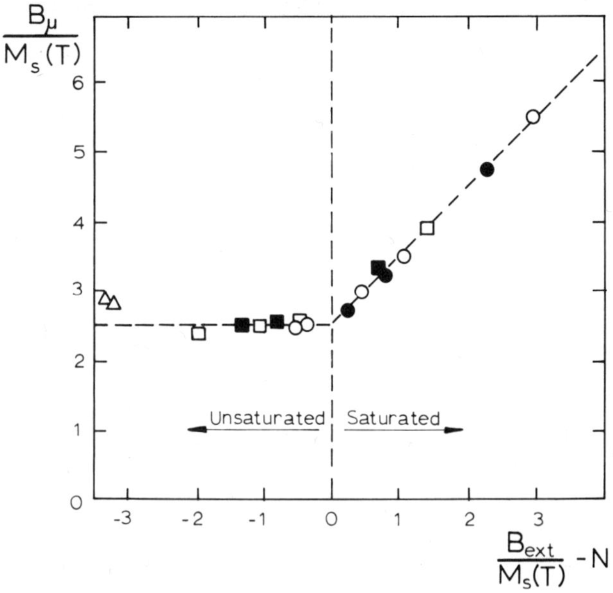

Fig. 26. Dependence of the local field B_{loc} on the external field in ferromagnetic Ni [51].

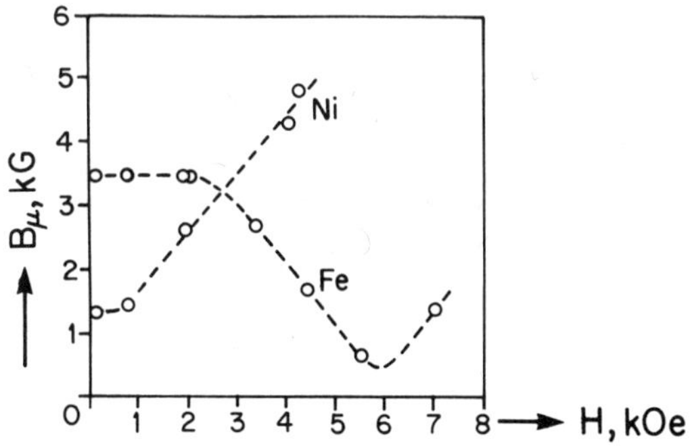

Fig. 27. Dependence of the local magnetic field B_{loc} on the external field in ferromagnetic Fe (and Ni) [58].

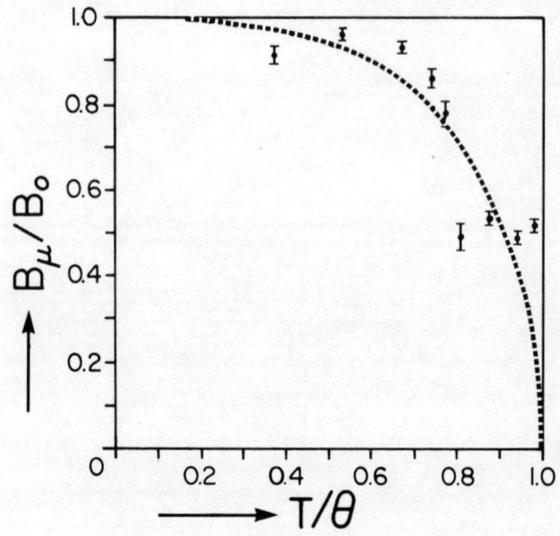

Fig. 28. Temperature dependence of the local magnetic field B_{loc} at the muon in Gd [55]. The data points roughly follow a Brillouin function with a saturation field 0^oK of 1760 gauss.

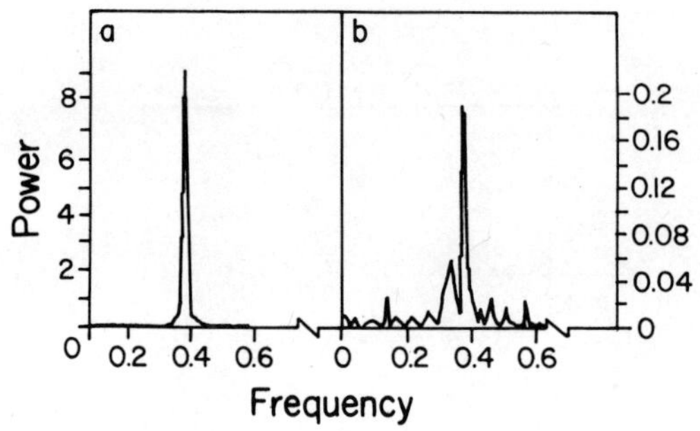

Fig. 29. Fourier spectrum of the muon precession frequencies in paramagnetic (a) and ferromagnetic (b) Dysprosium [54].

dip near T = 273°K is also related to a strange minimum in the polarization P. The authors of [55] are speaking of the possibility of a phase transition. From the dependence of the local field on the external field: (increases with increasing external field) one can infer again that $\vec{B}_{loc}$ and $\vec{B}_{ext}$ are directed parallel to each other.

<u>5.) Dy</u> Only some very sketchy information is available from measurements by Kossler et al. [54]. Fig. 29a shows a Fourier spectrum of the rate histogram at a temperature above the Curie temperature. Only one line is visible corresponding to a frequency determined by the external magnetic field. Fig. 29b shows the Fourier spectrum at T = 77°K in the ferromagnetic region. Several lines seem to be present now, not too far away however from the one line in the ferromagnetic region. Further detailed measurements are necessary before any meaningful conclusions can be drawn. In any case, Fig. 29b gives an impression of the wealth of phenomena that are to be expected in the course of future μSR-work.

Table 5 summarizes the results discussed so far and gives information on some relevant constants.

<u>Discussion and results</u> From the independence of the local field from the external field at low external fields for Ni and Fe we conclude that

$$B_{ext} - (N)\vec{M}_s(T) \cong 0 \qquad (6.3)$$

and it follows immediately that

$$\vec{B}_{hf} + \vec{B}_{dip} = \vec{B}_{loc} - \vec{B}_L \qquad (6.4)$$

The local field in cobalt has only been measured for $B_{ext} = 0$ and we assume that eq. (6.4) applies equally well. The fitted saturation field B_{loc} (0°K) = 1760 gauss in Gd is not too far away from the measured B_{loc} (93°K) = 1460 gauss in zero external field and we assume likewise the applicability of eq. (6.4). The Lorentz field can be calculated from the known macroscopic saturation magnetization constants and is listed for the investigated metals in Table 5. Applying eq. (6.4) we then calculate the sum $\vec{B}_{hf} + \vec{B}_{dip}$ whose values are also listed in Table 5, column 6.

We would like to know B_{hf} and the problem is to calculate B_{dip}. In order to calculate B_{dip}, we have to make an assumption as to where the muon will be located. In fcc nickel the available interstitial lattice sites are the octahedral and tetrahedral positions. Both have cubic symmetry with respect to the neighbor Ni cores (Fig. 16) and consequently $\vec{B}_{dip} = \sum_i \vec{b}_i$ will vanish. The octrahedral

Table 5. Experimental results for $\vec{B}_{loc}$ and $\vec{B}_{dip} + \vec{B}_{hf}$ in Ni, Co, Fe and Gd.

	T_c (°K)	crystal structure	$B_L = \frac{4\pi}{3} M_s$ (kGauss)	$\vec{B}_{loc}$ (0°K) [gauss]	$\vec{B}_{dip}+\vec{B}_{hf}$ [gauss]
Ni	630°	fcc	2.14*	+1480*	- 660
Fe	1045°	bcc	7.2**	(4100)** -3400	-10600
Co	1390°	hcp	5.9**	± 858**	-5040 or -6760
Gd	289°	hcp	8.4*	-1760*	-6640

* 0°K, ** room temperature

position is the more spacious one and it appears reasonable to locate the muon there (this is also suggested by neutron diffraction measurements in hydrogen charged Ni [59]. If the muon, however, is not localized, but rather diffuses through the crystal, it will probably jump from one interstitial octahedral site to the other. Assuming that the jump time itself is a very small ($\sim 10^{-12}$ sec) compared to the residence time in the interstitial positions, the dipole contributions will again vanish and we are left only with the hyperfine or contact field. Thus in the case of Ni one can be rather confident to have measured the pure contact field due to the conduction electrons in the assumed interstitial position.

In bcc-Fe the situation is much more complicated. Again we have interstitial sites in the octahedral and tetrahedral position (s. Fig. 30). However, for both positions we do not have a cubical symmetry with respect to the neighbor Fe-cores and the dipole fields consequently do not cancel. If the unit cell is oriented such that the [100] direction is parallel to the external field or the magnetization, we expect dipolar fields of +18.8 kGauss and -9.4 kGauss for the octahedral position and -5.2 kGauss and +2.6 kGauss for the tetrahedral position, respectively, depending on the orientation of the closest neighbor Fe-cores.

In a polycrystal sample with random orientations of the unit cell a huge field inhomogeneity is thus to be expected. If the muon were indeed localized in an interstitial site a very fast loss of phase coherence should happen, prohibiting any observation of a spin rotation. The observation of a single sharp precession frequency clearly contradicts this expectation. A way out of this dilemma is to assume that the muons diffuse very fast, jumping from one interstitial position to the next, hereby averaging over the dipolar

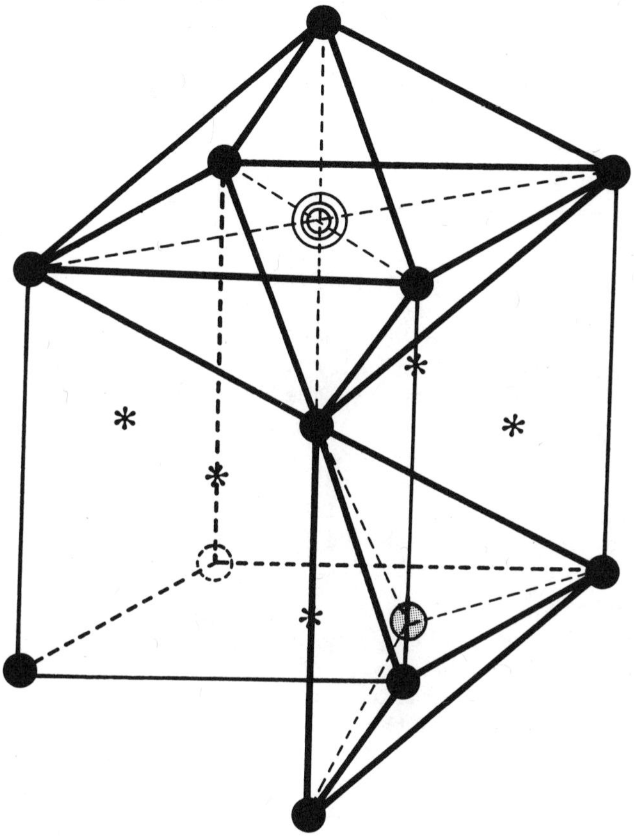

Fig. 30. Schematic structure of a body centered cubic (bcc) crystal. Indicated are the octahedral and tetrahedral interstitial positions.

field components by the motional narrowing mechanism. Indeed, there are twice as many sites with B_{dip} = -9.4 kGauss and +2.6 kGauss than with B_{dip} = +18.8 kGauss and -5.2 kGauss (see Fig. 30), so that a fast jumping will lead to an average of

$$\langle B_{dip} \rangle_{av.} = 2(-9.4 + 2.6) + (18.8 - 5.2) \simeq 0$$

It is known [60] that hydrogen diffusion in Fe at room temperature proceeds with a jump rate of about 10^9 sec^{-1}. The muon will probably diffuse even faster. If we take the average rms width of the field distribution in a polycrystal sample to be of the order of 1 kGauss, i.e. $\langle \Delta\omega^2 \rangle^{\frac{1}{2}} \simeq 10^8$ rad sec^{-1} we obtain the relation

$$\langle \Delta\omega^2 \rangle^{\frac{1}{2}} \tau_c \lesssim 0.1$$

which implies that at room or higher temperature the motional narrowing condition is indeed fulfilled. At lower temperatures, however, this may no longer be the case. And, indeed, Gurevich et al. [61] found a strong damping of the muon precession in Fe when decreasing the target temperature below $\sim$150°K (Fig. 31). Interestingly, as can be seen from Fig. 31, the damping rate in Ni stays independent of temperature, indicating that it does not matter whether the muon diffuses or not.

For the high temperature data it seems to be justified to assume that indeed $\langle B_{dip} \rangle_{av.} \cong 0$ and we interpret the measured internal field as originating only from the contact interaction with conduction electrons.

In hexagonal closed packed (hcp) Co and Gd two interstitial sites of local octahedral and tetrahedral symmetry are available (see Fig. 32). A slight deviation from cubic symmetry of the neighbor ion cores with respect to the interstitial positions leads to non-vanishing dipolar field contributions at the interstitial sites. For Co one calculates dipolar fields of -0.43 and +0.30 kGauss for the octrahedral and tetrahedral sites respectively, assuming a magnetization along the c-axis (easy axis of magnetization [52]). Corresponding numbers for Gd are not available. It appears that even a fast diffusion will not lead to a perfect averaging of the dipolar fields.

In their analysis Gurevich et al. [53,55] nevertheless assumed that B_{dip} can be neglected for both Co and Gd. In view of the smallness of the dipolar contribution involved, this assumption may not introduce too large a systematic error in determing B_{hf}. All values listed in Table 6, column 5 may thus be understood as originating solely from the contact interaction. Independent of some ambiguity in the determination of the sign of $\vec{B}_{loc}$ all extracted hyperfine fields carry a negative sign. This seems to indicate that the polarization of the conduction electrons in the considered interstitial sites is negative.

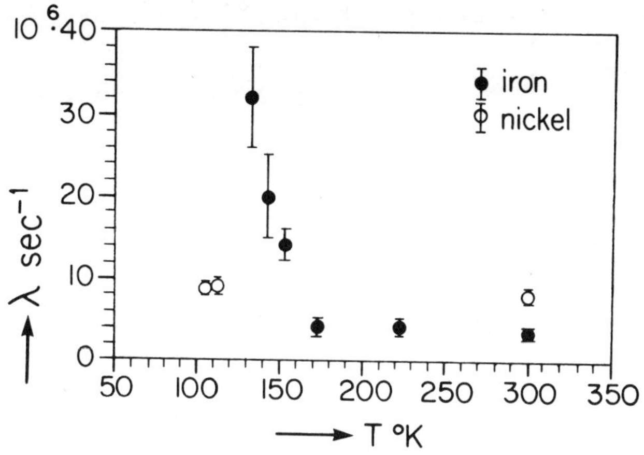

Fig. 31. Temperature dependence of the damping rate of the muon precession signal in ferromagnetic Fe and Ni [61].

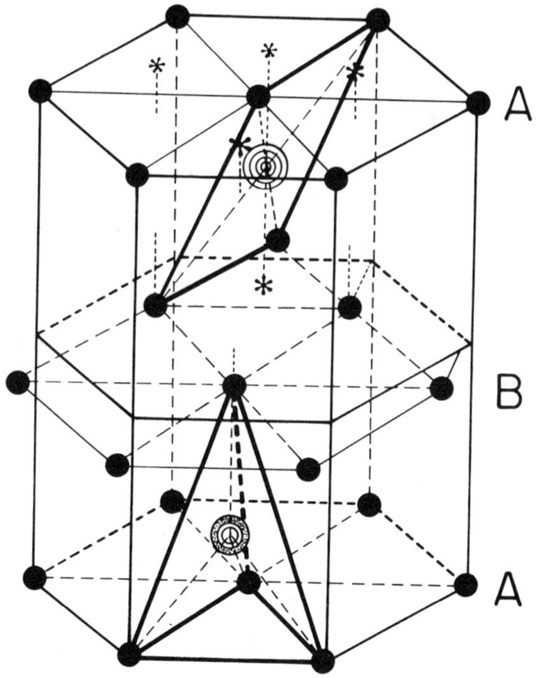

Fig. 32. Schematic structure of a hexagonal close-packed (hcp) crystal. Indicated are the tetrahedral and octahedral interstitial positions (◎ and ◎).

Table 6. Comparison of neutron and μSR data on interstitial magnetization.

	$\mu_B/Å^3$	M_{loc} (kGauss)	$B_{hf}=\frac{8\pi}{3} M_{loc}$ (kGauss)	$B_{hf,exp}$ (kGauss)	$\frac{\mu_{Ion}}{\mu_B}$
Ni	-0.0085	-0.079	-0.66	-0.66	0.6
Co	-0.12	-1.11	-9.30	-(5.0..6.7)	1.7
Fe	+0.106 -0.215	+1.0 -2.0	+8.38 -16.76	-10.6	2.2
Gd	-0.037	-0.343	-2.87	-6.6	7.1

Let us compare this result with the available neutron diffraction data on the magnetization in the interestitial region. Fig. 33 shows the magnetic moment distribution in the (100) plane in Ni, measured by Mook and Shull [62]. The magnetization in the interstitial volume is determined to be M_{loc} = 0.079 kGauss. The magnetization distribution in iron, determined also by Shull and Mook [63] is shown in Fig. 34. For the octahedral interstitial site one obtains

$$M_{loc}\ (\text{octa.}) = +\ 1.0(3)\ \text{kGauss}$$

and for the tetrahedral site

$$M_{loc}\ (\text{tetr.}) = -\ 2.0(4)\ \text{kGauss}$$

For Co Moon [64] has obtained a similar magnetization distribution as in Ni with a flat negative magnetization at and near the interstitial sites. The local magnetization is about

$$M_{loc} = -\ 1.11\ \text{kGauss}$$

For Gd the magnetization distribution was measured by Moon et al [65]. The projections of the spin density are shown in Fig. 35. For the neighborhood of the "c-site", which is the octahedral position, they obtained a magnetization of -0.343 kGauss.

The simplest comparison with the μSR-data is facilitated by assuming that the electron density at the considered interstitial sites is unchanged by the presence of the muon (no screening picture). The hyperfine field is then simply

$$\vec{B}_{hf} = \frac{8\pi}{3}\ \vec{M}_{loc}$$

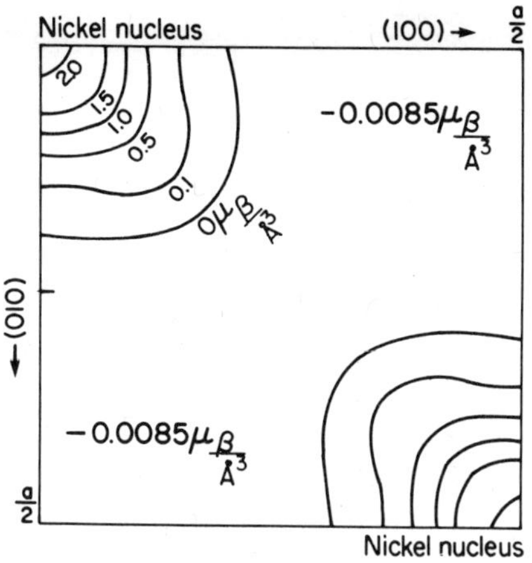

Fig. 33. Magnetic moment distribution in the [100] plane of Ni as obtained from neutron diffraction studies [62].

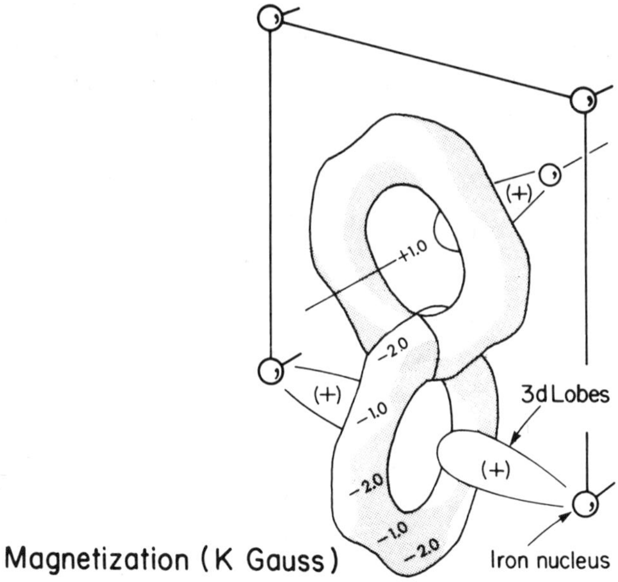

Fig. 34. Model of the magnetization distribution in Fe as suggested by neutron diffraction studies [63].

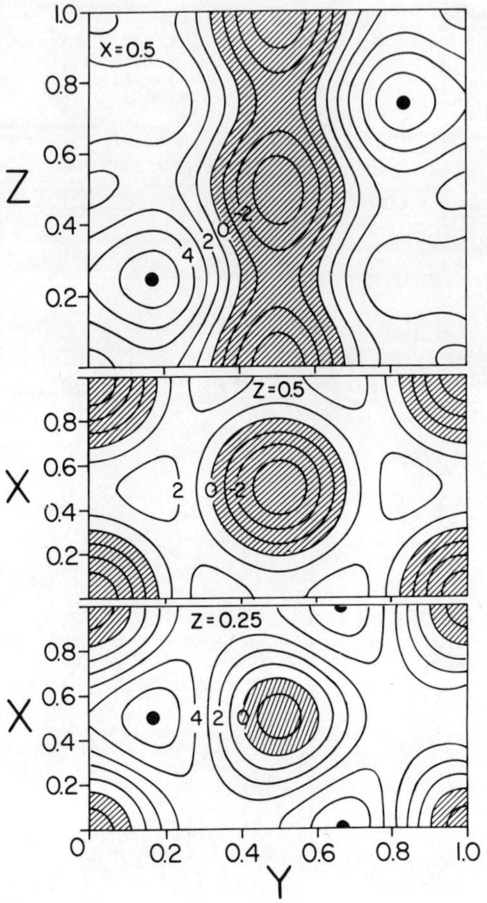

Fig 35. Magnetic moment distribution in Co projected on various planes in units of 0.01 $\mu_\beta/\mathring{A}^3$ as obtained from neutron diffraction studies [65]. The c-site has the position x = 0.5 and z = 0.5.

The results are shown in Tab. 6 together with the experimental hyperfine fields.

In the case of Ni one obtains perfect agreement, in the case of Fe and Co the numbers are also relatively close, while in Gd they are quite different. With the exception of Fe, the sign is determined by the two methods to be the same and always negative. This may be taken as evidence that in Fe, the muon is only sampling the magnetization in the tetrahedral interstitial positions, which is also the more spacious one. One also has to consider that due to a possible large vibration amplitude, the muon is in fact sampling the field over an extended range, which leads, in particular in Fe and Gd, to much smaller effective field values. Anyway, a comparison of the type just presented cannot be taken seriously. The negligence of screening effects is in no way justified and the coincidence of the numbers in the case of Ni has to be considered as fortuitous.

We turn now to the discussion of three attempts to interpret the μSR results in a more sophisticated way:

The first one is due to Patterson and Falicov [66], for Ni. A nickel atom has eight 3d- and two 4s-electrons. Their basic assumption is, that the 3d-electrons are well localized at the Ni-sites, while the 4s-electrons can be treated as forming a free electron gas. The total 4s-electron density is determined to be $n_o = 4.9\text{x}10^{22}$ cm^{-3}, assuming that each Ni-core contributes 0.6 electrons to the 4s-band. The 4s-electrons will then be responsible for the screening of the muon charge. The next assumption entering their theory is that the screening can be treated in linear response theory using the Thomas-Fermi model. The Thomas-Fermi screening length is

$$r_S = \left(\frac{E_F}{6\pi n_o e^2}\right)^{\frac{1}{2}} = 0.6\ \text{Å} \tag{6.5}$$

Since the muon-nearest neighbor nickel distance is 1.8 Å it seems to be justified to assume that only s-electrons are contributing to the screening. The enhancement of the electron density at the muon is now calculated, using the Lindhard expression for the free electron gas dielectric function [67]. Without going into the details of this calculation, we merely display the results in Fig. 36. Using the nickel conduction electron density as quoted above, one obtains from Fig. 36 a perturbed electron density of $n(o) = 5 \cdot n_o$ at the muon.

If we assume that the polarization of the conduction electrons in the screening cloud is the same as in the unperturbed case, as given by the neutron diffraction data, the hyperfine field should

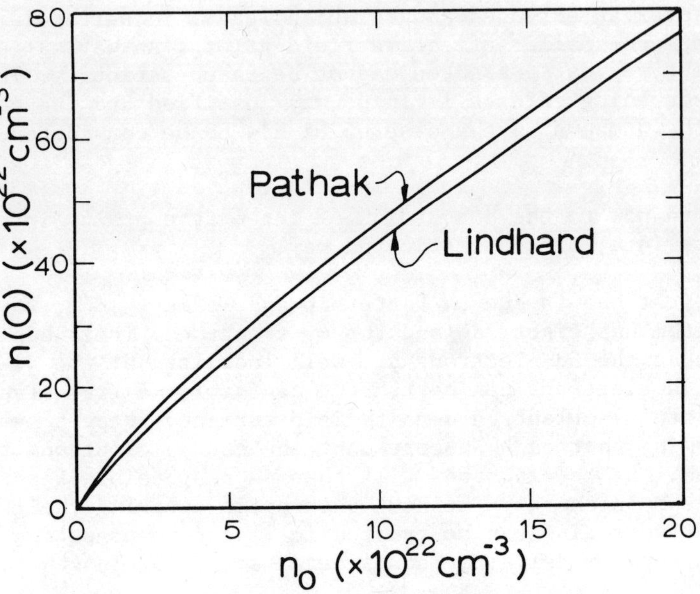

Fig. 36. Electron density dependence at the muon as a function of undisturbed electron density of a free electron gas in the linear response approximation. The calculation of Pathak includes electron-electron interactions.

have come out larger by a factor of 5. If one now believes that the neutron diffraction data are essentially correct, one has to find a mechanism by which the increase in charge density is accompanied by a decrease of the conduction electron polarization at the muon site. Patterson and Falicov tried to solve this problem by considering the total electron energy. For the unperturbed case this is

$$E_o = \frac{1}{V} \left(E_k + E_{sd} + E_{ss} + E_c\right) \tag{6.6}$$

with V - volume of the sample, E_k = kinetic energy, E_{sd} = s-d exchange energy, E_{ss} = s-s exchange energy, and E_c = correlation energy. The kinetic energy can be obtained by summing over all free electron states for the two spin directions up to the Fermi level

$$\begin{aligned} E_k &= \frac{1}{2} V \int d\varepsilon\, N(\varepsilon)\, \varepsilon \left(n^{\uparrow}(\varepsilon) + n^{\downarrow}(\varepsilon)\right) \\ &= \frac{3}{10} n_o E_F \left((1+\zeta)^{3/5} + (1-\zeta)^{3/5}\right) \end{aligned} \tag{6.7}$$

with $\zeta = (N\uparrow - N\downarrow)/n_o$

The s-d exchange energy term can be treated as a fictitious Zeeman interaction with an exchange field H_{sd}.

$$E_{sd} = \mu_B H_{sd} n_o \zeta$$

Neglecting the terms E_{ss} and E_c, H_{sd} can be determined by minimizing E_o with respect to ζ and setting the equilibrium value of the spin density equal to that given by the neutron diffraction results

$$\zeta_o \propto H_{sd} n_o^{-2/3} \quad \text{and} \quad \zeta_o = \frac{-0.85}{4.9} = -0.17$$

from which one obtains $H_{sd} \sim -10^8$ gauss. This field is of the order of normal Weiss fields.

The third and most critical assumption is now introduced by assuming that in the presence of the muon charge the magnetic and kinetic energy at some point x close to the muon will only depend on the charge and spin density at just this point. This is a local approximation.

It is immediately evident that this is certainly a reasonable approximation if the changes of n(x) and $\zeta(x)$ are slow over typical lattice distances. Due to the screening the charge density, however, changes drastically over distances of the screening length and the local approximation thus appears to be very badly justified. Nevertheless, the general idea of the approach of Patterson and Falicov appears to provide some intuitive understanding of how charge and spin density can be related to each other. In any case, the results that they obtained look very reasonable.

The local approximation implies that we can now write

$$E_o(x) = \frac{1}{V}(E_k(x) + E_{sd}(x) + E_{ss}(x) + E_c(x)) \qquad (6.8)$$

and $n_o \to n(x)$ $\zeta \to \zeta(x)$

As we are only interested in the density changes at x = o (muon site) we now minimize the total energy at x = o with respect to ζ. Introducing the relative spin and charge densities $\zeta_s = n(o)\ \zeta(o)/n_o\ \zeta_o$ and $\zeta_q = n(o)/n_o$ respectively, the result of the minimization procedure is shown in Fig. 37. The ss-exchange and correlation terms also have been taken into account according to ref. [66]. Fig. 37 shows that an increase of the charge density indeed leads to a decrease of the spin density, such that the product $n(o).\ \zeta(o)$ is almost constant in qualitative agreement with the Ni results. Besides the obvious shortcomings of the local approximation, the application of the Thomas-Fermi screening model as well as the treatment of the conduction electrons as a free electron gas is very questionable. In particular, the treatment of the screening in the framework of a linear response theory is probably not justified as the results of P.F. Meier do suggest. In the Meier-model the total charge density enhancement at the muon site is of the order of 30 (the total enhancement is not to be mixed up with $E(K_F)$ which describes the intensity enhancement for states close to the Fermi energy). In addition, it remains to be seen how reliable neutron diffraction can determine the true interstitial magnetization.

The second theoretical approach, due to Jena [68], is also based on the magnetization value for interstitial sites in Ni, obtained from [62]. The theory of Jena adopts the Daniel-Friedel model [69]. In this model the spin density is obtained in the framework of scattering theory. The model was specifically designed to explain the sign and magnitude of hyperfine fields due to conduction electrons at nonmagnetic substitutional impurities. This model describes correctly the observation that e.g. in Fe, nonmagnetic impurities in the first half of an sp series experienced a negative hyperfine field, while elements of the second half experienced a positive hyperfine field. The potential, introduced by the impurity, is approximated by a square well potential, which is not a bad picture for impurities having an extended electron core.

In a ferromagnet the conduction band is split due to the exchange interaction with the localized moments of the magnetic host atoms, leading to the spin polarization of the conduction electrons. The interaction is of the RKKY type, as already mentioned. For the 3d transition metals it is the s-d exchange, occurring also in the Patterson-Falicov model.

Treating the conduction electrons again as a free electron gas, the spin up and spin down bands are shifted with respect to each other

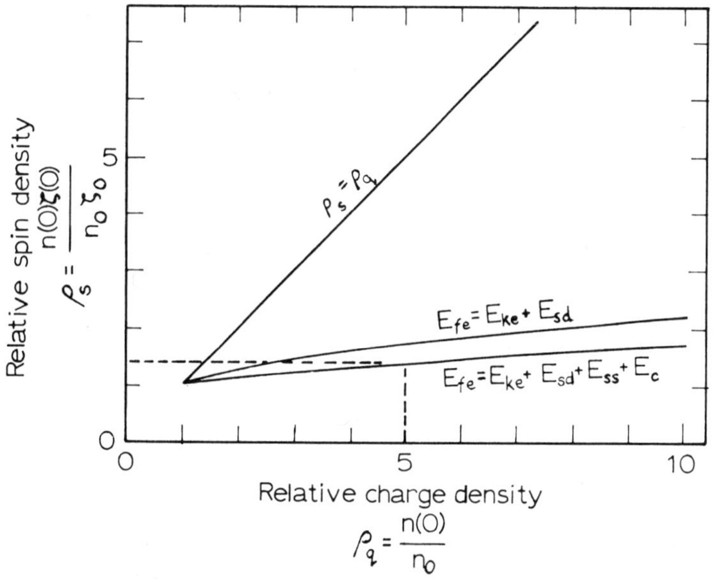

Fig. 37. Calculated relationships between the relative charge and spin densities (ρ_q and ρ_s) at the muon site. The straight line $\rho_s = \rho_q$ represents the "direct proportionality" hypothesis, while the other curves are calculations according to the minimization procedure [52].

in just the same fashion as already discussed for the electron paramagnetism. The spin split conduction band is characterized by the energy dispersion law:

$$(k_F^{\sigma})^2 + \sigma\Delta = k_F{}^2$$

The situation is sketched in Fig. 38. Near the impurity the total potential depth seen by the electrons depends now on the spin orientation ($\sigma = \pm 1$). For spin down electrons the impurity well looks less deep than for spin up electrons.

$$V^{\uparrow\downarrow} = -(V_o \pm \Delta) \tag{6.9}$$

Although there are more electrons with spin down, they are less attracted by the impurity than the spin up electrons. In the extreme i.e. for flat impurity wells, this may even lead to a reversal of the sign of the electron polarization at the impurity as compared with the undisturbed case [69]. For very deep impurity potential wells the band splitting will have not much effect and one expects the same spin polarization as in the absence of the impurity potential. These are in a rough way the essential features of the Daniel-Friedel model.

Jena now first determined Δ from the interstitial magnetization of Ni (as measured by neutron diffraction) in the free electron gas model

$$\begin{aligned} M_{loc} &= \mu_B\,(N\uparrow - N\downarrow) \\ &= \mu_B\;3\pi^2\left((k_F^{+1})^3 - (k_F{}^{-1})^3\right) \\ &\simeq \mu_B\;3\pi^2\left((k_F^{+1})^2 - (k_F^{-1})^2\right)\cdot k_F \simeq 6\pi^2\,\mu_B\,k_F\cdot\Delta \end{aligned} \tag{6.10}$$

To obtain the band splitting also for Fe and Co, he simply scaled Δ_{Ni} by the magnetic moments of Fe and Co [see Tab. 5]

The range of the square well potential was chosen to be the Thomas Fermi screening length, i.e. a = 0.6 Å in Ni, and nearly the same in Fe and Co. The depth of the unperturbed square wave potential $k_o = \sqrt{V_o}$ is determined self-consistently by satisfying the Friedel sum rule and the depth of the perturbed wells follow from

$$(k_F^{\sigma})^2 + (k_o^{\sigma})^2 = \text{const}$$

Following Daniel and Friedel [69] the scattering contribution to the electron density per spin for s-waves is given by

$$P_{sc}^{\sigma}(o) = \frac{1}{2\pi^2}\int_o^{k_F^{\sigma}} dk\;\frac{k^2}{1 - (k_o^{\sigma}/K^{\sigma})^2\,\sin^2(K^{\sigma}a)} \tag{6.11}$$

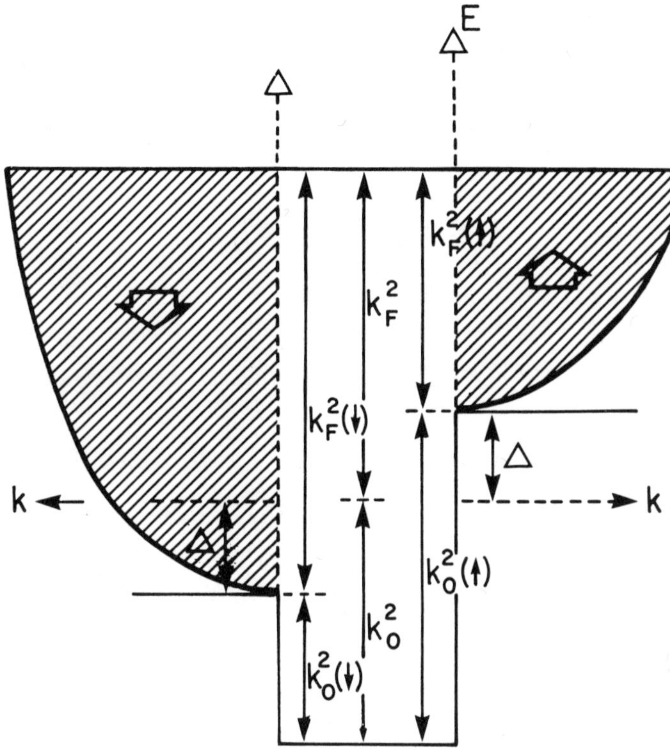

Fig. 38. Schematic plot of the effect of the relative shift of spin up and spin down conduction bands on the effective depth of an impurity potential in the Daniel-Friedel model.

with

$$(K^\sigma)^2 \equiv (k_o^\sigma)^2 + k^2$$

For a square wave potential with $k_o a > \frac{\pi}{2}$ there is also a contribution from the bound state in the $\ell = 0$ partial wave. This, if at all, leads to an additional contribution

$$P^\sigma_{bound}(o) = \frac{1}{2\pi} \frac{(\gamma^\sigma)^3}{\gamma^\sigma a - \tan(\gamma^\sigma a)} \tag{6.12}$$

with

$$(\gamma^\sigma)^2 \equiv (k_o^\sigma)^2 - (k_b^\sigma)^2$$

k_b^σ is the momentum of the electron in the bound state determined by the equation

$$k_b^\sigma = -\gamma^\sigma \cot(\gamma^\sigma a) \tag{6.13}$$

The total electron density per spin is then

$$P^\sigma(o) = P^\sigma_{sc}(o) + P^\sigma_{bound}(o) \tag{6.14}$$

From this Jena calculates the following hyperfine fields for Ni, Co and Fe

$$B_{hf}(\mathrm{Ni}) = -0.6 \text{ KGauss}$$

$$B_{hf}(\mathrm{Co}) = -2.5 \text{ KGauss}$$

$$B_{hf}(\mathrm{Fe}) = -5.6 \text{ KGauss}$$

Agreement with the measured value in Ni is good while this cannot be said for Co and Fe. Generally the right sign is obtained.

The following points need critical attention. First scaling the band splitting by the magnetic moments of the host atoms, starting from the magnetization data on Ni appears quite artificial. The magnetization data for Fe and Co, as obtained by neutron scattering do not display such a linear scaling behaviour. Secondly, the assumption of a square wave potential for the muon impurity needs further justification. Whatever the actual screened potential looks like, at the origin the potential can be expected to be quite deep, implying $\Delta/V_o \ll 1$. The scattering behaviour for spin up and spin down electrons should then be equal, as conjectured already above. The author mentions other points that need improvement:

1. Taking into account that a small percentage of d-electrons may be itinerant, i.e. being conduction electrons
2. Actually using a band model for the electrons instead of the free electron gas model.

Finally, Meier has also considered the application of his screening model for ferromagnetic metals [70]. If the difference in the spatial part of the electron wave function for spin up and spin down states can be neglected, and if the band splitting is small compared with the Fermi energy (i.e. applying a Knight Shift picture), we can directly apply eqs. (5.35), (5.20) in calculating the disturbed magnetization at the muon site

$$M(R_\mu)/M^o(R_\mu) \simeq E(k_F) \tag{6.15}$$

A difference in the spatial part would introduce a difference in the scattering pattern of the spin up and spin down electrons, leading also to a modified charge density of the muon for spin up and spin down electrons. In the Daniel-Friedel model this is accounted for by the different depths for the assumed square wave potential.

Meier comes to the conclusion that in Ni the actual band structure prohibits the naive application of his model. A detailed investigation would have to consider the s- and d-electron contributions to the screening explicitly.

In the case of Gadolinium, eq. (6.15) may be applied. With a Fermi momentum for Gd of about $1.4 \cdot 10^8 cm^{-1}$ one gets $E(K_F) \sim 10$ (see Fig. 22). The contact field in Gd was determined by μSR to ∿−6.6 KGauss, implying a magnetization density of $-0.09\ \mu_B/Å^3$ and hence an unperturbed density of $M_o(R_\mu) \cong -0.009\ \mu_B/Å^3$. A recent band structure calculation by Harmon and Freeman [71] gives a value of $M_o(c) \cong -0.002\ \mu_B/Å^3$ at the c-site. The spin density derived from the μSR measurement falls between the neutron diffraction result and the theoretical value, perhaps indicating that it comes closest to reality?

Finally, I would like to remark that most classical band structure calculations yield a negative spin polarization at the Fermi surface at least for strong ferromagnets [72]. The electrons at the Fermi surface are the most mobile ones and a naive picture would place these electrons in the interstitial region. This seems to be supported by the μSR measurements. Recent ESP photoelectron spin polarization measurements by Siegmann et al. [73] show however that in Ni, Co, Fe the electrons at the Fermi surface are positively polarized (see Fig. 39). The future will show how to solve this puzzle and it is hoped that the μSR technique will contribute in this fascinating task.

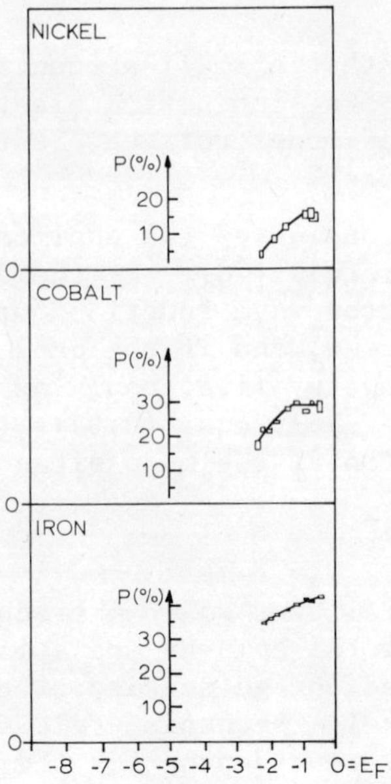

Fig. 39. Dependence of photoelectron spin polarization on the band energy in eV. The data have been obtained from cesiated Fe, Co and Ni-films [73].

I would like to close this chapter by mentioning the information which can be drawn from the two other parameters, determined by μSR, namely the asymmetry or precession amplitude and the damping constant. The asymmetry constant is a measure of the domain alignment [50]. For a complete random orientation of the magnetization of the domains, one would expect an asymmetry that is 2/3 of the maximum possible asymmetry. This should be the case if no external field is present [53]. In the instances where one has measured lower asymmetries one has to consider the possibility that for certain regions of the target sample larger field inhomogenities do exist, leading to a rapid loss of phase coherence of those muons that have been stopped there. The occurrence of a damping may have various origins [74]. The most attractive application results here from the possibility to observe critical fluctuations in the neighborhood of the Curie temperature or other phase transitions. The μSR applications look particularly promising as one can measure relaxation rates up to $10^8 sec^{-1}$, while e.g. NMR measurements are restricted to relaxation rates of up to $10^6 sec^{-1}$. Very promising first studies have been reported by Patterson et al. [75] for temperatures slightly above the Curie temperature in Ni.

VII. MUONIUM IN SOLIDS

We now turn to a quite different class of phenomena that is related to the presence of muonium atoms (μ^+e^-) or muonium-like centers in solids. As already mentioned in the introduction, for insulators one can generally count on the presence of muonium atoms as the state in which the implanted muons become thermalized. For metals, as we have seen, screening by the conduction electrons will prevent the existence of bound states for practically all cases. Only for metals with a very low conduction electron density as e.g. for Cs can bound states may exist. The binding energy will, however, be quite low and thermal ionization together with other disturbing effects may prevent the actual observation of such a weakly bound (μ^+e^-)- system. In any case, a careful search for weakly bound muonium states in certain metals is certainly indicated. Between the class of insulators and the class of metals (conductors) we have the semiconductors, which in principle allow to study the muonium behaviour in the presence of a widely ranging density of "mobile" electrons or holes.

Generally the behaviour of muonium, as evidenced through the evolution of the muons' polarization, is determined by

1. electrostatic interactions, changing the binding properties, and by
2. magnetic interactions, which, for all practical purposes, are those of the large magnetic moment of the electron ($\mu_e \sim 200 \cdot \mu_\mu$).

a) Muonium in Semiconductors

We will start by discussing some measurements that have been performed in the semi-conductors Ge [76] and Si [77] and which relate to the first category, i.e. the electrostatic interactions.

The first experiment that demonstrated the importance of the charge carrier concentration in Si and Ge was performed by Feher et al. [78]. They measured in a transverse field of 40 Gauss at room temperature the amplitude of the "free" muon precession as a function of carrier concentration by using differently doped p- and n-type silicon. The results are shown in Fig. 40. At a high concentration of free holes or free electrons the precession amplitude corresponds to essentially 100% polarization. At a free electron concentration of about $10^{14}/\text{cm}^3$ the amplitude drops to practically zero. An analogous result was obtained for strongly doped n-type Ge. At room temperature one finds a large precession amplitude, while at liquid nitrogen temperature (77°K) the amplitude has decreased considerably. In a subsequent work Eisenstein et al. [79] investigated the dependence of the longitudinal

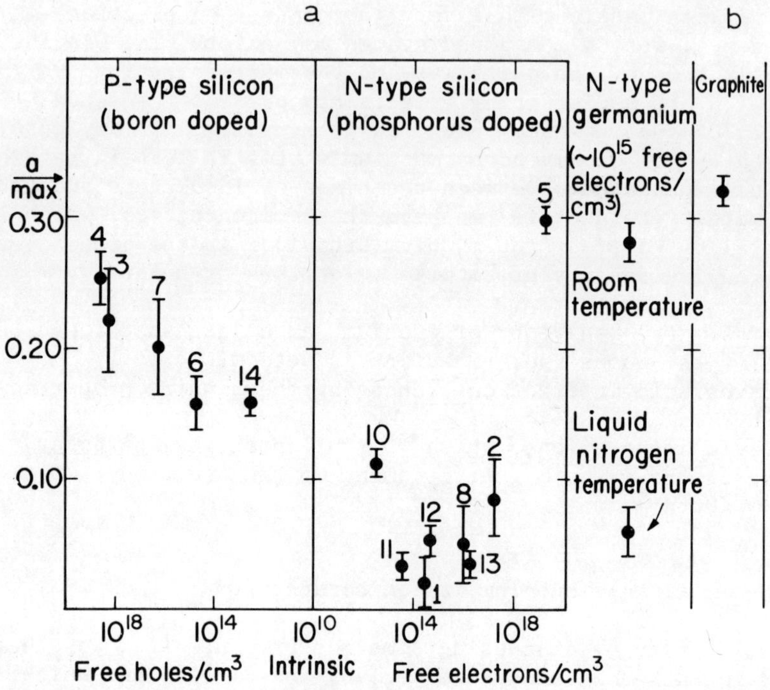

Fig. 40a. Dependence of the precession amplitude in Si on the electron and hole concentration.

Fig. 40b. Dependence of the precession amplitude in n-type Ge on temperature [78].

polarization on temperature and external field strength in Si with various donor concentrations. Evidence for weakly bound muonium centers (shallow donor muonium) was claimed, although no firm conclusion could be drawn from the statistically insufficient data. Later Adrianov et al. [80] measured the field dependence of the longitudinal polarization in a mildly p-type single crystal of Si at room temperature. The data are shown in Fig. 41. The quenching curve obtained can be fitted nicely by eq. (3.30), yielding an effective hyperfine frequency ω(Si) of

$$\frac{\omega_o(\mathrm{Si})}{\omega_o(\mathrm{Vac.})} = 0.405 \pm 0.026$$

As can be seen from the Fig. 41, in zero external field the residual polarization is larger than 0.5, indicating a fraction of "free" muons that are not bound to an electron.

After the first demonstration of the two frequency precession of muonium in quartz, things started to become exciting when Gurevich et al. [11] observed the two frequency precession at 77^oK in Ge [76]. The beating frequency was about twice as big as expected for muonium. Using eq. (3.37)

$$\omega_o(\mathrm{Ge}) = \frac{\omega_e^2}{\Omega}$$

one obtains an effective hyperfine frequency of

$$\frac{\omega_o(\mathrm{Ge})}{\omega_o(\mathrm{Vac.})} = 0.56 \pm 0.01$$

Subsequently Brewer et al. [77] observed the two precession frequency in p-type silicon at 77^oK. The Fourier spectrum of the rate versus time histogram, shown in Fig. 42, displayed a pair of frequencies with a splitting corresponding to an effective hyperfine field of

$$\frac{\omega_o(\mathrm{Si})}{\omega_o(\mathrm{Vac.})} = 0.45 \pm 0.02$$

This is in excellent agreement with the result of Adrianov et al. However, as can be seen from Fig. 42 there is a second pair of frequencies at much lower values, but still much larger than would correspond to a "free" muon precession in the applied field ($\sim$ 100 Gauss). This second pair of "anomalous frequencies" has so far been detected only in Si.

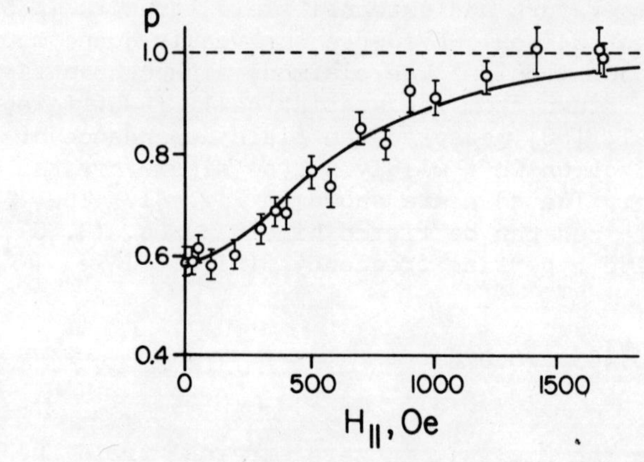

Fig. 41. Residual polarization versus magnetic field strength in Si [80].

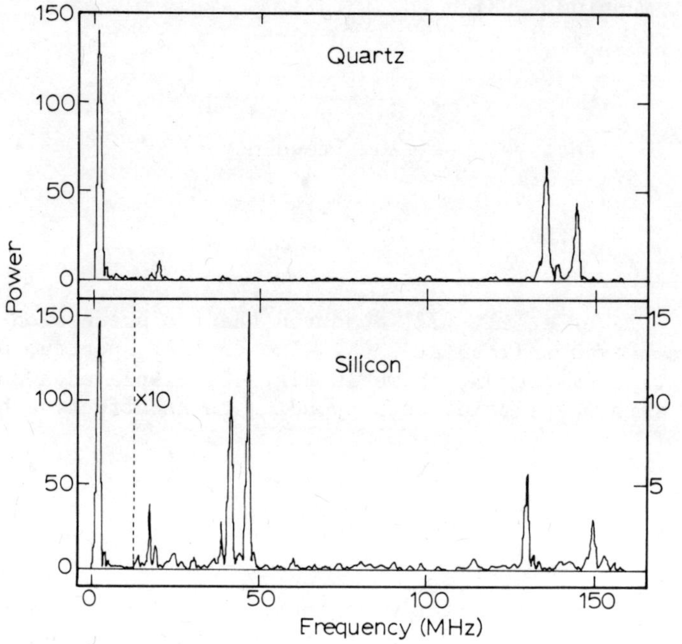

Fig. 42. Frequency spectra of muons in fused quartz at room temperature and in p-type silicon at 77°K in a transverse field of 100 gauss [77].

We will first discuss the physics behind the "normal" frequencies. As shown in Chapter III the hyperfine frequency ω_o is related to the Bohr radius a_o as follows:

$$\hbar\,\omega_o = \frac{8}{3}\, g_e\, \mu_B^e\, g_\mu\, \mu_B^\mu \frac{1}{a_o^3} = \frac{8\pi}{3}\, g_e\, \mu_B^e\, g_\mu\, \mu_B^\mu |\psi(0)|^2 \qquad (7.1)$$

A decreased hyperfine frequency thus corresponds to an increased size of the muonium atom, which in turn corresponds to a smaller binding energy. Inserting the numbers for the effective hyperfine frequency one obtains the following effective radii for muonium in Ge and Si

$$r_{Mu}(\mathrm{Ge}) = 0.645\ \text{Å} = 1.21\ a_o(\mathrm{Vac.}) \qquad a_o(\mathrm{Vac.}) = 0.532\ \text{Å}$$

$$r_{Mu}(\mathrm{Si}) = 0.719\ \text{Å} = 1.35\ a_o(\mathrm{Vac.})$$

The size of muonium in Ge and Si is still much smaller than the spacing between the Ge and Si cores in the crystals (lattice parameter ∿ 5 Å). Thus there is only a very small overlap of the muonium electron wave function and the core electron wave functions. It is hereby assumed that muonium will reside in an interstitial position. Such an impurity atom that fits well into some substitutional or interstitial volume is called a deep donor.

The electronic structure of "shallow" donors and acceptors in Si and Ge have e.g. been calculated by Luttinger and Kohn [81] within the framework of an effective mass theory, starting from crystal Bloch functions and treating the impurity potential as a perturbation. The donor electron is then described by a more or less localized wave function, expressed as a sum over crystal wave functions.

Such a treatment is, however, less suited to the case of deep donors, as has been recognized a long time ago by Reiss [82] and Kaus [83].

Kittel and Wang [84] have tried instead to explain the properties of deep donor muonium by a phenomenological model. In fact, two approaches were undertaken. The first one is the cavity model, the second one uses a wave vector dependent dielectric function $\varepsilon(q)$.

Cavity model. The cavity in which the muonium atom is enclosed is characterized by the following behaviour of the potential $V(r)$

$$V(r) = \begin{cases} -e^2/r + e\,(1 - 1/\varepsilon_o)R & r < R \\ -e^2/\varepsilon_o\, r & r > R \end{cases} \qquad (7.2)$$

Inside the cavity of radius R the potential is essentially the Coulomb potential of the muon. The second term is introduced to guarantee continuity at $r = R$. Outside the cavity the Coulomb potential is screened by the valence band electrons of the neighbouring silicon or germanium atoms. This is taken into account by introducing the dielectric function into the Coulomb potential and assuming that for $r > R$, $\varepsilon(r) = \varepsilon_o$, where ε_o is the macroscopic dielectric constant (Si: $\varepsilon_o = 12$; Ge: $\varepsilon_o = 15.8$). Inside the cavity the mass of the electron is given by the free electron mass, while outside the cavity one assumes that the mass of the electron is given by the effective mass of the conduction band electrons

$$m(r) = \begin{cases} m_o & r < R \\ m^* & r > R \end{cases} \qquad (7.3)$$

(Si: $m^* = 0.31\ m_o$; Ge: $m^* = 0.17\ m_o$).

Wang and Kittel have calculated the ionization energy as a function of cavity radius by solving the corresponding Schrödinger equation. The results are shown in Fig. 43. As can be seen there is quite a sudden decrease in ionization energy when the cavity radius is decreased below a certain value, indicating a transition from a deep donor state to a shallow donor state.

Si and Ge have a crystal structure of the diamond type. Two interstitial sites are available - one with hexagonal local symmetry and the other with tetrahedral local symmetry. The cavity radius was then calculated from the radius of the interstitial sphere that will fit inside the touching hard spheres of the lattice. The thus obtained radii are also indicated in Fig. 43 and it can be seen that in this model muonium has to belong in the class of deep donors.

Next, the probability density $|\psi(0)|^2$ was calculated numerically as a function of cavity radius. The results relative to $|\psi(0)|^2_{Vac}$ in vacuum are shown in Fig. 44. Using the calculated cavity radii one can then obtain theoretical values for $\omega_o(\text{Si, Ge})/\omega_o(\text{vacuum})$. The results are summarized in Table 7.

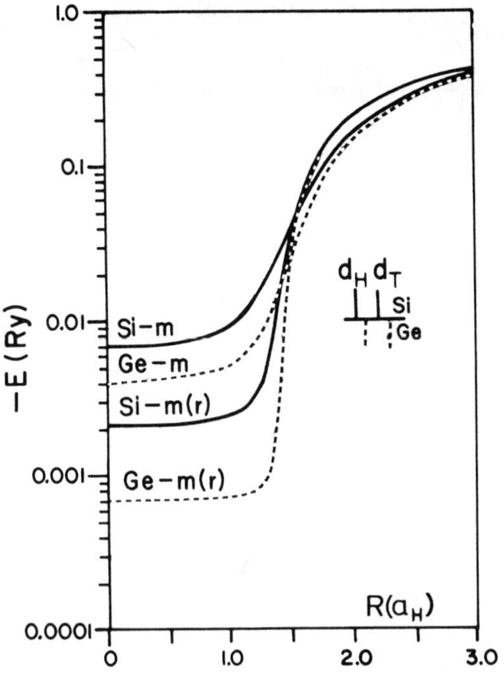

Fig. 43. Binding energy of muonium versus cavity radius [84].

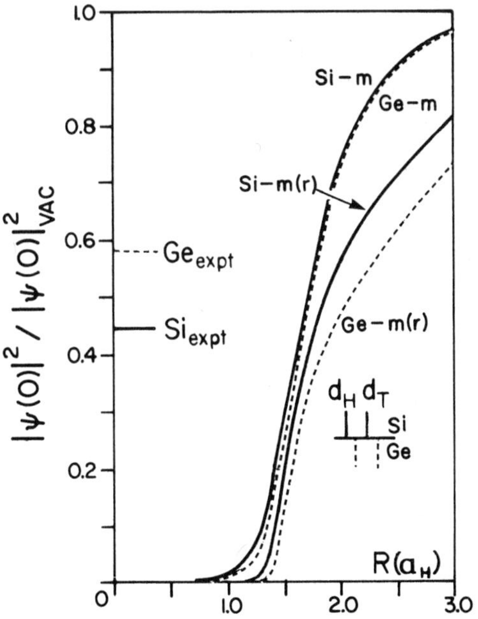

Fig. 44. Electron density $|\psi(0)|^2$ in muonium versus cavity radius [84].

Table 7

m(r)	R	$\|\psi(0)\|^2/\|\psi(0)\|^2_{vac}$ Si		Ge
$m(r) = \begin{cases} m, & r < R \\ m^*, & r > R \end{cases}$	d_H	0.578	>	0.506
	d_T	0.642	>	0.565
m	d_H	0.756	<	0.787
	d_T	0.837	<	0.860
Experimental		0.444	<	0.578

Comparison with the experimental data shows that the right order of magnitude is reproduced when the electron mass behaves as defined in eq. (7.3), except that the relative order is opposite to the one obtained experimentally. Assuming the electron mass to be equal everywhere to the free electron mass, one obtains the right relative order, but the electron density at the muon comes out too large.

In any case, the cavity model is able to principally explain the occurrence of a deep donor muonium state.

Dielectric function approach. Various authors [85] have calculated self-consistently the dielectric function $\varepsilon(q)$ from the actual band structure of Si and Ge. A model dielectric function can be defined as follows

$$\varepsilon(q) = \frac{q^2 + Q_D^2}{\varepsilon_o q^2 + Q_D^2} \tag{7.4}$$

where Q_D is a fitting or screening parameter. What we need is the dielectric function in r-space, which can be obtained from eq. (7.4) by a Fourier transformation

$$1/\varepsilon(r) = 1/\varepsilon_o + (1 - 1/\varepsilon_o)\, e^{-Q_D r} \tag{7.5}$$

Eq. (7.5) was fitted to the ε (q) values, calculated from the actual band structure to yield the parameter Q_D.

The potential was then defined as

$$V(r) = - {}^{e^2}/\varepsilon(r)r \tag{7.6}$$

and the Schrodinger equation was solved for this potential. The results are summarized in Table 8.

The right order of relative magnitudes is reproduced and also the absolute order of magnitude is close to the experimental values, in particular for Si.

We return now to the discussion of the anomalous precession frequencies detected in Si. Fig. 45 shows the field dependence of the pair of frequencies. The precise position of the frequency pair depends also on the orientation of the crystal with respect to the external field. Also there appear additional frequencies which have not found an explanation yet.

The dependence of the anomalous pair of frequencies on magnetic fields can be understood if one assumes that one is observing the transitions ω_{12} and ω_{34} (see Fig. 9). ω_{12} will start at zero, will have a maximum at some intermediate field value and will decrease again until it reaches the level crossing point. ω_{34} will be first equal to the hyperfine splitting frequency, will then decrease until it reaches a minimum and will then start to rise.

The anisotropy can be accommodated by using an "effective spin Hamiltonian" of the form

$$H = \vec{J}_e \cdot \underset{\sim}{A} \cdot \vec{S}_\mu + \mu_B^e \vec{B} \cdot \underset{\sim}{g}_e \cdot \vec{J}_e + g_\mu \mu_B^\mu \vec{S}_\mu \cdot \vec{B} \tag{7.7}$$

where $\underset{\sim}{A}$ and $\underset{\sim}{g}_e$ are tensors and $\vec{J}_e$ is an effective electron spin. This effective spin Hamiltonian is widely used in the analysis of ESR-spectra of paramagnetic impurities [86].

A good fit of the μSR results can be obtained by using this phenomenological Hamiltonian. The $\underset{\sim}{g}_e$-tensor turns out to be a scalar

$$(g_e)_{ij} = (13 \pm 3)\ \delta_{ij}$$

and the tensor $\underset{\sim}{A}$ displays a minimal anisotropy with symmetry about the (111) axis. $\underset{\sim}{A}$ is then diagonal with only two independent non zero elements:

Table 8

	Si					Ge				
	Q_D, in a_H^{-1}	ε_0	$-E$, in Ry	$\frac{\lvert\psi(0)\rvert^2}{\lvert\psi(0)\rvert^2_{vac}}$	$\langle r\rangle$, in a_H	Q_D, in a_H^{-1}	ε_0	$-E$, in Ry	$\frac{\lvert\psi(0)\rvert^2}{\lvert\psi(0)\rvert^2_{vac}}$	$\langle r\rangle$, in a_H
Walter and Cohen	0.9153	11.47	0.112	0.427	2.617	0.8702	14.00	0.116	0.453	2.526
Vinsome and Richardson	0.9221	10.53	0.116	0.429	2.593	0.8377	14.95	0.126	0.478	2.433

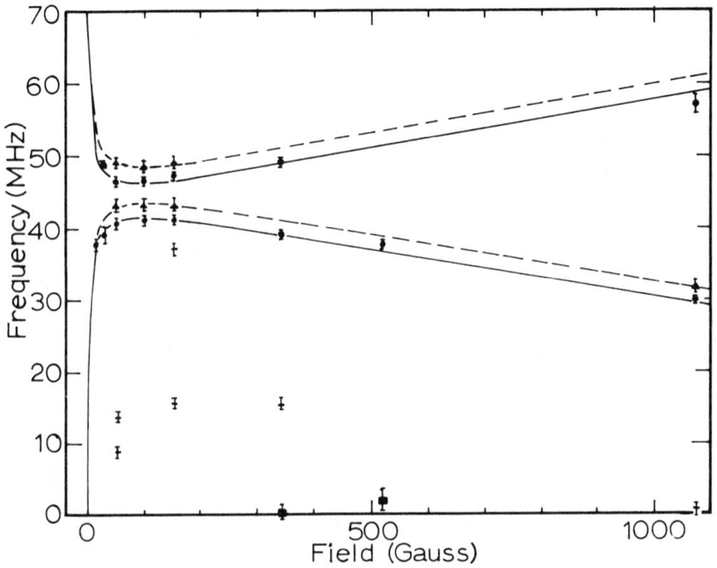

Fig. 45. Dependence of the "anomalous" precession frequencies in Si on magnetic field strength [77].

$$A_{33} = A_{11} = (0.0198 \pm 0.00002)\cdot A_o(\text{vac})$$

$$A_{12} = A_{22} = A_{\perp} = (1.035 \pm 0.02)\, A_o(\text{vac})$$

and

$$|\vec{J}_e| = \frac{1}{2}$$

According to this result, the object that produces the frequencies must be a spin ½-spin ½ system with a hyperfine interaction of about 1/50 of the vacuum muonium value and an effective electron g_e- factor of about 6 times the free electron g_e-factor.

The simplest interpretation of this object put forward so far assumes that it is a shallow donor muonium atom. The wave function now is spread over many lattice sites, whereas the entire deep-donor muonium fits into one interstitial site. However, the shallow state cannot be a pure s-state in view of the large effective g_e-factor and the anisotropy. To account for the first property a state with $\ell \neq 0$ is required. This allows interpretation of the g_e-factor as an orbital g-factor, involving a small effective electron mass. The origin of the anisotropy of the hyperfine interaction is less obvious and, in fact, no convincing explanation does exist. In any case, one may view the shallow donor object as some excited state of deep donor muonium. Because the overlap of an excited state with the valence band electrons looks different, an excited state may assume qualitatively different properties as compared to the deep donor ground state. In particular this may explain the life time of the excited state (e.g. a 2p-state) that is estimated from the measurements to be of the order of $\sim$ 300 nsec.

Still, this is not the whole story. We mentioned the unexplained additional frequencies showing up in the Fourier spectrum. There also appears to be a free muon precession signal, consisting of two components: one which is damped with a time constant of about 30 nsec [4] and a long lived one. In addition, both deep and shallow donor muonium were only observed at 77°K in p-type Si. In n-type Si at 77°K and in all samples at room temperature no signals were detected. In Ge no shallow donor muonium states were detected at all.

There remains a lot to be done, both theoretically and experimentally. Theoretical work in the frame of the Kohn-Luttinger formalism has been started [87].

b) Muonium in Insulators - Muonic U_2-centers

Direct muonium precession in insulators has only been seen in a very limited number of cases: in quartz, in ice and in frozen CO_2 [76, 88]. From the two frequency precession behaviour the hyperfine frequency ω_o was determined to be always compatible with ω_o(vac). The cavity radius in ice is approximately $4 \cdot a_o$ and, by looking at Fig. 43, it is obvious that muonium in ice should behave much like free muonium.

There is, however, much more evidence for the existence of muonium in insulating crystals. In ref. [4] a long list of asymmetries, determined by the "free" muon precession amplitude, in a large number of insulators is given. In all cases the asymmetry is lower than would correspond to the full polarization. The missing asymmetry is or was probably carried by those muons that are in the muonium state. For one or another reason a fast depolarization happened, prohibiting any direct observation of muonium precession.

Among the more likely reasons we mainly have to consider the effects of random local magnetic fields (RLMF) on the muonium precession. These will smear out the muonium precession in just the way as the "free" muon precession is effected by nuclear dipole fields. Sources of RLMF's may be paramagnetic impurities or nuclear dipole fields. We will come back to the importance of the latter ones in a while. In fact, in those cases where one has observed direct muonium precession, the precession pattern always displayed a damping behaviour [88]. Fig. 46 shows the muonium precession signal for two temperatures obtained in quartz. At the higher temperature (+30°) the damping is slower than at the lower temperature (-196°C). This is very reminiscent of the behaviour of the slow muon relaxation in copper. Perhaps the whole muonium atom is diffusing in the quartz sample and interactions with magnetic impurities are averaged to zero by a motional narrowing effect at higher temperatures.

Other reasons for fast depolarization may be chemical reactions of muonium or relaxation of the muonium electron itself, which in turn leads to a relaxation of the muon by way of the contact interaction. The latter mechanism may be important in paramagnetic substances and would be very interesting to study. A rich field of μSR-applications is still hidden there.

Whatever the reason may be that leads to a fast depolarization, restoration of polarization should always be observed in strong longitudinal magnetic fields due to the Paschen-Back effect, as shown in Chapter III. Indeed, much of the older work concentrates on measuring quenching curves. A restoration of polarization at

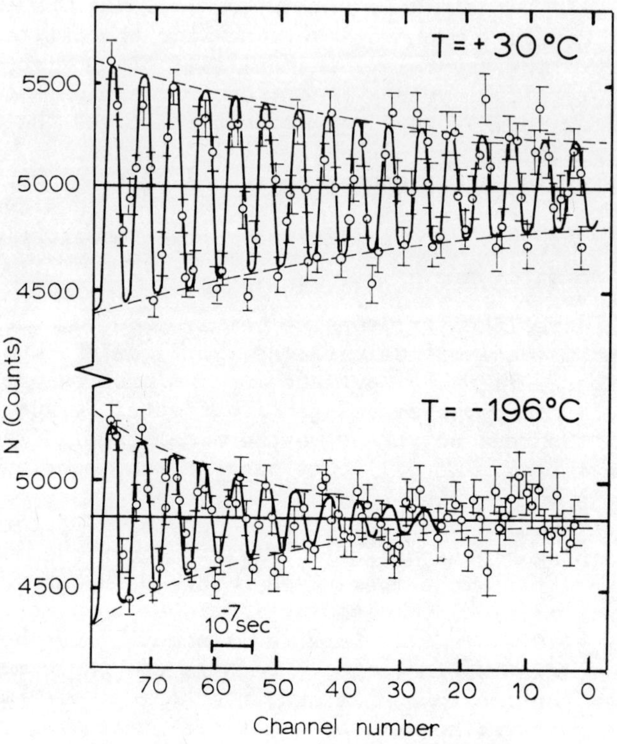

Fig. 46. Muonium precession at 30°C and at -196°C for a magnetic field of 7.2 gauss. The rate is corrected for the exponential decay of the muon [88].

strong external fields was always achieved, although details were not visible due to large statistical errors.

We will now discuss in somewhat more detail a very special coupling of muonium to nuclei that will strongly influence the evolution of the muons' polarization. Here it is appropriate to introduce a new concept, namely that of a muonic U_2-center [89].

A U_2-center is usually defined to be a paramagnetic hydrogen center i.e. atomic hydrogen, in an interstitial position. A muonium atom in an interstitial position is thus the complete analogue to a hydrogenic U_2-center. A muonic U_2-center will have the same properties as a hydrogenic U_2-center, except perhaps that, due to the lighter mass of muonium, a muonic U_2-center may vibrate with a larger amplitude. We will assume, however, that it is still a quite localized system.

Hydrogenic U_2-centers have been studied extensively by ESR and ENDOR techniques, particularly in alkali halides [90, 91]. These measurements show that quite a large hyperfine interaction exists between the electron of the U_2-center and the neighbor host nuclei. This interaction is partly of the nature of a dipole-dipole interaction, introducing spatial anisotropies, and partly of the nature of a contact interaction, due to the non-vanishing probability of finding the electron at the sites of the neighbor nuclei. This hyperfine interaction that exists in addition to the much stronger contact interaction between electron and proton is called a superhyperfine interaction (shf-).

It is to be expected that a muonic U_2-center will likewise be subject to a superhyperfine interaction (e.g. in alkali halides). The superhyperfine interaction acts coherently on all muonic U_2-centers, provided that one considers the effect in a single crystal. The presence of a shf-interaction thus does not lead to a simple smearing out effect of the muonium precession, but to a characteristic modulation that will in principle be visible in both longitudinal and transverse fields, if these fields are not too strong.

The magnetic interaction between the electron of a U_2-center and some neighbor nucleus with spin I and magnetic moment $g_I \mu_k$ is represented according to Chapter III eq. (3.7) by the Hamiltonian

$$H = g_e \mu_B^e g_I \mu_k \left(\frac{3(\vec{S}_e \cdot \vec{r})(\vec{I} \cdot \vec{r})}{r5} - \frac{(\vec{S}_e \cdot I)}{r3} + \frac{8\pi}{3} \vec{S}_e \cdot \vec{I}\, \delta(\vec{r}) \right) \tag{7.8}$$

Schumacher and Hall [92] and Seidel and Wolf [93] have derived from this expression the specific form of the shf-interaction Hamiltonian (notation is given in Fig. 47)

$$H_{shf} = (a - b)\vec{S}_e \vec{I} + 3b\, S_e^z I^z \tag{7.9}$$

with

$$a = \frac{8\pi}{3} g_e \mu_B^e g_I \mu_k |\psi(\vec{R}_I)|^2$$

$$b = \frac{1}{2} g_e \mu_B^e g_I \mu_k \int d^3r\, |\psi(r)|^2 \frac{3\cos^2\tau - 1}{r^3}$$

Choosing the direction of the applied field (z-axis) as the axis of quantization, we obtain

$$H = S_e^i T^{ij} I^j \tag{7.10}$$

with the matrix T given by

$$T = (a - b)\mathbb{1} + 3b \begin{pmatrix} \sin^2\theta & 0 & \sin\theta\cdot\cos\theta \\ 0 & 0 & 0 \\ \sin\theta\cos\theta & 0 & \cos^2\theta \end{pmatrix} \tag{7.11}$$

First we want to discuss some general features of the evolution of the muon's polarization in a muonic U_2-center, subject only to an isotropic shf-interaction with a single neighbor nucleus. No external field is assumed.

The initial state (following the treatment in Chapter III) is represented by the spin density matrix

$$\rho = \sum_i p_i |\chi_i\rangle\langle\chi_i| \tag{7.12}$$

with

$$|\chi_i\rangle = |\chi_\mu\rangle |\chi_e\rangle |\chi_I\rangle \tag{7.13}$$

and

$$p_i = \begin{cases} \dfrac{1}{2(2I+1)} & \text{if } \sigma_\mu^z|\chi_i\rangle = |\chi_i\rangle \\ 0 & \text{otherwise} \end{cases} \tag{7.14}$$

The muon polarization in z-direction evolves in time as

$$P_\mu^z(t) = \sum_n p_n \langle\chi_n| e^{iHt} \sigma_\mu^z e^{-iHt} |\chi_n\rangle \tag{7.15}$$

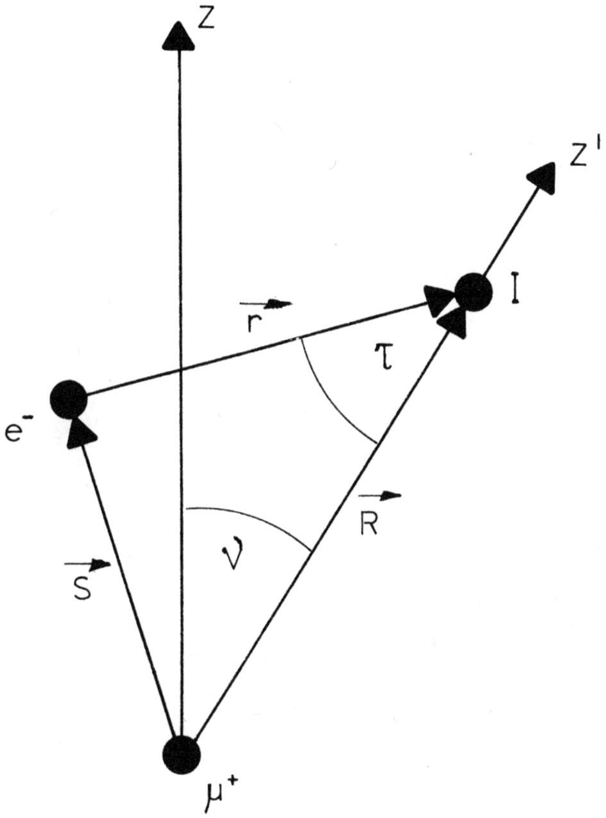

Fig. 47. Relative positions of muon spin and nuclear spin with regard to the z-direction defining angles ν and τ.

where the total Hamiltonian of the system muon-electron-nucleus is now given by

$$H = A \cdot \vec{S}_\mu \cdot \vec{S}_e + a \vec{S}_e \cdot \vec{I} \tag{7.16}$$

The eigenvalues of the Hamiltonian are found to be given by (if $I \geq 1$)

$$\begin{aligned}
\lambda_1 &= {}^A/4 + I\,{}^a/2 \\
\lambda_3 &= {}^A/4 - (I+1)\,{}^a/2 \\
\lambda_{2,4} &= -{}^A/4 - {}^a/4 \pm [(A - {}^a/2)^2 + I(I+1)a^2]^{\frac{1}{2}}/2
\end{aligned} \tag{7.17}$$

The level scheme is represented in Fig. 48. These eigenvalues coincide up to order $(a/A)^2$ with those of the Hamiltonian

$$\tilde{H} = A\vec{S} \cdot \vec{S}_e + {}^a/2(\vec{S}_\mu + \vec{S}_e) \cdot I \tag{7.18}$$

which describes a model where the muon and electron spins are coupled together and interact jointly with the nucleus. The eigenstates ϕ_n of $\tilde{H}$ can be expressed in terms of the χ_n by Clebsch-Gordon algebra: $\phi_n = C_{nn'}\chi_{n'}$. Following the procedure discussed in Chapt. III, the evaluation of eq. (7.15) leads by a lengthy but straightforward calculation to the result

$$\begin{aligned}
P^z_\mu(t) = \frac{1}{6}\Bigg\{ &\frac{I^2+I+1}{I^2+I} + \cos\omega_{24}t \\
&+ \frac{2I+3}{2I+1}\cos\omega_{14}t + \frac{2I-1}{2I+1}\cos\omega_{34}t \\
&+ \frac{2I^2+3I}{2I^2+3I+1}\cos\omega_{12}t + \frac{2I^2+I-1}{2I^2+I}\cos\omega_{23}t\Bigg\}.
\end{aligned} \tag{7.19}$$

Here ω_{ij} corresponds to the transition frequency between the energy levels λ_i and λ_j. Since in first approximation $\lambda_1 \simeq \lambda_2 \simeq \lambda_3 \simeq \omega_0/4$ and $\lambda_4 = -3\,\omega_0/4$, the transition frequencies ω_{14}, ω_{24}, and ω_{34} are of the order of ω_0 and are usually not resolvable in experiments. Thus the observable polarization is given by (assuming for simplicity a large value of I)

$$P^z_{\mu,obs.}(t) = \frac{1}{6}\{1 + \cos\omega_{12}t + \cos\omega_{23}t\} \tag{7.20}$$

Compared to the muonium case the constant term is reduced from 1/2 to 1/6. Such a reduction of the observed polarization below 1/2 is

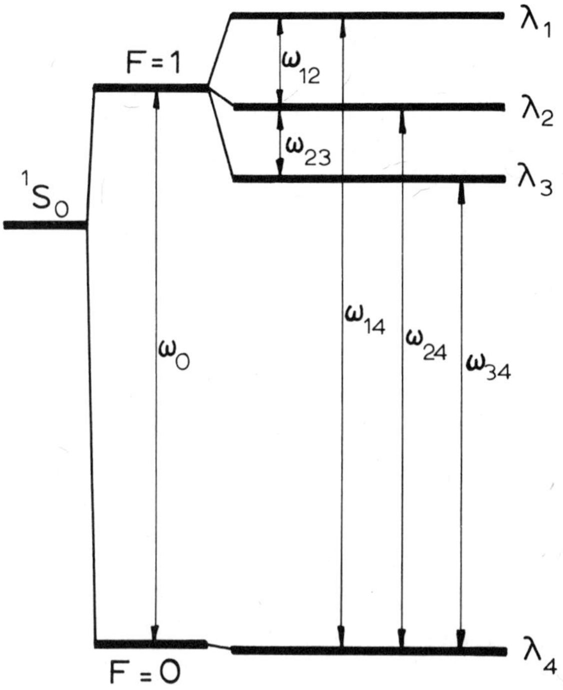

Fig. 48. Energy level scheme for a muonic U_2-center interacting isotropically with one neighbor nucleus with spin I in zero external magnetic field.

indeed seen in many experiments. Moreover, by including further super-hyperfine interactions with other nuclei (e.g. in the second neighbor shell or different species in the first neighbor shell) one gets a still smaller constant term and further oscillating terms.

We will now present a calculation on the evolution of the muon's polarization in a muonic U_2-center for some realistic cases in order to obtain an idea of the quality and magnitude of the observable effects [96].

In view of the extensive studies of hydrogenic U_2-centers in alkali halides, we will choose some of these substances as the host lattice for muonic U_2-centers. Only the evolution of the muon's polarization in longitudinal fields is considered.

The detailed ENDOR measurements of Spaeth and Sturm [90, 91] on U_2-centers in alkali halides show that the shf constants a and b for the halogen nuclei in the first neighbor shell are much greater than those of the alkali nuclei and those of nuclei farther away. Therefore, we shall confine ourselves in the following to a discussion of the influence of the four nearest halogen nuclei ($\alpha = 1,4$). For a general direction of the applied field the four shf tensors $T(\cos\theta_\alpha)$ are different. For certain directions symmetric with respect to the crystal axis the number of different tensors T reduce and the corresponding spins can be combined. This is illustrated in Fig. 49 with B along (1,0,0). In this particular case the total Hamiltonian reads

$$H = H_{hf} + H_Z + S_e^i\, T^{ij}\, (\cos\theta_1)\, (I_1^j + I_2^j) + S_e^i\, T^{ij}\, (\cos\theta_3)(I_3^j + I_4^j) \tag{7.21}$$

Again following the procedure outlined in Chapt. III, the evolution of the polarization in z-direction is given by eq. (3.27)

$$P_\mu^z(t) = \sum_{i,j,k,l,m} P_i\, U_{ik}\, U_{ij}\, U_{mk}\, U_{lj}\, e^{i(\lambda_k - \lambda_j)} \langle\chi_m|\sigma_\mu^z|\chi_e\rangle \tag{7.22}$$

where U is a unitary matrix that describes the transformation between the basis χ_i and the eigenvectors ϕ_i of the Hamiltonian eq. (7.21)

$$|\chi_i\rangle = \sum_j U_{ij}|\phi_j\rangle$$

and

$$H|\phi_j\rangle = \lambda_j|\phi_j\rangle \tag{7.23}$$

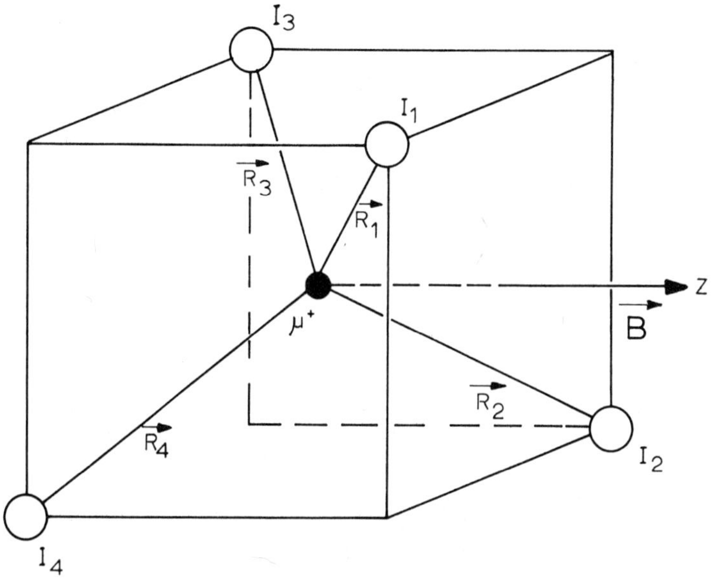

Fig. 49. Relative positions of the first four neighbor fluorine nuclei in KF with respect to the interstitial position of the muonic U_2-centers in the presence of a super-hyperfine interaction.

ENDOR measurements are usually performed in high static fields. The theoretical analysis of the shf interaction can then be based on perturbation theory. For muonic U_2-centers, however, the influence of the shf interaction is most pronounced in the low field region where the hyperfine interaction is dominant and the shf interaction competes with the Zeeman energy. Therefore one has to resort to a numerical evaluation of (7.23) and (7.22). This has been done for the particular case where the external field lies in the crystal (1,0,0) - and (1,1,1) - directions, respectively. Taking into account the rules for addition of spin operators we can then collect the four nuclear spins into two groups. This allows reduction of the dimensions of the matrices to a manageable degree.

Equation (7.22) for the polarization has been evaluated for various alkali halides using the hyperfine frequency for vacuum and the shf interaction parameters extracted from the ESR and ENDOR measurements of Spaeth and Sturm [91] (see Table 9). We split the longitudinal polarization into three parts:

$$P_\mu^z(t) = P^{res} + \sum_k g_k \cos \omega_k t + P_{high\ freq.} \qquad (7.24)$$

Table 7: Superhyperfine interaction parameters a and b for the halogen nuclei in the first-neighbor shell for hydrogenic U_2-centers taken from ref. [91].

Substance	I	a(MHz)	b(MHz)
KF	1/2	137.1	54.8
KBr	3/2	118.5	37.4
KCl	3/2	23.7	6.7
NaBr	3/2	200	50

i.e. the time independent residual part, contributions oscillating with low frequencies (< 1GHz), and a high frequency part (with frequencies of the order of ω_o). The latter contribution is uninteresting, since the fast oscillation cannot be observed. The variation of the residual polarization with the applied field is represented in Fig. 50 for some typical substances. With increasing field the muonium system first is decoupled from the nuclei (Paschen-Back effect for (μe)-N in fields where the Zeeman energy exceeds the shf interaction). This leads to the rise of P^{res} which approaches the curve of free muonium. The latter then exhibits the Paschen-Back effect in higher fields. The difference between the residual polarizations as calculated for B along (1,0,0) and (1,1,1) is very small and is only shown for NaBr.

The effects of crystal orientation however become clearly visible in the distribution of the low-frequency oscillations. In Fig. 51 we plot a histogram showing the weights g_k of the distribution of frequencies ω_k (channel width 10 MHz) for KF. The distribution of these frequencies and their weights strongly depends on the orientation and field strength. Since the sum of the total low-frequency contribution and the residual part equals the residual part of free muonium (see eq. (3.30))

$$P^{res} + \sum_k g_k = \frac{1 + 2x^2}{2 + 2x^2} \tag{7.25}$$

the g_k become small with increasing fields. For low fields, however, these oscillations should be observable.

In the above calculations only the shf interaction with the four nearest halogen nuclei has been taken into account. The shf parameters a and b for the nearest alkali nuclei (as measured with ESR and ENDOR [91]) are of the order of a few MHz. The corresponding additional interactions will modify the results for very low fields (of the order of a few gauss) in the following way. The residual polarization P^{res} will be further reduced and additional transition frequencies appear in the low frequency region and as satellites to the frequencies ω_k. Estimates show that the frequency distribution around each ω_k is of the order of the frequency bin width used in Fig. 51 and will not change the conclusion on the experimental observability of the frequencies ω_k.

These results show that interesting measurements of various properties of muonic U_2-centers are possible, provided that enough events are recorded to guarantee the necessary high statistics. An estimate shows that about 10^7 decay events have to be sampled. Although some experiments with muons in KCl [94] and in KCl and NA Cl [95] have already been performed, these investigations are not systematic enough to allow a clear analysis. The results of

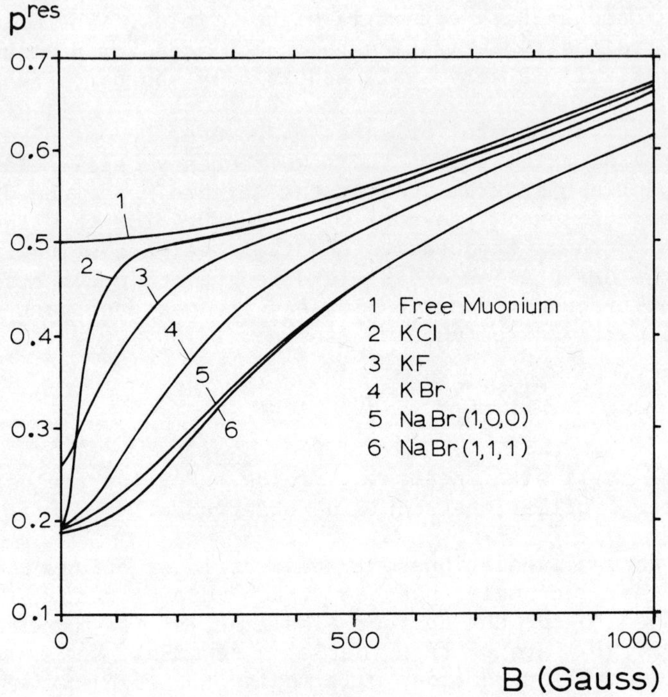

Fig. 50. Calculated quenching curves in various alkali halides for muonic U_2-centers in the presence of a superhyperfine interaction.

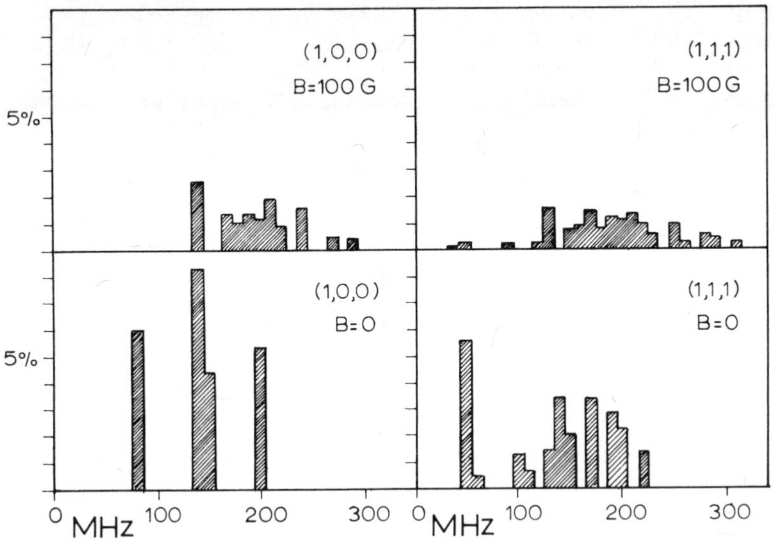

Fig. 51. Histogram of precession frequencies for muonic U_2-centers in KF for various conditions. Only the low frequency terms (<1 GHz) have been plotted in 10 MHz wide bins.

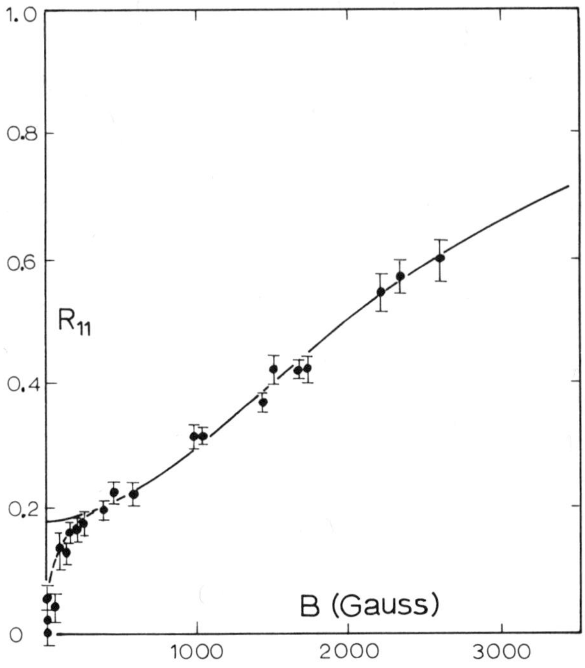

Fig. 52. Residual polarization versus magnetic field strength obtained in KCl at room temperature.

[95] indicate that muonium is chemically stable up to high temperatures (400°C) in KCl and NaCl, since no free muon signal has been observed in strong transverse fields. Ivanter et al. [94] have measured a quenching curve in a single crystal of KCl at room temperature. Their results are shown in Fig. 52. As can be seen, the experimental curve falls considerably below the calculated curve shown in Fig. 50. The steep rise at very small fields may be due to RLMF that are already quenched in very small longitudinal fields. The quenching curve at higher fields remains unexplained. Several additional processes may have to be taken into account, e.g. the lighter mass of the muonic U_2-center may lead to a larger vibration amplitude, increasing essentially the average contact interaction. (Using larger shf-constants, the experimental quenching curve can in fact be reproduced, except for the steep rise in low fields.) Also tunneling from one interstitial position to the next one cannot be excluded and may, if the rate is not too fast, constitute an efficient depolarization mechanism.

VIII. CONCLUDING REMARKS

Perhaps it is still too early to draw conclusions from the existing data about the analytical power of the μSR method in solid state physics. In a year or so, when experiments at TRIUMF and SIN and possibly CERN have really begun to accumulate data on a larger scale, and with the advancement of refined theories for interpreting the data, one will probably know more about it. I think, nonetheless, that already the body of experimental data and experience that is available today indicates that in fact μSR will find its place among other established methods like Mössbauer effect etc. The vast field of magnetism is certainly one of the areas where the μSR method is hoped to contribute to its investigation. In close connection with this, diffusion has been intensively studied, not only for basic reasons, but also as a necessity in analyzing data obtained in magnetic materials adequately. From a basic point of view, $\mu^{\pm}$ diffusion is also appearing to be a very interesting subject as it may allow a closer look at quantum rate processes.

The study and understanding of charge screening effects is likewise essential and a prerequisite to the analysis of μSR data obtained in ferromagnetic metals and alloys. The study of charge screening effects in metals by μSR appears to be very interesting in itself as it presents a kind of model situation, where a bare positive charge is implanted in an approximately free electron gas, thus perhaps allowing study of many body calculations concerning the response properties of the electron gas. Another field of μSR application, not treated in these lectures, is the study of critical phenomena or fluctuations in the neighborhood of magnetic phase transitions [75]. The ability of μSR to measure relaxation rates down to about 10^{-8} sec appears to make it an ideal tool for just that purpose. Likewise, the possibility of measuring electronic relaxation times down to $10^{-11} - 10^{-12}$ sec by the depolarization of muons in muonium should be exploited along these lines.

The study of muonium-like impurity centers in semiconductors and insulators, also magnetic ones, is hoped to provide information not only on the phenomenology of these centers, but vice versa on the properties of the host lattice that determine the properties of the impurity centers. With regard to this, it is interesting to note that, although hydrogen is the most common impurity in technically important semiconductors like Ge and Si, the only knowledge we have about the possible electronic structures of hydrogen impurities stem from the information on deep and shallow donor muonium states in these substances [85].

ACKNOWLEDGEMENTS

The preparation of these lectures has benefited greatly from the many discussions I had with colleagues and the information, help and advice I received from them. In particular I have to thank Drs. P.F. Meier, W. Rüegg, F.N. Gygax, M. Camani, B. Patterson and our graduate students H. Graf and H. Schilling (University of Zürich and ETH Zürich).

Last but not least I wish to express my thanks to Dave Garner, (Chemistry Department, UBC) for his fine job in correcting, proofreading and managing the final version of this paper.

REFERENCES

1) Proceedings of the Meeting on Muons in Solid State Physics (Bürgenstock, Switzerland 1971), two volumes (SIN 1971).

A. Schenck and K.M. Crowe in Proc. Topical Meeting on Intermediate Energy Physics, CERN-REPORT 74-8.

2) T. Yamazaki, K. Nagamine, S. Nagamiya, O. Hashimoto et. al., Physica Scripta. 11, 113 (1975).

S. Nagamiya, K. Nagamine, O. Hashimoto and T. Yamazaki, Phys. Rev. Lett. 35, 308 (1975).

3) R.L. Garwin, L.M. Lederman, and M. Weinrich, Phys. Rev. 105, 1415 (1957).

4) J.H. Brewer, K.M. Crowe, F.N. Gygax, and A. Schenck, in Muon Physics, eds. V.W. Hughes and C.S. Wu; to be published.

5) J.H. Brewer, K.M. Crowe, F.N. Gygax, R.F. Johnson, D.G. Fleming, and A. Schenck, Phys. Rev. A9, 495 (1974).

6) A. Abragam, The Principles of Nuclear Magnetism (Clarendon Press, Oxford 1970).

7) References are found e.g. in

Hyperfine Structure and Nuclear Radiations, eds. E. Matthias and D.A. Shirley (North Holland, Amsterdam 1968).

International Conference on Hyperfine Interactions studied in Nuclear Reactions and Decay (Uppsala, Sweden, 1974). Contributed Papers and Physics Scripta 11, No. 3 (1975) (invited papers).

8) J. Christiansen et al., Phys. Rev. C1, 613 (1970).

9) M. Camani et al., SIN-Proposal R-73-02.1.

10) E.V. Minaichev et al., Sov. Phys. JETP 31, 849 (1970).

11) I.G. Ivanter and U.P. Smilga, Sov. Phys. JETP 33, 1070 (1971).

12) A. Schenck, K.M. Crowe, Phys. Rev. Lett. 26, 57 (1971).

13) G.E. Pake, J. Chem. Phys. 16, 327 (1948).

14) I.I. Gurevich et al., Phys. Lett. 40A, 143 (1972).

15) See e.g.: International meeting hydrogen in metals, Jülich 1972, Jül-conf. 6 (1972) 2 volumes.
Ber. der Bunsengesellschaft 76, (1972).

16) H.C. Torrey, Phys. Rev. 92, 962 (1953).

17) W. Eichennauer et al., Z. Metallk. 56, 287 (1965).

18) L. Katz et al., Phys. Rev. B4, 330 (1971)

19) H. Schilling, Diplomarbeit ETH Zürich (1975) (unpublished).

20) A.T. Fiory et al., Phys. Rev. Lett. 33, 969 (1974).

21) T. Springer, Quasielastic Neutron Scattering for the Investigation of Diffusive Motion in Solids and Liquids, Springer Tracts 64 (1972).

22) G.H. Vineyard, J. Phys. Chem. 3, 121 (1957).

23) Y. Ebisuzaki et al., J. Chem. Phys. 46, 1373 (1967).

24) A.D. LeClaire, Phil. Mag. 14, 1271 (1966).

25) J.A. Sussmann, Ann. Phys. 6, 135 (1971).

26) P. Gosar, Nuovo Cim. 31, 784 (1964).

27) G. Blaeser and J. Peretti in "International Conference on Vacancies and Interstitials in Metals", Jülich 1968.

28) J.A. Sussmann and Y. Weissman, Phys. stat. sol. (b) 53, 419 (1972).

29) J.M. Guil et al., Proc. R. Soc. Lond. A 335, 141 (1973).

30) See also M. Abramowitz and I.A. Stegun, Handbook of Mathematical Functions, (Dover, New York, 1965).

31) A.M. Stoneham, Collective Phenomena 2, 9 (1975).

32) L.P. Flynn and A.M. Stoneham, Phys. Rev. B1, 3966 (1970).

33) J.A. Sussmann, Phys. Kondens. Mater. 2, 146 (1964).

34) Z.D. Popovic, J.M. Stott, Phys. Rev. Lett. 33, 1164 (1974).

35) C. Kittel, Introduction to Solid State Physics, (Wiley, New York, 1971).

36) A. Narath in Hyperfine Interactions, ed. A.J. Freeman, R.B. Frankel (Academic, New York, 1967).

37) W. Kohn and L.J. Sham, Phys. Rev. 140A, 1133 (1965).

38) L. Hedin, B.J. Lundquist, J. Phys. C : Proc. Phys. Soc. 4, 2064 (1971).

39) L. Hulthén, M. Sugawara, P.H. 39 (Springer 1957).

40) P.F. Meier, Helv. Phys. Acta. 48, 227 (1975).

41) J. Friedel, Adv. Phys. 3, 446 (1954), Nuovo Cim. Suppl. 7, 287 (1958).

42) C. Kittel, Quantum Theory of Solids, (Wiley, New York 1966).

43) R. Jost, Helv. Phys. Acta 20, 256 (1947).

44) See also: Goldberger and Watson, Collision Theory, (Wiley, New York 1967).

45) See e.g. F. Calogero, Variable Phase Approach to Potential Scattering, (Academic, New York 1967).

46) D.P. Hutchinson et al., Phys. Rev. 131, 1351 (1963).

47) See e.g. R.E. Watson in Hyperfine Interactions, loc. cit.

48) W. Rüegg, private communication.

49) W. Marshall, S.W. Lovesey, Theory of Thermal Neutron Scattering (Clarendon, Oxford 1971).

50) M.L.G. Foy et al., Phys. Rev. Lett. 30, 1064 (1973).

51) B.D. Patterson et al., Phys. Lett. 46A, 453 (1974).

52) B.D. Patterson, thesis (LBL-Berkeley 1975) (unpublished)

53) I.I. Gurevich et al., Sov. Phys. JETP 39, 178 (1974).

54) D.E. Murnick et al., in Contributed Papers, International Conference on Hyperfine Interactions [7].

55) I.I. Gurevich et al., ZhETF Pis. Red. 21, 16 (1975).

56) A.T. Fiory et al., BAPS 20, 319 (1975).

57) K. Nagamine et al., Proc. 14th Int. Conf. on Low Temp. Phys. Helsinki 3, 350 (1975).

58) I.I. Gurevich et al., ZhETF Pis. Red. 20, 558 (1974) (JETP Lett. 20, 254 (1974)).

59) E.O. Wollen et al., J. Phys. Chem. Solids 24, 1141 (1963).

60) See e.g. H.K. Birnbaum, C.A. Wert, Ber. d. Bunsengesellschaft 76, 806 (1972).

61) I.I. Gurevich et al., ZhETF Pis. Red. 18, 546 (1973).

62) H.A. Mook and C.G. Shull, J. Appl. Phys. 37, 1034 (1966).

63) C.G. Shull, H.A. Mook, Phys. Rev. Lett. 16, 184 (1966).

64) R.M. Moon, Phys. Rev. 136, A 195 (1964).

65) B.N. Harmon, A.J. Freeman, Phys. Rev. B10, 1979 (1974).

66) B.D. Patterson, L.M. Falicov, Solid State Comm. 15, 1509 (1974).

67) J. Lindhard, Kgl. Danske Videnskab. Selskab. Mat.-fys. Medd. 28, no. 8 (1954).

68) P. Jena, preprint, University of British Columbia (1975).

69) E. Daniel and J. Friedel, J. Phys. Chem. Solids 24, 1601 (1963).

70) P.F. Meier, Solid State Commun. 17, 989 (1975).

71) B.N. Harmon, A.J. Freeman, Phys. Rev. B10, 1979 (1974).

72) The situation may be illustrated by the following list of references:

Fe: pos. spin pol. at Fermi surface:
Duff, Das; Phys. Rev. B3, 192 (1971)
Tawil, Callaway, Phys. Rev. B7, 4242 (1973)
neg. spin pol. at Fermi surface:
M. Yasui et al., J. Phys. Soc. Japan 34, 396 (1973)
Wakoh, Yamashita, J. Phys. Soc. Japan 25, 1272 (1968)

Ni: neg. spin pol. at Fermi surface:
- Stoner-Wohlfarth model }
- Slater model } see e.g. C. Herring, in Magnetism IV Vol. IV, eds. G.T Rado and H. Suhl (Academic, New York, 1966)

- Wakoh, Yamashita loc. cit.
Connolly, Phys. Rev. 159, 415 (1967)

Co: neg. spin. pol. at Fermi surface:
Ballinger, Marshall, J. Phys. F : Metal Physics 3, 736 (1973).

73) H.C. Siegmann, Physics Rep. 17C, no. 2 (1975).

74) W.J. Kossler in "Proc. Topical Meeting on Intermediate Energy Physics", Zuoz, 1973 loc. cit.

75) B.D. Patterson et al., 20th Conf. Mag. Magnetic Materials, San Francisco (1974).

76) I.I. Gurevich et al., Sov. Phys. JETP 33, 253 (1971).

77) J.H. Brewer et al., Phys. Rev. Lett. 31, 143 (1973).

78) G. Feher, R. Prepost, A.M. Sachs, Phys. Rev. Lett. 5, 515 (1960).

79) B. Eisenstein, R. Prepost, A.M. Sachs, Phys. Rev. 142, 217 (1966).

80) D.G. Adrianov et al., Sov. Phys. JETP 31, 1019 (1970).

81) W. Kohn and J.M. Luttinger, Phys. Rev. 97, 869 (1955).

82) H. Reiss, J. Chem. Phys. 25, 681 (1956).

83) P.E. Kaus, Phys. Rev. 109, 1944 (1958).

84) J. Shy-Y. Wang, C. Kittel, Phys. Rev. B7, 713 (1973).

85) H. Nara, J. Phys. Soc. Japan 20, 778 (1965).
J.P. Walter, M.C. Cohen, Phys. Rev. B2, 1821 (1970).
P.K.W. Vinsome, D. Richardson, J. Phys. C4, 2650 (1971).

86) See e.g. B. Bleaney, in Hyperfine Interactions, loc. cit.

87) R.F. Johnson, private communication.

88) G.G. Myasishcheva et al., Sov. Phys. JETP 26, 298 (1968).

89) A. Schenck, Helv. Phys. Acta 47, 431 (1974).
P.F. Meier, A. Schenck, Phys. Lett. 50A, 107 (1974).

90) J.M. Spaeth, Z. Physik 192, 107 (1966).

91) J.M. Spaeth, M. Sturm, Phys. stat. sol. 42, 739 (1970).

92) J.L. Hall, R.T. Schuhmacher, Phys. Rev. 127, 1892 (1962).

93) H. Seidel, H.C. Wolf, Phys. stat. sol. 11, 3 (1965).

94) I.G. Ivanter et al., Sov. Phys. JETP 35, 9 (1972)

95) Unpublished results of the LBL group (Berkeley), 1973.

96) R. Beck, P.F. Meier, A. Schenck, Z. Physik B22, 109 (1975).

97) D. Lepski, Phys. Stat. Sol. 35, 697 (1969).

98) J. Lindhard, A. Winther, Nucl. Phys. A166, 413 (1971).

Warren

Q. There is presumably a stable negative muonium ion. Has the presence of this been detected in any of the muonium structures?

Schenck

A. No! In fact, it would be quite difficult to detect a negative muonium ion, because magnetically it very much resembles a free unbound μ^+ due to the diamagnetic structure of the negative ion and therefore it will precess with a frequency very close to the Larmor frequency of a free muon. There is however a small difference of the order of ppm's due to the diamagnetic shielding in $\mu^+e^-\uparrow e^-\downarrow$, which may be exploited in superprecise precession frequency measurements.

The question, whether one has to think of the μ^+ in certain metals as being quasi-free or in the state of a negative ion, is however a quite important one as it relates to the question of the electronic structure of dissolved hydrogen atoms in metals, which is still a matter of controversy, particularly in rare earth hydrides [15].

INTERACTIONS OF PIONS WITH NUCLEI

F. Scheck

SIN and ETH

Zurich, Switzerland

1. INTRODUCTION

With the advent of "pion factories" and of a number of pion spectrometer facilities with especially high resolution, pion-nucleus physics is entering a new era. Indeed, over the past fifteen years, theories and experiments on pion interactions with nuclei have been mostly of an exploratory nature only, and only rarely have they reached the precision and sophistication of experiments and theoretical analyses in, say, electron-nucleus, muon-nucleus or photo-nuclear physics. It is quite certain that this state of the field will change in the near future--mainly under the (expected) impetus of more refined and more precise experiments at the new accelerators. How rich the field of pion-nucleus physics actually is has been revealed, for instance, by the precision measurements of pion cross sections on nuclei by a CERN group [1], and of proton induced pion production near threshold by an Uppsala group [2]. Both these experiments which we quote as examples among other pioneering experiments, have revealed rather unexpected features of pion-nuclear interactions which subsequently have stimulated a great deal of theoretical activity. In particular, it has become clear that, potentially at least, the pion is indeed a good tool for the purpose of studying certain aspects of nuclear structure and of the behaviour of a hadron in nuclear matter--in a way which may well turn out to be rather complementary to the interaction of the nucleus with electromagnetic or weak probes.

At this turning point it may be good to ask where we actually stand in pion-nucleus physics. About the pion itself we have learned very little, in the past, from pion-nucleus experiments, and there is little chance that this will change in the future. The few exceptions are: the pion mass (π^-) whose most precise determination was done in pionic atoms; radiative corrections in pionic atoms (vacuum polarization in atoms with heavy nuclei) may be an interesting topic in this context, for the future. Otherwise, the intrinsic properties of the pion as a member of the spectrum of hadrons, its weak interactions, etc. are fairly irrelevant when it interacts with a nucleus.

On the other hand, we have not learned much about the nucleus, either, from pion-nuclear experiments. This was due to the fact that our theoretical descriptions of such experiments were too rudimentary and too crude so that, in most cases, the nuclear structure part had to be assumed known and had to be put into them. In fact, most investigations concentrated on the interaction mechanisms of the pion with the nuclear many body system, assuming the dynamics of the nucleus to be known from conventional, low-energy nuclear physics. The popular methods used to describe pion-nucleus interactions are, depending on the energy domain one is considering, optical potentials, eikonal methods or other intermediate theories of multiple scattering.

In these lectures we start out with a brief discussion of the pion-nucleon system, as far as it is relevant to the multiple scattering formalism discussed later on, and with a summary of definitions and basic features of pion-nucleus scattering and pionic bound states (Chapter 2). In Chapter 3 we formulate the basic equations for multiple scattering and derive and discuss the pion-nucleus optical potential, valid at low energies. We then go a little more into the fixed scatterer approximation and the theorem, due to Bég, on the pion-nucleus scattering amplitude for non-overlapping elementary interactions. This is the content of Chapter 4. Up to this point we take little care of the elementary pion-nucleon amplitudes assuming them to be smooth, well-behaved, short-ranged, etc. according to the needs. In reality, pion-nucleon scattering at intermediate energies is dominated by one resonating partial wave, the $J = I = 3/2$ N^* resonance, and, clearly, this must also show up in pion-nucleus scattering at appropriate energies. Chapter 5 is devoted to the first steps in the attempt of incorporating the N^* resonance into the multiple scattering formalism. We discuss, in particular, how and why the resonance upsets most of the static approximations of the previous sections, and we sketch a few of the qualitatively new phenomena which follow from the existence of a resonant pion-nucleon state. In Chapter 6, finally, we discuss the strong interaction hyperfine effects of pionic atoms with strongly deformed nuclei, in order to give at least one example where some new and interesting information on the nucleus itself can be obtained, in spite of the uncertainties in the optical model description of pionic atoms.

Since we intend here a pedagogical and mostly introductory treatment of the subject we do not show many data and detailed comparisons with specific models. As far as hadronic atoms are concerned, this omission is largely repaired by the experimental review of the topic by Dr. Koch at this School. For the rest, we refer to the review articles on pion-nucleus physics which have come out recently [3-5] and to recent reports at conferences on intermediate energy physics. Our list of references is far from complete and many more will be found, for example, in Hüfner's review in Physics Reports [3].

2. BASIC FEATURES OF PION-NUCLEON AND PION-NUCLEUS INTERACTIONS

The Pion-Nucleon Amplitudes

(a) Partial wave amplitudes. In this section we review briefly the main properties of pion-nucleon scattering, especially at low and intermediate energies. This serves the purpose of defining the notation and of providing us with the basic formulae to which we will be referring throughout these lectures.

Take any charge state $\pi^i N^j$ of pion and nucleon, and consider elastic scattering first. The scattering matrix is expanded, in the center-of-mass system, in terms of partial waves

$$F(E,\theta) = \sum_{\ell=0}^{\infty} (2\ell+1)[F_{\ell-}(E)\Pi_{j=\ell-\frac{1}{2}} + F_{\ell+}(E)\Pi_{j=\ell+\frac{1}{2}}]\, P_\ell(\cos\theta) \tag{2.1}$$

where

$$\Pi_{j=\ell-\frac{1}{2}} = \frac{\ell-\vec{\sigma}\cdot\vec{\ell}}{2\ell+1} \; ; \quad \Pi_{j=\ell+\frac{1}{2}} = \frac{\ell+1+\vec{\sigma}\cdot\vec{\ell}}{2\ell+1}$$

are projection operators for $j = \ell \mp 1/2$, respectively. The scattering amplitude is obtained by taking this matrix between initial and final Pauli spinors for the nucleon. Eq. (2.1) can be written alternatively as

$$F(E,\theta) = \sum_{\ell=0}^{\infty} [\ell F_{\ell-}(E) + (\ell+1)F_{\ell+}(E)]\, P_\ell(\cos\theta) \tag{2.2}$$

$$-i\vec{\sigma}\cdot\hat{q}\times\hat{q}' \sum_{\ell=0}^{\infty} [F_{\ell-}(E)-F_{\ell+}(E)]P'_\ell(\cos\theta); \quad (\hat{q}\equiv\frac{\vec{q}}{|\vec{q}|})$$

The unitarity relation for the partial wave amplitudes $F_{\ell+1}$ reads

$$\mathrm{Im}\, F_{\ell\pm}(q) \geq q|F_{\pm}(q)|^2 \tag{2.3}$$

q being the magnitude of the c.m. three momentum,

$$q = \frac{1}{2E}\sqrt{(E^2-M_\pi^2-M_N^2)^2 - 4M_\pi^2 M_N^2} \tag{2.4}$$

The inequality (2.3) is satisfied by the general form

$$F_{\ell\pm}(q) = \frac{\eta_{\ell\pm}(q)\, e^{2i\delta_{\ell\pm}(q)} - 1}{2iq} \tag{2.5}$$

with $0 \leq \eta_{\ell\pm}(q) \leq 1$ (inelasticity), $\delta_{\ell\pm}(q)$ being the real phase shift function. The optical theorem, finally, reads

$$F(E,\theta=0) = \frac{q}{4\pi}\sigma_{tot}(E) \tag{2.6}$$

where $\sigma_{tot}(E)$ is the total pion-nucleon cross section at the corresponding energy. A similar relation holds also for the partial wave amplitudes and the corresponding partial wave cross sections.

We will be dealing quite extensively with pion scattering at low energies. In favourable cases the expansion (2.2) might be restricted to s- and p-waves only. In an obvious notation we then have

$$F(E,\theta) \simeq F_{s\frac{1}{2}} + (F_{p\frac{1}{2}} + 2F_{p\frac{3}{2}})\cos\theta$$

$$- i\vec{\sigma}\,\frac{\hat{q}\times\hat{q}'}{|\hat{q}\times\hat{q}'|}\sin\theta\,(F_{p\frac{1}{2}} - F_{p\frac{3}{2}}) \tag{2.7}$$

with $\hat{q} = \vec{q}/|\vec{q}|$; $\vec{q}$ and $\vec{q}\,'$ being the pion momenta before and after the collision. Near threshold, in particular, we define the scattering lengths and volumes as usual by

$$\lim_{q\to 0}\left(\frac{F_{\ell j}}{q^{2\ell}}\right) = \lim_{q\to 0}\left(\frac{\delta_{\ell j}}{q^{2q+1}}\right) \equiv a_{\ell j} \tag{2.8}$$

(b) Isospin decomposition. For the sake of simplicity, we will give the isospin decomposition for the scattering lengths (2.8) only, the formulae being trivially extended to the more general scattering amplitudes. The standard notation is the following:

For s-Wave: a_{2T} with T the total isospin of the pion-nucleon system;

for p-Wave: $a_{2T,2j}$ with $j = 1/2$ or $3/2$

Thus, introducing the isospin dependence explicitly, eq. (2.7) takes the form

$$F(E,\theta) \simeq b_0 + b_1(\vec{t}\cdot\vec{\tau}) + [c_0 + c_1(\vec{t}\cdot\vec{\tau})]\vec{q}\cdot\vec{q}\,' - i\vec{\sigma}\cdot\vec{q}\times\vec{q}\,'[d_0 + d_1(\vec{t}\cdot\vec{\tau})] \qquad (2.9)$$

Here $\vec{t}$ denotes the isospin operators of the pion; $\vec{\tau}$ the ones of the nucleon. Using the relations

$$\langle T|\vec{t}\cdot\vec{\tau}|T\rangle = T(T+1) -2 -\frac{3}{4} \qquad \text{(T: total isospin)}$$

$$i\vec{\sigma}\vec{q}\times\vec{q}\,'\ P'_\ell(z) = \vec{\sigma}\cdot\vec{\ell}\ P_\ell(z)$$

$$\langle j|\vec{\sigma}\cdot\vec{\ell}|j\rangle = j(j+1) - \ell(\ell+1) - \frac{3}{4}$$

One has

$$b_0 = 1/3(a_1+2a_3); \quad b_1 = 1/3(a_3-a_1)$$

$$c_0 = 1/3[(a_{11}+2a_{31}) + 2(a_{13}+2a_{33})]$$

$$c_1 = 1/3[(a_{31}-a_{11}) + 2(a_{33}-a_{13})]$$

$$d_0 = 1/3[2a_{31}+a_{11}-2a_{33}-a_{13}] \qquad (2.10)$$

$$d_1 = 1/3[a_{31}-a_{11}-a_{33}+a_{13}]$$

We note here the experimental values for these quantities [6]

$$a_1 - a_3 = 0.288 \begin{matrix} +0.012 \\ -0.018 \end{matrix} \; m_\pi^{-1}$$

$$a_1 + 2a_3 = 0 \pm 0.04 \; m_\pi^{-1}$$

$$a_{11} - a_{31} = -0.045 \pm 0.006 \; m_\pi^{-3}$$

$$a_{13} - a_{33} = -0.243 \pm 0.007 \; m_\pi^{-3} \qquad (2.10')$$

$$a_{11} + 2a_{31} = -0.164 \pm 0.008 \; m_\pi^{-3}$$

$$a_{13} + 2a_{33} = 0.414 \pm 0.021 \; m_\pi^{-3}$$

We notice, in particular, the remarkable fact that b_0 is very much smaller in magnitude than a_1, a_3 or b_1. This implies that the s-wave interaction of a pion with an isospin zero nucleus which one naively would expect to dominate at low energies is, in fact, very small. This observation is crucial for the interaction of pions with nuclei at low energies. The s-wave term which normally is expected to be dominant, is almost absent, so that higher partial waves as well as renormalization effects in the nuclear medium, etc. become important.

Pion-Nucleus Scattering, Preliminaries

Before we turn to the theoretical description of pion-nucleus interactions we would like to recall briefly the main quantities that are accessible in experiments with pions and nuclei. Information about pion-nucleus interactions is obtained in a variety of ways, among which we will be concerned especially with pion elastic or inelastic scattering and with pionic atoms. Other aspects, such as pion production are treated in Dr. Reitan's lectures. In <u>scattering experiments</u> there are various quantities which can be determined: total cross sections, elastic differential cross sections, pion absorption with the observation of specific final states (such as neutron spectroscopy after pion absorption), pion single and double charge exchange, and so on. In all of these processes two phenomena are predominant:
(i) at intermediate energies, say, for $100 \leq T_\pi \leq 300$ MeV, pion-nucleus scattering is dominated by the $N^*(1236)$ resonance which appears in the p-wave pion-nucleon system with quantum numbers $j=I=3/2$. Fig. 1 shows the N^{*++} state in π^+p scattering, as an example.

The resonance appears at

$$T_\pi^{lab} = 185 \text{ MeV, i.e. } p_\pi^{lab} = 300 \text{ MeV/c} \qquad (2.11)$$

with a width of about $\Gamma_{N^*} \simeq 110$ MeV. The values of the total cross sections at resonance are

$$\sigma_{tot}(\pi^+ p) = 200 \text{ mb}; \quad \sigma_{tot}(\pi^- p) = 68 \text{ mb} . \qquad (2.12)$$

This resonance also shows up very strongly when we scatter pions off nuclei, though its width appears broadened and its position shifted towards lower energies by up to several tens of MeV [1]. This is illustrated by Figs. 2 and 3.

The occurrence of a strong resonance above threshold, and the predominence of one single partial wave is a new feature in hadron-nucleus scattering. We believe that this situation has not yet been fully exploited theoretically and that a lot more analysis will have to be done before we understand pion scattering on nuclei, in the presence of the N^* resonance. We will, therefore, not go into the rather crude models which have been put forward to explain the shift of the resonance in nuclei. These are briefly described in Hüfner's review article [3] where also many references can be found.

(ii) The second important and most interesting phenomenon is pion absorption. Simple kinematic arguments tell us that pion absorption on a single nucleon must be highly suppressed. The argument is that the momentum transfer to the absorbing nucleon is of the order of $(2m_n m_\pi)^{1/2} \simeq 525$ MeV/c and is far too large for nuclear dimensions: The corresponding high momentum Fourier components in the nucleon's shell model wave function are expected to be very small indeed. Pion absorption on clusters of nucleons, two at least, is much favoured, as the momentum transfer can now be shared between different partners in the nucleus. For example, if the pion is absorbed by a first nucleon which shares its momentum with a second nucleon, then the momentum transfer to each of them will be of the order of $(m_n m_\pi)^{1/2} \simeq$ 365 MeV/c. Here a clue comes from the observation of the decay products after pion absorption. We quote as an example the case of pion absorption in ^{165}Ho, where emission of up to eleven neutrons has been observed through the de-excitation X-rays of the ground state rotational band of the daughter nuclei $^{165-x}$Dy [9]. We must add, however, that pion absorption is a complicated process and the detailed mechanism which leads to the emission of these neutrons, for example, is far from understood.

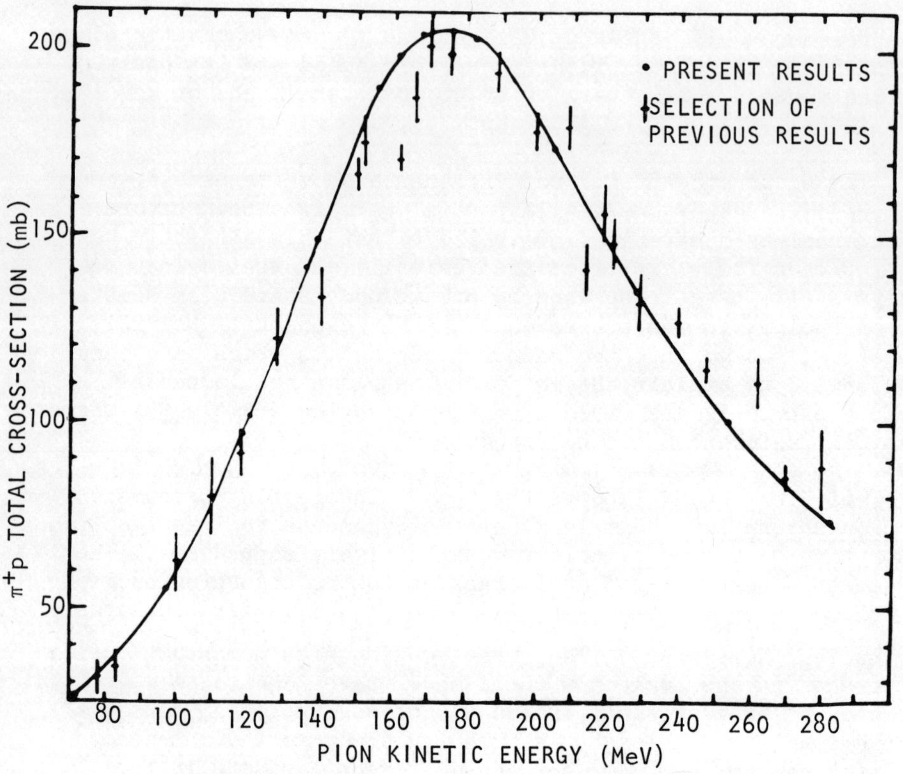

Fig. 1. The π^+p total cross section (Taken from A.A. Carter *et al.*, Ref. [7])

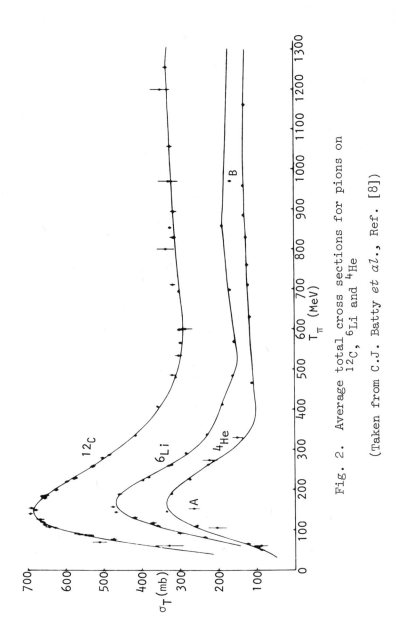

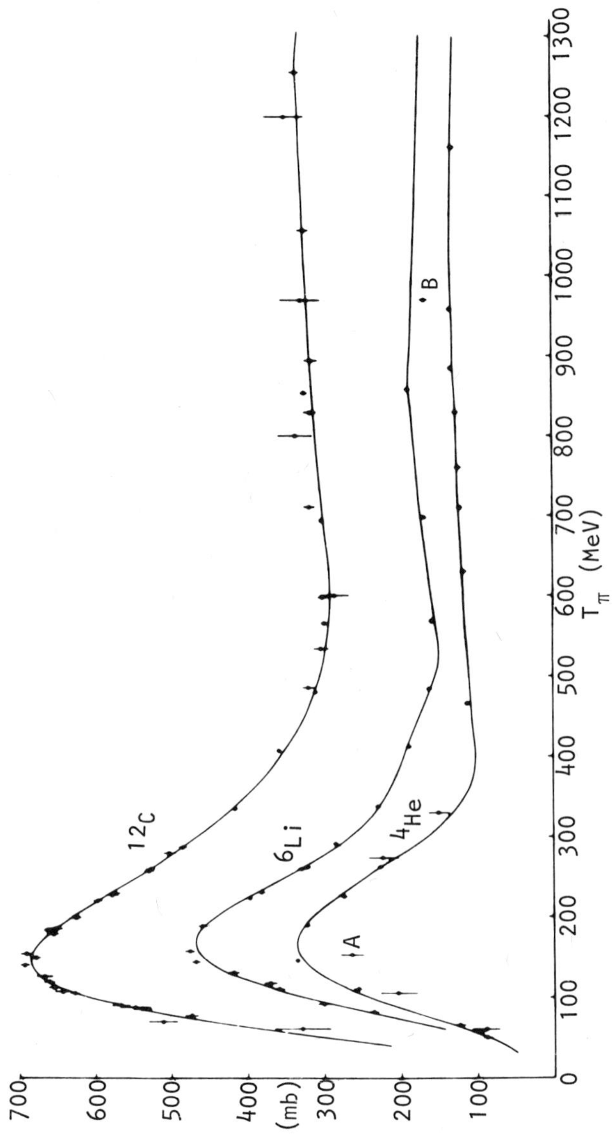

Fig. 2. Average total cross sections for pions on ^{12}C, ^{6}Li and ^{4}He

(Taken from C.J. Batty *et al.*, Ref. [8])

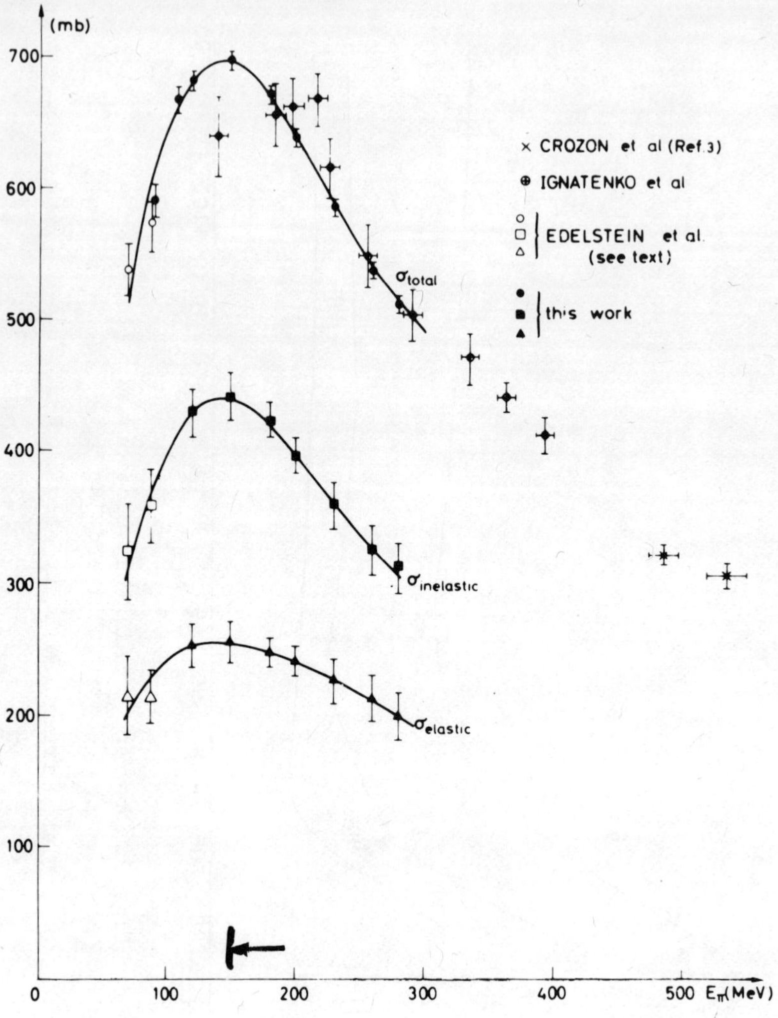

Fig. 3. π^- ^{12}C total cross section, total elastic cross section, and total inelastic cross section versus the π^- energy in the laboratory.

(Taken from F. Binon *et al.*, Ref. [1])

More generally, and beyond the two points discussed above, we may say that pion reactions on nuclei are particularly interesting because the charge states of the pion offer more degrees of freedom with which one can play than, e.g., in nucleon-nucleus scattering. Indeed, the pion can undergo single and double charge exchange in nuclei and thus can connect rather different states in neighbouring nuclei.

Pionic Bound States

The negative pion can be trapped in the Coulomb field of a nucleus. The system then forms a hydrogen-like atom with dimensions characterized by the pion's Bohr radius

$$a_\pi = \frac{1}{m_\pi \cdot Z\alpha} = (194/Z)\mathrm{fm} \tag{2.13}$$

The cascade from the initial atomic orbit into which the pion is captured ($n \simeq 15 - 20$), down to the level from which the pion eventually disappears through strong absorption in the nucleus, takes about 10^{-15} to 10^{-18} sec (depending on Z, the nuclear charge number). This is a very short time as compared with the pion's lifetime $\tau_\pi = 2.6 \times 10^{-8}$ sec. The cascade proceeds, for low transition energies (below ~ 100 keV) mainly through Auger transitions, for higher transition energies ($\gtrsim$ 100 keV) predominantly through emission of electric dipole X-rays. It is these X-rays which are normally being recorded and analyzed theoretically [10,11]. The strong interaction of the pion with the nucleus manifests itself through a (real) shift $\varepsilon(n,\ell)$ of the atomic energy levels and an additional level broadening $\Gamma_{abs}(n,\ell)$ due to the strong absorption of the pion. The shift, specifically, is defined as the difference of the energy of the same level (n,ℓ) in the field of a point-like Coulomb field with charge Z, and the actual position of the level which is due to the finite size of the nuclear charge distribution and to the strong interaction,

$$\varepsilon(n,\ell) \equiv E^{\mathrm{Point}}_{n,\ell} - E_{n,\ell} \text{ (finite size + strong interaction included)} \tag{2.14}$$

with

$$E^{\mathrm{Point}}_{n,\ell} = \bar{m}\left\{1 + \left(\frac{Z\alpha}{n-\ell-\frac{1}{2}+\sqrt{(\ell+\frac{1}{2})^2-(Z\alpha)^2}}\right)^2\right\}^{-\frac{1}{2}} \tag{2.15}$$

(energy eigenvalue of Ze/r potential for Klein-Gordon equation), $\bar{m}$ being the reduced mass of the pion and the nucleus. Due to its short range, the effects of the strong interaction will, very roughly, be proportional to the overlap of the pion's wavefunction and the nuclear mass density $\rho_A(\vec{r})$. This explains at once two facts: Firstly, in the lower part of its cascade the pion runs predominantly through circular orbits ($n = \ell + 1$). For fixed main quantum number n, these orbits have the least overlap with the nucleus, owing to their centrifugal tail r^{ℓ} near the origin. Inner orbits have a larger overlap with the nucleus and the pion will quickly be absorbed from them. Secondly, and more importantly, this estimate shows that $\Gamma_{abs}(n,\ell)$ (and similarly $\varepsilon(n,\ell)$) will depend little on n, but rather strongly on ℓ. From more realistic calculations, one finds, indeed, that $\Gamma_{abs}(n,\ell)$ behaves thus

$$\frac{\Gamma(n-1,\ell)}{\Gamma(n,\ell)} \sim 3 \text{ to } 5 \text{ but } \quad \frac{\Gamma(n,\ell-1)}{\Gamma(n,\ell)} \sim 10^2 \text{ to } 10^3 \qquad (2.16)$$

When proceeding from one circular orbit (n,n-1) to the next lower one (n-1, n-2) the absorption width thus increases by a factor between 10^2 and 10^3. In contrast to this, the radiative width Γ_{rad} (E1 transitions mainly) increases much more slowly, by about a factor 2 to 3 for the same step. This implies that the atomic cascade terminates rather abruptly at a critical level (n_c, $\ell_c = n_c - 1$) for which Γ_{abs} is larger than Γ_{rad}, and from which the pion is absorbed with high probability, having little chances to survive another E1 transition. (Remark that the next lower level is already so much broadened that it would not be possible to identify the X-ray leading to it, even if it still had a measurable intensity). Above this critical level, however, the atomic states of the pion will be hydrogenic, to a very good approximation, with no other modification of the simple hydrogen atom than through radiative corrections (vacuum polarization by e^+e^- -pairs mainly), and possibly small screening effects due to the electronic shells of the host atom.

In practice one can measure actually three quantities: shift and width of the critical level are read off directly from the X-ray corresponding to the transition $(n_c+1, n_c) \rightarrow (n_c, n_c-1)$; the width of the circular level just above the critical level is obtained from the intensity of the same X-ray. We thus have

$$\varepsilon(n_c);\ \Gamma_{abs}(n_c) \quad \text{and} \quad \Gamma_{abs}(n_c+1) \qquad (2.17)$$

at our disposal if we want to learn something about the pion-nucleus interaction from a pionic atom. Which level actually is the critical one in a given element, depends obviously on its charge number Z. Roughly, we have

$$\begin{array}{ll} 1s: & Z \lesssim 12 \\ 2p: & 13 \lesssim Z \lesssim 30 \\ 3d: & 30 \lesssim Z \lesssim 60 \\ 4f: & 60 \lesssim Z \lesssim 92 \end{array} \qquad (2.18)$$

As far as the theoretical analysis of pionic atoms is concerned, the situation sketched above has one important consequence: For all those E1 transitions which can actually be measured, Γ_{abs} is either very much smaller than, or at most comparable with Γ_{rad}. Therefore, for these orbits the Coulomb field remains the dominant term in the pion's equation of motion. The strong interaction, in some sense, appears only as a perturbation to this. Note, in particular, that

$$|\varepsilon(n_c)|,\ \Gamma(n_c) \ll |E_{n_c+1}-E_{n_c}| \tag{2.19}$$

This would not be true any more for the next level, just below the critical level (provided we could see it in an experiment). Here the situation is reversed, as the strong interaction is the predominant feature and the Coulomb field may even be negligible. This observation forms the basis for our semi-static treatment of the pion's energy levels, even though the pion appears as a highly unstable particle due to its strong absorption whenever it dips into the nucleus.

3. THE OPTICAL POTENTIAL IN THE STATIC APPROXIMATION

In what follows we will assume that the pion-nucleon interaction can be described by means of potentials $v_{\pi N}$. Even though we will try to eliminate these potentials from the final expressions for the pion-nucleus amplitudes, and to replace them by suitable pion-nucleon scattering amplitudes as much as this is possible, we will nevertheless have to rely on the potential picture as the theoretical basis of our developments. This is the basic assumption for virtually all theoretical descriptions of multiple scattering of pions in nuclei.

Starting from a potential picture it is relatively easy to write down exact equations for the description of multiple scattering in nuclear matter or in finite nuclei. It is much less easy, we will see, to actually solve these equations in even a very crude and approximate form.

Let us define

t_i : t-matrix describing pion scattering on a free nucleon, labelled i, and with suitable spin and isospin indices.

τ_i : t-matrix for the scattering of the pion on a bound nucleon, again labelled i.

Let T_π be the kinetic energy of the pion, H_A the nuclear Hamiltonian which describes the motion of the nucleons. Then the t-matrices defined above satisfy the Lippmann-Schwinger equations*.

$$t_i(E) = v_i + v_i \frac{1}{E-T_\pi+i\varepsilon} t_i(E) \tag{3.1}$$

$$\tau_i(E) = v_i + v_i \frac{1}{E-H_A-T_\pi+i\varepsilon} \tau_i(E) \tag{3.2}$$

If $T_A(E)$ denotes the pion-nucleus scattering matrix, then this matrix is a solution of the L.S. equation

$$T_A(E) = \sum_{i=1}^{A} v_i + \sum_{i=1}^{A} v_i \frac{1}{E-H_A-T_\pi+i\varepsilon} T_A(E) \tag{3.3}$$

It is convenient to introduce the auxiliary t-matrices

$$T_A^i(E) = v_i + v_i \frac{1}{E-H_A-T_\pi+i\varepsilon} T_A(E) \tag{3.4}$$

in terms of which

$$T_A(E) = \sum_i T_H^i(E)$$

Making use of eqs. (3.2) one can recast eq. (3.3) into a set of coupled equations for these auxiliary t-matrices. From eq. (3.4)

$$\left(1 - v_i \frac{1}{E-H_A-T_\pi+i\varepsilon}\right) T_A^i = v_i \left[1 + \frac{1}{E-H_A-T_\pi+i\varepsilon} \left(T_A-T_A^i\right)\right]$$

we get

$$T_A^i(E) = \tau_i(E) \left[1 + \frac{1}{E-H_A-T_\pi+i\varepsilon} \sum_{j\neq i} T_A^j\right] \tag{3.5}$$

These equations are exact expressions of the multiple scattering of pions in nuclei. Of course, it is vain to hope that we could solve them in an exact manner. Actually, this is not what we really would like to do as these equations contain by far too much information for situations of practical interest. Indeed, one must keep in mind that the t-matrices defined above contain all pionic and nucleonic variables and thus describe all kinds of elastic and inelastic processes between the incident pion and the nucleus. In practice, we will always try to simplify these equations by introducing further approximations, depending on the

---

*Note that τ is not defined uniquely. For instance, one could include a projection operator onto antisymmetric nucleon states in the Green's function in eq. (3.2).

situation we are considering. Before doing so, however, let us go one step further by writing down a formal iterative solution of eq. (3.5).

$$T_A = \sum_{i=1}^{A} T_A^i = \sum_i \tau_i + \sum_{i\neq j} \tau_i\, G^+\tau_j + \sum_{\substack{i\neq j\\ j\neq k}} \tau_i G^+\tau_j\, G^+\tau_k + \ldots \tag{3.6}$$

where

$$G^+ \equiv \frac{1}{E-H_A-T_\pi+i\varepsilon}$$

Eq. (3.6) expresses T_A, the pion-nucleus scattering matrix, as a series of single, double,...,multiple scattering of the pion on the nucleons bound in the nucleus. As such, this formal series is a good starting point for many approximate solutions of the multiple scattering problem. In the following subsections we will discuss some of the popular approximations by means of which the full complexity of eqs. (3.5) or (3.6) is reduced to a, hopefully, solvable problem.

Coherent Approximation

The so-called coherent approximation consists in assuming that the nucleus remains in its ground state whenever the pion has interacted with any one of the nucleons. In its multiple scatterings through the nuclear medium the pion does not excite the nucleus in intermediate steps. Thus, when taking the expectation value of the series (3.6) in the nuclear ground state $|\Omega\rangle$ we get approximately

$$\langle\Omega|\tau_i\, G^+\tau_j G^+\ldots\tau_m|\Omega\rangle \simeq$$
$$\langle\Omega|\tau_i|\Omega\rangle G_\Omega^+\langle\Omega|\tau_j|\Omega\rangle G_\Omega^+\ldots\langle\Omega|\tau_m|\Omega\rangle \tag{3.7}$$

with

$$G_\Omega^+ = \frac{1}{E-T_\pi+i\varepsilon}$$

When inserted into eq. (3.6) this gives

$$T_A^0 \equiv \langle\Omega|T_A|\Omega\rangle = A\langle\tau\rangle + A\langle\tau\rangle G_\Omega^+(A-1)\langle\tau\rangle +$$
$$+ A\langle\tau\rangle G_\Omega^+(A-1)\langle\tau\rangle G_\Omega^+(A-1)\langle\tau\rangle + \ldots$$

where the antisymmetry of $|\Omega\rangle$ in the A nucleon variables has been used. Thus we obtain

$$\left(\frac{A-1}{A} T_A^0\right) = U_A + U_A G_\Omega^+ \left(\frac{A-1}{A} T_A^0\right) \tag{3.8}$$

with

$$U_A = (A-1)\langle\tau\rangle \tag{3.9}$$

The many body multiple scattering series has been reduced to an effective one body problem, eq. (3.8). The quantity U_A is an optical potential which depends on the pion coordinates only. The factor (A-1)/A is, of course, close to one for medium-weight and heavy nuclei and is often omitted. The still rather symbolic expression (3.9) for the optical potential needs a little more explanation. The symbol $\langle\tau\rangle$ implies an average over spin and isospin degrees of freedom, as well as an average over the nuclear ground state wavefunction, i.e. an average over the Fermi motion of the nucleons in the nucleus. Note, however, that τ is still the pion-bound nucleon t-matrix which may or may not be known. As a first approximation we might replace it by t, the free pion-nucleon t-matrix--this is the standard impulse approximation. This last step can be justified in the limit where the nuclear density is low. We then obtain, in coordinate space, the optical potential

$$U_A^0(\vec{r}) = -\frac{4\pi}{2\bar{m}} \bar{F} \rho_A(\vec{r}) \left(\frac{A-1}{A}\right) \tag{3.10}$$

$\bar{m}$: reduced mass pion-nucleus

$\bar{F}$: averaged pion-nucleon amplitude

This is the first term in any density expansion of the optical potential in nuclear matter [3].

Optical Potential in Second Order

To go beyond the simple approximation (3.8) is not a simple matter and we will not have time to go into much detail here. A good starting point is the multiple scattering equations (3.3) and (3.5) expressed in terms of incident and scattered waves,

$$\phi = \psi + \frac{1}{E-H_A-T_\pi+i\varepsilon} \sum_{i=1}^{A} \tau_i \psi_i \tag{3.11}$$

$$\psi_i = \psi + \frac{1}{E-H_A-T_\pi+i\varepsilon} \sum_{j\neq i}^{A} \tau_j \psi_j \tag{3.12}$$

where

ψ : pion wave in the absence of the pion-nucleon interaction ("incident wave")

ψ_i: partial wave incident on nucleon number i

ϕ ; wavefunction of the pion in the nuclear medium

We make use again of the FSA and use closure in the nuclear states whereby

$$(E-H_A - T_\pi+i\varepsilon)^{-1} \rightarrow (E-T_\pi+i\varepsilon)^{-1}$$

and project eqs. (3.11) and (3.12) onto the nuclear ground state. The details are worked out in Fäldt's paper [12] for example. Let $\phi(\vec{r})$ be the pion wavefunction in coordinate representation

$$\phi(\vec{r}) = \langle\vec{r},\Omega|\phi\rangle$$

and let us neglect the recoil of the nucleons, i.e.

$$\langle\vec{r}',\{\vec{x}'\}|\tau_i|\vec{r},\{\vec{x}\}\rangle \simeq \prod_{\mu=1}^{A} \delta(\vec{x}'_\mu-\vec{x}_\mu)\langle\vec{r}'-\vec{x}_i|\tau_i|\vec{r}-\vec{x}_i\rangle \tag{3.13}$$

where $\{\vec{x}\} = \{\vec{x}_1, \vec{x}_2, \ldots ,\vec{x}_A\}$ denote the nucleon coordinates, and $\vec{r}$ the pion coordinates. Instead of τ we may write the (off-shell) scattering amplitude $\overline{F}$

$$\langle\vec{r}'-\vec{x}_i|\tau|\vec{r}-\vec{x}_i\rangle = -\frac{2\pi}{\overline{m}}\overline{F}(\vec{r}'-\vec{x}_i;\vec{r}-\vec{x}_i)$$

Then, with the definitions

$$\rho(\vec{x}_i)\psi(\vec{r};\vec{x}_i) \equiv \langle\vec{r}\Omega| \sum_{\mu=1}^{A} \delta(\vec{x}_\mu-\vec{x}_i)\psi_\mu|\Omega\rangle \tag{3.14}$$

and $\rho(\vec{x}_i)\rho(\vec{x}_j)[1 + C(\vec{x}_i,\vec{x}_j)]\psi(\vec{r};\vec{x}_i,\vec{x}_j) \equiv$

$$\langle\vec{r}\Omega| \sum_{\mu\neq\nu} \delta(\vec{x}_\mu-\vec{x}_i)\delta(\vec{x}_\nu-\vec{x}_j)\psi_\mu|\Omega\rangle \tag{3.15}$$

Eqs. (3.11) and (3.12) simplify to the approximate equations

$$\phi(\vec{r}) = \psi(\vec{r}) + \int d^3r'\, g(\vec{r}-\vec{r}\,') \int d^3x\rho(\vec{x}) \int d^3r''\, F(\vec{r}\,'-\vec{x};\vec{r}\,''-\vec{x})\psi(\vec{r}\,'';\vec{x}) \tag{3.16}$$

$$\psi(\vec{r};\vec{x}) = \psi(\vec{r}) + \int d^3r'\, g(\vec{r}-\vec{r}\,') \int d^3x'\, \rho(\vec{x}\,')[1 + C(\vec{x},\vec{x}\,')]$$

$$\int d^3r''\, F(\vec{r}\,'-\vec{x};\vec{r}\,''-\vec{x})\, \psi(\vec{r}\,';\vec{x};\vec{x}\,') \tag{3.17}$$

Here $C(\vec{x},\vec{y})$ is the two-body correlation function, defined through

$$<\Omega| \sum_{\mu\neq\nu} \delta(\vec{x}-\vec{x}_\mu)\, \delta(\vec{y}-\vec{x}_\nu)|\Omega> = \rho(\vec{x})\, \rho(\vec{y})[1 + C(\vec{x},\vec{y})] \tag{3.18}$$

This function has the properties

$$\int d^3x\, \rho(\vec{x})\, C(\vec{x},\vec{x}\,') = 0; \quad C(\vec{x},\vec{y}) \to 0 \quad \text{for } |x-y| \to \infty$$

and for antisymmetric states (or repulsive correlations between neutrons and protons)

$$C(\vec{x},\vec{y}) \to -1 \quad \text{for } |\vec{x}-\vec{y}| \to 0$$

Finally,

$$g(\vec{r}-\vec{r}\,') = e^{iq|\vec{r}-\vec{r}\,'|}/|\vec{r}-\vec{r}\,'|$$

q being the pion's momentum.

Eqs. (3.16) and (3.17) are still rather complicated and, in particular, do not close yet. The function $\psi(\vec{r},\vec{x},\vec{y})$ which appears on the right hand side of eq. (3.17) is determined by still another equation, which contains the three-body correlation function, and so on. The two equations above are only part of a system of coupled equations for multiple scattering of the pion [12]. This system of equations may be solved, however, by successive approximations. For example, neglecting two-nucleon correlations altogether, we might set

$$\psi(\vec{r},\vec{x}) \simeq \phi(\vec{r}) \tag{3.19}$$

i.e. assume that the wave incident on nucleon i is identical with the average wave in the medium. At this level of approximation the nucleus is still treated as a homogeneous medium. In a next step, taking two-body correlations into account only, we would set

$$\psi(\vec{r};\vec{x},\vec{y}) \simeq \psi(\vec{r};\vec{x}) \tag{3.20}$$

thus obtaining a set of two coupled equations. These equations simplify considerably if the elementary pion-nucleon amplitudes have zero range, and if we can use the on-shell threshold values (2.9), i.e.

$$F(\vec{r}\,'-\vec{x};\vec{r}-\vec{x}) = b_0\ \delta(\vec{r}\,'-\vec{x})\ \delta(\vec{r}-\vec{x}) + c_0(\vec{\nabla}_{r'}\delta(\vec{r}\,'-\vec{x}))(\vec{\nabla}_r\ \delta(\vec{r}-\vec{x})) + \text{spin and isospin dependent terms} \tag{3.21}$$

With the first approximation (3.19), eq. (3.16) is equivalent to the following Klein-Gordon equation

$$(\Delta + E^2-\bar{m}^2)\phi(\vec{r}) = 2\bar{m}\ U^K_{opt}(\vec{r})\phi(\vec{r}) \tag{3.22}$$

$$\text{with}\quad U^K_{opt} = -4\pi(b_0\rho(\vec{r}) - c_0\vec{\nabla}\rho(r)\vec{\nabla}) \tag{3.23}$$

The second approximation (3.20) leads to the Lorentz-Lorenz effect [13]. This effect gives rise to a nonlinear density dependence of the optical potential and is most important for the p-wave interaction where it amounts to replace

$$\vec{\nabla}\ c_0\ \rho(\vec{r})\vec{\nabla}\quad \text{by} \sim \vec{\nabla}\ \frac{c_0\rho(\vec{r})}{1-\frac{1}{3}\xi\ c_0\rho(\vec{r})}\ \vec{\nabla} \tag{3.24}$$

with $\xi = -C(0)$ in the case of zero range forces. For the p-wave term the derivation of this effect is rather involved and we shall sketch it only for the s-wave part of the optical potential (where it is numerically unimportant, though). Eqs. (3.16) and (3.17) can be combined to

$$\psi(\vec{r};\vec{x}) = \phi(\vec{r}) + b_0\int d^3r'g(\vec{r}\,'-\vec{r})\rho(\vec{r}\,')C(\vec{r},\vec{r}\,')\psi(\vec{r}\,';\vec{x})$$

$$\simeq \phi(\vec{r}) + b_0\rho(\vec{r})\eta\ \psi(\vec{r};\vec{x})$$

$$\text{with}\quad \eta = \int d^3r\ C(\vec{r})\ \ g(\vec{r})$$

Here we have assumed the correlations to be of short range and to depend on the relative coordinate only $C(\vec{r},\vec{r}\,') \simeq C(\vec{r}-\vec{r}\,')$. Thus

$$\psi(\vec{r};\vec{x}) \simeq \frac{1}{1-b_0\rho(\vec{r})\eta}\ \phi(\vec{r})$$

When inserted into eq. (3.16) this gives

$$\phi \simeq \psi + \int d^3r'\ g(\vec{r}'-\vec{r})\ \frac{b_0\rho(\vec{r}')}{1-\eta b_0\rho(\vec{r}')}\ \phi$$

or, in differential form,

$$(\Delta + E^2 - \bar{m}^2)\phi(r) = 2\bar{m}\ U^{LL}_{opt}\ \phi \tag{3.25}$$

with
$$U^{LL}_{opt} = -4\pi\ \frac{b_0\rho(r)}{1-\eta b_0\rho(r)} \tag{3.26}$$

and similarly for the p-wave part. We thus have obtained the (real part of) the Kisslinger potential

$$2\bar{m}\ \mathrm{Re}(U^{LL}_{opt}) = -4\pi\left(\bar{b}_0\rho - \vec{\nabla}\ \frac{c_0\,\rho}{1-\frac{1}{3}\xi c_0\rho}\ \vec{\nabla}\right) \tag{3.27}$$

The imaginary part is added by hand in the usual manner. Assuming that the absorption on two nucleons is predominant, we would expect the corresponding absorptive part of U_{opt} to be proportional to the square of the density [13]. Thus

$$2\bar{m}\ U^{LL}_{opt} = [A_0 + iB_0\rho(\vec{r})]\rho(\vec{r}) - \vec{\nabla}\ \frac{\alpha(\vec{r})}{1-\frac{1}{3}\xi\ \alpha(\vec{r})}\ \vec{\nabla} \tag{3.28}$$

with $\alpha(\vec{r}) = [A_1 + i\ B_1\,\rho(\vec{r})]\rho(\vec{r})$

$$A_0 = -4\pi\left(1+\frac{m_\pi}{m_n}\right)\left(b_0 + \frac{A-2Z}{A}\,b_1\right)$$

$$A_1 = -4\pi\left(1+\frac{m_\pi}{m_n}\right)^{-1}\left(c_0 + \frac{A-2Z}{A}\,c_1\right)$$

$$B_0 = -4\pi\left(1+\frac{m_\pi}{2m_n}\right)(\mathrm{Im}\ B_0);\quad B_1 = -4\pi\left(1+\frac{m_\pi}{2m_n}\right)^{-1}(\mathrm{Im}\ C_0) \tag{3.28'}$$

This is the form of the optical potential which has been used extensively in the analysis of pionic atom data*. For a detailed comparison we refer to Dr. Koch's lectures.

The Coulomb potential is added onto this by the replacement

$$E \to E - V_c(r)$$

in the Klein-Gordon equation (3.25), where $V_c(r)$ is determined by the nuclear charge distribution,

$$V_c(r) = -4\pi e^2\{\frac{1}{r}\int_0^r \rho(r')r'^2dr' + \int_r^\infty \rho(r')r'dr'\} \tag{3.29}$$

One last remark concerns the spin dependent terms in the amplitude (2.9) which give rise to hyperfine effects in the strong interaction. T. and M. Ericson, in their original paper on the subject, have estimated these effects to be small, of the order 1/A [13]. Recently, Partensky and Thévenet have calculated these effects for the case of pionic ^{17}O. They find that the strong hyperfine effects are of the same order of magnitude as the electromagnetic ones and, in particular, that the absorption width of the 2p level is changed from 11 eV (without spin dependent terms) to about 15 eV (with these terms included) [14].

4. FIXED SCATTERER APPROXIMATION, OFF-SHELL EFFECTS AND BEG'S THEOREM

In most approaches to the optical potential one relies on the approximation of fixed scatterers in the nucleus. This means that the scattering of the pion off the nucleus is calculated by taking first the nucleons in fixed positions, then taking the average over the wavefunction of the nuclear ground state. This also means that the recoil of the nucleons is neglected (See, for example, eq. (3.13) above). Under which circumstances this approximation is a good one, is difficult to state in a general way. Scattering of pions at <u>high</u> energy may be a good case, under certain conditions, as, indeed, a nucleus with fixed scatterers (or "frozen nucleus") means, in fact, that nuclear excitation in intermediate steps can be neglected. This applies if the kinetic energy of the pion is much larger than typical excitation energies of the nucleus

$$T_\pi \gg |E_n - E_0| \tag{4.1}$$

---

*The mass dependent factors keep track, in a rough approximation, of the transformation from the pion-nucleon to the pion-nucleus center of mass systems.

and if no high-lying nuclear state is strongly excited by the pion.

In the FSA (fixed scatterer approximation) the multiple scattering equation (3.3) reduces to an effective one-body L.S. equation which may be solved by standard methods. This is shown and worked out in the article by Foldy and Walecka (for separable potentials) [15] and discussed briefly by Hüfner [3]. A typical example of the FSA is the Glauber eikonal theory for multiple scattering at high energies--but we do not wish to enter this subject here. At low and medium energies the FSA is more questionable. In fact, we shall see below that in the region of the N^* resonance the supposition of fixed scatterers is badly violated and the FSA is not applicable.

At very low energies, around threshold, the FSA may become applicable again, provided the elementary pion-nucleon amplitudes are not strongly momentum dependent. For what follows, we shall assume that this is the case.

We now want to turn to an interesting theorem, due to Bég, which holds when the FSA is applicable and when the pion-nucleon potentials v_i are not overlapping [16-19]. Let us define the range of v_i (which is not necessarily local) as the distance R for which

$$v_i(\vec{r}-\vec{x}_i;\ \vec{r}\,'-\vec{x}_i) = 0 \quad \text{when}$$

$$\max\ (|\vec{r}-\vec{x}_i|,\ |\vec{r}\,'-\vec{x}_i|) \geqq R \tag{4.2}$$

The theorem then says this: If the potentials v_i created by the nucleons at fixed positions $\vec{x}_i$ do not overlap, i.e. if

$$|\vec{x}_i-\vec{x}_j|>2R \qquad \forall\ i,j \tag{4.3}$$

then the pion-nucleus scattering matrix $T_A(E)$ depends only on the on-shell elementary pion-nucleon t-matrices.

The theorem was demonstrated by Bég for the case of double scattering in 1961. It was rediscovered by Gal and Agassi [17] and was used by various authors mostly in connection with scattering at high energies. Hüfner has attempted a general proof by way of induction [13]. The proof of the theorem is rather technical and we will not try to repeat it here. We only mention that the assumption of strictly finite range is essential (as was shown in Bég's original work): as a consequence the pion propagates freely, i.e. on its mass shell in between the scattering potentials, What is surprising about this theorem is the statement that the distance between two neighbouring potential domains can be arbitrarily small. The pion's scattered wave need not become asymptotic before it hits the next scatterer.

In actual pion-nucleus scattering the condition of non-overlapping potentials holds true if nucleon-nucleon correlations are repulsive and if the correlation distance r_c is larger than twice the range R. The theorem evidently plays a crucial role in the discussion of off-shell effects in the pion-nucleon interaction. C. Wilkin and I have considered the example of the optical potential for pionic atoms [18]. Suppose that instead of the expression (2.9) we take the following form of the elementary pion-nucleon amplitude (neglecting the isospin terms for the moment)

$$F = b_0 + c_0[(E^2 - m_\pi^2) - \frac{1}{2}\vec{Q}^2] \qquad (4.4)$$

$\vec{Q}$ being the momentum transfer $\vec{Q} = \vec{q} - \vec{q}\,'$.

Since on the mass shell $\vec{Q}^2 = 2 \cdot q^2 - 2\vec{q} \cdot \vec{q}\,' = 2(E^2 - m_\pi^2) - 2\vec{q} \cdot \vec{q}\,'$,

eq. (4.4) is identical to eq. (2.9). Off the mass shell they are different; the optical potential of first order in the density which corresponds to the amplitude (4.4), reads

$$U^L_{opt} = -4\pi\, b_0 \rho(\vec{r}) + c_0[(E^2 - m_\pi^2)\rho(\vec{r}) + \frac{1}{2}\Delta\rho(\vec{r})]\} \qquad (4.5)$$

When we calculate pionic shifts and widths from U^L_{opt} we get rather different results than from U^K_{opt}, eq. (3.2). This is illustrated in Table I which shows the shift and width of the 1s-level in pionic oxygen.

Table I

	ε(1s)	Γ(1s)
U^K_{opt}	-17.5	6.07
U^L_{opt}	- 5.2	15.6
U^L_{opt} + L.L.	-16.5	8.2

Comparison of 1s shift and width in pionic oxygen, calculated from the Kisslinger potential, eq. (3.28) (first line); from the local potential eq. (4.5) (second line) and from the local potential to which its L.L. effect has been added (third line).

However, there should not really be any such off-shell effect, in view of Bég's theorem. Indeed, we started from fixed scatterers (cf. eq. (3.13)) and we used an interaction of zero range,(see eq. (3.21)). So something is inconsistent in our approach. It turns out, in fact, that neither U^K_{opt}, eq. (3.23) nor U^L_{opt}, eq. (4.5) fulfill Bég's theorem (as can be seen, e.g. by calculating the pion-nucleus double scattering term). If one goes to the next order in the density expansion (Order ρ^2), correction terms appear in both optical potentials, which tend towards a cancellation of off-shell effects. These corrections turn out to be precisely the Lorentz-Lorenz effect (3.24) for U^K_{opt} and an analogous effect for U^L_{opt}. When they are added, one gets the results in the third line of Table I which are much more alike than to first order in $\rho(\vec{r})$, in agreement with Bég's theorem.

The conclusion is that Bég's theorem is an important consistency check if one goes to the limit of non-overlapping potentials. Further, the practical importance of the Lorentz-Lorenz effect in pionic atoms depends crucially on what form of the on-shell pion-nucleon amplitude one starts from. When things are done properly there should be no (or very little) dependence on off-shell pion-nucleon scattering.

5. THE N^* RESONANCE IN PION-NUCLEUS INTERACTION

The static approximations in the discussion of pion-nucleus interactions are adequate for high kinetic energies of the pion and, possibly, may also be applicable in the very low energy domain. We now turn to a rather different situation: pion-nucleus scattering at intermediate energies, say $T^{lab} = 100 \to 300$ MeV. In this energy range the N*-resonance will be seen to play an important role (as expected) and to invalidate most of the assumptions made before (FSA, locality of pion-nucleus potential, etc.). For the discussion of the new features in the pion's multiple scattering, due to the N^* resonance, we will closely follow a recent paper by F. Lenz [20]. We refer to this paper for more details as well as for an extensive bibliography about the attempts to incorporate the resonance into pion-nucleus interactions.

As a main result one finds that the presence of the resonance invalidates the fixed scatterer approximation. However, the multiple scattering formalism, contained in our eqs. (3.1) - (3.3), still remains a useful concept, provided care is taken in the treatment of the detailed pion-nucleon dynamics. For example, it is essential to separate relative and center-of-mass motion of the πN system, such that the resonance appears in the right (relative) variables. Also, the intermediate propagation of the resonance must be taken care of in the dynamics of the problem. In incorporating the resonance dynamics no new assumptions or any other new ingredients are needed. The multiple scattering equations, the resonance-dominated pion-nucleon amplitude and the nuclear Hamiltonian are sufficient to describe pion propagation in the nucleus, and the coupling of the pion's motion to (N^*-nucleon hole) states and to pure nuclear states.

As another consequence of the presence of a strong πN resonance, the pion's optical potential becomes non-local. This non-locality is related to the energy dependence of the elementary pion-nucleon amplitude.

We start again from the L.S. equations (3.1) and (3.2) for the free scattering matrix and the bound nucleon scattering matrix τ. Since nucleon i is singled out in these equations, we write the nuclear Hamiltonian as

$$H_A = H_{A-1} + T_i + W_i \qquad (5.1)$$

H_{A-1}: Hamiltonian of the residual nucleus

T_i : kinetic energy of nucleon i

$$W_i(\vec{x}_i) = \sum_{j=1(\neq i)} V(\vec{x}_i - \vec{x}_j) \qquad (5.2)$$

We separate the kinetic energies of the pion and nucleon i into the kinetic energies of c.m. and relative motion

$$T_\pi + T_i = T_{c.m.} + T_r = \frac{P^2}{2M} + \frac{\kappa^2}{2\mu} \qquad (5.3)$$

with

$$M = m_N + m_\pi \quad ; \quad \mu = \frac{m_\pi m_N}{m_\pi + m_N}$$

$$\vec{P} = \vec{k} + \vec{K}, \text{ total momentum} \qquad (5.4a)$$

$$\vec{\kappa} = \frac{1}{M}(m_N\vec{k} - m_\pi\vec{K}) \text{ , relative momentum} \qquad (5.4b)$$

and where $\vec{k}$ and $\vec{K}$ are the laboratory momenta of the pion and nucleon i, respectively*. Finally, we need the corresponding

---

*We will neglect the difference between the lab. system in which the nucleus is at rest, and the pion-nucleus c.m. system.

c.m. and relative coordinates

$$\vec{R} = \frac{m_\pi}{M}\vec{r} + \frac{m_N}{M}\vec{x}_i \;;\quad \vec{\rho} = \vec{r} - \vec{x}_i$$

Then, from eqs. (3.1) and (3.2)

$$\tau_i = t_i(E-H_{A-1}-T_{cm})\{1 + \left[\frac{1}{E-H_{A-1} - T_{cm}-T_r-W_i + i\varepsilon} - \frac{1}{E-H_{A-1} - T_{cm}-T_r + i\varepsilon}\right]\tau_i\} \tag{5.5}$$

Many of the qualitatively new features can be discussed in the limit of weak binding

$W_i \simeq 0$, so that

$$\tau_i \simeq t_i(E-H_{A-1} - T_{cm}) \tag{5.6}$$

Taking τ_i between momentum eigenstates, we have

$$\langle\vec{k}',\vec{K}'|\tau_i|\vec{k},\vec{K}\rangle = \gamma\ \delta^3(\vec{k}+\vec{K}-\vec{k}'-\vec{K}') \; t_i(E-H_{A-1} - \frac{(\vec{k}+\vec{K})^2}{2M}\;;\ \vec{\kappa}',\vec{\kappa}\ ) \tag{5.7}$$

with $\vec{\kappa}'$ and $\vec{\kappa}$ given in terms of final and initial lab. momenta, respectively, by eq. (5.4b)**.

If we wish to apply the FSA we would replace t_i in eq. (5.7) by

$$t(E-H_{A-1} - \frac{(\vec{k}+\vec{K})^2}{2M}\;;\ \vec{\kappa}',\vec{\kappa}) \simeq t(E_{cm};\ \frac{m_N}{M}\vec{k}',\ \frac{m_N}{M}\vec{k}) \tag{5.8}$$

with

$$E_{cm} = E - \frac{p^2}{2M}$$

p being the free pion momentum. When is this approximation justified? Lenz gives the following criteria, together with numerical estimates (in parentheses) for scattering in the resonance region and for nuclear matter parameters:

**For non-relativistic scattering the extra factor γ is one. For relativistic kinematics it deviates from one and is given in eq. (A9) of ref. [20]. At the same time the replacements $M \to M = m_N + E_\pi$; $\frac{m_\pi}{M} \to \frac{m_\pi}{M}$, with E_π the total pion energy, must be made.

(i) The off-shell momentum dependence of t should not be modified by Fermi motion. Therefore,

$$\frac{E}{M}\frac{\langle K\rangle}{k} \ll 1 \qquad (0.1) \qquad (5.9a)$$

unless t depends on the momentum transfer only, but not on $\vec{\kappa}$ and $\vec{\kappa}'$ separately.

(ii) The energy dependence of t should be weak, i.e.

$$\left|\frac{\partial t}{\partial E}\frac{\langle K^2\rangle}{2M}/t\right| \ll 1 \qquad (0.2) \qquad (5.9b)$$

$$\left|\frac{\partial t}{\partial E}\frac{k\langle K\rangle}{M}/t\right| \ll 1 \qquad (0.7) \qquad (5.9c)$$

$$\left|\frac{\partial t}{\partial E}\frac{k^2-p^2}{2M}/t\right| \ll 1 \qquad (1.0) \qquad (5.9d)$$

where k is a typical value of the pion momentum in the nuclear medium. It is seen that in the resonance region these conditions are badly violated. Therefore, one has to try to incorporate the recoil term

$$\frac{(\vec{k}+\vec{K})^2}{2M}$$

of eq. (5.7) into the optical potential. It is precisely this dependence on the pion's off-shell momentum which invalidates the static approximation and gives rise to a number of new phenomena. If one assumes that the pion-nucleon interaction is dominated by the partial wave J = I = 3/2, then t is seen to be given by

$$t_{33}(E,\kappa',\kappa) = t^0_{33}\,\frac{h(\vec{\kappa}')h(\vec{\kappa})}{D(E)} \qquad (5.10)$$

where D(E) is the resonance denominator and where $h(\vec{\kappa})$ are (off-shell) form factors. For a nucleus with spin and isospin zero, the spin-isospin average of the pion-nucleon scattering matrix $T_{\beta\alpha}(E,\vec{\kappa}',\vec{\kappa})$, ($\beta,\alpha$ isospin indices referring to the pion), has to be taken

$$\langle T_{\beta\alpha}\rangle = \frac{4}{g}\,t_{33}(E,\vec{\kappa}',\vec{\kappa})(\hat{\kappa}\cdot\hat{\kappa}')\delta_{\alpha\beta}$$

Finally, using a shell model description of the nuclear ground state with single particle states $|\psi_n\rangle$, one obtains the following optical potential

$$U_{opt}(E,\vec{k}',\vec{k}) = \sum_{n=1}^{A} \langle\Omega,\vec{k}'|\tau_n(E)|\Omega,\vec{k}\rangle =$$

$$\frac{4}{9}\gamma\, \overset{0}{t}_{33} \sum_{n=1}^{A} \int d^3K \int d^3K'\ \psi_n^*(\vec{K}')\psi_n(\vec{K})\ \delta^3(\vec{k}'+\vec{K}'-\vec{k}-\vec{K})$$

$$(\hat{\kappa}'\hat{\kappa})\ h(\vec{\kappa}')\ h(\vec{\kappa})\ D^{-1}(E+\varepsilon_n - \frac{(\vec{k}+\vec{K})^2}{2M}) \qquad (5.11)$$

where

$$\vec{\kappa} = \frac{m_N}{M}\vec{k}' - \frac{E}{M}\vec{K}' \qquad (5.12)$$

$$\vec{\kappa}' = \frac{m_N}{M}\vec{k}' - \frac{E}{M}\vec{K}'$$

and where ε_n is the single particle energy pertaining to the state $|\psi_n\rangle$.

Note that the expression (5.11) contains a number of effects which we have been treating only very roughly or which we neglected altogether previously: Fermi motion is taken into account; the proper transformation to c.m. and relative motion is included; some off-shell dependence is taken care of by the form factor $h(\vec{\kappa})$; finally, the energy denominator contains the correct recoil term. This recoil term gives rise to a genuine non-locality of the optical potential.

Lenz then discusses the properties and the consequences of the non-local potential (5.11) in various simplified situations. In particular, the non-locality of U_{opt} implies a multiple eigenmode propagation of pions in nuclear matter. We do not have time to elaborate on this much further and we refer to the original paper for the details. We can only sketch some of the most striking results.

Let us assume the 33- amplitude to be given by a Breit-Wigner formula

$$F = F_0\ \frac{-\Gamma/2}{E-R+i\ \Gamma/2} \qquad (5.13)$$

with R = 297 MeV and $\Gamma/2$ = 55 MeV. F_0 is determined from the spin-isospin averaged total cross section $\bar{\sigma}$ at resonance through the optical theorem

$$F_0 = \frac{p_c\ \bar{\sigma}}{4\pi}$$

with p_c the relative momentum. Eq. (5.11) then simplifies to

$$U_{opt} \simeq t_0 \sum_n \int d^3K \, \psi_n^*(\vec{K}-\vec{k}'+\vec{k}) \frac{-\Gamma/2}{E+\varepsilon_n-R-\frac{(\vec{k}+\vec{K})^2}{2M} + i\,\Gamma/2} \times \tag{5.14}$$

$$\times \, \Psi_n(\vec{K})(\hat{\kappa}'\hat{\kappa})$$

with $$t_0 = -\frac{1}{(2\pi)^2} \frac{p}{p_c} \frac{1}{E} F_0$$

(p,p_c: free pion momenta in the lab. and c.m. system, respectively). If one neglects the angle dependence due to the factor $(\hat{\kappa}'\hat{\kappa})$, eq. (5.14) can be integrated and expressed in coordinate space. With $\vec{Q} = \vec{k} + \vec{K}$

$$U_{opt} \simeq t_0 \sum_n \int d^3Q \, \psi_n^*(\vec{Q}-\vec{k}') \frac{-\Gamma/2}{E+\varepsilon_n-R - \frac{\vec{Q}^2}{2M} + i\,\Gamma/2} \psi_n(\vec{Q}-\vec{k})$$

and

$$U_{opt}(E,\vec{r}',\vec{r}) = \frac{1}{(2\pi)^3} \iint d^3k' d^3k \, e^{-i\vec{k}'\vec{r}'} \, U_{opt}(E,\vec{k}',\vec{k}) \, e^{i\vec{k}\vec{r}}$$

Now, with

$$\frac{1}{(2\pi)^3} \int d^3k \, e^{i\vec{k}\vec{r}} \psi_n(\vec{k}) = \psi_n(\vec{r})$$

and

$$\int d^3Q \, \frac{e^{-i\vec{Q}(\vec{r}'-\vec{r})}}{\kappa^2-\vec{Q}^2} = -2\pi^2 \, \frac{e^{i\kappa|\vec{r}-\vec{r}'|}}{|\vec{r}-\vec{r}'|}$$

we obtain

$$U_{opt}(E,\vec{r}',\vec{r}) \simeq -(2\pi)^2 \, M \, t_0 \sum_{n=1}^{A} \psi_n^*(\vec{r}') \frac{e^{i\kappa_n|\vec{r}-\vec{r}'|}}{|\vec{r}-\vec{r}'|} \psi_n(\vec{r}) \tag{5.15}$$

with κ_n given by

$$\kappa_n = \sqrt{2M(E + \varepsilon_n - R + i\Gamma/2)} \tag{5.16}$$

This expression (5.15) displays very clearly the origin of the non-locality: The resonating pion-nucleon system propagates from $\vec{r}$ to $\vec{r}'$ with the characteristic wave number κ_n.

In the FSA this intermediate propagation is neglected and one would approximate

$$\frac{e^{i\kappa_n|\vec{r}-\vec{r}'|}}{|\vec{r}-\vec{r}'|} \sim \delta^3(\vec{r}-\vec{r}')$$

in which case U_{opt} becomes proportional to the ground state density

$$\delta(r) = \sum_1^A \psi_n^*(\vec{r})\, \psi_n(\vec{r}) \ .$$

The non-locality of the optical potential and the intermediate propagation of the pion-nucleon system in the form of a N^* - resonance have interesting consequences for the propagation of the pion in nuclear matter. For the Fermi gas model

$$\sum_n |\psi_n\rangle\langle\psi_n| \rightarrow (2\pi)^3 \frac{3\rho}{4\pi\, p_f^3} \int^{p_f} d^3Q\, |\vec{Q}\rangle\langle\vec{Q}|$$

$$\varepsilon_n \rightarrow \frac{\vec{K}^2}{2m_N} \ ; \quad \rho = \frac{2p_f^3}{3\pi^2}$$

The energy denominator in (5.14) is

$$E + \varepsilon_n - R + i\Gamma/2 - \frac{1}{2M}(\vec{k}+\vec{K})^2 \simeq$$

$$E - R + i\Gamma/2 - \frac{\vec{k}^2}{2M} - \frac{\vec{k}\vec{K}}{M} + \alpha\frac{\langle\vec{K}^2\rangle}{2m_N} \tag{5.17}$$

where $\vec{K}^2$ is replaced by its ground state expectation value, and where $\alpha \equiv \frac{E}{M}$.

$$U_{opt}(E,\vec{k}',\vec{k}) = \delta^3(\vec{k}'-\vec{k})U_0\, u(E,k) \tag{5.18}$$

$$U_0 = t_0(2\pi)^3\rho$$

$$u(E,k) = -\frac{3}{4\pi\, p_f^3}\, M\Gamma \int d^3K\, \frac{1}{\kappa_k^2 - 2\vec{k}\cdot\vec{K}} \tag{5.19}$$

$$\kappa_k^2 = \kappa^2 - k^2$$

$$\kappa^2 = 2M(E-R + i\Gamma/2) + \alpha \frac{M}{m_N} <\vec{K}^2>$$

The integral in u(E,k) can be worked out and one finds

$$u(E,k) = \frac{3}{8p_f^3} \frac{M\Gamma}{k} \left[\frac{\kappa_k^2 \, p_f}{k} + \left(p_f^2 - \frac{\kappa_k^4}{4k^2}\right) \ln \frac{\kappa_k^2 + 2p_f k}{\kappa_k^2 - 2p_f k} \right] \tag{5.20}$$

whereas in the static approximation, eq. (5.19) gives

$$u^{FSA}(E) = - \frac{M\Gamma}{\kappa^2 - p^2} = \frac{-\Gamma/2}{E-R - p^2/2M + i\Gamma/2} \tag{5.21}$$

Equipped with these expressions for the non-local optical in nuclear matter, we can now briefly touch upon the discussion of pion propagation in nuclear matter.

The Schrödinger equation in momentum space representation reads

$$(k^2 - p^2)\psi(\vec{k}) = -2E \int d^3k' \; U(E,\vec{k},\vec{k}')\,\psi(\vec{k}') \tag{5.22}$$

When using the form (5.18) this simplifies to

$$(k^2 - p^2)\psi(k) = -2E \; U_0 u(E,k)\psi(k) \tag{5.23}$$

The pion eigenmodes follow from the dispersion equation

$$k^2 - p^2 = c \; u(E,k) \tag{5.24}$$

$$c = \overline{p\sigma}\rho$$

With the FSA solution (5.21) for example, one has

$$k^2 - p^2 = - \frac{c\Gamma/2}{E-R + i\Gamma/2 - \frac{k^2}{2M}} = -2E \; u(E,k)$$

whose solutions are

$$k_{\pm}^2 = \frac{1}{2}[p^2 + 2M(E-R + i\frac{\Gamma}{2}) \pm \sqrt{(p^2 - 2M(E-R + i\frac{\Gamma}{2}))^2 + 4Mc\Gamma} \tag{5.25}$$

In order to identify the physics of these two eigenmodes, one discusses the solutions (5.25) first in the limit of low densities. The relevant parameter is

$$\gamma \equiv \frac{4c}{M\Gamma} = 4\,\frac{\ell^*}{\ell^\pi}$$

if $\ell^\pi = \frac{2}{\sigma\rho}$ and $\ell^* = \frac{k}{M}\frac{2}{\Gamma}$ are defined to be mean free paths of pion and N^*, respectively. If $\gamma \ll 1$, in which case the FSA holds, then k_+ is given approximately by

$$k_s^2 - p^2 = \frac{-c\Gamma/2}{E-R + i\Gamma/2 - p^2/2M} \qquad (5.26)$$

This is called the static eigenmode of the pion. The solution k_-, on the other hand, describes the pion-nucleon center-of-mass motion in the low density limit,

$$k_-^2 = 2M(E-R + i\Gamma/2) \qquad (5.27)$$

This mode is called $(N^*\text{-h})(=N^*$-nucleon hole) mode. For increasing density, the two modes are mixed more and more strongly and can no longer be identified as pion and (N^*h) modes. For example, right at the resonance, i.e.

$$E - \frac{p^2}{2M} = R,$$

there is a critical density

$$\rho_c = \frac{M\Gamma}{4p\sigma}$$

at which the two modes become identical.

$$k_\pm^2 = \frac{1}{2}\left[2p^2 + iM\Gamma \pm M\Gamma\sqrt{-1 + \rho/\rho_c}\,\right] \qquad (5.28)$$

For $\rho < \rho_c$ the damping of the pion mode (k_+) increases, the damping of the (N^*h)-mode decreases, with increasing density. For $\rho > \rho_c$ $\mathrm{Im}(k_\pm^2)$ are independent of the density and are equal, while $\mathrm{Re}(k_\pm^2)$ split in two modes. The coupling of these eigenmodes can also be described by means of the pion's Green's function in the nuclear medium. It is then seen that for low densities, in particular, the pion mode and the (N^*h) mode are represented diagrammatically by

pion mode

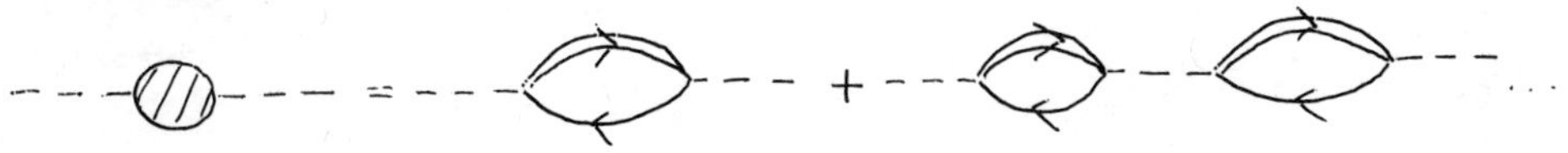

N^*-hole mode

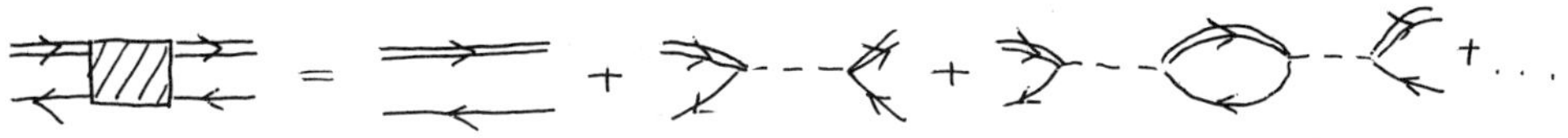

There are many more refinements to these simple estimates, such as the additional propagation modes which are due to the off-shell from factors in eq. (5.10), to the Fermi motion of the nucleons (which we have neglected here, cf. eq. (5.21)), and to the binding corrections (i.e. taking account of W_i in eq. (5.5)) and N^*-nucleus interaction. We refer to Lenz's paper for a detailed study of the more general situation. It is found, in particular, that the equations for the (N^*h) states, for the pion-nuclear states and for their coupling are determined unambiguously from the multiple scattering equations, with the resonant pion-nucleon amplitude and the nuclear Hamiltonian as only input.

It is clear that from these general and basic considerations to actual applications to specific pion-nuclear processes, there is still a long way to go. It is evident that a calculation for any specific reaction, on the basis of this new type of optical potential, is considerably more involved than with most of the standard (but often unjustified) approximate methods we have been used to in the past. Nevertheless, such improved calculations do not seem hopeless, and they open the way to the understanding of a wealth of new phenomena in pion-nucleus scattering around the resonance.

6. NUCLEAR PHYSICS FROM PIONIC ATOMS

In this last section we would like to discuss one example where interesting information about the structure of the nucleus can be extracted from pionic atoms--even though the optical potential describing the pion-nucleus strong interaction is still uncertain and poorly known. This is the quadrupole hyperfine structure in strongly deformed nuclei which is due to the strong interaction [21].

Consider a pionic atom with a strongly deformed nucleus whose spin J is greater than, or equal to one. Examples we will be studying are:

$^{165}_{67}Ho$	$J = \frac{7}{2}$	$Q = 3.47$ b
$^{175}_{71}Lu$	$J = \frac{7}{2}$	$Q = 3.49$ b
$^{235}_{92}U$	$J = \frac{7}{2}$	$Q = 4.0$ b

For these nuclei the critical pionic orbit is the 4f state. Due to the nonspherical electric and strong interaction, this level is split into a multiplet of states labelled by the total angular momentum F (vector sum of nuclear spin J and orbital angular momentum ℓ_π of pion). The electric quadrupole energies are given by the well-known formula

$$\langle (J\ell)F|V_C^{E2}|(J\ell)F\rangle = A_2 C(J,\ell,F) \tag{6.1}$$

where the coefficient $C(J,\ell,F)$ depends on the angular momenta only.

$$C(J,\ell,F) = \frac{3X(X-1)-4J(J+1)\ell(\ell+1)}{2J(2J-1)\ell(2\ell-1)} \tag{6.2}$$

and $X \equiv J(J+1)+\ell(\ell+1)-F(F+1)$

Since the 4f-level is still hydrogen-like, to a rather good approximation, the quadrupole constant A_2 is proportional to the spectroscopic quadrupole moment Q and to the expectation value of $\frac{1}{r^3}$ in the 4f-state,

$$A_2 \simeq e^2\, Q\, \frac{\ell}{2\ell+3}\, \frac{k^3}{S(S+1)(2S+1)} \tag{6.3}$$

with
$$k = \bar{m}\, \frac{Z\alpha}{\sqrt{(n-\ell+S)^2+(Z\alpha)^2}} \tag{6.4}$$

$$S = -\frac{1}{2} + \sqrt{(\ell+\frac{1}{2})^2 - (Z\alpha)^2} \tag{6.5}$$

These formulae hold for relativistic hydrogen wave functions; the small corrections due to the finite size of the nucleus and to the distortion of the pion's wave function due to strong interaction are easily added onto this [22,23].

The strong interaction will be described by the optical potential, eqs. (3.28), with the parameters b_i, c_i, as determined from the values (2.10') and with [24]

$$(\mathrm{Jm}\ B_0) = 0.040\ m_\pi^{-4};\ (\mathrm{Jm}\ C_0) = 0.080\ m_\pi^{-6} \tag{6.6}$$

As the deformation of the rare earth and the transuranium nuclei is known to be predominantly of quadrupole shape, we may expand the nuclear mass density in the lab. system

$$\rho_A(\vec{r}) \equiv \langle J,M=J|\sum_{i=1}^{A} \delta(\vec{r}-\vec{r}_i)|J,M=J\rangle \simeq$$

$$\rho_0(r) + \sqrt{\frac{5}{16\pi}}\,\rho_2(r)\ Y_{20}(\hat{r}) \tag{6.7}$$

and stop after the quadrupole term. When inserted into the optical potential, eq. (3.28), this gives a similar expansion in terms of strong multipole potentials, the quadrupole part of which will give an additional hyperfine splitting. As a consequence of the assumption (6.7) the strong interaction shift and width of a multiplet member with total angular momentum F have the general form

$$\varepsilon(F) = \varepsilon_0 + \varepsilon_2\ C(J,\ell,F) \tag{6.8a}$$

$$\Gamma(F) = \Gamma_0 + \Gamma_2\ C(J,\ell,F) \tag{6.8b}$$

The common shift ε_0 and common width Γ_0 are determined by the monopole part of the optical potential; the quadrupole constants ε_2 and Γ_2 are determined by the quadrupole part of the optical potential, while $C(J,\ell,F)$ is the same angular momentum factor as above, eg. (6.2)

With this result, the pattern of the quadrupole hfs remains unchanged. The electric quadrupole constant A_2, eqs. (6.1) and (6.3), is replaced by an effective constant

$$A_2^{\mathrm{eff}} = A_2 - \varepsilon_2 \tag{6.9}$$

(As to the sign in eq. 6.9, recall the definition of the shift in Section 2.3). These results are illustrated in Figs. 4 and 5, for pionic Lutecium and Uranium. Suppose now that we already have a precise value for the spectroscopic quadrupole moment Q, say, from the corresponding muonic atom. Then we can compute A_2, the electric quadrupole hf constant, from eq. (6.3), corrected for the (small and well-known) finite size and distortion effects.

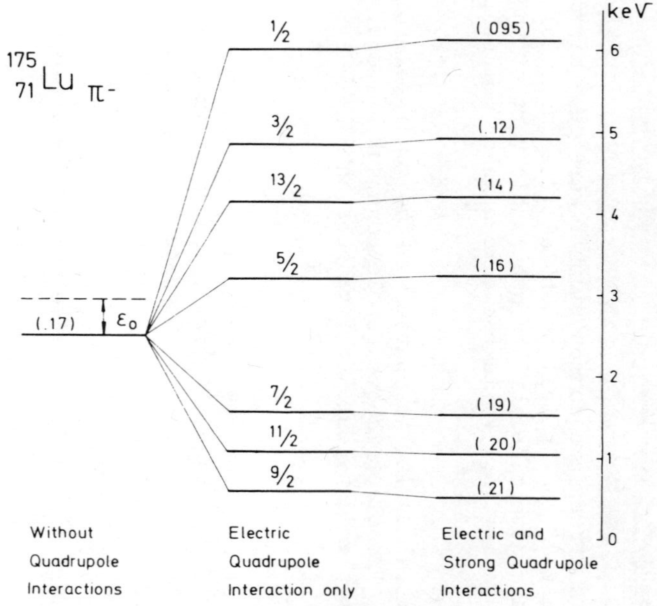

Fig. 4. Level scheme in pionic Lutecium. The widths of individual states are given in parentheses (units keV).

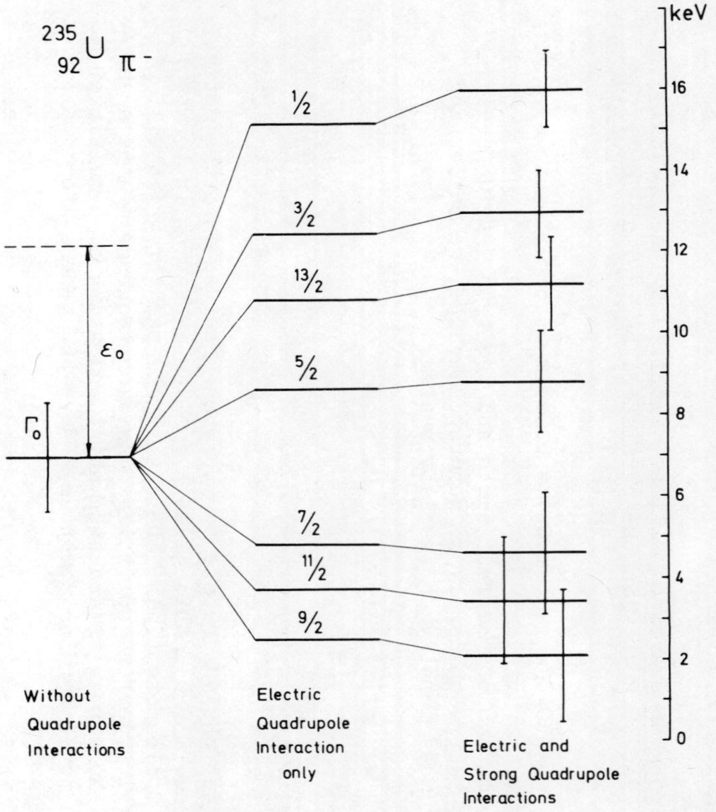

Fig. 5. Level scheme in pionic Uranium. Widths of individual states are displayed.

Comparing this value with the measured A_2^{eff} we can then extract ε_2, via the defining equation (6.9). ε_0 and Γ_0, on the other hand are obtained directly, as usual, from the (5g-4f) transition, while Γ_2 is probably more difficult to extract from the pictures shown in Figs. 4 and 5.

The important result now is that the ratios

$$\varepsilon_2/\varepsilon_0 \text{ and } \Gamma_2/\Gamma_0$$

to a fair approximation depend only on the nuclear mass density (6.7) in the nuclear surface region, and are independent, to a large extent, of the precise values of the parameters in the optical potential, eq. (3.28). To see this we must briefly describe how these new quantities are calculated in practice. (i) The monopole shift ε_0 and width Γ_0 are obtained from the monopole part of the optical potential, by integrating the Klein-Gordon equation numerically. (ii) The quadrupole constants ε_2 and Γ_2 are calculated perturbatively, with the (distorted) wave functions of the pion as they are obtained from (i). The formula is a little lengthy due to the gradient potential whose treatment requires some angular momentum algebra (gradient formula). One finds

$$\varepsilon_2 + i\frac{\Gamma_2}{2} = R + I \tag{6.10}$$

with

$$R = \frac{5\ell}{16\pi\bar{m}(2\ell+3)}\left\{A_0\int_0^\infty r^2dr\, R^2_{n\ell}(r)\rho_2(r) + A_1\int_0^\infty r^2dr\,\rho_2(r)\times\right.$$
$$\left[\frac{(\ell+2)(2\ell-1)}{(\ell+1)(2\ell+1)}F_1^2(r) + \frac{(\ell-1)(2\ell+3)}{\ell(2\ell+1)}F_2^2(r)\right.$$
$$\left.\left.+\frac{6}{(2\ell+1)\sqrt{\ell(\ell+1)}}F_1(r)F_2(r)\right]\right\}\times\left(\int_0^\infty r^2dr\,R^2_{n\ell}(r)\right)^{-1} \tag{6.11}$$

I is given by the same expression if A_0 and A_1 are replaced by B_0 and B_1, respectively, and if $\rho_2(r)$ is replaced by the corresponding expansion quantity for the squared density. Finally,

$$F_1(r) = \sqrt{\frac{\ell+1}{2\ell+1}}\left(\frac{d\,R_{n\ell}}{dr} - \ell\,\frac{R_{n\ell}}{r}\right) \tag{6.12a}$$

$$F_2(r) = \sqrt{\frac{\ell}{2\ell+1}}\left(\frac{d\,R_{n\ell}}{dr} + \frac{\ell+1}{r}R_{n\ell}\right) \tag{6.12b}$$

$R_{n\ell}(r)$ being the pion's wave function.

In order to see the physics contained in these new quantities, let us simplify matters still a little further, for a moment, as follows. Firstly, we estimate ε_0 by calculating it perturbatively as well, instead of integrating the Klein-Gordon equation numerically. Then

$$\varepsilon_0 \simeq -\frac{1}{2m}\left\{ A_0 \int_0^\infty \rho_0(r)\, R_{n\ell}^2\, r^2 dr + A_1 \int_0^\infty \rho_0(r)[F_1^2(r) + F_2^2(r)]r^2 dr \right\} \tag{6.13}$$

Secondly, one verifies by direct calculation that both expressions (6.11) and (6.13) are dominated by the p-wave term (proportional to A_1). The s-wave term, proportional to A_0 contributes only about 20% of the p-wave term. So, for our estimate, we neglect the s-wave term. Finally, as the critical pionic orbit is still far outside the nucleus, we might as well approximate

$$R_{n\ell}(r) \sim r^\ell$$

Then, from eqs. (6.12),

$$F_1(r) \simeq 0 \; ; \quad F_2(r) \alpha\, r^{\ell-1}$$

and we obtain

$$\left.\frac{\varepsilon_2}{\varepsilon_0}\right|_{\text{estimated}} \simeq -\frac{5}{8\pi}\,\frac{\ell-1}{2\ell+1}\,\frac{\int_0^\infty r^{2\ell}\,\rho_2(r)dr}{\int_0^\infty r^{2\ell}\,\rho_0(r)dr} \tag{6.14}$$

In this approximation the ratio $\varepsilon_2/\varepsilon_0$ (and similarly, the ratio Γ_2/Γ_0), is independent of the parameters of the optical potential. It probes the quadrupole mass density in the extreme nuclear surface region (recall that ℓ=3!).

These conclusions are found to be still approximately valid in a more realistic calculation, along the lines sketched above. For these we need a model for the mass density $\rho_A(\vec{r})$. We assume

(i) homogeneous mixture of protons and neutrons, that is

$$\rho_p(\vec{r}) : \rho_n(\vec{r}) : \rho_A(\vec{r}) = Z : N : A \tag{6.15}$$

(ii) the rotator model to describe the nuclear ground state. There is then a body-fixed system, with respect to which the mass density is axially symmetric and of quadrupole shape,

$$\bar{\rho}(r) = \bar{\rho}_0(r) + \sqrt{\frac{5}{16\pi}}\,\bar{\rho}_2(r)\,Y_{20}(\hat{r}) \qquad (6.16)$$

The intrinsic densities $\bar{\rho}_0$ and $\bar{\rho}_2$ are related to the densities (6.7) in the laboratory system by

$$\rho_0(r) = \bar{\rho}_0(r) \qquad (6.17a)$$

$$\rho_2(r) = \frac{J(2J-1)}{(J+1)(2J+3)}\,\bar{\rho}_2(r) \qquad (6.17b)$$

(iii) The intrinsic mass density is given by

$$\bar{\rho}(r) = N\{1 + \exp\,(X\,(\vec{r})4\,\ell n\,3)\}^{-1} \qquad (6.18)$$

with $X(\vec{r}) = \frac{1}{t}[r-c(1+\beta Y_{20})]$ (6.19)

N being the normalization constant which ensures that

$$\int d^3r\,\bar{\rho}(\vec{r}) = A\ .$$

The parameters in eq. (6.19) are known to be

$$c = 6.248\ \text{fm};\ t = 2.07\ \text{fm};\ \beta = 0.31 \qquad (6.20)$$

The densities $\bar{\rho}_0(r)$ and $\bar{\rho}_2(r)$, as well as the pionic and kaonic critical orbit wave functions $y_{n\ell} = r\,R_{n\ell}(r)$ are displayed in Fig. 6. We then obtain the theoretical numbers, for the case of Lutecium as an example, shown in Table 2. The Table also contains the experimental values for ε_0, Γ_0 and $\varepsilon_2/\varepsilon_0$. The latter quantity is obtained from the measured effective quadrupole hf constant, defined in eq. (6.9), and from the spectroscopic quadrupole moment of Lutecium, $Q = (3.49 \pm 0.02)$b, as it is obtained from an independent measurement in muonic Lutecium [25]. It is seen that the ratio $\varepsilon_2/\varepsilon_0$ is in fair agreement with the predicted value. Note, however, that while Γ_0 seems to agree with theory, ε_2 and ε_0, taken separately, are both predicted too small.

We close with a few remarks on these results.

(a) We could turn the analysis around, by assuming that $\varepsilon_2/\varepsilon_0$ can be taken from the theory. The hfs in the pionic atom can then be analyzed in terms of the spectroscopic quadrupole moment. This was first done for ^{165}Holmium [22] where it was found

$$Q = 3.47 \pm 0.11\ \text{b}$$

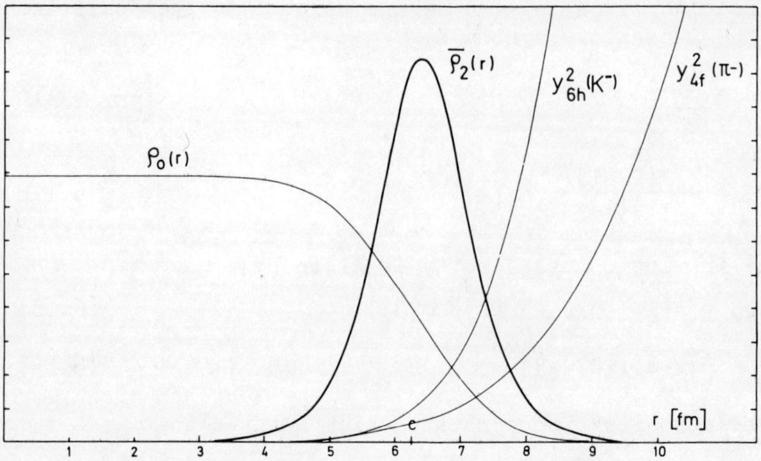

Fig. 6. Monopole density $\rho_0(r)$ and intrinsic quadrupole density $\overline{\rho_2}(r)$, eq. (6.17) for Lutecium, in arbitrary units. For comparison, the normalized and squared radial wave functions of the pionic and kaonic critical levels are also displayed.

Table II

	Theory	Experiment
ε_0	353 eV	(670 ± 70) eV
ε_2	-48 eV	$-(89 \pm 29)$ eV
Γ_0	201 eV	(230 ± 70) eV
Γ_2	-52 eV	
$\varepsilon_2/\varepsilon_0$	-0.14	-0.133 ± 0.045
Γ_2/Γ_0	-0.26	

Comparison of theoretical and experimental results for strong interaction quadrupole effects in pionic ^{175}Lu. Input: $Q = (3.49 \pm 0.02)$b from muonic ^{175}Lu gives experimental value for ε_2.

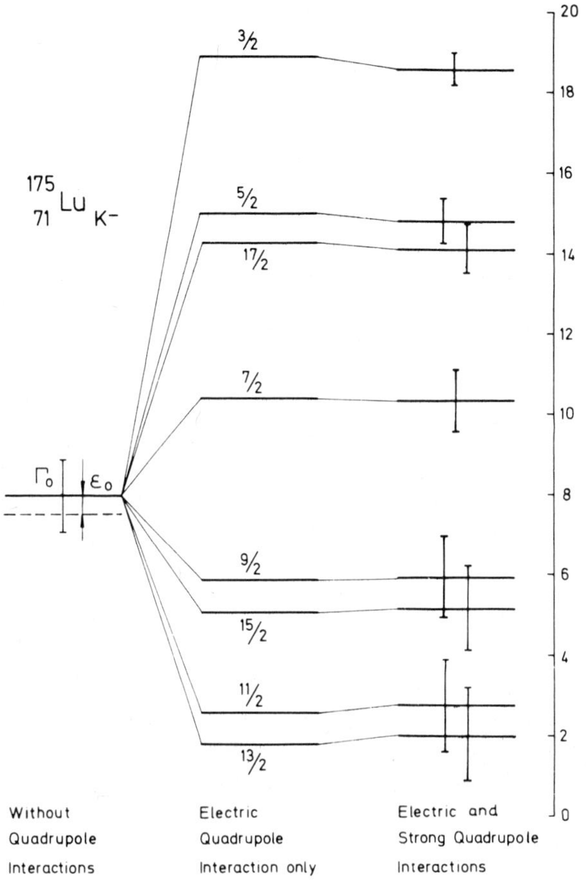

Fig. 7. Level scheme in kaonic Lutecium. Monopole shift ε_0 and Γ_0 are obtained from monopole part $V_s^{(0)}$ of optical potential; ε_2 and Γ_2 are calculated perturbatively from distorted wave functions.

(b) Notice that $A_1\ \rho_A(\vec{r})$, eq. (3.28), should be written more precisely as

$$A_1\ \rho_A(\vec{r}) \rightarrow -4\pi(1+\frac{m_\pi}{m_N})\{(c_0-c_1)\rho_p(\vec{r}) + (c_0+c_0)\rho_n(\vec{r})\} \tag{6.21}$$

when the assumption (6.15) is not valid. However, from the numerical values (2.10') of the scattering lengths we have

$$c_0 = 0.24\ m_\pi^{-3};\quad c_1 = 0.18\ m_\pi^{-3} \tag{6.22}$$

that is, the quantities we are considering here and which depend mainly on the term (6.21) are determined primarily by the neutron distribution $\rho_n(\vec{r})$ in the nuclear surface region.

(c) Finally, we remark that the same quadrupole effects have also been studied in kaonic atoms [21], where they are expected to be even larger than in pionic atoms. Fig. 7 shows, as an example, the case of kaonic Lutecium. Here the critical level is the 6h-level and we find

$$\frac{\varepsilon_2}{\varepsilon_0} \simeq -0.4\ ;\ \frac{\Gamma_2}{\Gamma_2} \simeq -0.3\ .$$

In summary, these examples may have shown to you that, under favorable conditions, we may extract useful information about the nucleus (here: the nuclear mass density in the surface region)--even though the optical potential (3.28) is not too well known in its details.

REFERENCES

[1] F. Binon, P. Duteil, J.P. Garron, J. Görres, L. Hugon, J.P. Peigneux, C. Schmit, M. Spighel and J.P. Stroot, Nucl. Phys. B17 (1970) 168.
[2] S. Dahlgren, B. Hoistad, P. Grafström and A. Åsberg, Phys. Letters 35B (1971) 219; Nucl. Phys. A211 (1973) 243.
[3] J. Hüfner, *Pions Interact with Nuclei*, Physics Reports (in press).
[4] D.S. Koltun in *Advances in Nuclear Physics*, M. Baranger and E. Vogt, eds., Plenum Press, N.Y. 3 (1969) 71.
[5] M.M. Sternheim and R.R. Silbar, Ann. Rev. Nucl. Sci. 24 (1974).
[6] H. Pilkuhn, *et al.*, Nucl. Phys. B65 (1973) 460, and Springer Tracts in Modern Physics 55.
[7] A.A. Carter, J.R. Williams, D.V. Bugg, P.J. Bussey and D.R. Dance, Nucl. Phys. B26 (1971) 445.
[8] C.J. Batty, G.T.A. Squier and G.K. Turner, Nucl. Phys. B67 (1973) 492.
[9] P. Ebersold, B. Aas, W. Dey, R. Eichler, H.J. Leisi, W.W. Sapp and H.K. Walter, to be published.
[10] G. Backenstoss, Ann. Rev. Nucl. Sci. 20 (1970) 467.
[11] M. Krell and T.E.O. Ericson, Nucl. Phys. B11 (1969) 521.
[12] G. Fäldt, Nucl. Phys. A206 (1973) 176.
[13] M. Ericson and T.E.O. Ericson, Ann. Phys. 36 (1966) 323.
[14] A. Partensky and C. Thévenet, preprint Lyon University, LYCEN 7507.
[15] L.L. Foldy and J.D. Walecka, Ann. Phys. 73 (1972) 525.
[16] M.A.B. Bég, Ann. of Phys. 13 (1961) 110.
[17] D. Agassi and A. Gal, Ann. of Phys. 75 (1973) 56.
[18] F. Scheck and C. Wilkin, Nucl. Phys. B49 (1972) 541.
[19] J. Hüfner, Nucl. Phys. B58 (1973) 55.
[20] F. Lenz, MIT preprint 461 (1975). See also F. Lenz and E.J. Moniz, MIT preprint (1975) 455.
[21] F. Scheck, Nucl. Phys. B42 (1972) 573.
[22] P. Ebersold, B. Aas, W. Dey, R. Eichler, J. Hartmann, H.J. Leisi and W.W. Sapp, Phys. Letters 53B (1974) 48, and further references therein.
[23] P. Ebersold, *et al.*, to be published.
[24] L. Tauscher, *Review on Pionic Atoms*, Strasbourg Meeting on Pion-Nucleus Interactions, University of Strasbourg (1971).
[25] W. Dey, Thesis ETH, Zurich (1975), and W. Dey, *et al.*, to be published.

Remarks by Dr. A. N. Kamal

Remark 1. The position of the (3,3) peak in pion-nucleus collisions is shifted relative to that in the free pion-nucleon case due to (i) the smearing due to the Fermi motion which tends to lower the resonance and broaden the peak; (ii) the dispersive effect of the medium on the pion momentum. This can lead to a lowering of the resonance position if the real part of the pion momentum in nuclear matter is enhanced over the value in vacuum; (iii) Pauli blocking has the effect of moving the peak away from threshold and at the same time making it sharper, that is the resonance becomes more stable as Pauli exclusion principle blocks some states to which (3,3) could decay.

Remark 2. In the calculations related to the problem of pion condensates in nuclear matter one generally uses the vacuum properties of (3,3) resonance i.e. the vacuum coupling constant and the position of the resonance. At high densities Pauli blocking should become stronger and it is not evident that the position and the width of the (3,3) may not be very different. Indeed the (3,3) resonance may not be there at all!

STUDIES OF LOW ENERGY π ELASTIC SCATTERING FROM LIGHT NUCLEI *

R.A. Eisenstein

Carnegie-Mellon University

Pittsburgh, Pennsylvania, U.S.A.

The elastic scattering of pions[1,2] from nuclei has been a topic of current interest for several years. Generally speaking, the results available[3,4] from the analysis of the limited amount of data that exists have indicated that optical model theories of suprising simplicity describe well the general features of the data, especially at energies above 100 MeV. The success in this energy region is probably due to the dominance of the (3,3) resonance that appears in π-N scattering at 180 MeV, which may mask more subtle details of the interaction. At low energies (0-75 MeV), where the dominance of this resonance is less pervasive,[2] the simple first-order models may be expected to do less well in describing the data. In particular, the use of the impulse approximation, the neglect of Fermi motion and neglect of two-nucleon absorption are less valid assumptions at these energies.

In an effort to learn more about these questions, The Carnegie-Mellon group[5] has studied the elastic scattering of π^+ from ^{12}C at 50 MeV using the low Energy Pion Channel at LAMPF. The channel slits were set such that $\Delta p/p$ varied between ±0.4 and ±1% while the solid angle setting was 17 msr. Graphite plates between 0.4 and 1.2 gm/cm^2 thickness were used as target material. The resolution of the experiment was principally determined by the momentum slit settings and target thickness effects; this varied between 0.5 and 1.5 MeV FWHM and was sufficient to separate elastic from inelastic scattering. The pions were detected using the Carnegie-Mellon dual-crystal intrinsic Ge spectrometer.[6] Pions of 50 MeV energy stop deep within the second crystal (see Fig. 1); this fact allows the use of particle identification

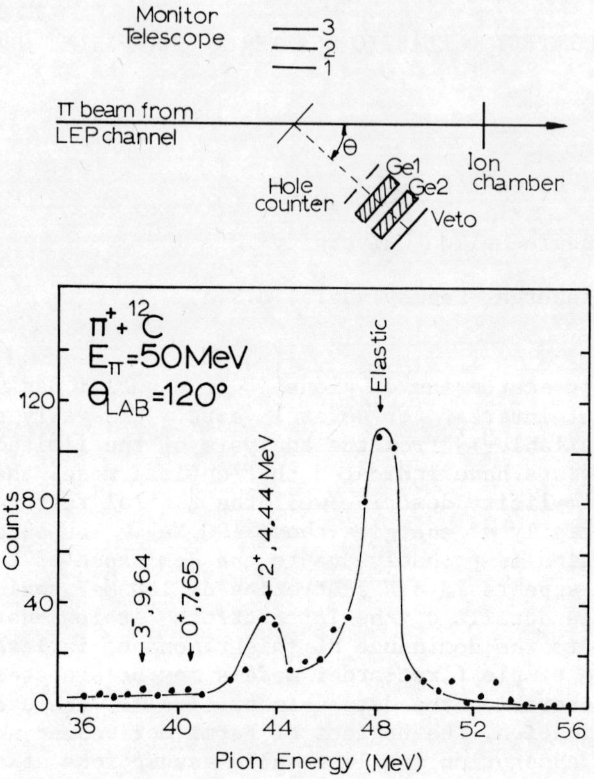

Fig. 1. Upper portion of figure shows a schematic diagram of the experimental set-up (see text). Lower portion shows pion scattering spectrum for 50 MeV π^+ at a lab. angle of 120°. Note the clean separation of the elastic peak from the first inelastic state (2^+, 4.44 MeV). Spectrum includes only pions stopping in the second Ge crystal.

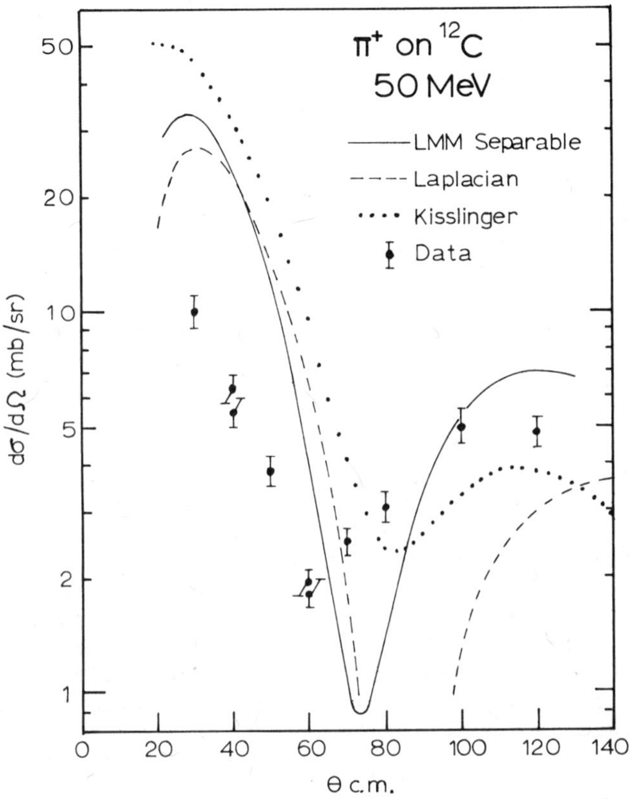

Fig. 2. The angular distribution measured in the experiment is compared to the predictions of three potentials, each obtained using free π-N phase shift information: the separable potential of Londergan McVoy and Moniz (ref. 8) (solid line), the local Laplacian potential (refs. 1,2) (dashed line), and the Kisslinger potential (refs. 1,2) (dotted line).

techniques to distinguish between π's and μ's. A scintillator with a 3/4" hole placed immediately in front of the first crystal determined the solid angle, while a large scintillator placed behind the spectrometer vetoed particles passing completely through it. The beam flux and beam-on-target were continuously monitored using an ionization chamber and a triplet scintillator telescope monitor. Typical beam fluxes were of order 2.5×10^5 π^+/sec average current. The angular distribution was normalized by scattering from polyethylene ($(CH_2)_x$) at 60° and scaling against the known hydrogen cross section.[7]

The angular distribution obtained is given in Fig. 2. The figure also shows the theoretical predictions made using free π-N phase shifts in three models of current interest: the Kisslinger (K) potential,[1,2] the local Laplacian (LL) potential[1,2] and the separable potential[8] of Londergan, McVoy and Moniz (LMM). The most striking feature of this comparison is that the minimum in the data is much farther forward (by as much as 20°) in scattering angle than predicted by any of the theories. In addition, the LL and LMM models are significantly deeper in the region of the minimum.

For the K and LL potential, very good fits to the data were obtained by allowing the (complex) parameters b_0 and b_1 to vary freely. The results of this search are shown in Fig. 3 and table I. Of interest is the fact that the magnitudes of $Re(b_0)$, $Im(b_0)$ and $Im(b_1)$ are much larger than the free values (see table I). In addition, the sign of $Im(b_0)$ is negative, which results in violations of unitarity in the K fit s-wave and production of π's (though no violation of unitarity) in the LL case. The fitting procedure establishes the real parts of b_0 and b_1 with better precision than the imaginary values. In these fits, the rms radius of ^{12}C was held fixed at the electron scattering value of 2.42 fm. Table I also lists the total cross section predicted by the best fit models.

Attempts to fit the data using restricted parameter spaces were made. In order to keep the potential manifestly unitary, the imaginary values of b_0 and b_1 were held fixed at the free π-N values and the real parts and the matter radius were varied. The resulting χ^2/N was 57.3/6; the rms radius found was 4.19 ± 0.1 fm. Other fits using other parameter spaces were tried with no better success. In addition, the "effective radius" model of Sternheim and Silbar[1] and the pionic atom potential of Ericson and Ericson[9] were tried using free πN and the best fit pionic atom values, respectively. In both cases, the minimum is at $\sim 100^o$ and the cross section has a value of ~0.1 mb/sr at the minimum.

Similar problems are found in a re-examination of the data

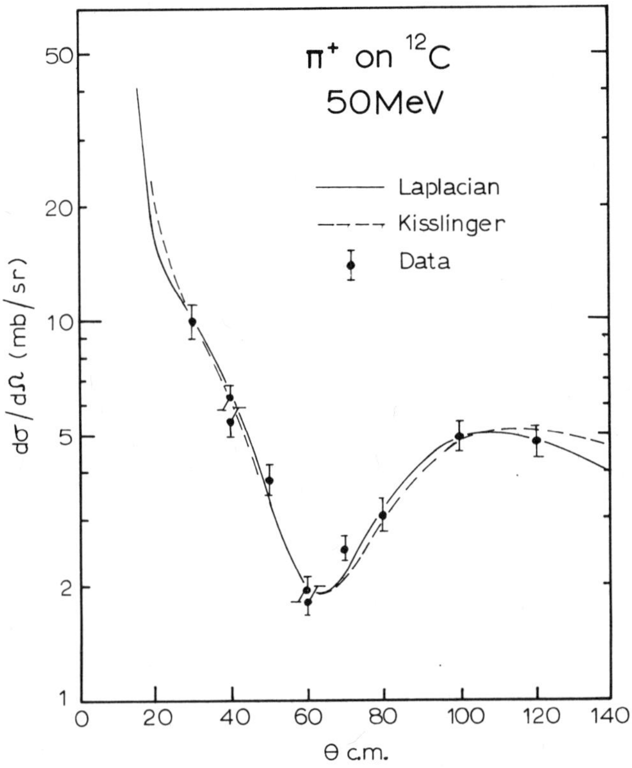

Fig. 3. The angular distribution measured for 50 MeV π^+ is compared to the best fit Kisslinger (dashed line) and local Laplacian (solid line) distributions. The values of (complex) b_0 and b_1 were varied freely to obtain these fits; these are tabulated in Table I. (See text.)

Table I

50 MeV $\pi^+ - {}^{12}C$

Model	χ^2/N	b_0		b_1		σ_{Tot}(mb)
K	7.2/6	-2.74 ± 2%	- 1.04 ± 13%	5.87 ± 2%	2.99 ± 10%	253
LL	6.3/6	-6.51 ± 5%	- 2.50 ± 22%	8.51 ± 5%	4.32 ± 11%	258
Free	---	-0.83	+ 0.51	7.89	1.04	507(K) 241(LL)

The best fit values of the complex parameters b_0 and b_1 are given with their percentage errors for the Kisslinger and local Laplacian models. The χ^2 per degree of freedom N is also given, as are the total cross sections corresponding to the parameter sets. The Kisslinger model violates unitarity in the $\ell = 0$ partial wave while the Laplacian model is unitary but produces pions in some regions of the nucleus. For comparison the last line gives the b_0 and b_1 values obtained using π-N phase shifts (see ref. 16).

of Marshall, et al.,[10] taken with π^+ at 30.2 MeV. Figure 4 shows these data as well as the prediction of the Kisslinger model using free π-N phase shift information. The minimum predicted is both deeper and further backward in scattering angle than the data. By allowing the (complex) parameters b_0 and b_1 to vary freely, good fits can be obtained (see figure 5). However, as in the 50 MeV case, the best fit parameters produce pathological potentials which violate unitarity slightly. (See Table II).

As evidence of the devastating effect such pion-producing potentials can have on other processes, we have used our 30 MeV best-fit (LL) potential in a calculation[11] of the (p,π^+) reaction[12] on ^{12}C for 185 MeV proton energy. Figure 6 compares our prediction using this potential with the data of Dahlgren et al.[12]. They could hardly look less alike.

Hence, we are presented with a situation in which none of the "simple" theories can describe the data, even with moderate excursions of the parameters from the values obtained from free π-N phase shifts. Though good fits can be obtained, the resulting potentials are not physically reasonable. I would remind you of the situation above 100 MeV, where the simple Kisslinger model, with moderate excursions of the parameters from free π-N values, gives good fits to the data.[3] (See figure 7). At the low energies studied here, the required inward shift of the position of the minimum seems to be the principal difficulty.

In a recent paper,[13] Cooper and Eisenstein show that this is indeed the case, at least for scattering from light nuclei. They investigated the existing π^{+}-^{12}C data at 30 and 50 MeV[5,10] and also the $\pi^\pm$ data on ^4He at 50 MeV,[14] where all of the problems mentioned above persist. Since the nuclear scattering at these energies will be dominated by the $\ell=0$ and 1 partial waves, we may write the elastic amplitude as

$$f(\theta) = \frac{i}{2k}\left[(1-\eta_0 e^{2i\delta_0}) + 3(1-\eta_1 e^{2i\delta_1})\cos\theta\right] \tag{1}$$

after neglecting the Coulomb amplitude. Here η_ℓ and $2\delta_\ell$ are the magnitude and phase of the ℓth scattering matrix element and θ is the scattering angle. The approximations used in writing (1) are most accurate for the ^4He case, but since the results are qualitative, they should be more general. The cross section obtained from (1) is parabolic in $\cos\theta$, and has a minimum at θ_M such that

$$\cos\theta_M = \frac{-[1-\eta_0\cos2\delta_0 - \eta_1\cos2\delta_1 + \eta_0\eta_1\cos2(\delta_1-\delta_0)]}{3(1-2\eta_1\cos2\delta_1 + \eta_1^2)} \tag{2}$$

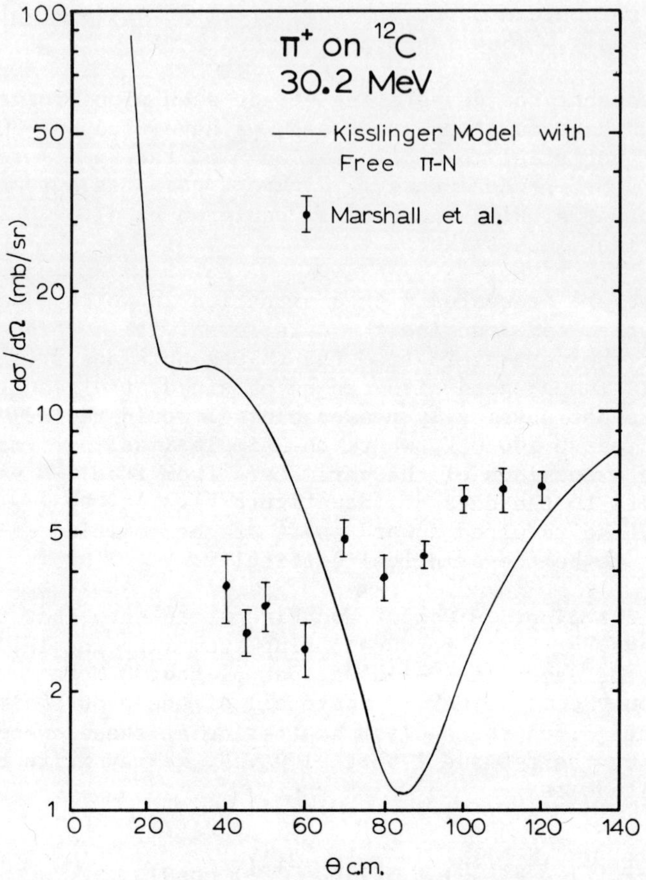

Fig. 4. The angular distribution measured in the experiment of Marshall et al. (ref. 10). The solid curve is the prediction of the Kisslinger model using parameters derived from free π-N phase shifts. (See text.)

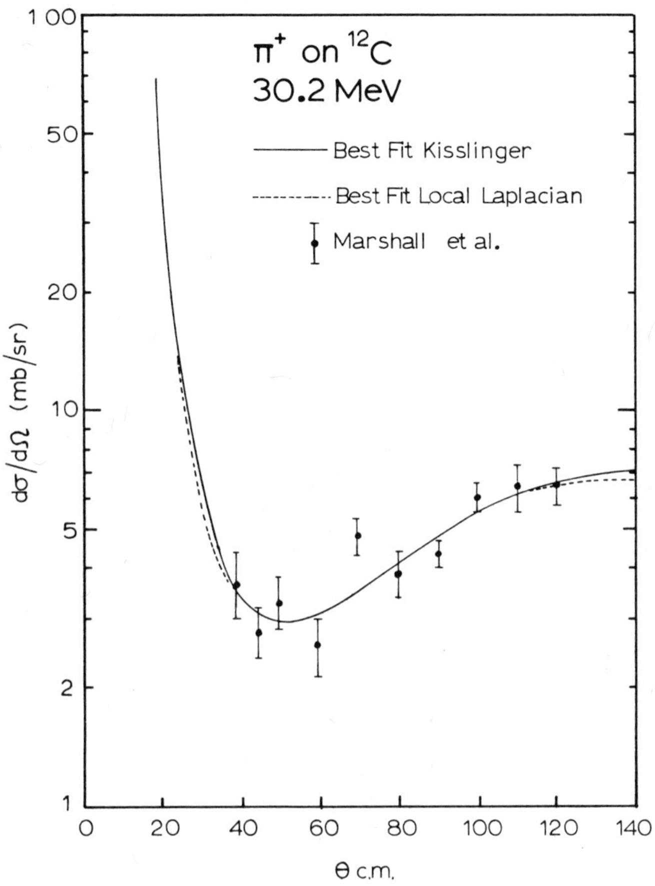

Fig. 5. The angular distribution measured by Marshall et al. is compared to the best fit (K) and (LL) potentials obtained by allowing (complex) b_0 and b_1 to vary freely. (See text.)

Table II

30.2 MeV $\pi^{\pm}$-^{12}C

Model	χ^2/N	b_0		b_1	
K	12.0/6	-4.10 ± 38%	-3.16 ± 84%	5.65 ± 17%	-1.31 ± 120%
LL	12.0/6	-12.46 ± 16%	8.75 ± 40%	9.66 ± 6%	-2.28 ± 58%
Free	---	-0.63	0.68	7.73	0.46

The best fit values of the complex parameters b_0 and b_1 are given with their percentage errors for the K and LL models. (See table I caption). Both of these potentials violate unitarity slightly.

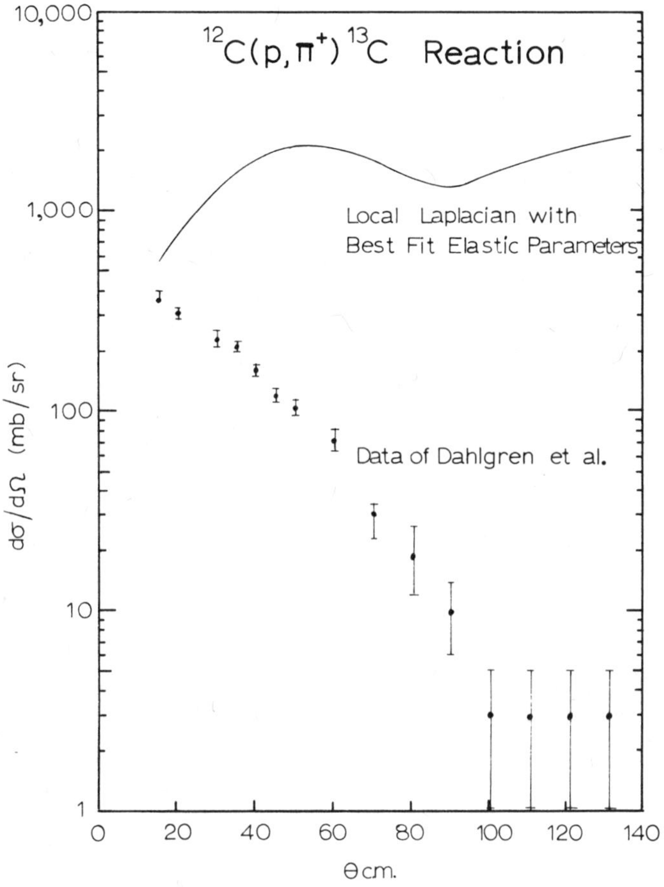

Fig. 6. The angular distribution measured by Dahlgren et al. in the $^{12}C(p,\pi^+)^{13}C$ reaction (ref. 12) compared to a calculation by G. Miller (ref. 11) made using the best fit LL potential for the Marshall 30 MeV π^+ elastic data to generate the π distortion. (See text.)

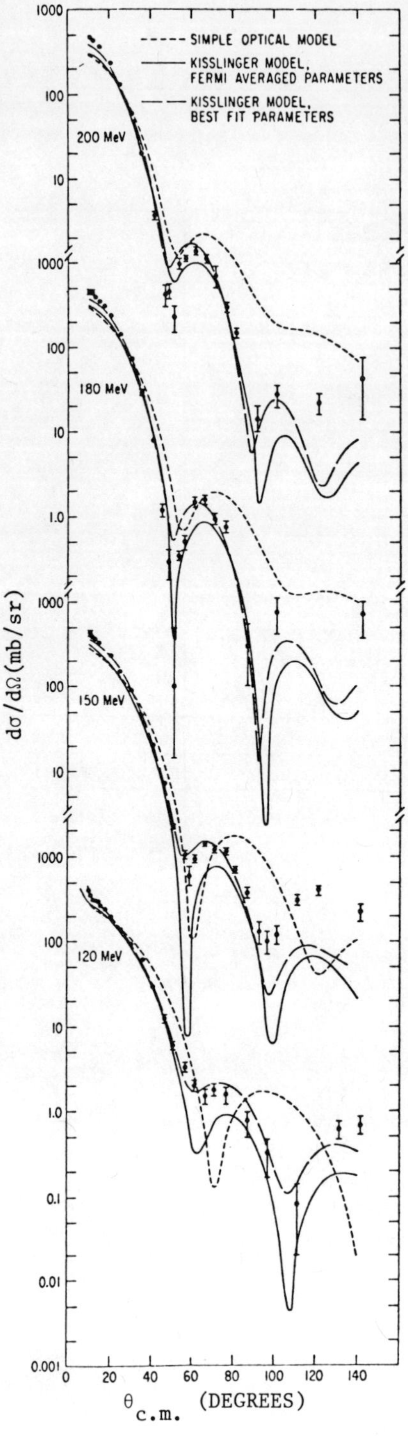
SIMPLE OPTICAL MODEL
KISSLINGER MODEL, FERMI AVERAGED PARAMETERS
KISSLINGER MODEL, BEST FIT PARAMETERS
200 MeV
180 MeV
150 MeV
120 MeV
dσ/dΩ (mb/sr)
θ c.m. (DEGREES)

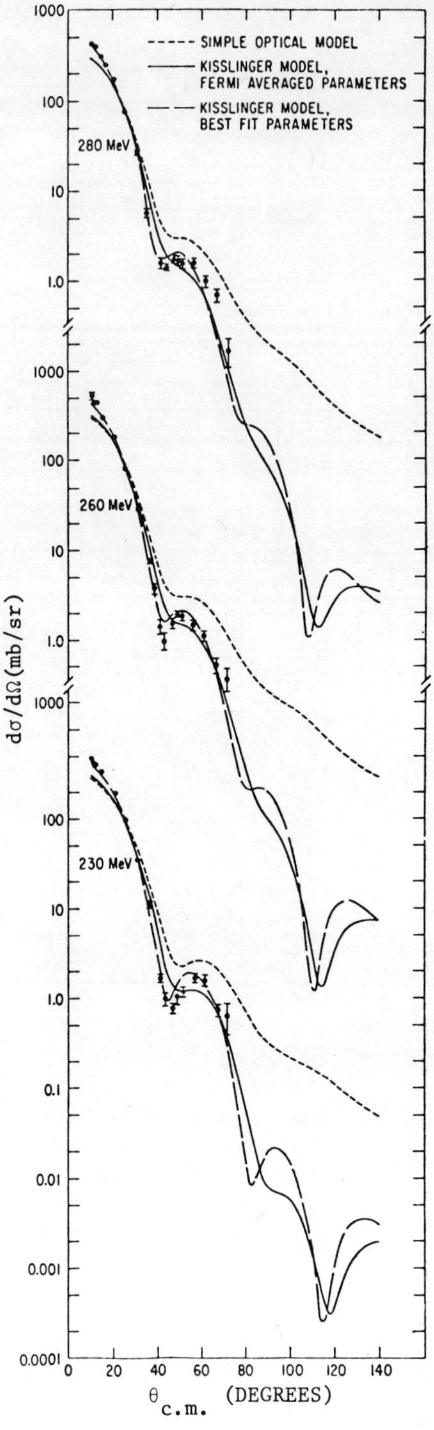
SIMPLE OPTICAL MODEL
KISSLINGER MODEL, FERMI AVERAGED PARAMETERS
KISSLINGER MODEL, BEST FIT PARAMETERS
280 MeV
260 MeV
230 MeV
dσ/dΩ (mb/sr)
θ c.m. (DEGREES)

We see that for θ_M to lie forward of 90°, $\cos\theta_M$ must be positive. Requiring unitarity ($\eta_\ell \leq 1$) and $\cos\theta_M > 0$, the following restrictions are found:

$$\delta_\ell \neq n\frac{\pi}{2}\ (n = 0,1,2)\ ; \quad \delta_0 \neq \delta_1 \quad ; \quad \eta_\ell \neq 0\ . \tag{3}$$

Eqs. (3) mean, in particular, that the small s-wave interaction given by free π-N phase shifts ($\delta_0 \approx 0$) cannot be correct in the nuclear case. This is confirmation of the situation in pionic atoms,[9] where a large s-wave π-nucleus parameter is also required to fit the data. Note that the prediction of eq. (3) is independent of the optical potential used. Other features predicted by eq. (3), such that the nucleus cannot be black ($\eta_\ell \neq 0$), and that the s- and p-wave interactions are likely to be opposite in sign, have been confirmed by investigation[13] with complete optical model codes.[15]

The above analysis shows that the location of the minimum places restrictions on the scattering matrix. But why is it that the simple (K) and (LL) models used to analyze the data want to violate unitarity? To see this, consider the following simple model for the scattering amplitude in plane wave Born approximation, neglecting nuclear partial waves $\ell > 1$ and the Coulomb amplitude. Then both the (K) and (LL) models (on-shell) have the form

$$f(\theta) \sim (b_0 + b_1 \cos\theta)\ \rho(q) \tag{4}$$

where b_0 and b_1 are the usual s- and p-wave parameters and $\rho(q)$ is the Fourier transform of the nuclear density. For a uniform nuclear density normalized to unity, $\rho(q)$ takes the form of a spherical Bessel function:

$$\rho(q) \sim \frac{1}{qR}\ j_1(qR)$$

Fig. 7. The angular distributions obtained by Binon et al. compared to theoretical calculations of Sternheim and Auerback (ref. 3). Note the good quality of the fits obtained using the Kisslinger optical potential with Fermi-averaged parameters. The "best fit" parameters represent only minor excursions from the free π-N values. None of the pathologies mentioned in the text are present. The curve labelled "simple optical model" is obtained from a local density proportional potential.

with q the momentum transfer and R the nuclear radius. By examining the argument of j_1, we can see that for the energies of interest here ($E < 75$ MeV) and for light nuclei ($A \sim 12$) j_1 is a slowly varying function of θ with no zeros in the range $0 < \theta < \pi$. Therefore, any minimum in the cross section will not be due to diffraction but to interference between the s- and p-wave pieces of eq. (4).

Let us now examine the cross section given by (4). We neglect $\rho(q)$ since it varies slowly with θ and obtain

$$\frac{d\sigma}{d\Omega}(\theta) \sim |b_0|^2 + 2\mathrm{Re}(b_0 b_1^*)\cos\theta + |b_1|^2\cos^2\theta \tag{5}$$

whose minimum occurs at

$$\cos\theta_M = -\frac{(b_{0R}b_{1R} + b_{0I}b_{1I})}{(b_{1R}^2 + b_{1I}^2)} \quad . \tag{6}$$

In the above, b_{0R} and b_{0I} are the real and imaginary parts of b_0, and similarly for b_1. Note that if all terms in (6) were positive, the minimum would be backward of 90°. In the free π-N case however (see tables I and II) b_{0R} is <u>negative</u> while the others are positive. Because b_{1R} is large the first term in (6) dominates and $\cos\theta > 0$ slightly. From (6) we see that if the minimum must be moved further forward, this can be done by: (a) making b_{0R} more negative (<u>increasing s-wave repulsion</u>); (b) reversing the sign of b_{0I} or b_{1I} (<u>thereby making some part, or all, of the amplitude non-unitary or pion producing</u>); (c) a combination of (a) and (b). (Another possibility is to allow b_{1R} to be negative, but we choose to ignore this possibility because b_{1R} is thought to be given reliably by the free π-N phase shifts. Our fits with full distortion and the Coulomb interaction included almost never choose to vary b_{1R} by more than $\pm$ 30%.)

If we now "normalize" eq. (5) by dividing through by $|b_0|^2$ and calculate the latus rectum (LR) of the resulting parabola when reduced to "standard form" we obtain

$$\mathrm{LR} = \frac{|b_0|^2}{|b_1|^2} \quad . \tag{7}$$

The LR is taken as a measure of the sharpness of the minimum. When we compare the latus recta computed from the best-fit solutions to those from free π-N, we find that the former are larger than the latter, in agreement with the predictions of eq. (6) and (7). Since the ratio $|b_0|/|b_1|$ has increased, the minima of the best fits are less sharp than in the free π-N case. If a limit is placed on the size of $|b_0|^2$ by the sharpness of the minimum in the data and by the value of b_{1R}, the difficulty in pushing the minimum far enough forward may lead to the violations of unitarity observed.

Although the above findings are modified slightly when a full distorted treatment (including the Coulomb amplitude) of the K or LL models is made, the conclusions reached are not substantially modified. It appears that: (a) the minimum in the low energy (∿75 MeV and lower) data on light nuclei (A ∿ 12) is due to s- and p-wave interference and not diffraction from the nuclear surface; (b) the forward position of the minima in the data sets analyzed requires a large repulsive π-nucleus phase shift δ_0; (c) the movement of the free π-N minimum to more forward angles and its broadening leads, within the accuracy of the Born approximation, to a large repulsive real s-wave parameter b_{0R} and to possible pion production in the nucleus or violations of unitarity.

Part (c) has been gleaned from a Born approximation analysis which shows clearly the connection between b_0, b_1, and the cross section. Although this simple connection is obscured in the exact distorted wave analysis by such complications as distortion and off-shell behavior of the models, the results of exact fitting procedures bear out qualitatively the conclusion.

Any potential whose scattering is dominated by a Born term of the form of eq. (3) will have similar difficulties if b_0 and b_1 are obtained from the free π-N phase shifts. Clearly some values of b_0 and b_1 exist for which these difficulties are not present. These values will be acceptable so long as the internal restrictions of the potential will let the parameters vary more freely than they can in the K and LL models. The derivation of these phenomenological values from a more fundamental basis will still remain a problem.

The above statements offer fairly convincing evidence that the simple K and LL models are inadequate descriptions of π elastic scattering phenomena at low energies. We can infer that "improved" potentials, which include effects due to, e.g., nuclear binding, nuclear density variations in b_0 and b_1, Fermi motion, and two-nucleon absorption will do much to mitigate the above problems and will improve agreement with experiment.

In closing I wish to emphasise the essential role of my colleagues J. Amann, P. Barnes, M. Doss, S. Dytman and A. Thompson in the data collection, and that of M. Cooper in the interpretation of the results. I would like also to thank the staff of LAMPF for their gracious hospitality during a year's leave of absence.

REFERENCES

[*] Work supported by the USERDA.
[1] M. Sternheim and R. Silbar, Ann. Rev. Nuc. Sci. 24 (1975) 249.
[2] D. Koltun, Adv. Nuc. Phys. 3 (1969) 71, Plenum Press.
[3] M. Sternheim and E. Auerbach, Phys. Rev. Letters 25 (1970) 1500.
[4] M. Blecher et al., Phys. Rev. C10 (1974) 2247.
[5] J.F. Amann, P.D. Barnes, M. Doss, S.A. Dytman, R.A. Eisenstein, A.C. Thompson (Present address: Lawrence Berkeley Laboratory, Berkeley, Ca. 94720) "Elastic Scattering of 50 MeV π^+ From ^{12}C" (Submitted to Phys. Rev. Letters).
[6] J. Amann et al., Nuc. Inst. Meth. (to be published).
[7] D. Dodder, J. Frank, R. Mischke, private communication.
[8] J. Londergan, K. McVoy and E. Moniz, Ann. Phys. 86 (1974) 147.
[9] M. Ericson and T.E.O. Ericson, Ann. of Phys. 36 (1966) 323.
[10] J. Marshall, M. Nordberg, R. Burman, Phys. Rev. C1 (1970) 1685.
[11] Dr. G.A. Miller kindly performed this calculation for us.
[12] S. Dahlgren et al., Nuc. Phys. A211 (1973) 243.
[13] M. Cooper and R. Eisenstein, (submitted to Phys. Rev. Letters).
[14] K. Crowe et al., Phys. Rev. 180 (1969) 1349.
[15] R.A. Eisenstein and G.A. Miller, J. Comp. Phys. 8 (1974)
[16] D. Roper, R. Wright, B. Feld, Phys. Rev. 138 (1965) B190. Solution 24 has been used to generate the parameters b_0 and b_1.

RADIATIVE PION CAPTURE AND MUON CAPTURE IN THREE-NUCLEON SYSTEMS

A.C. Phillips

Department of Physics

University of Manchester, Manchester, England

Radiative pion capture and muon capture in nuclei can be studied using two methods, the elementary particle method (EPM) and the impulse approximation method (IAM). The EPM is fully relativistic and incorporates the effects due to virtual mesons and baryon resonances within nuclei. The structure of the nuclei are described by form factors which are partially determined by electron scattering and β decay data. Unlike the EPM, the IAM can be applied to a variety of muon and pion capture reactions. In this method the nuclear structure is described by non-relativistic wave functions and the interaction of the nucleus is expressed in terms of the interaction of the constituent nucleons. Meson and baryon resonance effects are not included. Dynamical models for these effects give rise to 10% corrections to the IAM. Alternatively the EPM can be used to relate the meson interaction effects that occur in β decay, muon capture and radiative pion capture.

In this talk I shall describe some IAM and EPM calculations, done in collaboration with F. Roig and J. Ros, on the following reactions:

$$\mu^- + {}^3He \to {}^3H + \nu, \ nd + \nu \text{ and } nnp + \nu$$

$$\mu^- + {}^3H \to nnn + \nu$$

$$\pi^- + {}^3He \to {}^3H + \pi^0, \ {}^3H + \gamma, \ nd + \gamma, \ nnp + \gamma$$

and

$$\pi^- + {}^3H \rightarrow nnn + \gamma$$

The technical details are contained in refs. [1] and [2].

In the IAM the amplitude for reactions of this type is approximately of the form

$$A \sim a_N \int \psi_{E_f}(p_1,p_2,p_3)\, \psi_{E_i}(p_1+k,p_2,p_3)\, d\tau \qquad (1)$$

where E_i and E_f are the energies of the initial and final nuclear states, k is the momentum of the outgoing neutrino or photon and a_N is the nucleon capture amplitude. It is useful to consider the transition to a bound ^3H state and to a continuum 3-nucleon state separately.

TRANSITION FROM ^3He TO ^3H

In this case the energy transfer $E_f - E_i \sim o$ and the momentum transfer is $k \sim m_\pi$ or m_μ . Accurate impulse approximation calculations are possible, the main error arising from the uncertainty in the relative proportion of the S,S' and D-state components of the 3-nucleon bound states. In this situation the main topic of interest is the magnitude of the discrepancy between the IAM and experiment. For example it is found that the calculated and experimental values for the $\mu^- + {}^3He \rightarrow {}^3H + \nu$ rates differ by 10%. This discrepancy can be attributed to meson interaction effects which modify the axial vector current. A similar discrepancy occurs when ^3H β decay is considered: The impulse approximation for the Gamow-Teller matrix element is approximately

$$|M_{GT}|^2 \sim 3[P(S) - \tfrac{1}{3}P(S') + \tfrac{1}{3}P(D)]^2$$

For S' and D-state probabilities P(S') ∿ 1% and P(D) between 5 and 10%, the theoretical value for $|M_{GT}|^2$ is 8 to 13% smaller than the experimental value of 2.91 ± 0.05. In the EPM the axial vector form factor is normalized to the experimental value for M_{GT} and the calculated rate for $\mu^- + {}^3He \rightarrow {}^3H + \nu$ agrees with experiment.

Next consider the Panofsky ratio in ^3He. One expects

$$P = \frac{\pi^- + {}^3He \to {}^3H + \pi^o}{\pi^- + {}^3He \to {}^3H + \gamma} \sim \frac{\pi^- + p \to n + \pi^o}{\pi^- + p \to n + \gamma} \left[\frac{F(0.03fm^{-2})}{F(0.47fm^{-2})}\right]^2 \tag{2}$$

where $F(q^2)$ is a form factor which represents the probability amplitude that the 3-nucleon bound state remains bound after absorbing a momentum transfer q. The experimental value for $P = 2.68 \pm 0.13$ is roughly reproduced if one takes

$$F(q^2) \sim 1 - q^2R^2/6 \tag{3}$$

where R^2 is the RMS charge of ${}^3He$.

This rough estimate illustrates that the Panofsky ratio in ${}^3He$ can be accurately calculated since it is more or less proportional to the accurately measured Panofsky ratio in hydrogen. Here again the main topic of interest is the magnitude of the discrepancy between the IAM and experiment. Table 1 indicates that there is a discrepancy of 4% which increases to 10% as the assumed D-state probability density of ${}^3He$ is increased from 5 to 10%.

It is possible to estimate the expected discrepancy by exploiting the EPM relation between the radiative pion capture amplitude and the axial vector current that occurs in β decay and muon capture.

Consider the PCAC hypothesis in the presence of the electromagnetic field $A_\alpha(x)$ [4]:

$$[\partial_\alpha + ie\, A_\alpha(x)]\, J_\alpha^A(x) = -m_\pi^2\, f_\pi\, \phi_\pi(x) \tag{4}$$

where J_α^A is the weak axial current, $f_\pi = .95\, m_\pi$ is the pion decay constant and ϕ_π is the pion field operator. Take the matrix element between the states $|{}^3He\rangle$ and $|{}^3H,k\lambda\rangle$ and explicitly separate out the pion pole terms. The L.H.S. of eqn. (4) becomes

$$\frac{-f_\pi q^2}{q^2 + m_\pi^2} \langle {}^3H,k\lambda|j_\pi(o)|{}^3He\rangle - iq_\alpha \overline{\langle {}^3H,k\lambda|J_\alpha^A(o)|{}^3He\rangle}$$

$$+ ie\, \varepsilon_\alpha(\lambda)\, \langle {}^3H|J_\alpha^A(o)|{}^3He\rangle \tag{5}$$

Table 1 : Theoretical Results for Panofsky ratio in ^3He,P.

^{3}He D-State Probability	5%	10%
P	2.79	2.98
Discrepancy with expt. P=2.68 ± 0.13	(4 ± 4)%	(10 ± 4)%
"Expected Discrepancy"	(8 ± 2)%	(13 ± 3)%

The R.H.S. becomes

$$\frac{f_\pi m_\pi^2}{q^2 + m_\pi^2} \langle ^3H,k\lambda | j_\pi(o) | ^3He\rangle \qquad (6)$$

Here j_π is the pion source, i.e.

$$(\Box^2 - m_\pi^2)\ \phi_\pi(x) = -\ j_\pi(x) \qquad (7)$$

Equating the expressions (5) and (6) and taking the soft pion limit $q_\alpha \to o$, we obtain the relation

$$\varepsilon_\alpha(\lambda)\ \langle ^3H | J_\alpha^{em} | ^3He\pi\rangle \equiv \langle ^3H,k\lambda | j_\pi | ^3He\rangle$$

$$\to \frac{ie}{f_\pi}\ \varepsilon_\alpha(\lambda)\ \langle ^3H | J_\alpha^A | ^3He\rangle \qquad (8)$$

We see that in the soft pion limit the matrix element of the electromagnetic current that governs radiative pion capture is directly proportional to the matrix element of the axial vector current that enters into β decay and muon capture. This relation is true even in the presence of meson interaction effects. Thus one expects that the discrepancy between the IAM and experiment in radiative pion capture by ^{3}He to be similar to the 8 to 13% discrepancy found in the analysis of the triton β decay and in muon capture by ^{3}He. This connection between the weak and radiative pion processes can be used to evaluate the "expected discrepancy" in the IAM calculation of the Panofsky ratio. The results are listed in table 1. We note that, for each choice of the ^{3}He D-states, the actual discrepancy and the expected discrepancy are similar in magnitude and sign. We conclude that the data is consistent with the soft pion relation between the weak and the pion capture processes.

TRANSITIONS FROM ^{3}He AND ^{3}H TO 3-NUCLEON CONTINUUM STATES

Provided the energy transfer, $E_f - E_i$, is small, the analysis of the reactions of the type

$$\pi^- + {}^3He \to nnp + \gamma,\ \mu^- + {}^3He \to nnp + \nu, \qquad \text{etc.}$$

using the IAM is straightforward. However, when the energy transfer of the order of m_π (or m_μ), the calculations are less reliable, because of sensitivity to high momentum components of the wave functions. (By inspection of eqn. (1), we see that when $E_f - E_i \sim m_\pi$ and $k \sim 0$, $p_1{}^2/M \sim m_\pi$; i.e. the reaction is sensitive to $\psi_{Ei}(p_1)$ at $p_1 \sim \sqrt{m_\pi M} \sim 3f_m{}^{-1}$.) In addition, we expect small corrections to IAM due to meson interaction effects. The discussion of β decay, muon capture and the Panofsky ratio suggests that these effects are likely to be of the order of 10%. However, as emphasised in Professor Primakoff's lectures, meson interaction effects could be more significant in situations where the IA is suppressed. This could occur in the reactions $\pi^- + {}^3H \rightarrow nnn + \gamma$ and $\mu^- + {}^3H \rightarrow nnn + \gamma$ where the production of the 3 neutron systems is inhibited by the Pauli principle.

Closure calc ulations suggest that final state interactions in the breakup reactions are large. Consider, for example,

$$\pi^- + {}^3He \rightarrow {}^3H + \gamma, \quad nd + \gamma \text{ and } \quad nnp + \gamma$$

The completeness relation for a system of two neutrons and a proton may be written in terms of the states for interacting nucleons,

$$1 = |{}^3H\rangle\langle{}^3H| + \sum|\psi_{nd}\rangle\langle\psi_{nd}| + \sum|\psi_{nnp}\rangle\langle\psi_{nnp}| \tag{9}$$

or in terms of the states for non-interaction nucleons

$$1 = \sum|\phi_{nnp}\rangle\langle\phi_{nnp}| \tag{10}$$

If the pion capture mechanism were independent of the type and of the energy of the final 3-nucleon system, the total rate for producing interacting systems consisting of ${}^3H$, nd and nnp states would be equal to the rate for the production of non-interacting nnp states. To reproduce such an effect, it is essential to use an "exact" model for the final 3-nucleon states. In table 2 we list the radiative pion capture rates for each spin-isospin channel. The capture rates were obtained using two representations for the final state: three nucleons interacting via separable potentials (i.e. Amado model) and three non-interacting nucleons. Note that the total rate obtained using interacting nucleons is comparable with nnp rates obtained when the final state interactions are neglected. Note also that when final state interactions are included, each spin-isospin channel is dominated by the contribution from the state with the lowest threshold; 94% of the S = 3/2, I = 1/2 channel is due to the nd state and 92% of the S = 1/2, I = 1/2

Table 2 : Radiative pion capture rates in ^{3}He for final states of definite spin and isospin

S	I	Interacting nucleons ^{3}H	Interacting nucleons nd	Interacting nucleons nnp	Non-interacting nucleons nnp
1/2, 3/2	3/2	-	-	0.5	0.3
3/2	1/2	-	2.1	0.1	1.7
1/2	1/2	3.7	0.1	0.2	2.9

channel is due to the ^3H state. A similar phenomena also occurs in muon capture and in the photodisintegration of ^3He.

The Amado model neglects several important features of the two-nucleon interaction, the most important being the tensor interaction. However, the closure approximation for the total capture rate can be used to determine how the results would vary when a more realistic model is used. It is also straightforward, to calculate the rate for the production of ^3H for a variety of bound state wavefunctions. Thus, the uncertainty of the calculated branching ratio for breakup can be found. The results agree with experiment: For pion capture

$$\frac{\pi^- + {}^3He \rightarrow nd\gamma \text{ and } nnp\gamma}{\pi^- + {}^3He \rightarrow {}^3H\gamma} = 1.10 \pm 0.15$$

The experimental value is 1.12 ± 0.05 [3].

For muon capture

$$\frac{\mu^- + {}^3He \rightarrow nd\nu \text{ and } nnp\nu}{\mu^- + {}^3He \rightarrow {}^3H\nu} = 0.44 \pm 0.06$$

The experimental value is 0.42 ± 0.10 [5].

Relativistic and meson interaction effects have been neglected in the calculation of these branching ratios and this may result in an error of the order of 10%.

Meson effects may have a more important role in pion and muon capture by ^3H since the IA for the capture rates is strongly suppressed by the Pauli principle. For capture from the 1s state the calculated rates are $W(\mu^- + {}^3H \rightarrow nnn + \nu) = 9.5\ sec^{-1}$, and $W(\pi^- + {}^3H \rightarrow nnn + \gamma) = 0.07 \times 10^{15}\ sec^{-1}$.

The meson corrections to these results could be as big as 40%. Unfortunately there is no data on muon capture in tritium. The recent experiment for pion capture in tritium gives [6]

$$R_\gamma = \frac{\pi^- + {}^3H \rightarrow nnn + \gamma}{\pi^- + {}^3H \rightarrow nnn \text{ and } nnn + \gamma} = 4.1 \pm 0.7\%$$

A clean comparison with theory is not possible because of the difficulties in calculating the non-radiative pion absorption rate.

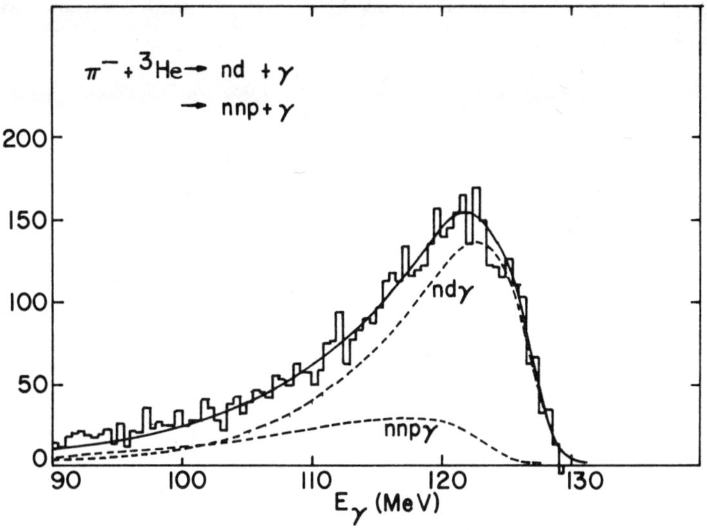
π⁻ + ³He → nd + γ
→ nnp + γ
200
150
100
50
0
ndγ
nnpγ
90
100
110
120
130
Eγ (MeV)

Fig. 1

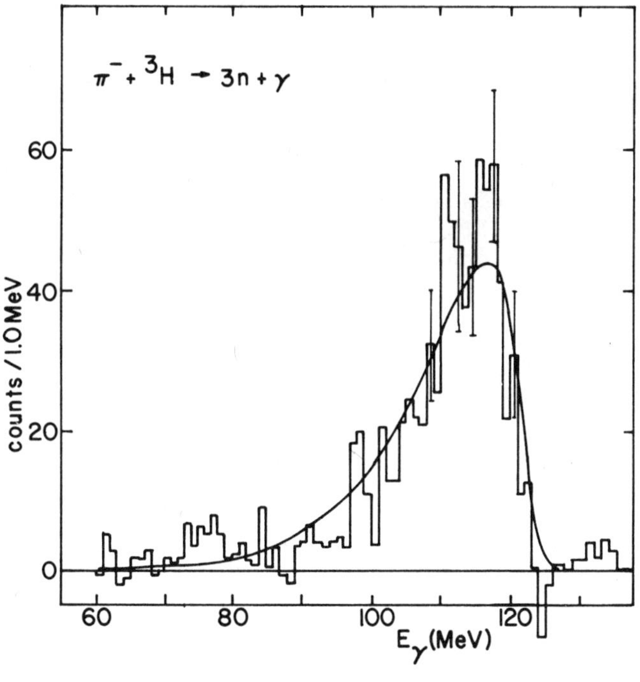
π⁻ + ³H → 3n + γ
60
40
20
0
counts / 1.0 MeV
60
80
100
120
Eγ (MeV)

Fig. 2

Nevertheless, using the two-nucleon absorption model to calculate the latter, we find

$$R_\gamma = 7 \pm 2\ \% .$$

One of the main motivations for the study of the radiative pion break-up reactions is to examine the dependence on the 3-nucleon continuum states and in particular to assess whether there is any evidence for the existence of 3-nucleon resonances. From the theoretical point of view, the only unambiguous information on the 3-nucleon continuum states is provided by calculations using the Amado model; no resonances are found. The photon spectrum in radiative pion capture is likely to be sensitive to resonant states. In fig. 1 we compare the experimental spectrum for

$$\pi^- + {}^3\mathrm{He} \rightarrow nd + \gamma \text{ and } nnp + \gamma$$

with the theoretical spectrum calculated using the (non-resonant) Amado model. The experimental spectrum shows no sign of a resonance.

In view of the close agreement obtained in $\pi^-\ {}^3\mathrm{He}$ any disagreement between theory and experiment in the pion capture in ^{3}H is likely to be a manifestation of peculiarities in the 3-neutron final state.

The theoretical and experimental [6] spectra are compared in fig. 2. Note that there are signs of structure at low 3-neutron energies. However, at this stage, the data is not precise enough to warrant any definite conclusions.

REFERENCES

[1] A.C. Phillips and F. Roig, Nucl. Phys. A234 (1974) 378.
[2] A.C. Phillips, F. Roig and J. Ros, Nucl. Phys. A237 (1975) 493.
[3] P. Truol, H.W. Baer, J.A. Bistirlich and K.M. Crowe, Phys. Rev. Lett. 32 (1974) 1268.
[4] See, for example, M. Ericson and M. Rho, Phys. Rep. 5 (1972) 57.
[5] L.B. Auerbach et al., Phys. Rev. 138 (1965) B127.
D. Clay et al., Phys. Rev. 140 (1965) B856.
O.A. Zaimidoroga et al., Phys. Lett. 6 (1963) 100.
[6] H.W. Baer, J.A. Bistirlich, S. Cooper, K.M. Crowe, J.P. Perroud, R.H. Sherman, F.T. Shively, P. Troul, Proceedings of the VI International Conference on High Energy Physics and Nuclear Structure, Santa Fe, 1975.

THE OMICRON SPECTROMETER

B.W. Allardyce

CERN

Geneve, Switzerland

INTRODUCTION

It is intended to build a spectrometer with a large solid angle and a large momentum acceptance at the reconstructed synchrocyclotron at CERN. This spectrometer will have an energy resolution of about 1 MeV for particles with momenta up to about 400 MeV/c.

The project has been approved at CERN and construction will take just over one year; it is hoped that physics runs will commence in the autumn of 1976. The cost of the project, excluding the magnet itself, is of order 400 K$ and this is being met by CERN and by the participating laboratories at Turin, Oxford, Birmingham, and IKO, Amsterdam. The Omicron collaboration consist of the people named in table 1. Further details may be found in the documents PH III 74/57 and 75/11.

The Omicron spectrometer will be sited at the improved SC at CERN. The accelerator's improvement programme has recently been completed and results so far are encouraging: high duty cycle extracted beams have been achieved: a proton extraction efficiency of 70% has been attained: internal beam currents of 600 na have been used at a reduced r.f. pulsing rate of 1 in 16, thus corresponding to the full design current of 10 μa if the r.f. were to be pulsed every cycle. The intention is to proceed to high intensity beams within the next few months.

TABLE 1

The Omicron Collaboration

G. Bonnazola T. Bressani E. Chiavasa S. Costa S. Dellacasa B. Gallio A. Musso G. Pasqualini	Turin
N. Tanner	Oxford
J. Davies	Birmingham
H. Arnold R. van Dantzig	Amsterdam
B. Allardyce	CERN

TECHNICAL FEATURES OF OMICRON

1. General

The spectrometer consists of a large magnet with roughly 1.7 m^3 of usable field volume, homogenious to about 10%. It is intended to place planes of multiwire chambers in this field in front of a target, followed by arrays of multiwire and drift chambers. The general philosophy is to detect inside the magnet both the incident particle and the one(s) leaving the target over a wide angular range. The trajectories of these particles will then be determined from the known magnetic field.

2. Magnet

A large magnet has been obtained on loan from RHEL. It weighs 189 tons and has external dimensions of 3m x 3m x 3m. This magnet was a bubble chamber magnet and consequently new poles have to be made. The poles are 2m x 1m in area and a gap of 85 cm has been chosen. With such a gap, and using flat poles with no shims, it has been calculated that the magnetic field will remain constant to within ± 10% over the whole volume which could be used for physics (i.e. the whole volume between the poles, apart from what is lost due to chamber frames, etc.). These calculations were performed at RHEL using the programme GFUN.

Various other minor modifications are to be made to the magnet and it will be sent to CERN around Christmas 1975. The magnet is shown in fig. 1.

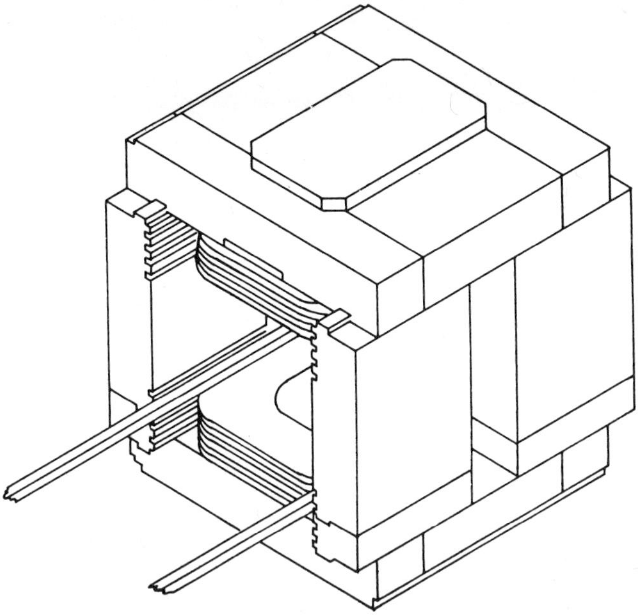

Fig. 1

Magnet

3. Location

The magnet will be located in the Proton Hall of the SC in such a way that it could in principle receive beams of π, μ, n and p. However, as will be shown later, most interest attaches to pions and muons at present. An arrangement for the backward scattering of low energy pions is shown in fig. 2.

The magnet will be mounted with its field vertical in such a way that rotation about a vertical axis is possible . This involves sinking the magnet in a pit in order to adjust the centre to the beam height at the SC of 1.25m.

4. Detectors

The incident beam (where fluxes up to 10^7/sec might be obtained), will be detected by planes of multiwire proportional chambers. These are being manufactured in Turin and will be of dimensions 12.5cm x 12.5cm. The chambers will have 1mm wire spacings, and will be thinner than normal chambers in order to reduce the multiple scattering. The specification for these chambers is as

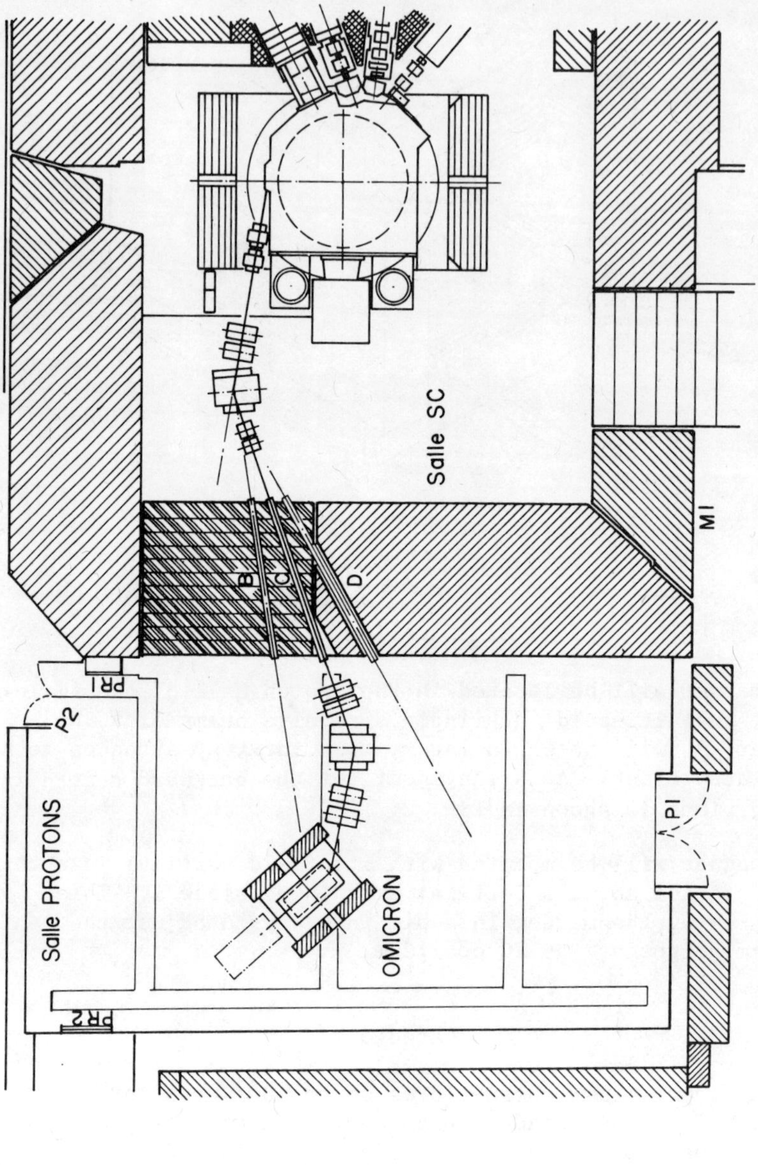

Fig. 2. An arrangement for the backward scattering of low energy pions

follows:

area: 12.5cm x 12.5cm

sense wires: 10μ gold-plated tungsten

wire spacing: 1 mm

sense wire to cathode gap: 4 mm

cathode wires: 50μ beryllium copper, perpendicular to sense wires

cathode wire spacing: 1 mm

windows: 10μ mylar

gas filling: 1 atm "magic" gas

A combination of chambers with vertical, horizontal and inclined wires will be used to eliminate ambiguities in track reconstruction.

After the target the first detector plane will also be mwpc for rate reasons, but following that there will be drift chambers arranged in "rings" around the target. Initially these "rings" will be polygons made of planar chambers, but one future possibility is to develop cylindrical chambers which might be suitable for certain experiments. The biggest drift chambers at present envisaged are 150 cm x 70 cm with 5 cm sense wire spacings.

The electronics for all these detectors will be of a standard CERN type.

5. Magnet Box

As mentioned earlier, the detectors as well as the target will be inside the magnet; in addition there will of course have to be various thin scintillators. The intention is to mount these components on a solid base plate which can be removed from the magnet on rails to an area behind the magnet. The positioning of target and detectors would be done here, and we are aiming at a positional accuracy of 1/10 mm. This base plate would then be pushed into the magnet to locate on prepositioned pins relative to the known magnet field.

Because multiple scattering is a serious problem at low energies, a helium atmosphere will be used to surround the detectors and target, so that the base plate will have a cover to retain this helium. Panels in the cover will carry the cable connections, light pipes, etc.

Calculations of the resolution including multiple scattering effects show that 1 MeV resolutions should be readily attainable.

6. Computing

An HP21 computer system will be used for on-line data acquisition. We hope to do some of the later off-line analysis on this system also. The system consists of two 32K HP21M computers with normal peripherals.

The translation of hits on wire planes into particle trajectories is a well-known problem with this sort of spectrometer. There is a good deal of expertise at CERN with Omega and SFM, but there are various differences in the case of Omicron. These are principally that

(a) the field is fairly homogenious

(b) the energy is low, so that multiplicity of tracks is low

(c) the background from particles striking windows, etc. is very much reduced because of the low energy.

However, in Omicron there will be serious multiple scattering, there is a lower redundancy of information along the tracks, and there might be problems from spiralling of low energy tracks.

It is intended to use the parametrisation technique for track finding if at all possible in order to reduce to a minimum the amount of computer time required. The field homogeneity of order 10% in Omicron will probably be adequate to allow such a technique to be used.

The programme chain will consist of the stages: simulation, track finding, geometry, and kinematics.

EXPERIMENTS

In early discussions about the possibility of constructing a spectrometer along the lines of what is now proposed, a long list of interesting experiments was drawn up. From this list we have recently concentrated attention on a smaller group of experiments, chosen as a balance between their physics content, their technical difficulty and the prejudices of the people concerned.

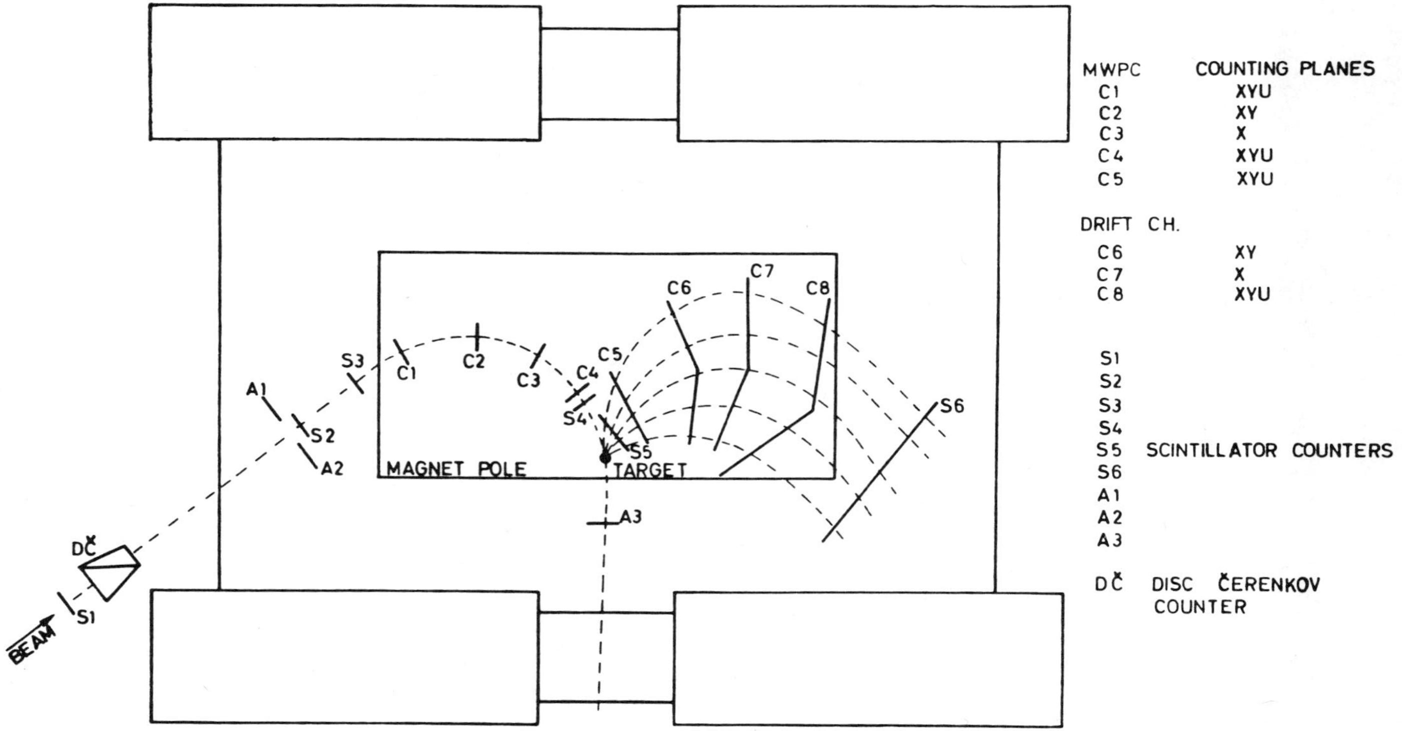

Fig. 3. Layout of the Omicron spectrometer for pion backward scattering measurement

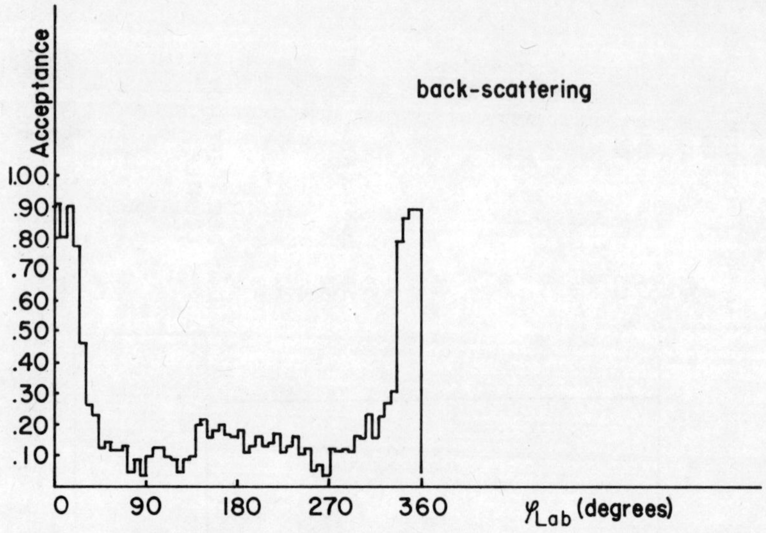

Fig. 4. Acceptance for the azimuthal angle ϕ, integrated over all the θ values, calculated by the Monte Carlo simulation of the experiment.

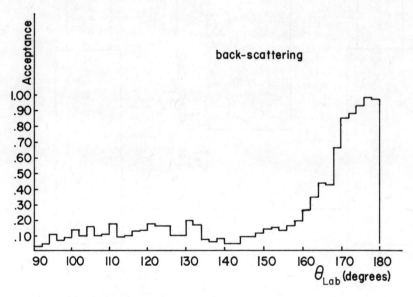

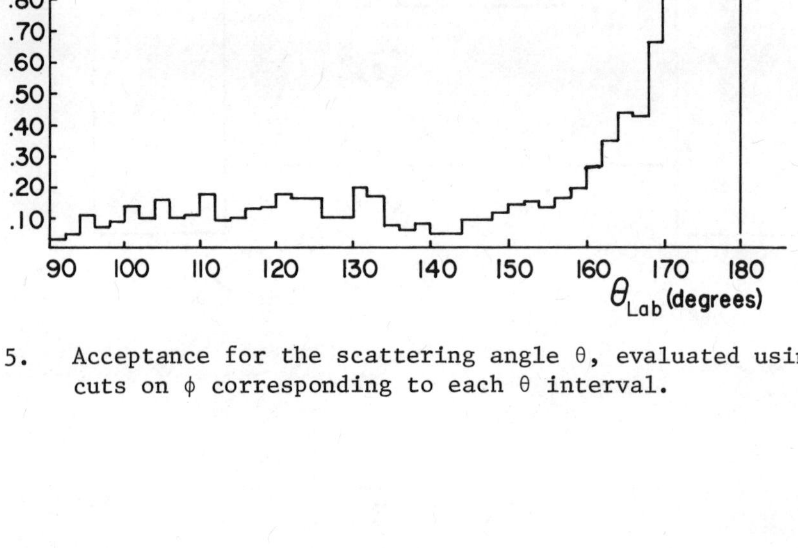

Fig. 5. Acceptance for the scattering angle θ, evaluated using the cuts on ϕ corresponding to each θ interval.

1. π,μ scattering at backward angles on light nuclei.

At present it is intended that this be the first experiment on Omicron. A good deal of theoretical interest attaches to a measurement of the elastic scattering of pions of low energy (up to 120 MeV) in the backward direction. Such measurements were requested by speakers at the Santa Fe conference in order to test the various models (e.g. optical models) which give very different predictions only at very backward angles.

The intention is to use a ^{12}C target because of the existence of the extensive data of Binon et al. on this nucleus. The arrangement of detectors within the magnet is shown in fig. 3. Monte-Carlo calculations of the acceptance of Omicron for such an arrangement are shown in fig. 4 and fig. 5 for the ϕ and θ angles; it will be seen that for $\theta > 160^o$ the acceptance is large.

It should be noted that since all pion beams are contaminated with muons, the above experimental arrangement would also allow a measurement of the muon backward scattering to be made. A DISC counter and/or time of flight will be used to distinguish π and μ.

Since low energy μ are essentially non-relativistic particles, the radiative effects which are present in electron scattering data are very much reduced here. Consequently some information on the radiative corrections to be applied in electron scattering may be obtained. In addition, it might be possible to compare the scattering of low energy μ^- and μ^+ in order to extract the electric polarizability of, for example, ^4He; differences in the shape of the angular distribution at back angles of order 1% are expected, so such an experiment would be difficult.

2. Double charge exchange reactions

Double charge exchange reactions have not received a great deal of attention in the past, due principally to rather low cross-sections. Various theoretical estimates have been made, the latest of which gives cross-sections which are fairly healthy. This may well be due to the favourable nuclear states involved (mirror states) in the calculation of the reaction $^{18}O\ (\pi^+,\pi^-)\ ^{18}Ne$. The result of this calculation is shown in fig. 6 for 450 MeV pions.

As will be seen, the forward direction out to $\sim 15^o$ is the only place where good cross-sections are to be found, and the experimental arrangement inside Omicron shown in fig. 7 will be used to study this reaction. Monte-Carlo calculations of the geometrical acceptance of this arrangement have been made with the results shown in

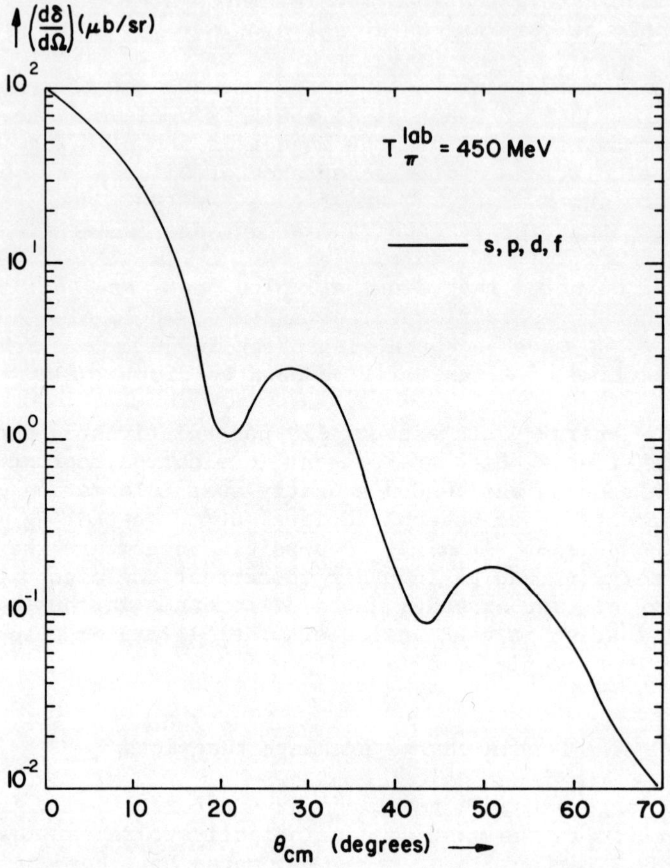

Fig. 6. Differential cross-section for the $\pi^+ + {}^{18}O \rightarrow \pi^- + {}^{18}N$ (g.s.) reaction at T_π = 450 MeV, following the calculation of Lin and Franco.

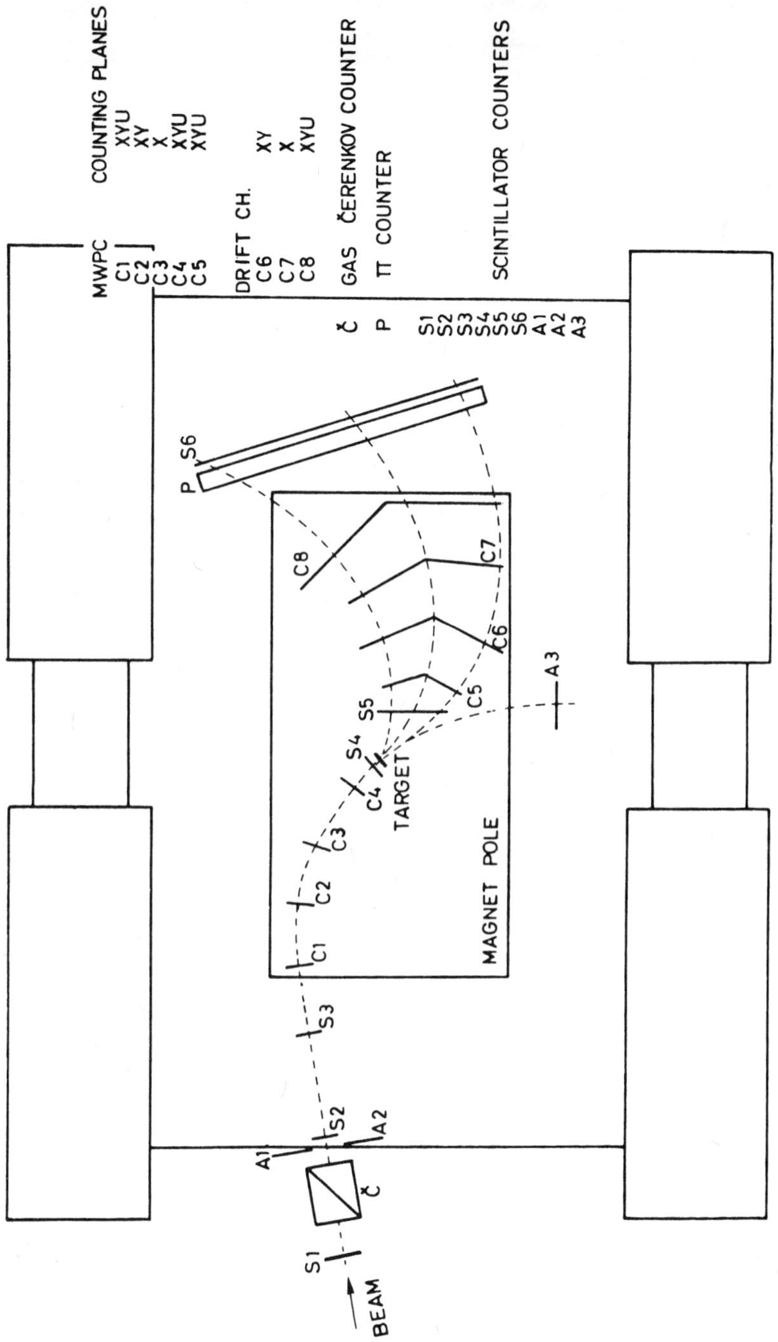

Fig. 7. Experimental layout of the Omicron spectrometer to measure double charge exchange reactions.

figs. 8 and 9 for the ϕ and θ angles. With the expected pion fluxes at the SC2, anticipated count rates of order 1 per minute will be obtained corresponding to the worst case of the recent calculations.

In addition to an ^{18}O target there is interest in the d.c.e. reaction on ^{7}Li and ^{9}Be for which various groups disagree over the formation of bound states in $^{7}B^{*}$ and $^{9}C^{*}$. Also by using a ^{12}C target we might be able to investigate double analogue states in nuclei via $\pi^- + {}^{12}C \rightarrow \pi^+ + {}^{12}Be$ and $\pi^+ + {}^{12}C \rightarrow \pi^- + {}^{12}O$.

Finally, if the apparatus works superbly well, the experimental arrangement for d.c.e. might allow a measurement of the π^0 lifetime, a quantity known only to the 10 - 15% level to date. This would utilise the reactions $\pi^- + p \rightarrow \pi^0 + n$ at 300 MeV/c to form the π^0, followed by $\pi^0 + p \rightarrow \pi^+ + n$. All the cross-sections are known, and the observed π^+ flux as a function of distance of π^0 travel would obviously yield the π lifetime. However this is a very difficult measurement, and requires excellent reconstructional accuracy on the observed tracks of the incident π^- and the final π^+.

3. Branching Ratio for the Rare Decay $\pi^0 \rightarrow e^+e^-$.

This rare decay mode of the π^0 has not yet been directly measured, but is of considerable theoretical interest, because calculations of the rate involve assumptions about the character of the interaction (e.g. vecter, axial vecter interaction and/or scaler, pseudoscaler) which will have repercussions elsewhere.

The present upper limit of 5 10^{-6} for the branching ratio is two orders of magnitude above the unitarity limit, and to measure to the 10^{-8} level obviously requires patience. Counting rate calculations for the direct observation of $\pi^0 \rightarrow e^+e^-$ give typically one count per hour with an arrangement as shown in fig. 10. This experimental arrangement has been optimised to accept events with momenta in the range 165 ± 55 MeV/c and with opening angles between the two electrons of $50^0 \pm 2^0$. The π^0 are produced by π^-p at 300 MeV/c. Extremely tight geometric and reconstructional constraints will have to be used to reject background events of which there are numerous sources, e.g. $\pi^0 \rightarrow \gamma e^+e^-$, electron contamination of the π^- beam, π^-p elastic scattering, $\pi^-p \rightarrow \gamma n$, $\pi^-p \rightarrow ne^+e^-$ by internal conversion. These sources of background have been extensively investigated by Davies.

The intention is to measure the number of events as a function of invariant mass of the e^+e^- system, in which the reaction $\pi^0 \rightarrow e^+e^-$ will appear as a small bump at a mass corresponding to the π^0 mass.

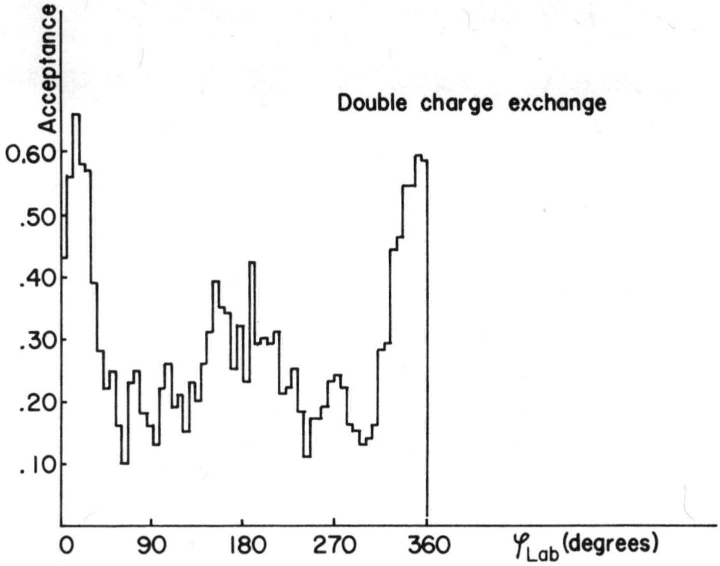

Fig. 8. Acceptance for the azimuthal angle ϕ, integrated over all the θ values.

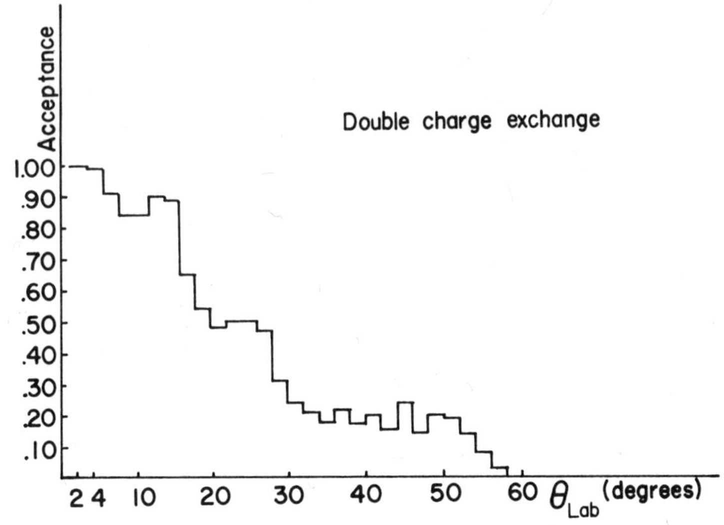

Fig. 9. Acceptance as a function of the emission angle θ, taking into account the cuts on ϕ in correspondence to each interval of θ.

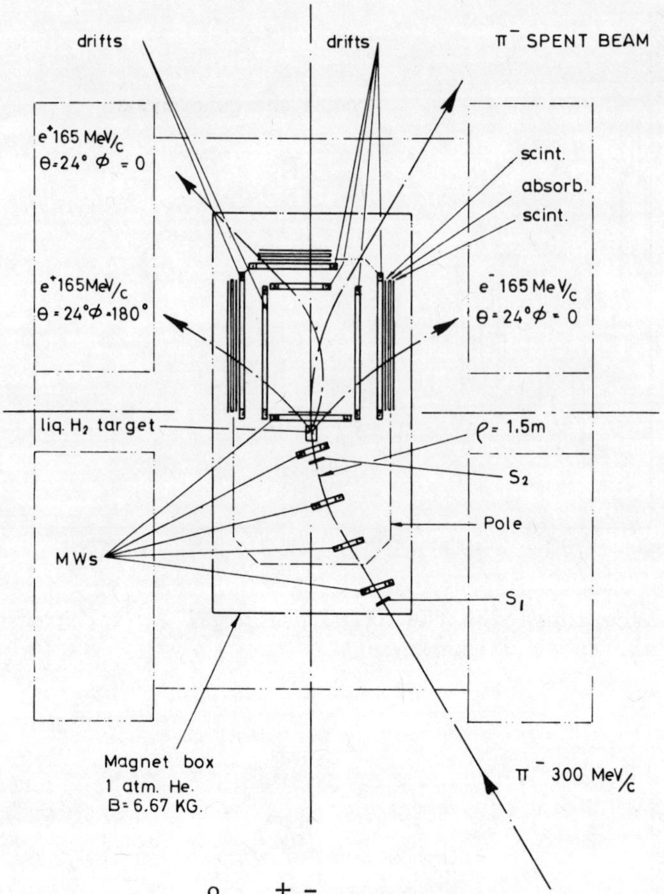

Fig. 10. Layout for $\pi^o \to e^+e^-$ in the Omicron spectrometer.

4. Other Experiments

Two other experiments are also included in the proposal for the Omicron spectrometer but will not be discussed in detail here because of the limitation of time. The first is the measurement of π-π scattering lengths by threshold pion production, $\pi p \to \pi\pi N$. The main problem here is the low energy of the two pions, where detection and track reconstruction will be difficult.

The other experiment is the measurement of the axial form factor from $\pi^- p \to n "\gamma"$, the virtual γ becoming an e^+e^- pair by internal conversion. This experiment will give information on the form factor as a function of momentum transfer, since the virtual γ is off shell. Hence the information will complement that from β decay and μ capture where the momentum transfer is 0 and M_μ respectively.

STATUS REPORT ON THE SIN UNIVERSAL SPECTROMETER INSTALLATION (SUSI)

E.T. Boschitz*

University of Karlsruhe

Germany

At an early stage of planning the experimental facilities at SIN it was decided to build a high resolution π-channel and a π-spectrometer for nuclear structure studies. For the π-channel an achromatic system with a large dispersion at the intermediate focus was designed in order to provide a versatile beam line for different kinds of experiments. The lay out is shown in Fig. 1. The π-channel consists of two ion optically almost symmetric sections. The first one (consisting of the quadrupoles Q_1 - Q_6, the separator S, the compensating magnets M_s, the bending magnet M1 and the beam blocker B) disperses the pion beam (produced at the target P) in the horizontal plane at M. At this point a horizontal focus is also required in order to select the pion momenta by means of a MWPC (C_1). The second section (consisting of the bending magnet M2 and the quadrupoles Q_7 - Q_{10} recombines the dispersed beam into an achromatic beam spot on the scattering target (1 cm x 2 cm). The second MWPC (C_2) determines the incident angle of the pions on the target.

The advantage of the small beam spot on the target is many fold. First of all it reduces considerably the design problems for the π-spectrometer. Then it may be very useful for experiments like (π^+,2p) using solid state counters for the detection of the protons. Gas targets as well as solid targets

---

*ETH-GRENOBLE-HEIDELBERG-KARLSRUHE-NEUCHÂTEL-SIN-COLLABORATION. (J. Arvieux, E.T. Boschitz, R. Corfu, J. Domingo, J.P. Egger, K. Gabathuler, P. Gretillat, Q. Ingram, C. Lunke, E. Pedroni, C. Perrin, J. Piffaretti, E. Schwarz, C. Wiedner, J. Zichy.)

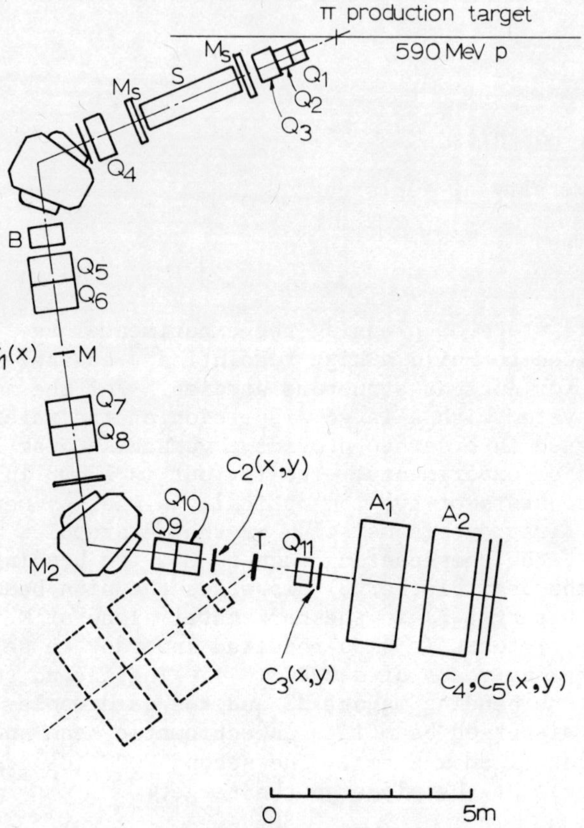

Fig. 1. Layout of High Resolution π-channel

consisting of rare isotopes can readily be used. The design of an auxiliary magnet for 180°-scattering becomes much simpler. Finally the scattering angle is well enough defined so that a special determination of the scattering center by the computer is not needed.

The disadvantage of this channel design is the use of MWPC's in the beam which limits the total pion flux which can be handled to 3×10^7/s and because of energy-straggling it limits the momentum resolution to about $dp/p = 5 \times 10^{-4}$.

The original design of the π-channel used standard beam handling components of SIN. More detailed calculations in connection with the π-spectrometer showed that the standard bending magnets were inadequate. New magnets with properties of spectrometer magnets have been designed and constructed.

The research program of the future users was reviewed in detail in order to decide upon the design criterea for the π-spectrometer. There were basically three groups of experiments:
1) elastic and inelastic pion scattering
2) knock out reactions (π,π'N), or more general (π,π'X)
3) double charge exchange reactions.

It was also stressed that the future program with the spectrometer should utilize the better duty cycle of the cyclotron, and be complimentary to the high resolution facility EPICS at LAMPF (i.e. give preference to coincidence experiments). These considerations suggested a broad range spectrometer of large solid angle with a momentum resolution matching the one of the channel.

Various options were considered for the spectrometer: commercially available spectrometers, special design with standard SIN components and existing designs from various laboratories. The performance - price - time - comparison gave preference to the design of the SACLAY SPES II spectrometer (P. BIRIEN, J. THIRION). Fig. 2 shows the schematic layout of this spectrometer. It consists of one quadrupole singlet followed by two dipole magnets. There is a detector D located between the quadrupole and the first magnet. It records the x and y coordinates of a passing particle. Two more detectors at the focal plane F and 50 cm behind produce information on x and y, and on the corresponding angle.

The optics of the system has the following features
<u>In the radial (deflexion) plane</u>:
1. Point to point imaging between D and F
2. Between the two dipole magnets the trajectories of particles with the same momenta are parallel
3. At F the dispersed trajectories are parallel.

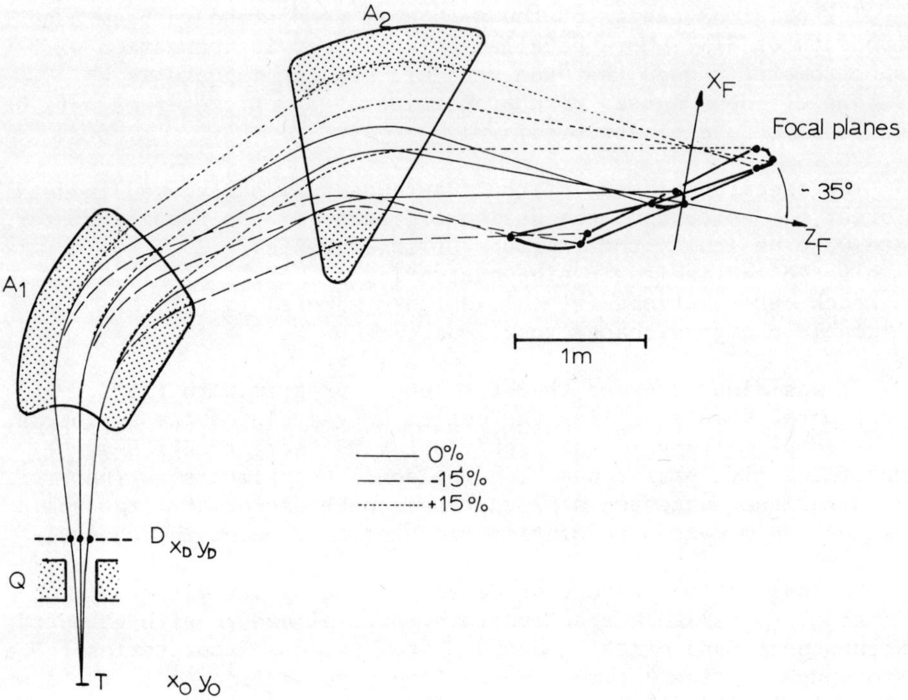

Fig. 2 . Layout of π-Spectrometer

The focal lines for a given (x_D, y_D) at D are thus almost straight.

In the axial plane:

Point to point imaging between T and F.

The principle advantage of this system is the following: whereas a spectrometer with a large figure of merit (product of solid angle, momentum acceptance and momentum resolution) requires a careful reduction of spherical and chromatic aberrations up to fifth order with the aid of complicated curvatures at magnet edges, H_t windings, multipole elements etc., these aberrations are only important to second order in this design. The determination of the coordinates (x,y) at D reduces the basic solid angle to 50 μsr (due to target size and straggling effects). Of course this advantage has to be paid for: Each point at D has "its own focal plane" which has to be determined and stored in the computer for on-line analysis. The intersects of these focal planes with the particle trajectories measured by detector C4 and C5 yield the particle momenta (Fig. 3). In addition to determining the focal planes the coordinates x_D and y_D measure the angle of scattering from the target.

In the following the specifications for the π-channel and the π-spectrometer are listed.

Specifications for	π-channel	π-spectrometer
momentum resolution $\frac{dp}{p}$	5×10^{-4}	5×10^{-4}
solid angle Ω [msr]	6	16
momentum acceptance	± 1 %	± 18 %
$p_{min.} - p_{max.}$ $\left[\frac{MeV}{c}\right]$	100 - 450	100 - 450
dispersion D [cm/%]	7.5 (first half) 10^{-5} (second half)	5.5 (along the focal plane)
inclination of focal plane	80°	35°
angular range without - with auxiliary magnet assembly		0 - 140° 140°-180° vertical

A detailed examination of the fabrication drawings for the Saclay spectrometer led to the conclusion that major redesign work had to be done. The magnetic field needed for deflecting particles of p = 450 MeV/c is B = 11.5 kG whereas the SACLAY version was

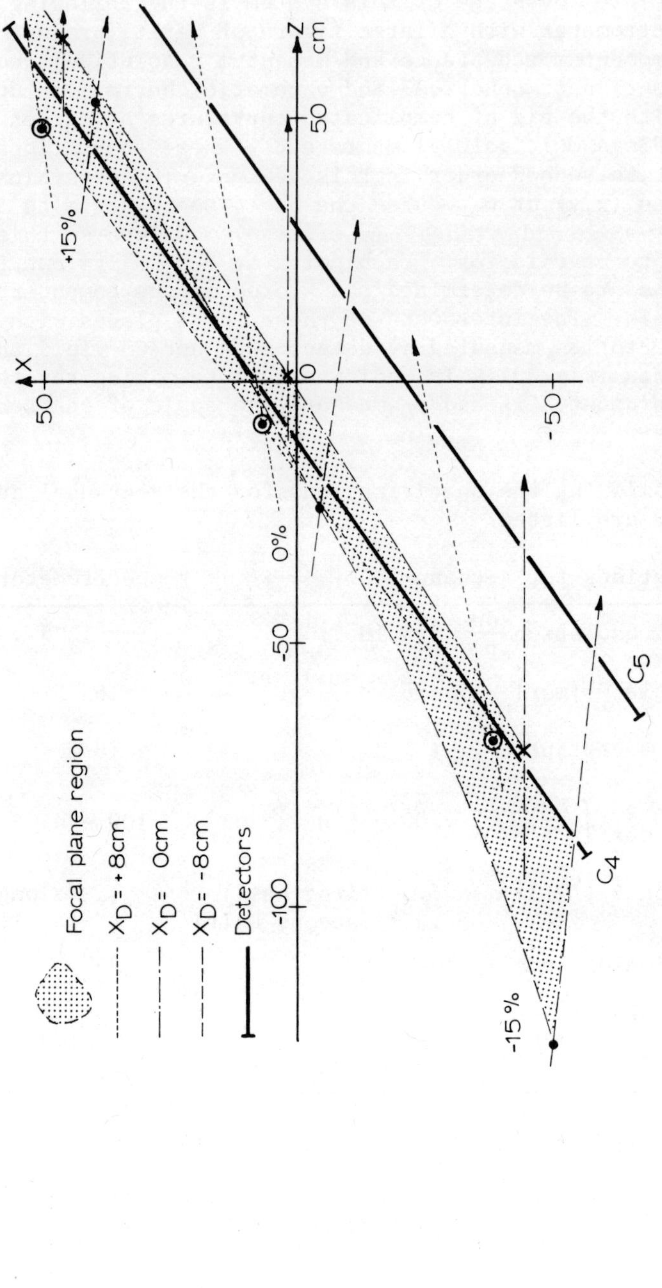

Fig. 3. Situation in Focal Plane Region

designed for 18 kG. Consequently the coil and yoke configuration had to be changed. Also technical concessions had to be made to the manufacturer of the yoke and the pole pieces in order to save time and money. For easy assembly the vacuum chambers were redesigned. The magnets were manufactured in less than 9 months. After having adopted a system of 130 flip coils which has been used for mapping the magnetic field of the SIN accelerator magnets extensive field measurements were performed for both spectrometer magnets in and out of the mid plane for several fields. These data are being evaluated using the ray trace program SGOUB1 from SACLAY. After the final assembly with the vacuum chambers the magnets will be installed in the spectrometer frame in the next weeks. (Fig. 4).

It is clear from the principle of operation of the π-channel and the π-spectrometer that the performance of the entire system relies heavily on the quality of the detectors, on a fast read out system and on the efficient use of the on-line computer. Great effort has been put into these three parts of the system.

The main quality required for C_1 and C_2, in addition to minimum thickness, is the ability to handle a particle flux up to 3×10^7/s; more specifically, these detectors must be able to distinguish events separated by 20 ns (microstructure of the primary proton beam at SIN). In principle there is no high rate requirement for C_3, C_4, C_5, but the need for good spacial resolution (less than 1 mm) and minimum thickness. Various MWPC's have been developed at the University of Neuchâtel. In addition to standard chambers (following a recent design at CERN) special chambers have been produced to achieve the good time resolution. These chambers consist of 2 standard anode planes of 1mm wire spacing shifted by 0.5 mm. In these chambers it was possible to limit the electron collection around the wire within a cylinder of about 0.3 mm radius by increasing the amount of freon to 6% without losing much efficiency. The time fluctuation of the electron collection was then greatly reduced. The time spectrum obtained during the tests for any triggered wire with respect to a scintillator had a FWHM of 5.4 ns and was almost completely contained in 20 ns. The high voltage was 5.6 kV and the efficiency 95%.

For the particle detection in the focal plane of the spectrometer one MWPC has been built (the size is 1040 mm x 260 mm). The principle of induced charge is used for the readout with delay lines. A fast coincidence and readout system has been developed for the MWPC's C_1, C_2, C_3*.

*A fast readout system for multiwire proportional chambers, R. Foglio, C. Perrin, J. Pouxe, U. Bart, E. Schwarz, Proceedings of the 2nd Ispra Nuclear Electronics Symposium, May 30, 1975.

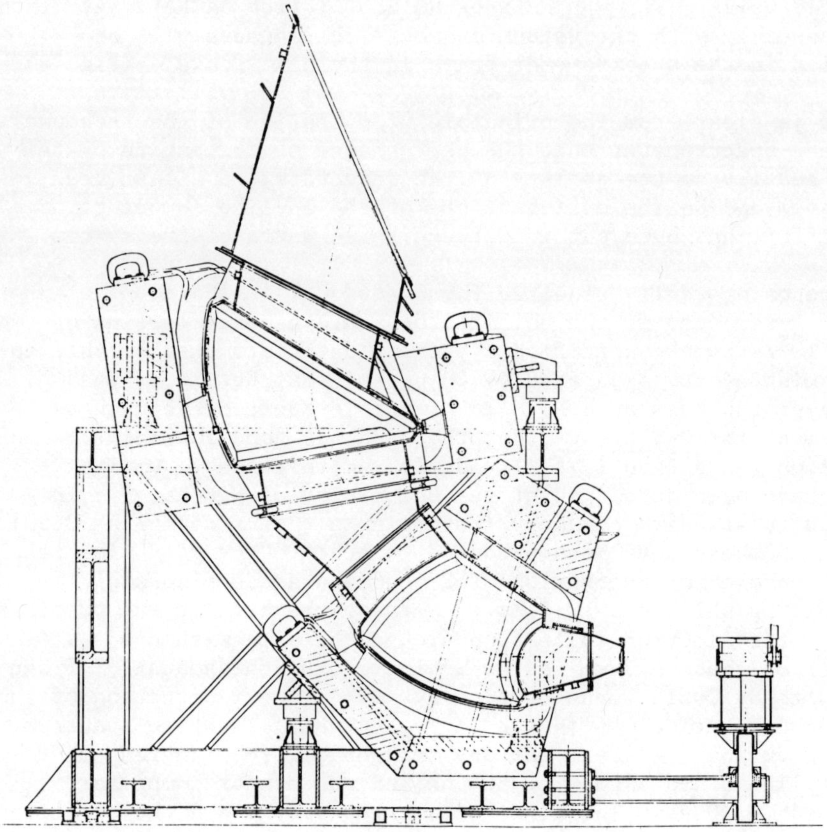

Fig. 4. π-Spectrometer in frame

It consists of the following elements: preamplifiers, fast or, fast coincidence registers, crate controllers, coding and fast rejection unit and a camac buffer connected to a PDP11/45 computer. The computer contains 28 k memory, 2 magnetic tape units, 1 RK05 Disk, 1 paper tape punch/reader, 1 DECwriter LA-30 terminal, 1 Tektronix 4010 Display terminal connected to CAMAC through a dedicated CC11 crate controller. The on-line computing is handled by a system based on DEC's RSX-11A, modified and improved to suit our purposes. This permits up to 36 separate "tasks" (i.e. user programs) to be used to run the experiment.

The present schedule calls for assembly of channel and spectrometer until end of August 1975 in order to start the first tune-up tests in September 1975.

Acknowledgement:

The dedicated effort of many members of the technical services at SIN has been essential for the rapid progress of this project.

PION PRODUCTION ON NUCLEI

A. Reitan

University of Trondheim

Trondheim, Norway

1. THE EXPERIMENTAL SITUATION

To characterize the experimental situation on π production on complex nuclei in a few words, one might say that while more experiments certainly are needed if one wishes to understand the details of the reaction mechanism, the quality of the experiments is better and the number of cases studied larger than in some other important areas of pion-nuclear physics. This has, above all, come about through the efforts of the Uppsala [1-3] group, that has studied (p,π^+) and (p,π^-) reactions on a number of targets (^9Be, ^{10}B, ^{12}C, ^{13}C, ^{16}O, ^{28}Si and ^{40}Ca for π^+, ^9Be and ^{13}C for π^-) at 185 MeV incident proton energy. What makes these experiments so important is first of all the fact that the angular distributions have been measured, with rather high accuracy (at least in the π^+ case); furthermore, it has been possible to separate the cross sections leading to the various states of the final nucleus, as long as these states are not too closely spaced. Generally speaking, the results of these experiments are that a) the cross section for π^+ production is much larger and shows more variation with angle than that for π^- production, and that b) the angular distribution and absolute cross section is very noticeably dependent on which final nuclear state is reached in the reaction. A typical example is shown in fig. 1.

Other $(p,\pi^{\pm})$ experiments have been performed by Domingo et al. [4-6] (d, ^3He, ^4He, ^9Be, ^{12}C, ^{13}C, ^{14}N (p,π^+), ^9Be (p,π^-), 600 MeV, 0°), Le Bornec et al. [7] (^{10}B, ^{14}N, ^{32}S, ^{40}Ca (p,π^+), 154 MeV, angular distribution to ground states), Cochran et al. [8] (d, Be, C, $A\ell$, Ti, Cu, Ag, Ta, Pb, Th $(p,\pi^{\pm})$, 730 MeV, angular distribution summed over final states), and Dollhopf et al. [9] ($d(p,\pi^+)^3H$, 470 and 590 MeV, angular distribution). There are also data by Amato et al. [10] on the inverse (π^+,p) reaction for 70 MeV pions (^6Li, ^9Be, ^{12}C, ^{16}O). As far as more exotic π-production experiments on complex nuclei are concerned, Wall et al. [11] have studied the $(^3He,\pi^0)$ reaction on ^{12}C and Pb at 180 and 200 MeV.

2. QUALITATIVE FEATURES

One of the important features of the (p,π) experiments that have been performed is the pronounced momentum mismatch between the incident and produced particle. To take a specific example, a proton with lab. kinetic energy T_p = 185 MeV incident on ^{12}C has a CM momentum k_p = 562 MeV/c; with the final nucleus ^{13}C in its ground state the CM momentum for the emitted π^+ is k_π = 102 MeV/c; the momentum transfer $q = k_p - k_\pi$ being q_{min} = 460 MeV/c at 0^0 and q_{max} = 664 MeV/c at 180^0. At the incident energy T_p = 600 MeV the momentum transfer ranges from q_{min} = 561 MeV/c to q_{max} = 1592 MeV/c. With these numbers in mind it comes as no surprise that

a) The $A(p,\pi^+)B$ cross section is small compared to the total reaction cross section for protons on the same nucleus and at the same energy. At 185 MeV the (p,π^+) cross section on ^{12}C, leading to the first few levels in ^{13}C, is $\sigma_{\pi^+} \approx 5$ μb, whereas the total reaction cross section for protons on ^{12}C is $\sigma_R \approx 200$ mb [12]. Thus, $\sigma_{\pi^+}/\sigma_R \sim 10^{-5}$.

b) The (p,π^+) differential cross sections in the forward direction are not all that different at 600 MeV and at 185 MeV. For $^{12}C(p,\pi^+)^{13}C$ g.s. we have at 0^0 $d\sigma_{\pi^+}/d\Omega \approx 0.40$ μb/sr [2] for T_p = 185 MeV and $d\sigma_{\pi^+}/d\Omega \approx 0.25$ μb/sr [4,6] for T_p = 600 MeV.

c) The differential cross section drops much more rapidly with increasing angle at 600 MeV than it does at 185 MeV (or, any backward peak appearing in the 185 MeV differential cross section should be strongly suppressed at 600 MeV). As it stands, this is a prediction; however, that kind of behaviour is seen in the $d(p,\pi^+)t$ data of ref. [9] when one compares the cross sections at 470 and 590 MeV.

One of the reasons given most often for doing pion-nuclear physics is the possibility of using such reactions to extract nuclear-structure information. In the simplest picture of the (p,π^+) reaction, the so called one-nucleon mechanism (fig. 2), one imagines that after the proton has emitted the pion, the resulting neutron gets captured directly into an available single-nucleon state. In the most simple-minded version of this

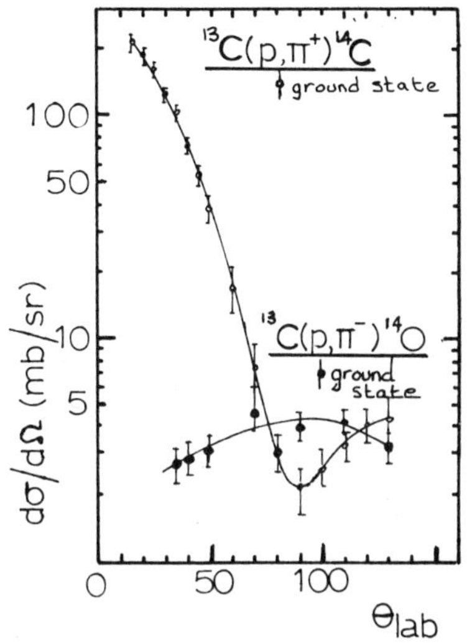

Fig. 1. Angular distributions for $^{13}C(p,\pi^{+})^{14}C$ and $^{13}C(p,\pi^{-})^{14}O$.

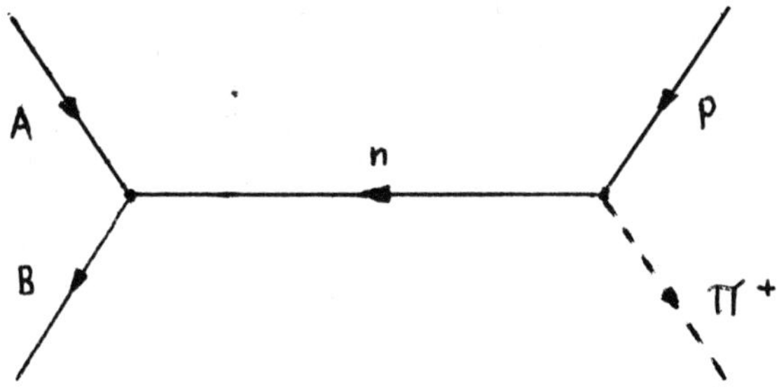

Fig. 2. The one-nucleon mechanism.

model, the plane-wave approximation, this takes place with no participation of the other nucleons in the target, except for the fact that these make up the potential well that determines the properties of the state in question. The differential cross section then becomes proportional to the square of the single-particle wave function in momentum space, evaluated at the (large) momentum transfer $\underline{q}$. When the first 0° differential cross sections at 600 MeV became known it was thus hoped [4] that (p,π^+) reactions could be used as a simple, sensitive and direct measure of the tail end of nuclear wave functions in momentum space. This has since turned out to be too optimistic. Likewise, when the first angular distributions to separate final states appeared, it was hoped that other properties of such states could be deduced in a simple manner by comparing the different angular distributions. At present, this also seems a bit too much to hope for. Nevertheless, the variations in absolute magnitude and angular dependence that one sees in these cross sections indicate a strong influence of nuclear properties, and, generally speaking, it is safe to state that

d) The (p,π^+) reactions can indeed be used as a tool for obtaining nuclear-structure information. The point that should be made, however, is that in pionic nuclear physics one should no more expect to get something for nothing than in more conventional branches of nuclear physics. Nuclear structure is a messy subject, and the complexities that exist are bound to show up also in any realistic treatment of pionic reactions with nuclei, the only simplification being that arising from the fact that the pion has zero spin. In conventional nuclear reaction theory it was realized quite early that any realistic model of a reaction should at the very least treat the geometry (i.e. the angular momenta) of the problem properly; any approximation in which angular momentum is disregarded can at most be qualitatively correct and be used only to understand how a certain reaction comes about, not to extract detailed information about nuclear properties.

In the remarks above only (p,π^+) reactions have been mentioned. About the even rarer (p,π^-) reactions one can first of all make the obvious observation that they cannot arise from the simple one-nucleon mechanism of fig. 2, at least two nucleons have to be involved in the process. Since the angular distributions of π^- generally seem to show less structure than those of π^+, it seems likely that from a nuclear-physics point of view there is less to be learned from (p,π^-) than from (p,π^+) reactions. However, it seems equally likely that the flat part of the (p,π^+) angular distributions at backward angles is due to the same sort of processes which cause the (p,π^-) reaction at these and other angles; in other words, one can hope to use π^- data to disentangle part of the problem involving the π^+ reaction.

We shall now try to give some insight into some of the theoretical approaches to $(p,\pi^{\pm})$ processes, leaving alone cute reactions such as $(^3He,\pi^0)$, where at low energies the production of pions must take place through a coherent effort of the nucleons in the projectile, and where the cross section not only is small, but minute (~ 10pb/(sr·MeV) [11]).

3. THE ONE-NUCLEON MODEL

Let us, then, take a closer look at the process represented by fig. 2. As indicated before, the idea is to describe the reaction as a one-nucleon problem, in which the initial and final nucleon states are $|p, \underset{\sim}{k}_p, \mu_p\rangle$ and $|n, \psi_{NLJ}^{J_z}\rangle$, μ_p being the z-component of the proton spin and $\psi_{NLJ}^{J_z}$ a bound neutron state; we are for the time being considering the case of a spin zero target and a final nucleus consisting of the same spin zero core plus a neutron. In terms of the interaction Hamiltonian $\underline{H}_{\pi^+N}$ the CM differential cross section is

$$\frac{d\sigma}{d\Omega} = (2\pi)^{-2}(E_pE_AE_\pi E_B/s)(k_\pi/k_p)^{\frac{1}{2}} \sum_{\mu_p} \sum_{J_z} | \langle n,\psi_{NLJ}^{J_z}|H_{\pi^+N}|p,\underset{\sim}{k}_p,\mu_p\rangle|^2 ,$$

$$s = (m_A + m_p)^2 + 2m_AT_p = (E_A + E_p)^2 = (E_B + E_\pi)^2 , \qquad (1)$$

$$k_p^2 = \tfrac{1}{4}s^{-1} [(s - m_A^2 - m_p^2)^2 - 4m_A^2 m_p^2],$$

$$k_\pi^2 = \tfrac{1}{4}s^{-1} [(s - m_B^2 - m_{\pi^+}^2)^2 - 4m_B^2 m_{\pi^+}^2].$$

a) The Interaction Operator

Already at this stage there are some uncertainties in the problem, connected with the interaction operator $\underline{H}_{\pi N}$. One can use the relativistic pseudoscalar interaction $H'_{\pi N} = \underline{i}\ \underline{g}\ \gamma_5\ \underset{\sim}{\tau}\cdot\underset{\sim}{\phi}$ as a starting point, where $\underset{\sim}{\phi}$ is the pion field and $\underset{\sim}{\tau}$ is the nucleon isospin operator, and where $H'_{\pi N}$ is to be applied between Dirac spinors for the nucleon. We look at the unphysical single-pion absorption and emission by a free nucleon, described by plane-wave initial and final states $|\underset{\sim}{k}\rangle = \exp(\underline{i}\ \underset{\sim}{k}\cdot\underset{\sim}{r})$ and $|\underset{\sim}{k}'\rangle = \exp(\underline{i}\ \underset{\sim}{k}'\cdot\underset{\sim}{r})$. We have for both cases

$$H'_{\pi N} = \overline{\psi}_f H'_{\pi N} \psi_i = \psi_f^\dagger \gamma_4 H'_{\pi N} \psi_i$$

$$= ig\,[(E'+m)/2m]^{\frac{1}{2}}\,[(E+m)/2m]^{\frac{1}{2}}\,\langle \underset{\sim}{k}'|\underset{\sim}{\tau}\cdot\underset{\sim}{\phi}|\underset{\sim}{k}\rangle$$

$$\times\, \chi_f^\dagger \begin{pmatrix} 1 \\ \dfrac{\underset{\sim}{\sigma}\cdot\underset{\sim}{k}'}{E'+m} \end{pmatrix}^\dagger \begin{pmatrix} 1 & 0 \\ 0 & -1 \end{pmatrix} \begin{pmatrix} 0 & -1 \\ -1 & 0 \end{pmatrix} \begin{pmatrix} 1 \\ \dfrac{\underset{\sim}{\sigma}\cdot\underset{\sim}{k}}{E+m} \end{pmatrix} \chi_i \qquad (2)$$

$$= -ig\,[(E'+m)(E+m)]^{-\frac{1}{2}}\,\langle \underset{\sim}{k}'|\underset{\sim}{\tau}\cdot\underset{\sim}{\phi}|\underset{\sim}{k}\rangle$$

$$\times\, \chi_f^\dagger\, \underset{\sim}{\sigma}\cdot\left[\frac{E'+E+2m}{4m}(\underset{\sim}{k}-\underset{\sim}{k}') + \frac{E'-E}{4m}(\underset{\sim}{k}+\underset{\sim}{k}')\right]\chi_i\,,$$

where χ is a Pauli spinor and where we have taken $\underline{m}_p = \underline{m}_n = \underline{m}$, similarly $\underline{m}_{\pi^\pm} = \underline{m}_{\pi^0} = \mu$. If we express this in terms of the energies and momenta of the pion ($\underline{E}_\pi$, $\underset{\sim}{k}_\pi$) and the $\begin{smallmatrix}\text{initial}\\ \text{final}\end{smallmatrix}$ nucleon for $\begin{smallmatrix}\text{absorption}\\ \text{emission}\end{smallmatrix}$ $\left(\underline{E}'', \underset{\sim}{k}'' = \begin{smallmatrix}\underline{E},\underset{\sim}{k}\\ \underline{E}',\underset{\sim}{k}'\end{smallmatrix}\right)$ we get to terms of the order $\underline{E}_\pi/\underline{m}$

$$H'_{\pi N} = \pm(4\pi)^{\frac{1}{2}} if\,\mu^{-1}\,\langle \underset{\sim}{k}'|\underset{\sim}{\tau}\cdot\underset{\sim}{\phi}|\underset{\sim}{k}\rangle\,\chi_f^\dagger\,\underset{\sim}{\sigma}\cdot\left[\left(1-\tfrac{1}{2}\frac{E_\pi}{E''+m}\right)\underset{\sim}{k}_\pi - \frac{E_\pi}{E''+m}\underset{\sim}{k}''\right]\chi_i$$

$$= \pm(4\pi)^{\frac{1}{2}} if\,\mu^{-1}\,\langle \underset{\sim}{k}'|\underset{\sim}{\tau}\cdot\underset{\sim}{\phi}|\underset{\sim}{k}\rangle\,\chi_f^\dagger\,\underset{\sim}{\sigma}\cdot\left(\underset{\sim}{k}_\pi - \lambda\,\frac{2m}{E''+m}\,\frac{E_\pi}{m}\,\frac{\underset{\sim}{k}+\underset{\sim}{k}'}{2}\right)\chi_i\,, \qquad (3)$$

$$f^2 = (4\pi)^{-1}\,(\mu/2m)^2\,g^2 \approx 0.08, \quad \lambda = \tfrac{1}{2}.$$

For plane-wave pions $\underset{\sim}{\phi} \propto \exp(\pm\,\underline{i}\,\underset{\sim}{k}_\pi\cdot\underset{\sim}{r})$ and $\underset{\sim}{\pi} = \partial\underset{\sim}{\phi}/\partial t = \mp\,\underline{i}E_\pi\underset{\sim}{\phi}$, where $\underset{\sim}{\pi}$ is the conjugate field. The non-relativistic p-wave pion-nucleon interaction obtained here is therefore, with $\underline{E}'' = \underline{m}$,

$$H_{\pi N} = (4\pi)^{\frac{1}{2}}\,f\,\mu^{-1}\,\underset{\sim}{\sigma}\cdot\{\underset{\sim}{\nabla}_\pi(\underset{\sim}{\tau}\cdot\underset{\sim}{\phi}) + \frac{\lambda}{2m}[\underset{\sim}{p}(\underset{\sim}{\tau}\cdot\underset{\sim}{\pi}) + (\underset{\sim}{\tau}\cdot\underset{\sim}{\pi})\underset{\sim}{p}]\}, \qquad (4)$$

where the nucleon momentum operator $\underset{\sim}{p}$ is $i\underset{\sim}{\nabla}$ operating to the left and $-i\underset{\sim}{\nabla}$ operating to the right. More conveniently, we can write

$$H_{\pi N} = (4\pi)^{\frac{1}{2}} \; f \, \mu^{-1} \, \underset{\sim}{\sigma} \cdot [(1 \mp \lambda \frac{E_\pi}{2m}) \underset{\sim}{\nabla}_\pi \mp \lambda \frac{E_\pi}{m} \underset{\sim}{\nabla}_N] \, (\underset{\sim}{\tau} \cdot \underset{\sim}{\phi}), \tag{5}$$

in which $\underset{\sim}{\nabla}_N$ operates to the right on nucleon coordinates only.

The interaction operator we are dealing with here is equivalent to the one derived in a similar way by Miller [13], except for the fact that he also includes terms of higher order in E_π/m; this may or may not be meaningful. It is not the so-called Galilean invariant operator $\propto \underset{\sim}{v}_\pi - \underset{\sim}{v}_N$, which would correspond to $\lambda = 1$ rather than $\lambda = \frac{1}{2}$ in the equations above, and which intuitively seems more pleasing. To arrive at that form of the non-relativistic πN interaction one has to proceed more carefully and use the entire field equations for the coupled fields of the pion and the nucleon, not just the interaction term by itself, see Cheon [14] and Barnhill [15]. However, it turns out that this only introduces $\lambda = 1$ as a possibility, not as a uniquely determined quantity, the point being that formally any change in λ can be compensated for by terms of higher order in the pion field. The fact therefore remains that the exact form of the pion-nucleon interaction to be used in a non-relativistic calculation, such as that of (p,π) reactions on nuclei, is not uniquely determined. This problem is not one of purely academic interest, since the term $\propto (E_\pi/m)\underset{\sim}{\nabla}_N$ is not always small numerically.

We should like to point out that to apply the operator $H_{\pi N}$ with $\lambda = \frac{1}{2}$ to a problem involving bound nucleons and nucleons distorted by a nuclear potential is equivalent to the following procedure, in which one pretends to know what a relativistic nucleon in a nuclear potential is: α) Expand the nucleon states in plane waves β) Force these plane waves to be relativistic by giving them a lower component related to the upper component in the same way as for free waves γ) Use the relativistic γ_5 pion-nucleon interaction. Such a procedure is to be preferred from a computational point of view for more complicated problems involving both relativistic and non-relativistic particles [16].

As our excuse for having spent so much time talking about the interaction operator $H_{\pi N}$ we mention in conclusion that there has been a tendency not to show enough care in handling the ± signs appearing in eq. (5). For the processes $|i\rangle + \pi \rightleftarrows |f\rangle$, where $|i\rangle$ and $|f\rangle$ are nucleon states, we do not get the required relation $\langle f|H_{\pi N}|i\rangle = \langle i|H_{\pi N}|f\rangle^*$ unless the signs are changed in accordance with the prescription given here; by partial integration we find indeed

$$\langle f|\widetilde{H}_{\pi N}|i\rangle = \int d^3r\psi_f^*(\underset{\sim}{r})[(1-\lambda\frac{E_\pi}{2m})\underset{\sim}{\nabla}_\pi - \lambda\frac{E_\pi}{m}\underset{\sim}{\nabla}_N]\ \psi_i(\underset{\sim}{r})\ e^{i\underset{\sim}{k}_\pi\cdot\underset{\sim}{r}} \tag{6}$$

$$= \langle i|\widetilde{H}_{\pi N}|f\rangle^* = \{\int d^3r\psi_i^*(\underset{\sim}{r})\ [(1+\lambda\frac{E_\pi}{2m})\underset{\sim}{\nabla}_\pi + \lambda\frac{E_\pi}{m}\underset{\sim}{\nabla}_N]\ \psi_f(\underset{\sim}{r})\ e^{-i\underset{\sim}{k}_\pi\cdot\underset{\sim}{r}}\}^*.$$

b) The Plane-Wave Approximation

We take here plane waves for the wave functions of the incident proton and the emitted pion. The part of $\underset{\sim}{\tau}\cdot\underset{\sim}{\phi}$ that corresponds to π^+ production is then $-(2E_\pi)^{-\frac{1}{2}}\ \tau_-\exp(-i\ \underset{\sim}{k}_\pi\cdot\underset{\sim}{r})$, where $\tau_-|p\rangle = 2^{\frac{1}{2}}|n\rangle$, and we have

$$H_{\pi+N}|p,\underset{\sim}{k}_p,\mu_p\rangle = -(4\pi)^{\frac{1}{2}}f\ \mu^{-1}(E_\pi)^{-\frac{1}{2}}|n\rangle\underset{\sim}{\sigma}\cdot[(1+\lambda\frac{E_\pi}{2m})\underset{\sim}{\nabla}_\pi +$$

$$\lambda\frac{E_\pi}{m}\underset{\sim}{\nabla}_N]e^{-i\underset{\sim}{k}_\pi\cdot\underset{\sim}{r}}\ e^{i\underset{\sim}{k}_p\cdot\underset{\sim}{r}}\ \chi_{\frac{1}{2}}^{\mu_p}$$

$$= (4\pi)^{\frac{1}{2}}if\ \mu^{-1}(E_\pi)^{-\frac{1}{2}}|n\rangle\underset{\sim}{\sigma}\cdot\underset{\sim}{Q}\ e^{i\underset{\sim}{q}\cdot\underset{\sim}{r}}\ \chi_{\frac{1}{2}}^{\mu_p}, \tag{7}$$

$$\underset{\sim}{q} = \underset{\sim}{k}_p - \underset{\sim}{k}_\pi,\quad \underset{\sim}{Q} = (1+\lambda\frac{E_\pi}{2m})\ \underset{\sim}{k}_\pi - \lambda\frac{E_\pi}{m}\underset{\sim}{k}_p.$$

The final single-nucleon state is

$$|n,\psi_{NLJ}^{J_z}\rangle = |n\rangle R_{NLJ}(r)\sum_M\sum_{\mu_n}\langle L\tfrac{1}{2}M\mu_n|JJ_z\rangle\ Y_L^M(\underset{\sim}{\hat{r}})\ \chi_{\frac{1}{2}}^{\mu_n},$$

$$\int_0^\infty r^2\ dr\ R_{NLJ}^2(r) = \int d^3r\langle n,\psi_{NLJ}^{J_z}|n,\psi_{NLJ}^{J_z}\rangle = 1, \tag{8}$$

and the cross section (1) becomes

$$\frac{d\sigma}{d\Omega} = [J]\ \pi\ E_p(E_A E_B/s) f^2 (k_\pi/k_p)(Q/\mu)^2 |\phi_{NLJ}(q)|^2, \tag{9}$$

where we use the notation $[a] = 2a + 1$, $[a/b] = (2a + 1)(2b + 1)^{-1}$ etc., and where

$$\phi_{NLJ}(q) = (2/\pi)^{\frac{1}{2}} \int_0^\infty r^2 dr\ R_{NLJ}(r)\ j_L\ (qr) \tag{10}$$

is the single-nucleon wave function in momentum space. As it has been noted by Domingo et al. [4] one should also take recoil effects into consideration when calculating the form factor (10). The correct way to do this is to think of R_{NLJ} as the wave function describing the motion of the captured neutron relative to the core A, and replace $\underset{\sim}{q}$ by the corresponding relative momentum

$$\tilde{\underset{\sim}{q}} = \underset{\sim}{q}_{nA} = m_{nA}(\underset{\sim}{v}_n - \underset{\sim}{v}_A) = m_{nA}[(\underset{\sim}{q}/m_n) + (\underset{\sim}{k}_p/m_A)]$$

$$= \underset{\sim}{k}_p - \frac{m_A}{m_A + m_n}\underset{\sim}{k}_\pi \approx \underset{\sim}{k}_p - \frac{A}{A+1}\underset{\sim}{k}_\pi . \tag{11}$$

As indicated earlier, if the cross section were given with sufficient accuracy by eq. (9) (with q replaced by $\tilde{q}$) it could be used to determine the form factor $\phi_{NLJ}(\tilde{q})$ at large momentum transfer in a very direct way. E.g., it was shown in ref. [4] that at 600 MeV and 0° the cross section differs by about a factor 10^4 if one goes from harmonic-oscillator to Fermi-distribution wave functions.

We also note the appearance of the momentum $\underset{\sim}{Q}$ in the expression for the cross section. The factor Q^2 can in fact have a large influence on the plane-wave result. The point is that while near threshold the pion energy E_π is small compared to the nucleon mass m, the pion momentum k_π is also small compared to the proton momentum k_p. Thus, it can easily happen that Q gets very close to zero in the forward direction, as observed by Eisenberg et al. [17, 18] for the opposite reaction (π^+,p) at 50 MeV when using the Galilean-invariant interaction with $\lambda = 1$. Authors who have considered the plane-wave (p,π^+) reaction directly have preferred to use a plus sign in the last term of $\underset{\sim}{Q}$ and thus do not experience this cancellation effect.

c) Distorted Waves

Calculations on the (p,π^+) reaction in the DWBA approximation have been reported by Eisenberg et al. [18, 19], Keating and Wills [20], Rost and Kunz [21], Höistad et al. [22] and Miller et al. [13, 23]. Instead of using plane-wave pion and proton wave functions one now makes the substitutions

$$e^{i\underset{\sim}{k}_\pi\cdot\underset{\sim}{r}} \Rightarrow \psi_\pi(\underset{\sim}{r}) = \int d^3K\ \phi_\pi(\underset{\sim}{K})\ e^{i\underset{\sim}{K}\cdot\underset{\sim}{r}},$$

$$e^{i\underset{\sim}{k}_p\cdot\underset{\sim}{r}} \Rightarrow \psi_p(\underset{\sim}{r}) = \int d^3P\ \phi_p(\underset{\sim}{P})\ e^{i\underset{\sim}{P}\cdot\underset{\sim}{r}}, \tag{12}$$

where we have disregarded the spin-orbit term in the proton distortion. In eq. (9) the distortions then correspond to the replacement

$$Q^2|\phi_{NLJ}(q)|^2 \Rightarrow F(\theta) = 2(4\pi)^2\sum_\Lambda\sum_\nu |W(1\tfrac{1}{2}LJ:\tfrac{1}{2}\Lambda)|^2|I_{NLJ\Lambda}^{\nu}|^2,$$

$$I_{NLJ\Lambda}^{\nu} = \int d^3K\ d^3P\ \phi_{NLJ}(q')\ \phi_\pi^*(\underset{\sim}{K})\ \phi_\pi(\underset{\sim}{P})\ Q'\ Y_{L1\Lambda}^{\nu *}(\underset{\sim}{q}', \underset{\sim}{Q}'), \tag{13}$$

$$Y_{L1\Lambda}^{\nu}(\underset{\sim}{q}', \underset{\sim}{Q}') = \sum_M\sum_m \langle L1Mm|\Lambda\nu\rangle\ Y_L^M(\underset{\sim}{q}')\ Y_1^m(\underset{\sim}{Q}'),$$

$$\underset{\sim}{q}' = \underset{\sim}{P} - \underset{\sim}{K},\quad \underset{\sim}{Q}' = (1 + \lambda\frac{E_\pi}{2m})\underset{\sim}{K} - \lambda\frac{E_\pi}{m}\underset{\sim}{P},$$

this is reduced to the previous result if $\phi_\pi(\underset{\sim}{K}) = \delta(\underset{\sim}{K} - \underset{\sim}{k}_\pi)$ and $\phi_p(\underset{\sim}{P}) = \delta(\underset{\sim}{P} - \underset{\sim}{k}_p)$.

Even at 185 MeV it is a fairly good approximation to use the eikonal form of the proton wave function,

$$\psi_p(\underset{\sim}{r}) = e^{i\underset{\sim}{k}_p\cdot\underset{\sim}{r}} \exp\{-i \frac{E_p}{kp} \int_{-\infty}^{z} V_p(\underset{\sim}{r}') \, dz'\}, \tag{14}$$

$$\underset{\sim}{r} = \underset{\sim}{b} + z\underset{\sim}{e}_z, \; \underset{\sim}{r}' = \underset{\sim}{b} + z'\underset{\sim}{e}_z, \; \underset{\sim}{b} \perp \underset{\sim}{e}_z = \underset{\sim}{k}_p/k_p,$$

where V_p is the optical potential. At 600 MeV it is also reasonable to treat the pion wave function in the same way, i.e.

$$\psi_\pi(\underset{\sim}{r}) = e^{i\underset{\sim}{k}_\pi\cdot\underset{\sim}{r}} \exp\{i \frac{E_\pi}{k_\pi} \int_{z}^{\infty} V_\pi(\underset{\sim}{r}') \, dz'\}. \tag{15}$$

On the assumption that the amplitude for scattering of a proton or a pion by a nucleon varies slowly with momentum transfer compared to the nuclear form factor one can use Glauber theory to find the optical potentials

$$V_j(r) = -\tfrac{1}{2}Ai \frac{k_j}{E_j} \sigma_j(1 - i\alpha_j) \, \rho(r), \; \int d^3r \, \rho(r) = 1, \tag{16}$$

where $\rho(\underset{\sim}{r})$ is the nuclear density, σ_j is the total cross section for scattering of particle j by a nucleon, and α_j is the ratio between the real and the imaginary part of the forward scattering amplitude. Eisenberg et al. [19] used this approximation at 600 MeV for comparison with the CERN [4-6] data and obtained quite good agreement with the measured 0° cross sections, taking reasonable wave functions for the captured neutron.

At 185 MeV, where one has whole angular distributions to fit, the situation is so far not very clear. The proton distortions can be treated by using well-established optical potentials derived from elastic scattering experiments. It turns out that while the proton distortions give a significant reduction in the cross section as compared with the plane-wave case, and tends to improve the agreement with the experiments, the calculated cross section is not terribly sensitive to minor variations in the optical potential; also, the spin-orbit part of the potential has only a small influence on the (p,π^+) cross section [22]. The sensitivity to the choice of wave function for the captured neutron is smaller than in the case of plane waves. However, enough of it remains that the observed cross section could indeed be used as a good means of determining the high-momentum part of this wave function, provided that one could be sure that the rest of the problem was treated correctly.

The main uncertainty in this respect is connected with the optical potential that one uses to calculate the pion distortions. For 185 MeV protons the pion kinetic energy is only some 30 MeV, and one does not like to resort to the eikonal approximation to obtain the distorted pion wave functions. One uses instead the optical potential $\underline{V}_\pi$ in the Klein-Gordon equation; this can be written as a Schrödinger equation with an optical potential $\underline{V}'_\pi$,

$$(\nabla^2 + k_\pi^2 - 2\mu V'_\pi)\psi_\pi(\underset{\sim}{r}) = 0,$$

$$V'_\pi = (E_\pi/\mu)V_\pi[1-(V_\pi/2E_\pi)]. \tag{17}$$

As for the proton case, it is required that the potential should successfully describe the elastic scattering, but this does not determine the potential uniquely. The first thing to be noted about the application of pion optical potentials to (p,π^+) reactions is that the popular Kisslinger (ξ=0) and Kroll-Kisslinger (ξ=1) potentials

$$2\mu V'_\pi = A\{-b_o\, k_\pi^2\, \rho(r) + b_1 \underset{\sim}{\nabla}\, \rho(r)\, [1+(\xi/3)b_1\rho(r)]^{-1} .\underset{\sim}{\nabla}\} \tag{18}$$

give cross sections that are 1-2 orders of magnitude too large compared with the experimental results [20, 21, 13], as long as the parameters $\underline{b}_j$ are fitted to the elastic-scattering data subject to the condition that their signs should be the same as those arising from free pion-nucleon scattering phase shifts. The local counter part

$$2\mu V'_\pi = A\{(-b_o + b_1)\, k_\pi^2\, \rho(r) + \tfrac{1}{2}b[\nabla^2\, \rho(r)]\} \tag{19}$$

of the Kisslinger potential suffers from the same deficiency [22].

The difficulty with these potentials as applied to the (p,π^+) reaction is that the $\nabla\rho\cdot\nabla$ or $\nabla^2\rho$ term gives too much weight to the high-momentum components of the pion wave function, as displayed more clearly by the undamped $\mathbf{k}\cdot\mathbf{k}'$ or $k_\pi^2 - \frac{1}{2}(\mathbf{k}-\mathbf{k}')^2$ term in the scattering amplitude that is used to derive them. Miller [13] was able to suppress the high-momentum components sufficiently by changing the relative magnitudes of the parameters b_0 and b_1 in the Kisslinger potential and also changing the sign of both, while still fitting the pertinent elastic-scattering data; this procedure seems, perhaps, a bit artificial. More importantly, however, and more pleasing from a theoretical point of view, Miller and Phatak [23] used a separable πN T-matrix with damping factors in the $\mathbf{k}\cdot\mathbf{k}'$ term to generate an optical potential, and were able to obtain good agreement with the $^{12}C(p,\pi^+)^{13}C$ angular distributions at 185 MeV, at the price of a somewhat poorer fit to the $\pi^{-12}C$ elastic scattering data. Furthermore, we note that it is quite possible to fit both the (p,π^+) and at least the small-angle pion elastic-scattering cross section by using a simple square-well potential [22].

In all these distorted-wave calculations the question of choosing the correct interaction operator $H_{\pi N}$ is implicitly there as an open problem. The value of λ is rather unimportant (i.e. the ∇_N part of the interaction does not contribute much) for the case in which the Kisslinger type potentials are used and the cross section comes out much too large anyway. For pion optical potentials that describe the (p,π^+) reaction better, the calculated cross section can change significantly as one goes from the static $\lambda = 0$ interaction through the intermediate $\lambda = \frac{1}{2}$ case to the Galilean-invariant $\lambda = 1$ operator.

In conclusion, provided that the simple one-nucleon mechanism gives the correct way of describing the (p,π^+) reaction, there are still at least three unknown quantities that go into a calculation of the cross section, the pion-nucleon interaction, the pion optical potential, and the bound-state neutron wave function at high momenta. In a single experiment, reliable information about one of these unknowns can only be extracted if one can learn enough about the others from different sources of information.

4. SOME REACTION THEORY

As indicated earlier, it is our belief that pionic nuclear reactions should be treated on the same footing as nuclear reactions involving other hadrons. Not only will this enable one to treat the complexities arising from the nuclear-physics part of the problem in a systematic manner, it will also make it easier to make meaningful comparisons between pionic and other reactions. For example, in the single-nucleon picture the similarity between the $A(p,\pi^+)B$ reaction and the stripping reaction A(d,p)B is apparent, pictorially all one has to do is to replace the proton and the pion in fig. 2 by a deuteron and a proton, respectively. For this reason the (p,π^+) reaction is also known as "pionic stripping", the proton then being considered as a bound state of a neutron and a positive pion; it is then a question how literally one is allowed to take that description. Let us for the moment take it quite literally and look at the more general reaction A(a,b)B. In fig. 3, diagram I with an exchanged neutron corresponds to that in fig. 2. Diagram II in fig. 3 is another possible one-particle-exchange diagram, the exchanged particle is in this case a complex nucleus. This diagram is known to give important contributions in the backward direction for (d,p) reactions, hopefully it may also explain at least part of the backward (p,π^+) cross section. In addition, we can here also obtain a (p,π^-) cross section by a simple mechanism. That can also be obtained by a more exotic version of diagram I, where the exchanged particle is the Δ resonance, Δ^0 for the (p,π^+) case and Δ^{++} for the (p,π^-) one. Another, but probably not very important, one-particle-exchange diagram is shown as diagram III in fig. 3.

In order that we may be able to include even more diagrams in a systematic manner let us expand the A(a,b)B differential CM cross section

$$\frac{d\sigma}{d\Omega} = (2\pi)^{-2} (E_a E_A E_b E_B/s)(k_f/k_i)[J_A J_a]^{-1} \sum_{\mu_a} \sum_{\mu_A} \sum_{\mu_b} \sum_{\mu_B} |M|^2, \quad (20)$$

$$M = \sum_n M_n, \quad \underset{\sim}{k}_i = \underset{\sim}{k}_a = -\underset{\sim}{k}_A, \quad \underset{\sim}{k}_f = \underset{\sim}{k}_b = -\underset{\sim}{k}_B,$$

where $\underline{M_n}$ is the matrix element for diagram $\underline{n}$, in multipoles $\underset{\sim}{j} = \underset{\sim}{J}_B - \underset{\sim}{J}_A$, $\underset{\sim}{s} = \underset{\sim}{J}_b - \underset{\sim}{J}_a$, $\underset{\sim}{\ell} = \underset{\sim}{j} + \underset{\sim}{s}$. We define the multipole amplitudes $\underline{G}$ so that

$$M_n = [J_A J_a]^{\frac{1}{2}} (m_{aA} m_{bB})^{-1} \sum_j \sum_s \sum_\ell i^\ell [sj]^{\frac{1}{2}} \sum_{\mu_j} \sum_{\mu_s} \sum_{\mu_\ell} \quad (21)$$

$$\times < J_A j \mu_A \mu_j | J_B \mu_B > < J_a s \mu_a \mu_s | J_b \mu_b > < s j \mu_s \mu_j | \ell \mu_\ell > G_{njs\ell}^{\mu_\ell},$$

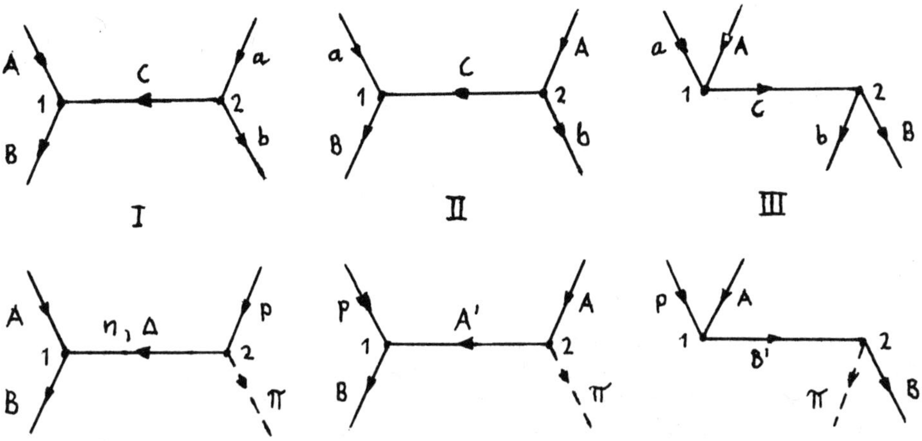

Fig. 3. Pole diagrams.

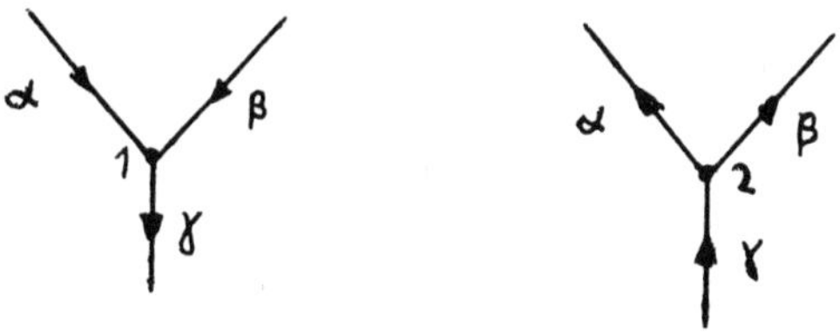

Fig. 4. 3-prong vertices.

where $\underline{m}_{aA}$ and $\underline{m}_{bB}$ are reduced masses. The differential cross section (for unpolarized particles) is then

$$\frac{d\sigma}{d\Omega} = (2\pi)^{-2}(E_{aA}/m_{aA})(E_{bB}/m_{bB})(m_{aA}m_{bB})^{-1}(k_f/k_i)[J_BJ_b]\Big(\sum_n S_n + \sum\sum_{nn'>n} S_{nn'}\Big),$$

$$S_n = \sum_j \sum_s \sum_\ell \sum_{\mu_\ell} |G_{njs\ell}^{\mu_\ell}|^2, \quad S_{nn'} = 2\sum_j \sum_s \sum_\ell \sum_{\mu_\ell} \mathrm{Re}\Big\{G_{njs\ell}^{\mu_\ell *} G_{n'js\ell}^{\mu_\ell}\Big\}. \tag{22}$$

In the usual manner the amplitude for a given diagram is obtained by associating an amplitude with each vertex and a propagator with each exchanged particle. We shall in the present exposition do this on the assumption that all particles are non-relativistic, and what we are doing in the following then largely amounts to a re-formulation and an application of the methods of Shapiro et al., see e.g. ref. [24]. One uses here plane waves for all unbound particles, and distortions are introduced as higher-order diagrams.

For the (2 in + 1 out) and (1 in + 2 out) vertices of fig. 4 we write the amplitudes as

$$M_1 = -2\pi\, m_{\alpha\beta}^{-\frac{1}{2}}\, (2\eta_{\alpha\beta\gamma})^{\frac{1}{2}} \sum_L \sum_M i^L F_L^{\mu_\alpha \mu_\beta \mu_\gamma M*}(\zeta_{\alpha\beta})\, Y_L^{M*}(\hat{\underset{\sim}{\zeta}}_{\alpha\beta}),$$

$$M_2 = -2\pi\, m_{\alpha\beta}^{-\frac{1}{2}}\, (2\eta_{\alpha\beta\gamma})^{\frac{1}{2}} \sum_L \sum_M i^{-L} F_L^{\mu_\alpha \mu_\beta \mu_\gamma M}(\zeta_{\alpha\beta})\, Y_L^{M}(\hat{\underset{\sim}{\xi}}_{\beta\alpha}), \tag{23}$$

$$\eta^2_{\alpha\beta\gamma} = 2(m_\alpha + m_\beta - m_\gamma)/m_{\alpha\beta},\ \underset{\sim}{\zeta}_{\alpha\beta} = \underset{\sim}{q}_{\alpha\beta}/m_{\alpha\beta},\ \underset{\sim}{q}_{\alpha\beta} = m_{\alpha\beta}[(\underset{\sim}{k}_\alpha/m_\alpha) - (\underset{\sim}{k}_\beta/m_\beta)].$$

Depending on the coupling scheme we write

$$F_L^{\mu_\alpha \mu_\beta \mu_\gamma M}(\zeta_{\alpha\beta}) = \sum_J \sum_{\mu_J} < J_\alpha L \mu_\alpha M | J \mu_J > < J_\beta J \mu_\beta \mu_J | J_\gamma \mu_\gamma > f_{LJ} \mathcal{F}_{LJ}(\zeta_{\alpha\beta})$$

$$= \sum_{J'} \sum_{\mu_J'} < J_\alpha J_\beta \mu_\alpha \mu_\beta | J' \mu_J' > < J' L \mu' M | J_\gamma \mu_\gamma > f_{LJ'} \mathcal{F}_{LJ'}(\zeta_{\alpha\beta}), \tag{24}$$

where we have the relationship

$$f_{LJ} \mathcal{F}_{LJ}(\zeta_{\alpha\beta}) = \sum_{J'} (-1)^{2J+J_\beta-J'-J_\alpha} [JJ']^{\frac{1}{2}} W(J_\alpha J_\beta L J_\gamma : J'J) f_{LJ'} \mathcal{F}_{LJ'}(\zeta_{\alpha\beta}). \tag{25}$$

The quantum number $\underline{L}$ describes the relative motion of the particles α and β, and it is implied that parity is conserved in the vertex, i.e. $\underline{P_\alpha P_\beta P_\gamma} = (-1)^{\underline{L}}$. The first eq. (25) corresponds, for instance, to the motion of a nucleon α with quantum numbers $(\underline{J_\alpha L}\ \underline{J}) = (\frac{1}{2}\ \underline{L}\ \underline{J})$ relative to a core β, the last eq. (25) is the so-called channel spin representation. We use here the former of these expressions. The amplitudes $\underline{f_{LJ}}$ are defined so that $|\underline{f_{LJ}}|^2 = \underline{S_{LJ}}$, where $\underline{S_{LJ}}$ is the spectroscopic coefficient, normalized in accordance with ref. [25], and the form factors are

$$\mathcal{F}_{LJ}(\zeta_{\alpha\beta}) = (2\eta_{\alpha\beta\gamma})^{-\frac{1}{2}} (\eta^2_{\alpha\beta\gamma} + \zeta^2_{\alpha\beta}) \int_0^\infty y^2_{\alpha\beta}\, dy_{\alpha\beta}\, R_{LJ}(y_{\alpha\beta})\, j_L(\zeta_{\alpha\beta} y_{\alpha\beta}),$$

$$\int_0^\infty y^2_{\alpha\beta}\, dy_{\alpha\beta}\, R^2_{LJ}(y_{\alpha\beta}) = 1, \qquad y_{\alpha\beta} = m_{\alpha\beta}\, r_{\alpha\beta}, \tag{26}$$

where R_{LJ} is the radial wave function for the relative motion of α and β, chosen to be positive and real at large distances. The normalization is such that $F_{oJ_\alpha} \equiv 1$ for an s-state interaction with zero range.

If we now look at diagram I in fig. 3, the matrix element

$$M_I = -(2i/m_{CA})(\eta_1^2 + \zeta_1^2)^{-1} \sum_{\mu_C} M_1 M_2 \tag{27}$$

can be written in a re-coupled form compatible with eq. (22), such that the multipole amplitude becomes

$$G_{Ijs\ell}^{\mu_\ell} = (4\pi)^2 \frac{m_{aA} m_{bB}}{m_{CA}^{3/2} m_{bC}^{1/2}} (\eta_1\eta_2)^{\frac{1}{2}} [J_A J_b]^{-\frac{1}{2}} (-1)^{J_a - J_b + J_C + j - s} (\eta_1^2 + \zeta_1^2)^{-1}$$

$$\times \sum_{L_1} \sum_{L_2} i^{L_2 - L_1 - \ell} W(L_1 j L_2 s : J_C \ell) f_{L_1 j}^{*} f_{L_2 s} F_{L_1 j}(\zeta_1) F_{L_2 s}(\zeta_2) Y_{L_1 L_2 \ell}^{\mu_\ell *}(\hat{\zeta}_1, \hat{\zeta}_2),$$

$$\eta_1 = \eta_{CAB}, \quad \eta_2 = \eta_{Cba}, \quad \zeta_1 = \zeta_{CA}, \quad \zeta_2 = \zeta_{bC}, \tag{28}$$

$$Y_{L_1 L_2 \ell}^{\mu_\ell}(\hat{\zeta}_1, \hat{\zeta}_2) = \sum_{M_1} \sum_{M_2} \langle L_1 L_2 M_1 M_2 | \ell \mu_\ell \rangle \, Y_{L_1}^{M_1}(\hat{\zeta}_1) \, Y_{L_2}^{M_2}(\hat{\zeta}_2).$$

It is implied that to get the total transition amplitude from all diagrams of type I (different particles C, different states of the same particle) one should sum coherently over the contributions from the various diagrams. For the (p,π) case ($J_a = 1/2$, $J_b = 0$) the multipoles are $\underset{\sim}{j} = \underset{\sim}{J}_B - \underset{\sim}{J}_A$, $s = 1/2$, $\ell = j - 1/2$, and for the (p,π^+) reaction with neutron exchange we have $J_C = J_2 = 1/2$, $L_2 = 1$. By considering diagram I alone and adjusting the neutron wave function by a Butler [26] cut-off procedure it is possible to fit the Uppsala angular distributions in the region where the cross section is falling, using a constant pion form factor and not too unreasonable spectroscopic coefficients in vertex 1 [27]. In a more honest approach one would use pion form factors as given e.g. by Afnan and Thomas [28], and spectroscopic coefficients and neutron form factors determined as accurately as possible from other sources of information, trying to obtain agreement with experiments by including other diagrams. In this way one would then hope to be able to say something definite about the reaction mechanism for (p,π) as well as for e.g. (d,p) reactions at various energies.

As far as diagrams II and III in fig. 3 are concerned the amplitudes are

$$G^{\mu_\ell}_{IIjs\ell} = (4\pi)^2 \frac{m_{aA} m_{bB}}{m_{Ca}^{3/2} m_{bC}^{1/2}} (\eta_1\eta_2)^{\frac{1}{2}} [sj]^{\frac{1}{2}} [J_a J_b]^{-\frac{1}{2}} (\eta_1^2+\zeta_1^2)^{-1} \sum_{L_1}\sum_{J_1}\sum_{L_2}\sum_{J_2} i^{L_2-L_1-\ell}$$

$$\times \; [J_1 J_2]^{\frac{1}{2}} (-1)^{J_B+J_C-J_A+\ell+s-J_1} \, W(L_1J_1L_2J_2:J_C\ell) \begin{Bmatrix} s & j & \ell \\ J_a & J_B & J_1 \\ J_b & J_A & J_2 \end{Bmatrix}$$

$$\times \; f^*_{L_1J_1} f_{L_2J_2} \mathcal{F}_{L_1J_1}(\zeta_1) \mathcal{F}_{L_2J_2}(\zeta_2) Y^{\mu_\ell *}_{L_1L_2\ell}(\hat{\underset{\sim}{\zeta}}_1, \hat{\underset{\sim}{\zeta}}_2), \qquad (29)$$

$\eta_1 = \eta_{CaB}$, $\eta_2 = \eta_{CbA}$, $\underset{\sim}{\zeta}_1 = \underset{\sim}{\zeta}_{Ca}$, $\underset{\sim}{\zeta}_2 = \underset{\sim}{\zeta}_{bC}$, and

$$G_{IIIjs\ell}^{\mu_\ell} = (4\pi)^2 (m_{bB}/m_{aA})^{\frac{1}{2}} (\eta_1\eta_2)^{\frac{1}{2}} [J_C][sj]^{\frac{1}{2}} [J_A J_B J_a J_b]^{-\frac{1}{2}} (\eta_1^2+\zeta_1^2)^{-1}$$

$$\times \sum_{L_1}\sum_{J_1}\sum_{L_2}\sum_{J_2} i^{L_2-L_1-\ell} [J_1 J_2]^{\frac{1}{2}} (-1)^{J_B+J_C+J_1+L_2+s}$$

$$\times W(J_A J_B J_1 J_2 : jJ_C) \begin{Bmatrix} s & j & \ell \\ J_a & J_1 & L_1 \\ J_b & J_2 & L_2 \end{Bmatrix} \tag{30}$$

$$\times f^*_{L_1J_1} \; f_{L_2J_2} \; \mathcal{F}_{L_1J_1}(\zeta_1) \; \mathcal{F}_{L_2J_2}(\zeta_2) \; Y_{L_1L_2\ell}^{\mu_\ell *}(\hat{\zeta}_1, \hat{\zeta}_2) \; ,$$

$$\eta_1 = \eta_{aAC}, \quad \eta_2 = \eta_{BbC}, \quad \underset{\sim}{\zeta}_1 = \underset{\sim}{\zeta}_{aA} = \underset{\sim}{k}_i/m_{aA}, \quad \underset{\sim}{\zeta}_2 = \underset{\sim}{\zeta}_{bB} = \underset{\sim}{k}_f/m_{bB} .$$

We note that a diagram of type II gives an angular distribution which is roughly

$$\frac{d\sigma}{d\Omega} \propto [1 + (a/2)\,(3\cos^2\theta - 1)]\,(1 + b\cos\theta)^{-2},$$

$$-1 < a < 1 \;, \quad b > 0 \;. \tag{31}$$

This is already not too unlike the observed (p,π^-) cross sections.

To get a more complete picture of the (p,π) reaction mechanism one will now have to include more complicated diagrams. As far as triangle diagrams are concerned we can at least think of those displayed in fig. 5, where we have refrained from including diagrams involving Δ's or other resonances. The diagrams that are included in the distorted-wave one-nucleon model are those denoted by IV and VI, whereas diagrams V and VII are pre- and rescattering versions of diagram II. In diagram VIII the particle D is a deuteron or a singlet deuteron, and IX and X are pion-exchange diagrams.

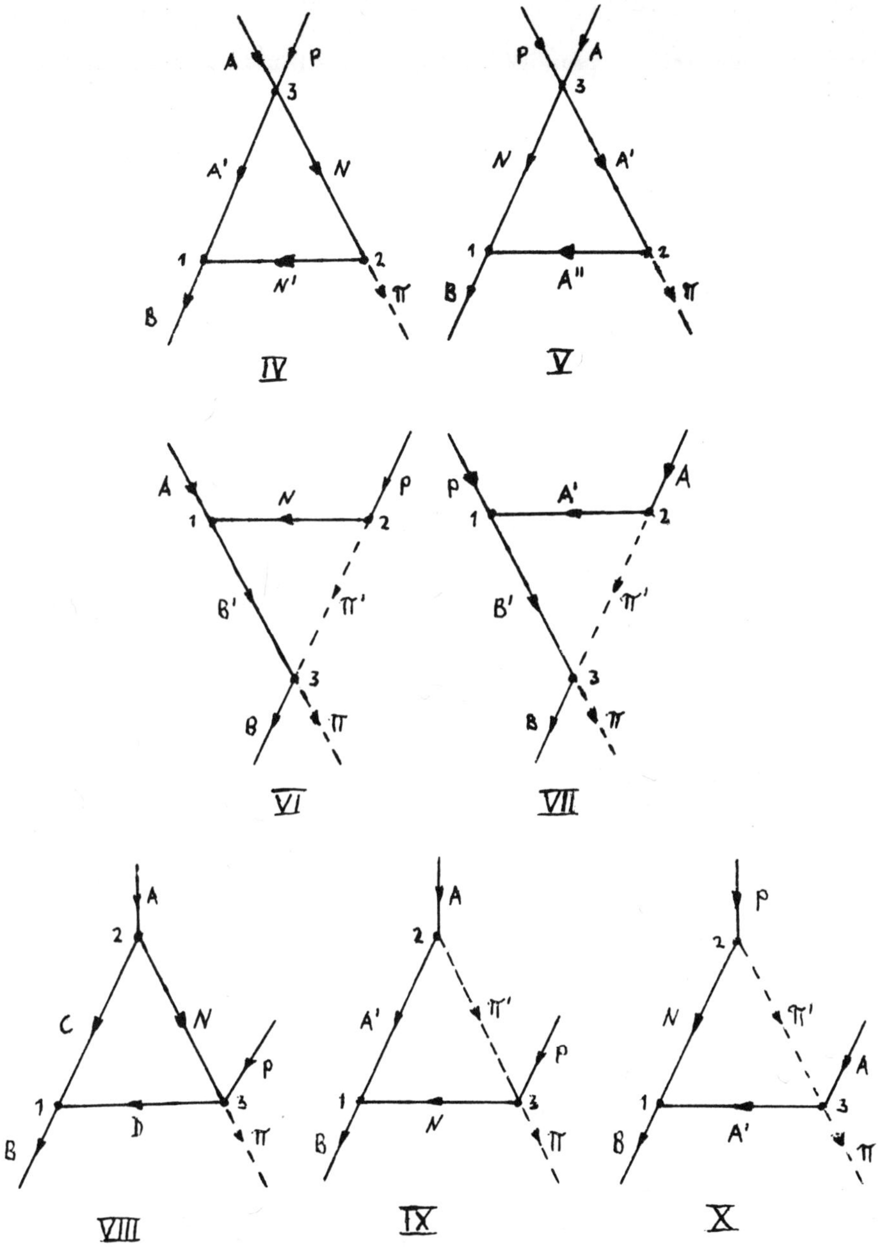

Fig. 5. Triangle diagrams for the reaction A(p,π)B.

The matrix elements for these diagrams can be obtained from those of the more general ones in fig. 6. These have been numbered in accordance with fig. 5, in that diagram IV and diagram V of fig. 5 both are of type IV as far as fig. 6 is concerned; similarly for diagrams VIII and IX of fig. 5 and diagram VIII of fig. 6.

For the 4-particle vertices $\alpha\beta \to \gamma\delta$ we use another multipole expansion of the amplitude,

$$M_3 = [J_\alpha J_\beta]^{\frac{1}{2}} (m_{\alpha\beta} m_{\gamma\delta})^{-1} \sum_J \sum_S \sum_L i^L [SJ]^{\frac{1}{2}} \sum_{\mu_J} \sum_{\mu_S} \sum_{\mu_L} \tag{32}$$

$$\times < J_\beta J \mu_\beta \mu_J | J_\delta \mu_\delta > < J_\alpha S \mu_\alpha \mu_S | J_\gamma \mu_\gamma > < SJ\mu_S \mu_J | L\mu_L > G_{JSL}^{\mu_L}(\underset{\sim}{\zeta}_{\alpha\beta}, \underset{\sim}{\zeta}_{\gamma\delta}),$$

$$G_{JSL}^{\mu_L}(\underset{\sim}{\zeta}_{\alpha\beta}, \underset{\sim}{\zeta}_{\gamma\delta}) = \sum_{L_3} \sum_{L_4} i^{L_3 - L_4 - L} g_{JSLL_3L_4}(\zeta_{\alpha\beta}, \zeta_{\gamma\delta}) \, Y_{L_3L_4L}^{\mu_L *}(\hat{\underset{\sim}{\zeta}}_{\alpha\beta}, \hat{\underset{\sim}{\zeta}}_{\gamma\delta}).$$

We introduce the notation

$$F_n \equiv F_{nL_1J_1L_2J_2JSLL_3L_4L'\ell}^{\mu_\ell}(\underset{\sim}{\zeta}_1, \underset{\sim}{\zeta}_2, \underset{\sim}{\zeta}_3, \underset{\sim}{\zeta}_4)$$

$$= f_{L_1J_1}^{*} f_{L_2J_2} \mathcal{F}_{L_1J_1}(\zeta_1) \mathcal{F}_{L_2J_2}(\zeta_2)$$

$$\times g_{JSLL_3L_4}(\zeta_3, \zeta_4) Y_{L_3L_4L, L_1L_2L', \ell}^{\mu_\ell *}(\hat{\underset{\sim}{\zeta}}_3, \hat{\underset{\sim}{\zeta}}_4; \hat{\underset{\sim}{\zeta}}_1, \hat{\underset{\sim}{\zeta}}_2),$$

$$Y_{L_3L_4L, L_1L_2L', \ell}^{\mu_\ell}(\hat{\underset{\sim}{\zeta}}_3, \hat{\underset{\sim}{\zeta}}_4; \hat{\underset{\sim}{\zeta}}_1, \hat{\underset{\sim}{\zeta}}_2) = \sum_M \sum_{M'} < LL'MM' | \ell\mu_\ell > \times$$

$$Y_{L_3L_4L}^{M}(\hat{\underset{\sim}{\zeta}}_3, \hat{\underset{\sim}{\zeta}}_4) Y_{L_1L_2L'}^{M'}(\hat{\underset{\sim}{\zeta}}_1, \hat{\underset{\sim}{\zeta}}_2),$$

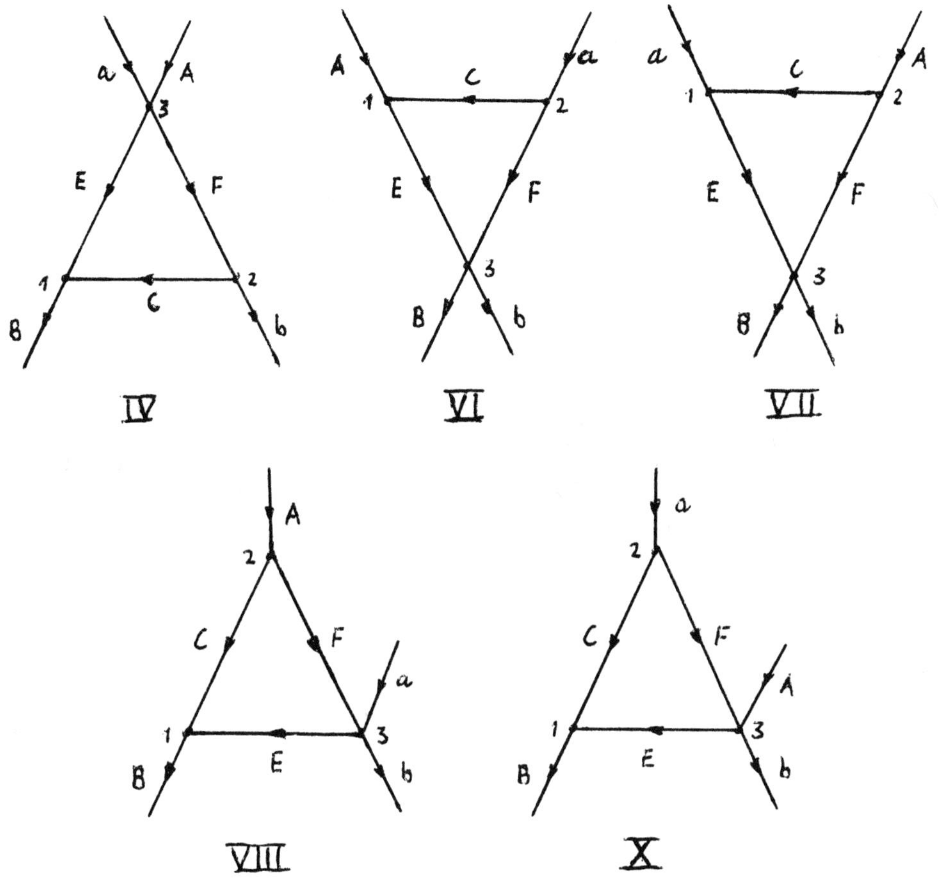

Fig. 6. Triangle diagrams for the reaction A(a,b)B.

$$I_n \equiv I^{\mu_\ell}_{nL_1J_1L_2J_2JSLL_3L_4L'\ell} \quad ,$$

$$A_n \equiv A_{nJ_aJ_bJ_AJ_BJ_CJ_EJ_FL_1J_1L_2J_2JSLL'js\ell} \quad ,$$

$$\eta^2_{\alpha\beta,\gamma\delta} = 2(m_\alpha + m_\beta - m_\gamma - m_\delta)/(m_{\alpha\beta}m_{\gamma\delta})^{\frac{1}{2}} \quad , \tag{33}$$

$$\begin{Bmatrix} J_1 & J_2 & J_3 & J_4 & J_5 \\ J_6 & J_7 & j_1 & j_2 & j_3 \\ j_4 & j_5 & j_6 & j_7 & j_8 \end{Bmatrix} = \sum_{ab}\sum [ab](-1)^{b+j_7+j_8} W(j_1j_2j_3j_4:aJ_6)W(j_5j_6j_7j_8:bJ_7)$$

$$\times \begin{Bmatrix} J_1 & a & b \\ J_2 & j_1 & j_8 \\ J_3 & j_2 & j_7 \end{Bmatrix} \begin{Bmatrix} J_1 & a & b \\ J_4 & j_4 & j_6 \\ J_5 & j_3 & j_5 \end{Bmatrix} \quad ,$$

$$\begin{bmatrix} J_1 & J_2 & J_3 & J_4 & J_5 \\ J_6 & J_7 & j_1 & j_2 & j_3 \\ j_4 & j_5 & j_6 & j_7 & j_8 \end{bmatrix} = \sum_a \sum_b [ab] \; W(j_1j_2j_3j_4:aJ_6) \; W(j_5j_6j_7j_8:bJ_7)$$

$$\times \begin{Bmatrix} J_1 & a & b \\ J_2 & j_1 & j_7 \\ J_3 & j_2 & j_8 \end{Bmatrix} \begin{Bmatrix} J_1 & a & b \\ J_4 & j_4 & j_6 \\ J_5 & j_3 & j_5 \end{Bmatrix} \quad ,$$

and we can write the multipole for the diagrams of fig. 6 as

$$G_{njs\ell}^{\mu\ell} = \frac{4}{\pi} \nu_n (\eta_1\eta_2)^{\frac{1}{2}} \sum_{L_1} \sum_{J_1} \sum_{L_2} \sum_{J_2} \sum_{JSLL_3} \sum_{L_4} \sum_{L'} i^{L_1-L_2+L_3-L_4-\ell} A_n I_n \ . \tag{34}$$

The quantities ν_n depend on the masses of the particles, in that

$$\begin{aligned}
\nu_{IV} &= (m_{EF}/m_{CE})^{3/2}\, m_{bB}(m_{aA}m_{bC})^{-\frac{1}{2}}, \\
\nu_{VI} &= (m_{EF}/m_{FC})^{3/2}\, m_{aA}\,(m_{CA}m_{bB})^{-\frac{1}{2}}, \\
\nu_{VII} &= (m_{EF}/m_{FC})^{3/2}\, m_{aA}(m_{Ca}m_{bB})^{-\frac{1}{2}}, \\
\nu_{VIII} &= (m_{CE}/m_{FC})^{3/2}\, m_{aA}\, m_{bB}(m_{aF}m_{bE})^{-1}, \\
\nu_{X} &= (m_{CE}/m_{FC})^{3/2}\, m_{aA}\, m_{bB}\,(m_{AF}m_{bE})^{-1}.
\end{aligned} \tag{35}$$

The geometrical factors are

$$A_{IV} = [J_F]\,[J_E J S L L' J_1 J_2 j s]^{\frac{1}{2}}\,[J_b]^{-\frac{1}{2}}$$

$$\times\,(-1)^{2J_E+(J_C+J_2)+(J_b-J_a-s)+(J_A+J_F-J)+(j+s-\ell)+(L_1+L_2+L')}$$

$$\times\,W(L_1 J_1 L_2 J_2 : J_C L')\begin{Bmatrix} J_b & J_F & J_2 & J_a & s \\ J_B & L & J_A & J_1 & j \\ J_E & \ell & S & L' & J \end{Bmatrix},$$

$$A_{VI} = [J_E]\,[J_F J S L L' J_1 J_2 j s]^{\frac{1}{2}}\,[J_A]^{-\frac{1}{2}}$$

$$\times\,(-1)^{(J_C+J_2)+(J_a-J_F-J_2)+(J-S)+(j-s)+(L_1+L_2)}$$

$$\times\,W(L_1 J_1 L_2 J_2 : J_C L')\,W(J_A J_B J_1 J : j J_E)\,W(J_a J_b J_2 S : s J_F)\begin{Bmatrix} L & \ell & L' \\ J & j & J_1 \\ S & s & J_2 \end{Bmatrix},$$

$$A_{VII} = [J_E]\,[J_F J S L L' J_1 J_2 j s]^{\frac{1}{2}}\,[J_a]^{-\frac{1}{2}}$$

$$\times\,(-1)^{2J_A+2J_1+(J_C+J_2)+(J_F-S+J_b)+(J_a+s-J_b)+(S-j)+(j-s)+(L_1+L_2)}$$

$$\times\,W(L_1 J_1 L_2 J_2 : J_C L')\begin{Bmatrix} J_b & J_F & S & J_a & s \\ J_B & L' & J_A & J & j \\ J_E & \ell & J_1 & L & J \end{Bmatrix},$$

(36)

$$A_{VIII} = \delta_{sS}\ [J_E J_F J S L L' J_1 J_2 j]^{\frac{1}{2}}\ [s]^{-\frac{1}{2}}$$

$$\times\ (-1)^{2J_A + 2J_F + (J_C + J_2) + (L+J-S) + (\ell+j-s) + (L_1+L_2+L')}$$

$$\times\ W(L_1 J_1 L_2 J_2 : J_C L')\ W(LL'sj:\ell J) \begin{Bmatrix} L' & j & J \\ J_1 & J_B & J_E \\ J_2 & J_A & J_F \end{Bmatrix} ,$$

$$A_X = [J_E J_F J S L L' J_1 J_2 j s]^{\frac{1}{2}}$$

$$\times\ (-1)^{2J_a + (J_C - J_1) + (J_E - J_F - J) + (J_A + J_B - j) + (L_1 + L_2)}$$

$$\times\ W(L_1 J_1 L_2 J_2 : J_C L') \begin{bmatrix} L & S & J & L' & \ell \\ J_a & J_B & J_b & J_F & s \\ J_2 & j & J_1 & J_A & J_E \end{bmatrix} ,$$

and the dynamics is contained in the integrals $\underline{I_n}$, where

$$I_{IV} = \int d^3\zeta_4\ (\zeta_1^2 + \eta_1^2 - i\delta)^{-1}\ (\zeta_4^2 - \zeta_3^2 + \eta_3^2 - i\delta)^{-1}\ F_{IV} ,$$

$$\eta_1 = \eta_{CEB} ,\quad \eta_2 = \eta_{CbF} ,\quad \eta_3 = \eta_{aA,EF} ,$$

$$\underset{\sim}{\zeta}_1 = \underset{\sim}{\zeta}_{CE} ,\quad \underset{\sim}{\zeta}_2 = \underset{\sim}{\zeta}_{bC} ,\quad \underset{\sim}{\zeta}_3 = \underset{\sim}{\zeta}_{aA} ,\quad \underset{\sim}{\zeta}_4 = \underset{\sim}{\zeta}_{EF} ,$$

$$I_{VI} = \int d^3\zeta_3\ (\zeta_2^2 + \eta_2^2 - i\delta)^{-1}\ (\zeta_3^2 - \zeta_4^2 - \eta_3^2 - i\delta)^{-1}\ F_{VI}\ ,$$

$$\eta_1 = \eta_{CAE}\ , \quad \eta_2 = \eta_{CFa}\ , \quad \eta_3 = \eta_{FE,bB}\ ,$$

$$\underset{\sim}{\zeta}_1 = \underset{\sim}{\zeta}_{CA}\ , \quad \underset{\sim}{\zeta}_2 = \underset{\sim}{\zeta}_{FC} \quad \underset{\sim}{\zeta}_3 = \underset{\sim}{\zeta}_{FE}\ , \quad \underset{\sim}{\zeta}_4 = \zeta_{bB}\ ,$$

$$I_{VII} = \int d^3\zeta_3\ (\zeta_2^2 + \eta_2^2 - i\delta)^{-1}\ (\zeta_3^2 - \zeta_4^2 - \eta_3^2 - i\delta)^{-1}\ F_{VII}\ ,$$

$$\eta_1 = \eta_{CaE}\ , \quad \eta_2 = \eta_{CFA}\ , \quad \eta_3 = \eta_{FE,bB}\ ,$$

$$\underset{\sim}{\zeta}_1 = \underset{\sim}{\zeta}_{Ca}\ , \quad \underset{\sim}{\zeta}_2 = \underset{\sim}{\zeta}_{FC}\ , \quad \underset{\sim}{\zeta}_3 = \zeta_{FE}\ , \quad \underset{\sim}{\zeta}_4 = \zeta_{bB}\ , \qquad (37)$$

$$I_{VIII} = \int d^3\zeta_1\ (\zeta_1^2 + \eta_1^2 - i\delta)^{-1}\ (\zeta_2^2 + \eta_2^2 - i\delta)^{-1}\ F_{VIII}\ ,$$

$$\eta_1 = \eta_{CEB}\ , \quad \eta_2 = \eta_{CFA}\ , \quad \eta_3 = \eta_{aF,bE}\ ,$$

$$\underset{\sim}{\zeta}_1 = \underset{\sim}{\zeta}_{CE}\ , \quad \underset{\sim}{\zeta}_2 = \underset{\sim}{\zeta}_{FC}\ , \quad \underset{\sim}{\zeta}_3 = \underset{\sim}{\zeta}_{aF}\ , \quad \underset{\sim}{\zeta}_4 = \underset{\sim}{\zeta}_{bE}\ ,$$

$$I_X = \int d^3\zeta_1\ (\zeta_1^2 + \eta_1^2 - i\delta)^{-1}\ (\zeta_2^2 + \eta_2^2 - i\delta)^{-1}\ F_X\ ,$$

$$\eta_1 = \eta_{CEB}\ , \quad \eta_2 = \eta_{CFa}\ , \quad \eta_3 = \eta_{AF,bE}\ ,$$

$$\underset{\sim}{\zeta}_1 = \underset{\sim}{\zeta}_{CE}\ , \quad \underset{\sim}{\zeta}_2 = \underset{\sim}{\zeta}_{FC}\ , \quad \underset{\sim}{\zeta}_3 = \underset{\sim}{\zeta}_{AF}\ , \quad \underset{\sim}{\zeta}_4 = \underset{\sim}{\zeta}_{bE}\ .$$

It is to be understood that the relative momenta that are not integrated over, should be expressed in terms of the external momenta and the integration variable.

It is our hope that some eager person will carry through a detailed calculation in this spirit e.g. for the reactions $d(p,\pi^+)t$ and $d(d,p)t$. Subject to the availability of more data than what exists at the present time, it should then be possible to make rather definite statements about the respective reaction mechanisms at different energies.

5. THE TWO-NUCLEON MODEL

By the two-nucleon model for (p,π) reactions we mean a mechanism where the large momentum transfer $\underset{\sim}{q} = \underset{\sim}{k}_p - \underset{\sim}{k}_\pi$ is shared among two (or more) nucleons in the target, rather than being adsorbed by a single nucleon. Within this loosely defined framework there is room for a great variety of approaches when it comes to making detailed calculations, as demonstrated in a number of papers in the last few years. The difference between the various treatments is first of all due to the fact that the authors have included different sub-processes (diagrams) in their models; secondly, even for a given basic picture of the reaction mechanism there is a great variety of methods and approximations available when doing the actual calculations.

Let us, first of all, again emphasize that when one does a distorted-wave calculation in the framework of the one-nucleon model one has already included some contributions from the two-nucleon model (diagrams IV and VI of fig. 5) in an average way, by using optical potentials. Thus, the importance of distortions in itself proves the importance of the two-nucleon mechanism.

Among the calculations done directly in the two-nucleon model we mention first of all those where one uses information on the reaction $pp \to d\pi^+$ as one's basic input. This amounts to taking the point of view that diagram VIII of fig. 5 dominates the process and limiting the exchanged particle D to be the physical deuteron. Such a calculation was done by Ingram et al. [29], and later Fearing [30] has improved this model; the latter author also includes distortive corrections to this diagram and has in particular been interested in the processes $d(p,\pi)\tau$, with $\tau = {}^3H$ or ${}^3He$. More generally, one can use partial-wave amplitudes for the process $NN \to NN\pi$ as input for (p,π) computations on complex nuclei [31], the calculations then become very complicated and suffer from a rather disturbing lack of transparency. It is often said that this type of model also includes the one-nucleon mechanism; this is true in the sense that the basic processes

$pp \rightarrow d\pi^+$ or $NN \rightarrow NN\pi$ in their turn can be described in terms of diagrams one of which corresponds to the one-nucleon case. However, we do not believe that there is any double counting involved in the sense that diagram VIII also includes diagram I, as long as the proper spectroscopic coefficients are used in each vertex.

As far as the reaction $d(p,\pi)\tau$ is concerned, Locher and Weber [32] have looked at the contributions to the cross section from processes corresponding to diagrams VIII and X in fig. 5. It was found that the backward peak in the $d(p,\pi^+)t$ cross section at 470 MeV [9] can largely be explained as an effect of the interference between the amplitudes for these two diagrams. The importance of diagram X has previously been stressed by other authors [33, 34].

There are also several other calculations that include or are solely concerned with the two-nucleon mechanism for (p,π) reactions. Calculations where pion-exchange contributions are treated explicitly are reported in ref. [16, 35]; in this category falls also the impressive paper by Locher et al. [36]. The basic reaction $NN \rightarrow NN\pi$ is here assumed always to involve the Δ as an intermediate state, and the interaction is then described by an equivalent potential. The approximations that are made, however, limit the applicability of the method to the region near threshold.

Finally, we mention that Huber et al. [37, 38] have studied the (p,π) reaction by a method which involves the plane-wave one-nucleon mechanism, but with short-range nucleon-nucleon correlations included. It is our feeling that what these authors call short-range correlations, is perhaps more properly described as a manifestation of the importance of distortions and other two-nucleon effects, the correlation function then parametrizes these effects in a simple manner.

The overall picture which emerges from the various calculations based on the two-nucleon model, is that it is quite possible to explain the existing data within that model, and without including the pre- and rescattering diagrams that appear as part of the distortions in calculations within the one-nucleon model. Since the latter method by itself also seems to represent a quite satisfactory approach, there are clearly some problems that have to be solved before we can reach a satisfactory understanding of these reactions.

NOTE ADDED IN PROOF

At the Santa Fe conference (June, 1975) we learned that

a) Le Bornec et al.* have measured (p,π^+) angular distributions at 154 MeV on ^{13}C, ^{25}Mg and ^{28}Si, in addition to the targets mentioned in section 1.

b) Bauer et al.† have measured (p,π^+) angular distributions at 600 MeV on CD_2, ^6Li and ^7Li. The differential cross section drops by a factor ~30 in the region from 0° to 40°, at 185 MeV the corresponding ratio is typically ~2-3.

---

*Y. Le Bornec, B. Tatischeff, L. Bimbot, I. Brissaud, H.D. Holmgren, F. Reide and N. Willis, University of Maryland report IPNO/Ph N/ 75-15.

†T. Bauer, R. Beurtey, A. Boudard, G. Bruge, A. Chaumeaux, P. Couvert, H.H. Duhm, D. Garreta, M. Matoba, Y. Terrien, L. Bimbot, Y. Le Bornec, B. Tatischeff, E. Aslanides, R. Bertini, F. Brochard, P. Gorodetzky and F. Hibou, Sixth international conference on high energy physics and nuclear structure, Santa Fe, 1975, Abstract IV.C.24.

REFERENCES

[1] S. Dahlgren, B. Höistad and P. Grafström, Phys. Lett. 35B (1971) 219.

[2] S. Dahlgren, P. Grafström, B. Höistad and A. Åsberg, Nucl. Phys. A204 (1973) 53; Phys. Lett. 47B (1973) 439; Nucl. Phys. A211 (1973) 243; Nucl. Phys. A227 (1974) 245.

[3] S. Dahlgren and P. Grafström, Physica Scripta 10 (1974) 104.

[4] J.J. Domingo, B.W. Allardyce, C.H.Q. Ingram, S. Rohlin, N.W. Tanner, J. Rohlin, E.M. Rimmer, G. Jones and J.P. Girardeau-Montaut, Phys. Lett. 32B (1970) 309.

[5] K. Gabathuler, J. Rohlin, J.J. Domingo, C.H.Q. Ingram, S. Rohlin and N.W. Tanner, Nucl. Phys. B40 (1972) 32.

[6] J. Rohlin, K. Gabathuler, N.W. Tanner, C.R. Cox and J.J. Domingo, Phys. Lett. 40B (1972) 539.

[7] Y. Le Bornec, B. Tatischeff, L. Bimbot, I. Brissaud, J.P. Garron, H.D. Holmgren, F. Reide and N. Willis, Phys. Lett. 49B (1974) 434; University of Maryland Annual Report (1974).

[8] D.R.F. Cochran, P.N. Dean, P.A.M. Gram, E.A. Knapp, E.R. Martin, D.E. Nagle, R.B. Perkins, W.J. Shlaer, H.A. Thiessen and E.D. Theriot, Phys. Rev. D6 (1972) 3085.

[9] W. Dollhopf, C. Lunke, C.F. Perdrisat, W.K. Roberts, P. Kitching, W.C. Olsen and J.R. Priest, Nucl. Phys. A217 (1973) 381.
[10] J. Amato, R.L. Burman, R. Macek, J. Oostens, W. Shlaer, E. Arthur, S. Sobottka and W.C. Lam, Phys. Rev. C9 (1974) 501.
[11] N.S. Wall, J.N. Craig, R.E. Berg, D. Ezrow and H.D. Holmgren, Proc. Fifth int. conf. on high-energy physics and nuclear structure (Almqvist & Wiksell International, Stockholm, 1974), p. 279.
[12] P.U. Renberg, D.F. Measday, M. Pepin, P. Schwaller, B. Favier and C. Richard-Serre, Nucl. Phys. A183 (1972) 81.
[13] G.A. Miller, Nucl. Phys. A224 (1974) 269.
[14] Il-T. Cheon, Prog. Theor. Phys. Suppl. Extra No. (1968) 146.
[15] M.V. Barnhill, Nucl. Phys. A131 (1969) 106.
[16] A. Reitan, Nucl. Phys. B29 (1971) 525.
[17] J. LeTourneux and J.M. Eisenberg, Nucl. Phys. 87 (1966) 331.
[18] W.B. Jones and J.M. Eisenberg, Nucl. Phys. A154 (1970) 49.
[19] J.M. Eisenberg, R. Guy, J.V. Noble and H.J. Weber, Phys. Lett. 45B (1973) 93.
[20] M.P. Keating and J.G. Wills, Phys. Rev. C7 (1973) 1336.
[21] E. Rost and P.D. Kunz, Phys. Lett. 43B (1973) 17.
[22] B. Höistad, S. Dahlgren, P. Grafström and A. Åsberg, Physica Scripta 9 (1974) 201.
[23] G.A. Miller and S.C. Phatak, Phys. Lett. 51B (1974) 129.
[24] I.S. Shapiro, in Proc. Int. School of Physics "Enrico Fermi" Course 38 (Academic Press, London, 1967).
[25] N. Austern, Direct nuclear reaction theories (Wiley-Interscience, New York, 1970).
[26] S.T. Butler, Proc. Roy. Soc. (London) A208 (1951) 559.
[27] A. Reitan, Nucl. Phys. A237 (1975) 465.
[28] I.R. Afnan and A.W. Thomas, Phys. Rev. C10 (1974) 109.
[29] C.H.Q. Ingram, N.W. Tanner and J.J. Domingo, Nucl. Phys. B31 (1971) 331.
[30] H.W. Fearing, Phys. Lett. 52B (1974) 407; Phys. Rev. C11 (1975) 1210; Phys. Rev. C11 (1975) 1493; Univ. of Alberta preprint UAE-NPL-1072.
[31] A. Reitan, Nucl. Phys. B50 (1972) 166.
[32] M.P. Locher and H.J. Weber, Nucl. Phys. B76 (1974) 400.
[33] G.W. Barry, Phys. Rev. D7 (1973) 1441.
[34] V.S. Bhasin and I.M. Duck, Phys. Lett. 46B (1973) 309.
[35] B.R. Wienke, Prog. Theor. Phys. 49 (1973) 1220.
[36] Z. Grossman, F. Lenz and M.P. Locher, Ann. Phys. 84 (1974) 348.
[37] M. Dillig, H.M. Hofmann and M.G. Huber, Phys. Lett. 44B (1973) 484.
[38] M. Dillig and M.G. Huber, Nuovo Cim. Lett. 11 (1974) 728.

QUASI-FREE SCATTERING AND NUCLEAR STRUCTURE

LECTURE 1

Th.A.J. Maris

Instituto de Fisica
Universidade Federal do Rio Grande do Sul
Porto Alegre - RS - Brasil

As you see from the title of my talks, I have narrowed down the general subject of "Nuclear Structure Effects with Proton Probes". I understand that Dr. Beurtey will among others discuss the elastic and inelastic scattering of medium energy protons; it seems to me that the quasi-free process is the simplest reaction occurring in this field and that it is the one which has given the most significant information on the structure of the nucleus. However this may be just my personal prejudice.

Of the two talks which I shall give, I have thought to spend the first one for a presentation of an overall review of the field of quasi-free scattering. This might be of most interest to those of you who work in other fields, though it may be boring for those who know quasi-free scattering well. In an attempt to give also something of interest to this last group, I will in my second talk leave out the detailed derivation of known formalism and jump to the other extreme of discussing problems which are open or debatable. But I think that also this part can be generally followed, because I will continuously keep the physics in view, as I believe one should do, at least in a subject as concrete as the present one. This first talk will consist of two parts. First I shall give a review of the experimental situation, then the main theoretical ideas will be discussed.

1. EXPERIMENTAL REVIEW

In a quasi-free (p,2p) process an incoming proton of medium

energy knocks a proton out of a nucleus without the incoming and the two outgoing protons having any other violent interaction with the nucleus. For brevity I shall speak of (p,2p) processes; if not explicitely stated otherwise, however, everything I shall say is also applicable to (p,pn) reactions. I will limit myself to processes with incoming energies not lower than 150 MeV and nuclei not lighter than ^{4}He, mainly because in these cases the theoretical description used is better.

Quasi-free processes were observed nearly 25 years ago [1] at the synchrocyclotron in Berkeley by Chamberlain and Segré as they bombarded a ^{7}Li-target with 340 MeV protons. They measured coincidences between outgoing proton pairs and found a strong angular correlation for such pairs. This correlation could be qualitatively understood from the assumption that the incoming proton had knocked out a moving nuclear proton, very much as if it were free. The momentum distribution of the nuclear proton necessary to give the observed smearing of the angular correlation around the usual angles for free target protons in rest was quite reasonable [2], if compared to the one expected for a Fermi gas of the nuclear dimension.

In about the same period theoretical work [3] cleared up some essential features of multiple scattering and the impulse approximation [4], which is the formal expression of the "quasi-free-ness" of the collision, was introduced. A simple estimate from the known total nucleon-nucleon cross sections shows that the mean free path of a nucleon of medium energy is of the order of 3.5 fm and therefore such quasi-elastic scattering events (as they were originally called) are reasonably probable. In the following I shall use the following notation: (Fig. 1).

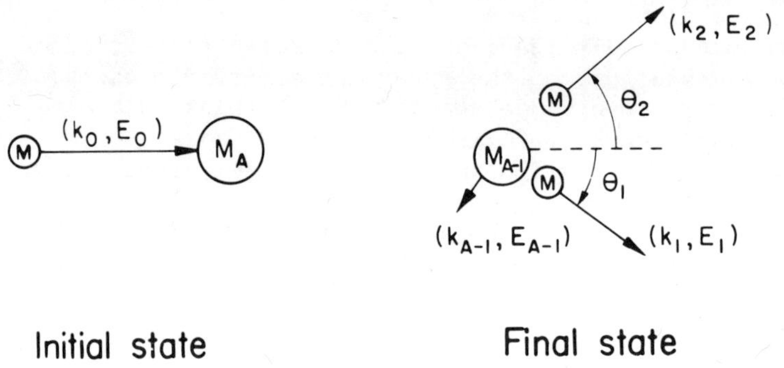

Fig. 1

One would expect for the cross section an expression of the type:

$$\frac{d\sigma}{d\underline{k}_1 d\underline{k}_2} = \text{factor} \times P_E(\underline{k}_3) \; \frac{d\sigma(pp)}{d\Omega} \qquad (1)$$

in which $P_E(\underline{k}_3)$ is the probability density to find a nuclear proton with momentum $\underline{k}_3 = \underline{k}_1 + \underline{k}_2 - \underline{k}_0$ and the separation energy $E = T_0 - (T_1 + T_2 + T_{A-1})$; $d\sigma(pp)/d\Omega$ is the free proton-proton cross section for the kinematics of the experiment (i.e. for a moving target proton) and therefore a Lorentz (or Galilean) transformation is required to connect it with the usual free proton-proton cross section.

In the Berkeley experiment only the directions and not the final energies T_1 and T_2 of the outgoing protons were measured. This means that the expression (1) was effectively integrated over $|\underline{k}_1|$ and $|\underline{k}_2|$ and that only $d\sigma/d\Omega_1 d\Omega_2$ was determined.

This was the state of affairs some 15 or 20 years ago at the time that H. Tyrén invited P. Hillman and myself (as house-theoretician) to work with him at the 185 MeV Synchrocyclotron in Uppsala. Since the Berkeley work had been done, an essential aspect of nuclear structure had become established, namely, the surprisingly good validity of the single particle shell model. This would lead one to expect that different energies are required to knock out protons from the various nuclear shells and that it therefore might be interesting not to integrate eq. (1) over the energies; i.e., to measure also the kinetic energies of each of the emerging coincident particles. In this way one would be able to determine $P_E(\underline{k}_3)$ and the shell structure should come out [5] as peaks in the energy dependence of $P_E(\underline{k}_3)$.

We decided to try the experiment. The equipment consisted of two range telescopes which were chosen to be symmetrically located about the direction of the incident beam and coplanar with it. Coincidences were measured, always choosing $|\underline{k}_1| = |\underline{k}_2|$ so that in effect the nuclear proton had its momentum in or opposite to the incident beam direction. $P_E(\underline{k}_3)$ was measured by varying the values of $|\underline{k}_1|$ $(=|\underline{k}_2|)$ keeping the angles fixed, just by placing absorbers in front of the range telescopes. As the relative variation of the outgoing momenta necessary to scan the interesting part (0–60 MeV) of the separation energy spectrum is rather small, $\underline{k}_3$ varies not very much and the intensity variations found should reflect the E-dependence, i.e. the nuclear shell structure. Fig. 2 shows the first spectrum [6], obtained for ^{12}C.

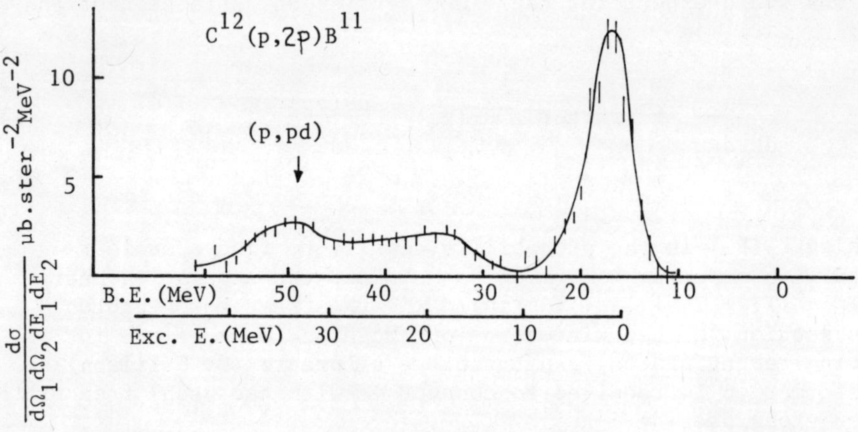

Fig. 2

In the shell model ^{12}C has 4p -protons bound by 16 MeV and 2 strongly bound 1s protons. The result was therefore encouraging. The very short life time of the highly excited 1s hole state is not at all surprising.

The ^{16}O spectrum is much more crucial, because in this case there should be an extra $p_{1/2}$-peak separated by the 6 MeV spin-orbit splitting from the $p_{3/2}$-peak observed in ^{12}C. We therefore took a water target and obtained the spectrum of Fig. 3.

At this point we believed the idea and measured several 1p-shell nuclei. Fig. 4 shows the measured separation energies.

The s and p shells are clearly separated and the spin-orbit splitting is visible.

Now one could measure the momentum distribution of any shell by choosing the sum of the outgoing energies appropriate for the separation energy of the shell. There is one qualitative prediction one can make. The quantity $P_E(\underline{k}_3)$ is just the momentum distribution for protons with separation energy E. A shell model state with non-zero angular momenta cannot have protons at rest, so $P_E(\underline{0})$ should vanish. On the other hand for an s-state there should be many protons at rest, so that $P_E(\underline{0})$ should be a maximum.

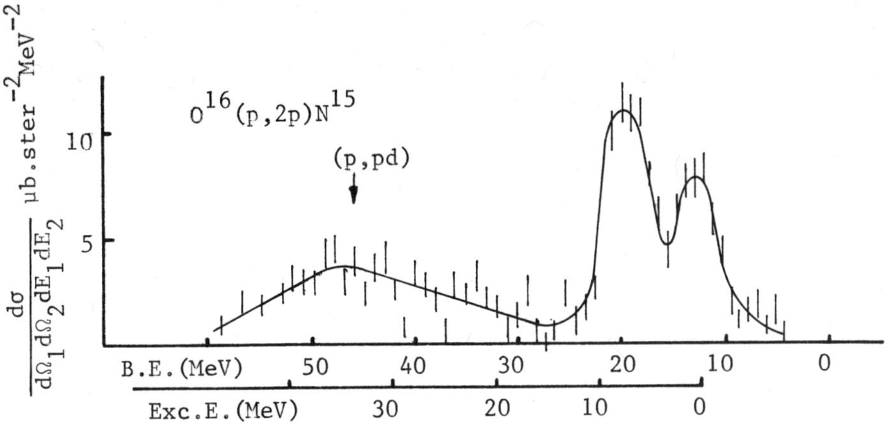
$O^{16}(p,2p)N^{15}$
(p,pd)
$\frac{d\sigma}{d\Omega_1 d\Omega_2 dE_1 dE_2}$ μb.ster^{-2}MeV^{-2}
10
5
B.E.(MeV)
50
40
30
20
10
0
Exc.E.(MeV)
30
20
10
0

Fig. 3

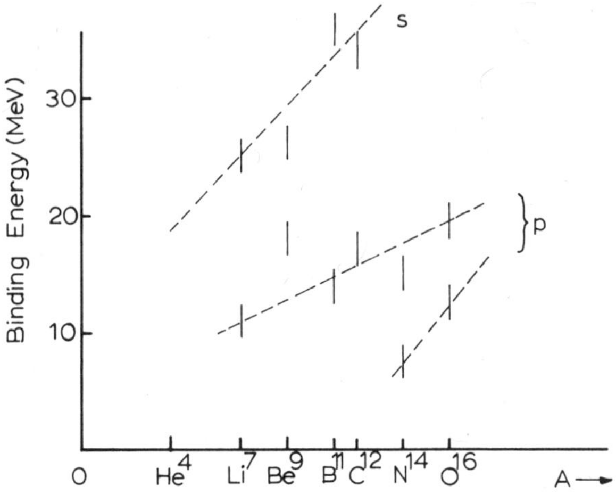
Binding Energy (MeV)
30
20
10
s
p
O
He^4
Li^7
Be^9
B^{11}
C^{12}
N^{14}
O^{16}
A

Fig. 4

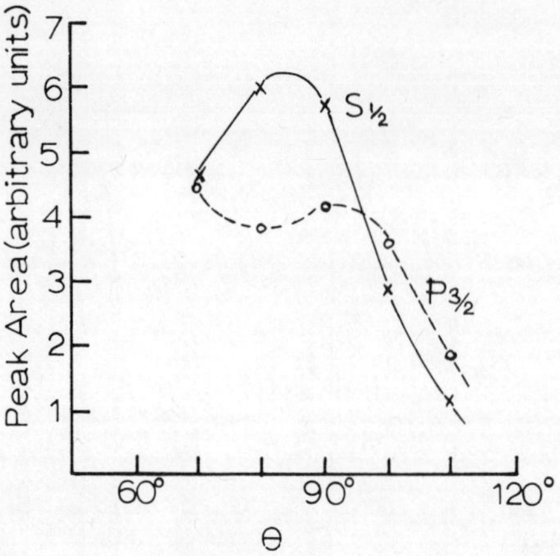

Fig. 5

Fig. 5 shows the first measurement of a momentum distribution [7], namely the one for ^{7}Li. The expected dip for vanishing target proton momentum in the p-shell and the maximum in the s shell distributions are visible. We shall later see why the p-shell minimum does not go to zero.

Briefly after this work a very extensive measurement on ^{12}C has been performed at Harwell and in the last decade or so the Uppsala measurements have been vastly extended and improved in an effort in which practically all medium energy proton accelerators have participated. For more complete results please see the reviews quoted at the end. Let me give some representative results.

The Chicago measurement [8] (Fig. 6) for 460 MeV incoming protons of the energy spectrum and momentum distribution of ^{4}He shows surprising agreement (for a light nucleus) with the shell model; i.e. one sharp separation energy peak for the two 1s protons and a typical s-state momentum distribution.
A Harvard measurement [9] of the momentum distribution of the p and s shell of ^{12}C, performed for asymmetric choices of the outgoing energies, is shown in Fig. 7.

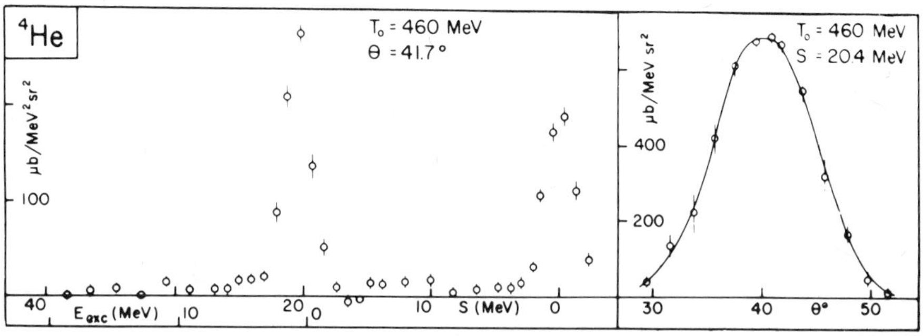

4He
T0 = 460 MeV
θ = 41.7°
μb/MeV² sr²
100
40
Eexc (MeV)
10
20
0
10
S (MeV)
0
T0 = 460 MeV
S = 20.4 MeV
μb/MeV sr²
400
200
30
40
θ°
50

Fig. 6

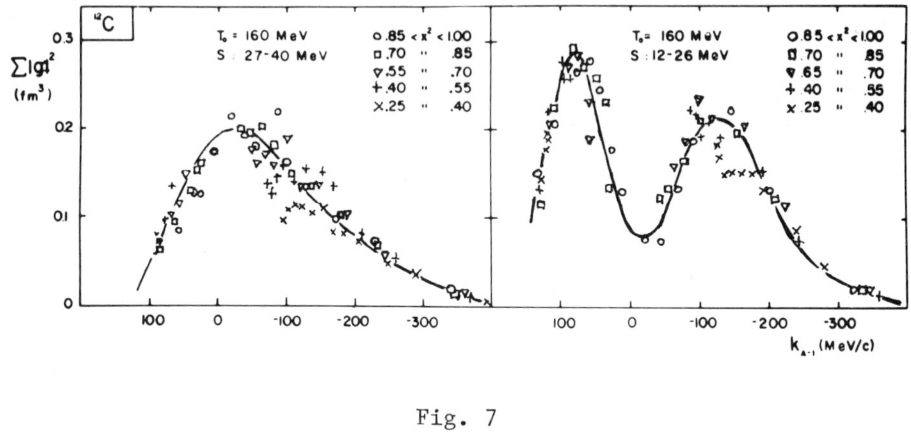

12C
T0 = 160 MeV
S : 27-40 MeV
T0 = 160 MeV
S : 12-26 MeV
Σ|g|²
(fm³)
0.3
0.2
0.1
0
100
0
-100
-200
-300
kA-1 (MeV/c)

Fig. 7

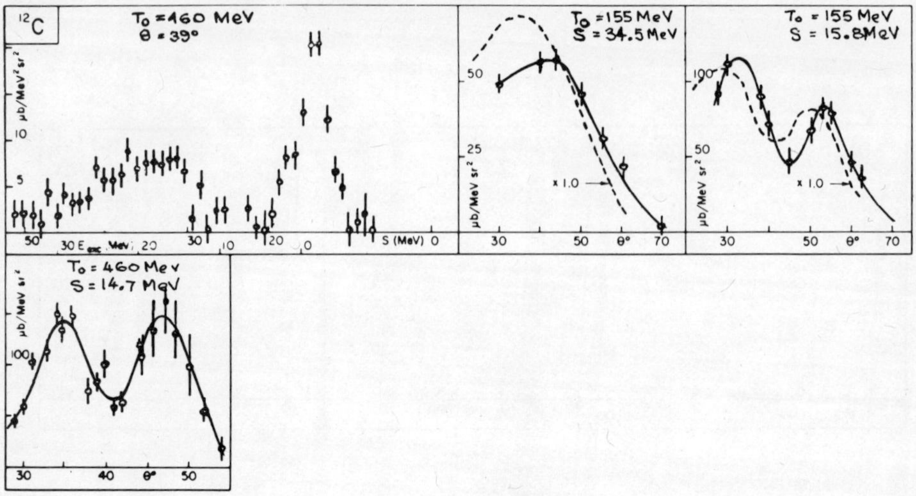

Fig. 8

An Orsay [10] and Chicago measurement of ^{12}C (Fig. 8).

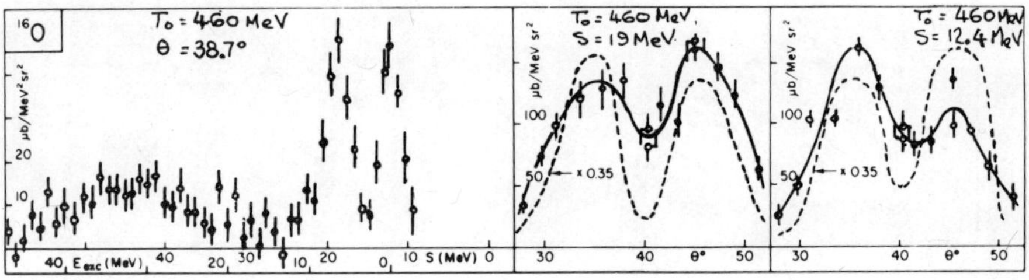

Fig. 9

Again a Chicago measurement for ^{16}O. Both p-states and the s-state are clearly visible (Fig. 9).

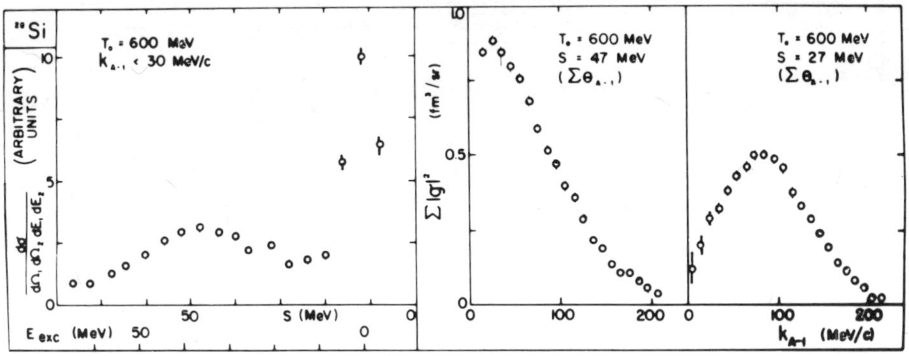

Fig. 10

A 600 MeV CERN measurement [11] of the spectrum of Si^{28} for small target momenta so that only the 1s and 2s states are seen; also shown are the corresponding momentum distribution (Fig. 10).

Some remarkable experiments have been performed by the Liverpool group [12] with 385 MeV incoming protons, where momentum distributions $P_E(\underline{k}_3)$ were determined for about 5 MeV energy intervals (Fig. 11).

Figure 11 shows the case of ^{40}Ca. It is impressive to see how strong the momentum distribution varies with separation energy and how well this dependence agrees with the shell model. In fact, the quasi-free experiments have given the first convincing experimental evidence that the shell model is not just applicable to the slightly bound nucleons of the upper shells, but that it is still a valid concept for the stronger bound particles. One might perhaps also maintain that these experiments form one of the most direct evidences that the nucleus consists of rather "intact" nucleons and not of some nucleon-meson soup.

There is another type of quasi-free experiment which has given quite similar information as the (p,2p) one, which I would

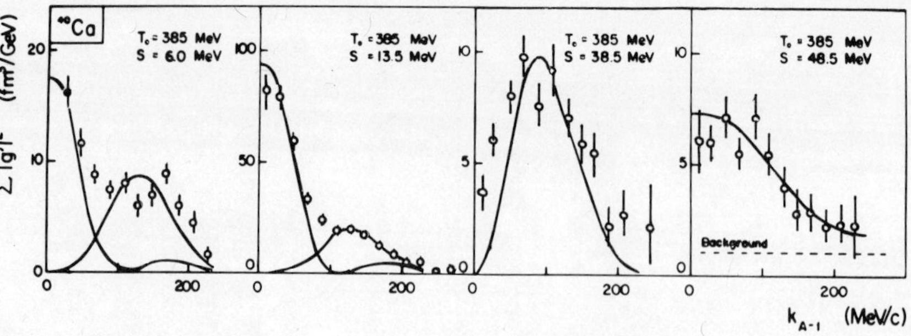

Fig. 11

like briefly to mention although it involves only the electromagnetic interaction. In the beginning of the '60s my colleague Gerhard Jacob came back to Porto Alegre from a visit to the USA and told me, that after he had given a talk McIntyre had asked him whether one could not do quasi-free scattering with electrons. Our first reaction was that this would not work, because of the large range of the Coulomb force, which would mainly have collective effects. We then realized however, that if a large momentum transfer would be selected, effectively only the short distance part of the interaction would contribute. In fact there are even large advantages in doing such (e,e'p) experiments, which compensate for the smallness of the electromagnetic cross section.
The nuclear absorption by multiple scattering (to which I shall come later) is much smaller because only one proton traverses the nucleus. In a typical case in which the (p,2p) cross section would be reduced by a factor of 8, the (e,e'p) cross section only suffers a reduction by a factor of $8^{1/3} = 2$. For the same reason the distortion of the shape of the angular correlation by multiple scattering is expected to be much smaller [13]. The DWIA, which here is practically a DWBA, should be very much better than in the (p,2p) case.

The first experiments showing the shell structure in ^{12}C were a few years later performed by the Sanitá Group in Rome. Since that time a considerable amount of work has been done by the same group and by Russian, Japanese, French and German laboratories.

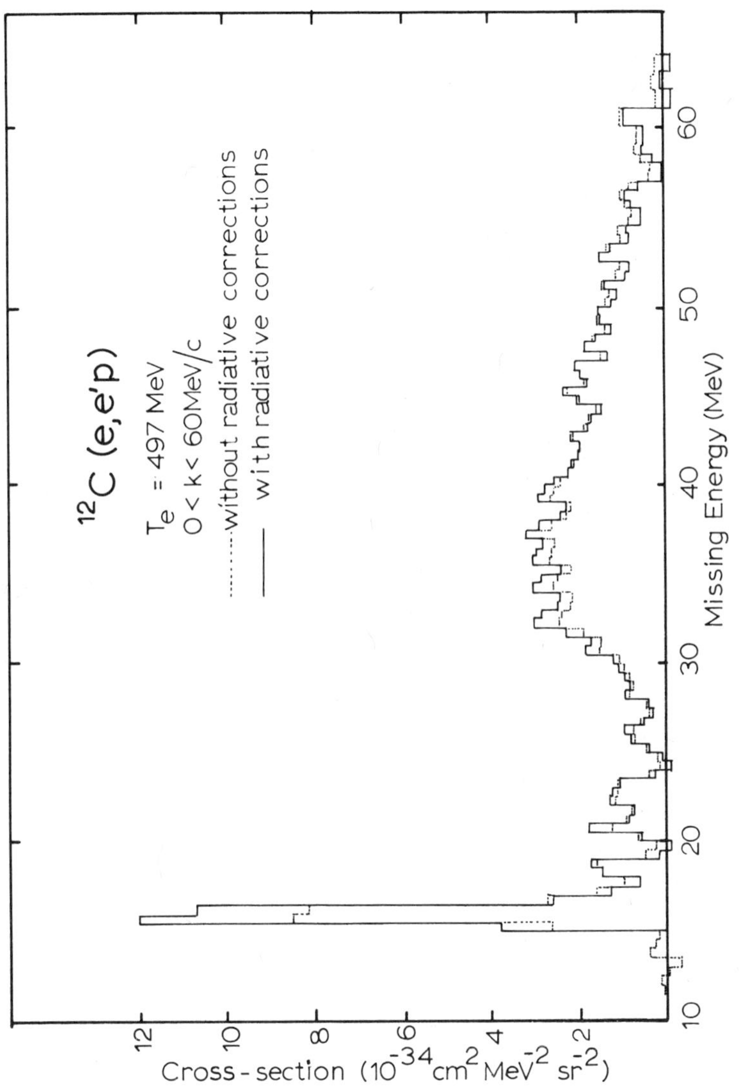

^{12}C (e,e'p)
T_e = 497 MeV
0 < k < 60MeV/c
without radiative corrections
with radiative corrections
Missing Energy (MeV)
Cross-section (10^{-34} cm^2 MeV^{-2} sr^{-2})
10
20
30
40
50
60
2
4
6
8
10
12

Fig. 12

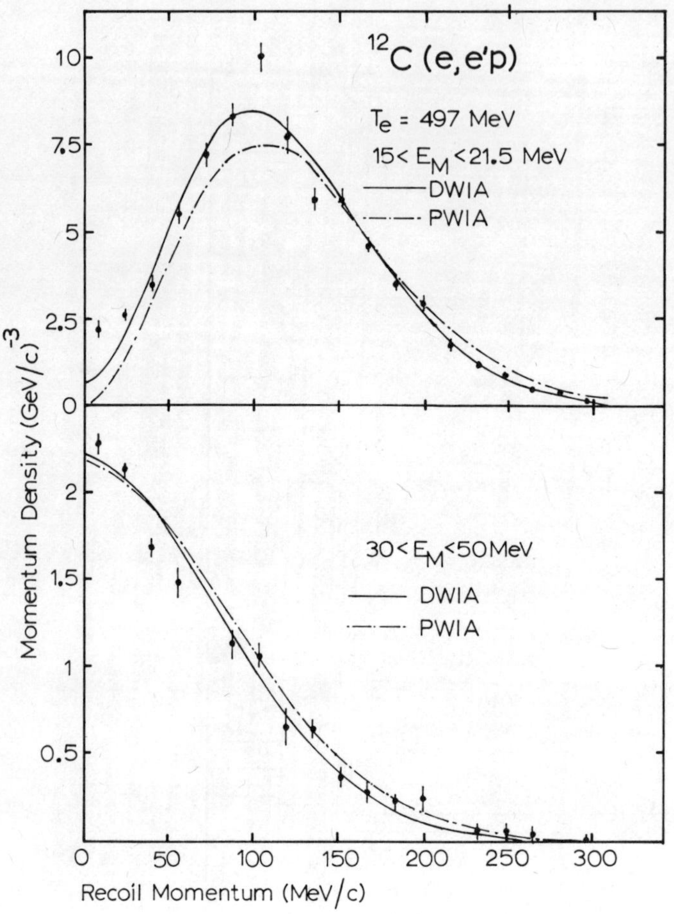

^{12}C (e,e'p)
T$_e$ = 497 MeV
15< E$_M$ <21.5 MeV
DWIA
PWIA
30< E$_M$ <50MeV
DWIA
PWIA
Momentum Density (GeV/c)$^{-3}$
Recoil Momentum (MeV/c)
0
2.5
5
7.5
10
0.5
1
1.5
2
0
50
100
150
200
250
300

Fig. 13

Fig. 14

Figures 13 and 14 shows a recent example of this work [14] (Saclay, 700 MeV electrons) which perhaps is the most extensive investigation of quasi-free scattering in ^{12}C. The general agreement between (p,2p) and (e,e'p) results is good. The expected small absorption and distortion in (e,e'p) experiments have been quantitatively verified.

To end the experimental part of this review and to give an impression of the progress made in the last fifteen years, Fig. 15 shows a recent compilation of separation energy data (Review "e").

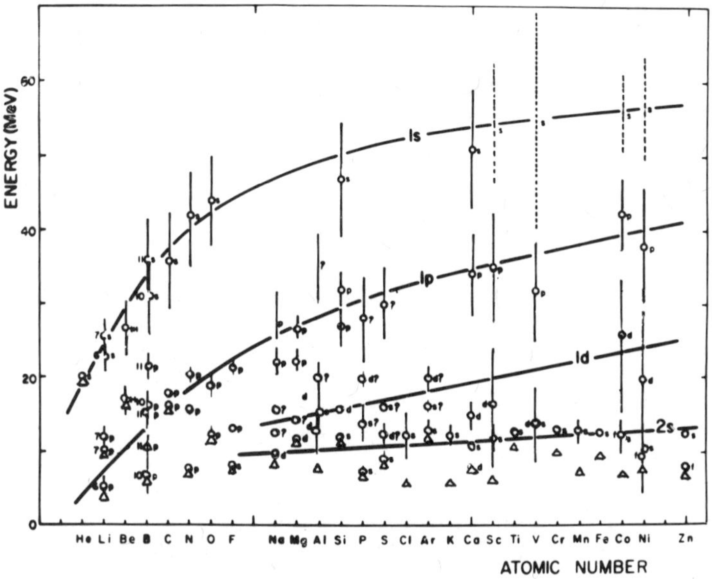

Fig. 15

This figure should be compared with the old Uppsala result of fig. 4.

2. THE MAIN LINES OF THE THEORY

What has been done with respect to the theoretical description of these experiments? The naive formula

$$\frac{d\sigma}{d\underline{k}_1 d\underline{k}_2} = \text{kinematical factor} \times P_E(\underline{k}_3) \cdot \frac{d\sigma(pp)}{d\Omega} \qquad (1)$$

clearly has to be improved. Its most obvious mistake is that multiple scattering has been neglected and this is a very serious error. Let us take the case of a light nucleus like ^{16}O. Its radius is about 3.5 fermis which just about equals the mean free path of a medium energy proton in the nucleus. For a knockout in the center of the nucleus the incoming and outgoing protons have together to traverse 3 nuclear radii, i.e. 3 mean free paths. The intensity reduction of the quasi-free process resulting from multiple scattering should then be of the order of exp (-3), that is by a factor of 20! The situation is in reality somewhat better because the quasi-free process can also occur at the nuclear surface but experiment and calculation give a reduction factor of the order 10. This means that the cross section is only 10% of the one given by our naive formula!

As we realized this, our first worry was the following. Due to multiple scattering, several protons may be knocked out in one process. Each pair of these protons is able to make coincidences; the energies of these coincidences will have nothing to do anymore with the shell model separation energies. As these "bad" events occur even in light nuclei with the overwhelming weight of 9:1 compared to the "good" quasi-free events, will they not completely mix-up our spectra?

The reason why, at the moment we saw this risk, we did not give up the whole experiment was the following wishful thought. If a multiple scattering has taken place, the angular correlation should be strongly dispersed over a large solid angle. A similar effect should occur for the separation energies. There is a broad continuum made up by possible energy losses once one admits multiple collisions and in general these losses will be large. We hoped, therefore, that the multiple scattering would give rise to a smooth background, rising towards increasing separation energy, and that over this background the shell model peaks caused by "good" quasi-free events would be visible. The experiments have shown this hope to be justified. Rather recently the Tokyo group [15] has for the case of (e,e'p) scattering, shown with a Monte Carlo Calculation that the expected background is indeed small. For the case of (p,2p) scattering, I know of no quantitative estimate. As long as this has not been made, one cannot be sure that certain "inner shell peaks" in (p,2p) scattering are not simulated by a structure in the multiple scattering background.

Besides the mentioned introduction of unwanted events, caused by inelastic multiple scattering, elastic multiple scattering will deflect the protons coming from the remaining interesting events. In general this influence is taken into account by distorting the plane proton waves by complex optical potentials, obtained from elastic proton scattering. The imaginary parts take care of the absorption. The replacement of the plane waves by distorted waves has the effect that in our naive cross section formula the momentum distribution $P_E(\underline{k}_3)$ is replaced by a calculable "distorted momentum distribution".

Multiple scattering demands the most drastic correction, but there are some more errors which I shall discuss in my next talk. Here I will just mention two of them. The free scattering matrix element is "off the energy shell". This just means that for the exact momenta of the incoming and outgoing protons no free scattering process exists, because the nucleus has taken away the separation energy. Consequently there is some arbitrariness in the choice of the appropriate free nucleon-nucleon cross section. In practice this means that one does well to avoid kinematical situations in which the free cross-section changes rapidly with energy or angle. Fortunately this is not very difficult for the (p,2p) case as this free cross section equals about 4mb/ster. independent of energy and angle for energies between 100 and 400 MeV.

Then there is the difficulty that the hole state of the nucleus, after the knockout process, is not an eigenstate of the residual nucleus. This is so because the single particle model is only a zero order approximation. There are in the first place the short range correlations predominantly caused by the hard core in nucleon-nucleon interaction, which creates high momentum components. If a high momentum nucleon is knocked out, the resulting recoil of the residual nucleus will mostly not be taken up by the whole nucleus, but by one or a few nucleons. This resembles a multiple scattering and will result in an immediate decay of the hole state. One may estimate the ensuing reduction to amount to something like 15%. There are also the long range correlations, which cause the splitting of the hole state strength over more than one state of the residual nucleus. In the case of bound upper shell states, this will result in a set of discrete excitations of one angular momentum around an average separation energy. For unbound inner shell states it is a mixing of continuum states and will cause a broadening of the peak, corresponding to a state of short lifetime. This we observed already in the spectrum of ^{12}C for the 1s state.

As a result of the short and long range correlations, the normalization of the nuclear form factor is often poorly known. The peak in the experimental spectrum over which one integrates may contain only a part of the strength of the hole state. This strength may be divided over several long lived states and continuum states. Also, since the calculated absorbtion of the traversing protons is very sensitively dependent on the chosen optical potentials, the normalization of the calculated angular correlations often differs from the experimental ones by a factor of two or more and can be easily adjusted by a slight change of parameters. But the shapes of the "distorted momentum distributions" are less sensitive to the correlations and absorbtion and have often been fitted well, resulting in meaningful information. The same may become true for the effective polarizations, to be discussed in my next talk. The most direct information is given by the energy spectra, which are not affected by the above mentioned uncertainties.

Finally I would like to mention some interesting processes which in a certain way are related to quasi-free (p,2p) scattering.

In the first place there are the pick-up experiments, like (p,d) and (D,^{3}H) processes which snatch a nucleon out of the nucleus and lead also to hole states [16]. Experimentally these measurements do not require coincidences and are therefore simpler and more precise. Theoretically their description is in my opinion even more approximate than the one of quasi-free scattering, because the distorted wave concept for a deuteron or triton moving through the nucleus is dubious. There seems to be a not yet fully understood depression for states of higher separation energies.

Then there are the quasi-free cluster knockouts like (p,pd), (p,pα), (e,eα) etc., which often are treated by a distorted wave formalism similar to the one of (p,2p) scattering. In this case there is, in addition to the distorted wave difficulty for complex particles, also the question in how far the quasi-free concept is valid. In contradistinction to the nucleon, whose structure inside the nucleus may not be very much affected by the other nucleons, one would think that a deuteron or α-particle would be strongly deformed, if not partially dissolved, by the nucleus. Also, multiple scattering [17] will cause new coherent more-step-contributions to the quasi-free cross section.

In spite of, or perhaps because of, these theoretical difficulties, measurements of these processes are interesting.

REVIEWS ON QUASI-FREE SCATTERING

a. M. Riou, Rev. Mod. Phys. 37 (1965) 375.
b. T. Berggren and H. Tyrén, Ann. Rev. Nucl. Sci. 16 (1966) 153.
c. The lectures of T. Berggren, S. Kullander and C.F. Perdrisat given at the S.I.N. Summer School in Leysin, Switzerland, September 1969.
d. D.F. Jackson, Ad. Nucl. Phys. 4 (1971) 1.

e. G. Jacob and Th.A.J. Maris, Rev. Mod. Phys. 38 (1966) 121 and 45 (1973) 6.

f. Th.A.J. Maris, Proceedings Fifth International Conference on High-Energy Physics and Nuclear Structure, Uppsala, 1973; edited by G. Tibell.

g. U. Amaldi Jr., Proceedings of the International School of Physics, Erice, 1970.

h. K. Nakamura, Review Talk presented at Gordon Photonuclear Conference, 1974.

REFERENCES

[1] O. Chamberlain and E. Segré, Phys. Rev. 87 (1952) 81.
J.B. Cladis, W.N. Hess and B.J. Moyer, Phys. Rev. 87, (1952) 425.

[2] P.A. Wolff, Phys. Rev. 87 (1952) 434.
J.M. Wilcox and B.J. Moyer, Phys. Rev. 99 (1955) 875.

[3] R. Serber, Phys. Rev. 72 (1947) 1114.
M.L. Goldberger, Phys. Rev. 74 (1948) 1269.
K.M. Watson, Phys. Rev. 89 (1953) 575.

[4] G.F. Chew, Phys. Rev. 80 (1950) 196.
G.F. Chew and M.L. Goldberger, Phys. Rev. 87 (1952) 778.

[5] Th.A.J. Maris, P. Hillman and H. Tyrén, Nucl. Phys. 7 (1958) 1.

[6] H. Tyrén, Th.A.J. Maris and P. Hillman, Nuovo Cimento 6 (1957) 1507.
H. Tyrén, P. Hillman and Th.A.J. Maris, Nucl. Phys. 7 (1958) 10.

[7] P. Hillman, H. Tyrén and Th.A.J. Maris, Phys. Rev. Letters 5 (1960) 107.

[8] H. Tyrén, S. Kullander, O. Sundberg, R. Ramachandran, P. Isacsson and T. Berggren, Nucl. Phys. 79 (1966) 321.

[9] B. Gottschalk, K.H. Wang and K. Strauch, Nucl. Phys. A90 (1967) 83.

[10] J.P. Garron, J.C. Jacmart, M. Riou, C. Ruhla, Y. Teillac and K. Strauch, Nucl. Phys. 37 (1962) 126.

[11] G. Landaud, J. Yonnet, S. Kullander, F. Lemeilleur, P.V. Renberg, B. Fagerström, A. Johanson and G. Tibell, Nucl. Phys. A173 (1971) 337.

[12] A.N. James, P.T. Andrews, P. Kirkby and B.G. Lowe, Nucl. Phys. A138 (1969) 145.

[13] G. Jacob and Th.A.J. Maris, Nucl. Phys. 31 (1962) 139 and 152.

[14] M. Bernheim, A. Bussière, A. Gillebert, J. Mougey, Phan Xuan Ho, M. Priou, D. Royer, I. Sick and G.J. Wagner, Phys. Rev. Lett. 32 (1974) 898.

[15] K. Nakamura et al., Univ. of Tokyo preprint.

[16] D.F. Jackson, Adv. Nucl. Phys. 4 (1971) 1.
G.J. Wagner, Habilitationsschrift, MPI (1970) unpublished.

[17] V.V. Balashov, invited talk at Second International Conference on Clustering Phenomena in Nuclei, Maryland, 1975.

LECTURE 2

As I have remarked last time, I will in this talk make a jump to the other extreme and discuss a series of theoretical points in quasi-free scattering which, it seems to me, are less well known. Several of these points have only very partially been investigated and might be cleared up in model calculations. I shall not go into detailed derivations as some of these easily could each take the whole hour. Instead, I shall try to make the used formulae plausible. What I would like to do is to look as much as possible at the physics of the problem, which is often neglected in the literature, although it is not only simpler to understand than the formalism but can also indicate the direction which this formalism should take.

There are three things which I shall do: a) Discuss the model of the reaction mechanism, and what it neglects: b) Discuss some nuclear structure aspects and show why sum rules go wrong; c) Discuss investigations with polarized protons.

Before going into this, allow me to briefly sketch a remark made many years ago by Hartree which is valid for all systems with many degrees of freedom, meaning for 90% of theoretical physics. Hartree considered the problem of the 26-electron configuration of an iron atom and remarked that in such a case a drastic approximation in which most degrees of freedom are neglected is unavoidable. Suppose one neglects the spin variables and that one would try to make a crude table of the total 26-electron wavefunction, taking only 10 values for each space argument. Then it is easy to see that the table would have $10^{3x26} = 10^{78}$ entries and Hartree remarked that the solar system does not contain enough atoms to supply the paper for such a table. Of course, we have even neglected the infinitely many degrees of freedom arising from the possibility of virtual particle creation!

This remark makes particularly clear how hopeless the solution of a many-body problem in general is. There exist, of course, approximation methods, but I do not know of any method in which a meaningful purely mathematical estimate of the committed errors can be given. Therefore, if by some expansion one finds a result which agrees with experiment, the conclusion one can make is not that one has derived the experimental result from first principles; one can only maintain that the used approximation method seems to work well. If the same or similar approximations give good results in many cases, one feels that the neglected terms are indeed small and that one has "understood" the problem.

The central point in the theory of many-body problems appears therefore to be the search for good approximation methods and it is clear that in this situation a close contact between theory and experiment is highly desirable. All the successful theories, as the shell model, the theory of multiple scattering, the BCS theory of superconductivity and the theory of critical phenomena have resulted from such an approach.

The lack of sufficient power to really attack the many body problem purely mathematically may seem deplorable, but on the other hand it opens the field for physical intuition which is one of the features which distinguishes us from an electronic computer. It is in this spirit that I think we should consider our special problem.

REMARKS ON THE REACTION MECHANISM

The cross section for the quasi-free process is given by

$$\frac{d\sigma}{d\underline{k}_1 d\underline{k}_2 d\underline{k}_{A-1}} = \text{kin. factor} \times \sum_f |t_{fi}|^2 \ \delta(\underline{k}_1+\underline{k}_2+\underline{k}_{A-1}-\underline{k}_0) \ .$$

$$\cdot \ \delta(T_1+T_2+T_{A-1}+E^f_{A-1}-T_0-E_A) \tag{1}$$

where t_{fi} is related to the S-matrix by

$$S_{fi} = \delta_{fi} - 2\pi i\delta(E_f-E_i)T_{fi}$$

$$T_{fi} = (2\pi)^{-1} \ t_{fi} \ \delta^3(\underline{k}_1+\underline{k}_2+\underline{k}_{A-1}-\underline{k}_0) \tag{2}$$

A perturbation expansion for T is

$$T_{fi} = \langle f|V + V(E_i - H_0 + i\varepsilon)^{-1} V + -- \ |i\rangle \ , \tag{3}$$

V being the interaction of the incoming proton with all the other ones.

In our sketch of a derivation we leave out all antisymmetrizations, which can be put in. The first Born approximation with

$$\langle \underline{r}|i\rangle = e^{i\underline{k}_0\cdot\underline{r}_0} \quad \psi_A(\underline{r}_1 --- \underline{r}_A)$$

$$\langle \underline{r}|f\rangle = e^{i(\underline{k}_1\cdot\underline{r}_1+\underline{k}_2\cdot\underline{r}_0)} \; \psi^f_{A-1}(\underline{r}_2 --- \underline{r}_A) \tag{4}$$

gives after some algebra:

$$T_{fi} \sim \int \psi^{*f}_{A-1}(\underline{r}_2---\underline{r}_{A-1})e^{i(\underline{k}_0-\underline{k}_1-\underline{k}_2)\cdot\underline{r}_1} \psi_A(\underline{r}_1---\underline{r}_{A-1})d\underline{r}_1---d\underline{r}_A \quad \cdot$$

$$\cdot \quad \int e^{i(\underline{k}_0-\underline{k}_2)\cdot\underline{r}_0} \, V(\underline{r})d^3r \quad , \tag{5}$$

i.e., the cross section factorizes in a free matrix element in Born approximation and an overlap integral. This overlap integral is evidently the probability amplitude to find in the ground state of the nucleus A_N a proton with the required momentum $\underline{k}_1+\underline{k}_2-\underline{k}_0$ (="$\underline{k}_3$") and the residual nucleus ^{A-1}N in the state f with the selected energy given by the δ-function in eq. 1. This overlap integral contains the momentum δ-function (which should occur in T) because

$$\psi^f_{A-1} \, (\underline{r}_2 --- \underline{r}_{A-1}) = e^{i\underline{k}_{A-1}\cdot\underline{R}_{A-1}} \, \psi^f(\underline{\rho}_1 --- \underline{\rho}_{A-2}) \tag{6}$$

where $\underline{R}_{A-1}$ is the centre of mass coordinate and ψ the internal wavefunction. Extracting this $\delta^3(\underline{k}_1+\underline{k}_2+\underline{k}_{A-1}-\underline{k}_0)$ function one finds t_{fi}, and $|t_{fi}|^2$ is consequently proportional to the above mentioned probability times a free cross section in the Born approximation.

The factorization works still if one improves the free nucleon-nucleon matrix element to its t-matrix element, resulting in the impulse approximation, which is much better than the Born approximation and which just means that one uses the experimental free cross section in formula (1). At this point we have arrived at the cross section formula which we guessed in my first talk. Of course one is making the well-known errors of the impulse approximation, which neglects the influence of the nuclear environment on the p-p collision.

If we take multiple scattering into account, improving the incoming and outcoming waves by an optical potential distortion, the factorization is no longer exact (see for example [1]). This is also physically clear because the distortion means that the protons before and after the quasi-free collision have been deflected by the nucleus and therefore the initial and final momenta do not determine anymore the momenta relevant to the elementary collision. There must occur a kind of averaging of the free matrix element over the values of the external momenta and that is just why one can no longer formally factorize the matrix element exactly.

Now one would like to use the factorization because then one can reduce the nucleon-nucleon matrix element to a measurable cross-section and the overlap integral is manageable. Without factorization, in principle one can still evaluate the complete integral using a phase shift analysis for the nucleon-nucleon interaction and for the three proton waves. Computer experts maintain that this calculation is formidable even for modern computers, whereas the factorized calculation I did fifteen years ago by hand. In addition one has to work with nucleon-nucleon phase shifts which are less well-known than nucleon-nucleon cross sections. So, if possible, one should work in a situation in which factorization is a good approximation and it is interesting to see when this will be so.

The factorization will be good if the mentioned averaging does not matter, i.e., if the matrix element varies slowly compared to the spread of momenta caused by the distortion. The matrix element is constant for a δ-function nucleon-nucleon force, but this is not a very realistic assumption. The proton-proton cross section is in the region of 100-400 MeV indeed very slowly varying with energy and angle. However there seems to occur a curious cancellation effect, because as we shall see in the polarization experiment, the separate spin dependent matrix elements vary considerably stronger. We must hope, therefore, that there will be only a small smearing of the momenta by the distortion. Qualitatively, this is just the essential condition allowing the use of the WKB method, which assumes that the optical potential affects only slightly the shape of the wavefront, i.e., supposing that the classical paths of the particles suffer only small changes.

Therefore if one calculates the overlap integral in the WKB approximation and also "exactly" by a good phase shift calculation and one finds very different results, then the factorization had probably not been allowed in the first place and one should go back to the non-factorized expression. If the mentioned difference is small, which is in our energy region often true [2], one may reasonably hope the factorization to be good.

Let me excuse myself for stating things, here and later, rather categorically. This I do for brevity; in reality there probably exists very different opinions from the ones I have.

Having discussed the factorization, let me go one step back and ask the question what type of things one neglects even if one does not make the factorization approximation. The point is that we take into account the complicated effects of the spectator nucleons only through the optical potentials. Here is the point where we reduce the number of degrees of freedom drastically and from Hartree's remark it is clear that it is difficult to estimate the committed error. In the language of multiple scattering expansions, we have neglected the possibility that the nucleus is excited and de-excited more than once, ending up in the considered final state.

From elastic scattering, where the optical model works well for not too large angles, one might hope this effect not to be serious if the considered final state has a well defined energy, i.e. a long lifetime. This is not necessarily true. In a model calculation Balashov et al [3] have shown for the case of (e,e'p) scattering that the dynamical effect of the outgoing proton can be considerable even for final nuclear states with well defined energies, by mixing the one particle-one hole states. In fact, taking a pure Coulomb interaction which naively should only knockout protons, these authors obtain at not too high energies an appreciable (e,e'n) contribution essentially by a p-n charge exchange. But this effect decreases with increasing energies. Model calculations for the (p,2p) case would be desirable.

For short lived states, as are the highly excited inner shell hole states, the traversing protons may have another serious effect. Physically the point is that the nucleus can decay before the protons have left the interaction region. The time dependence of the resulting influence will show up in a deformation of the peak in the energy spectrum. Some aspects of this effect have been treated in the literature. Brueckner et al remarked [4] that there are two extremes, the adiabatic and the sudden removal of the particle, where the lifetime of the residual hole state should be compared with the traversal times of the protons. By going from a sudden to an adiabatic removal, changing the bombarding energy, presumably a broad energy peak should deform so that its centre of gravity moves towards lower excitation energy. Another calculation is the one of Austern and Pittel, who showed [5] for the case of the 1s state of ^{12}C that the energy peak deforms because one of the main decays of the 1s hole-state is by a kind of Auger effect in the 1p-shell, leading to a 1 particle - 2 hole state. This same final state can be reached by a double scattering in the p-shell and there will be an interference between the two mechanisms, changing slightly the energy peak.

In all these cases the situation is more favourable in (e,e'p) scattering where there is only one traversing proton. Perhaps the comparison of (e,e'p) and (p,2p) measurements might eventually give experimental indications for the size of the effects. There is a need for more model calculations clarifying the quality of the DWIA for (p,2p) reactions.

I should mention an interesting calculation [6] in which the authors compare the DWIA for the knockout of a particle bound to a potential of finite mass with the results of an exact Fadeev calculation. The DWIA fares very well at medium energies, but of course again the residual nucleus has been represented only by a potential.

Closing this qualitative discussion of the reaction formalism, let me make one final remark on the evaluation of the factorized-out overlap integral. One might evaluate this integral by a phase shift method, but if the factorization is valid, the WKB method should also be reasonably good. It has the advantage of being simple and giving a physical picture. I wonder how, for example, one could physically understand the polarization phenomena, later to be discussed, in an angular momentum basis for the incoming and outgoing protons.

These remarks have been concerned with the reaction mechanism. Let me now have a short look at the final nuclear state.

NUCLEAR STRUCTURE

The nuclear structure part in the factorized cross section is given by the distorted overlap integral

$$|g'_{fi}(\underline{k}_3)|^2 = \Big|\int e^{i\underline{k}_0\cdot\underline{r}_1} D_0(\underline{r})\, e^{-\underline{k}_1\cdot\underline{r}_1} D_1(\underline{r}_1)\, e^{-\underline{k}_2\cdot\underline{r}_1} D_2(\underline{r}_1) \qquad (7)$$

$$\psi^{*f}_{A-1}(\underline{r}_2 \;---\; \underline{r}_A)\; \psi^{i}_{A}(\underline{r}_1 \;---\; \underline{r}_A)\, d^3 r_1 \;---\; d^3 r_A\Big|^2$$

To understand the meaning of this integral it is helpful to first neglect completely the multiple scattering, i.e., to put all D's equal to one. Then, as we have already seen last time, the $|\text{integral}|^2$ has a simple physical meaning which is also directly connected with the reaction: it is equal to the probability of finding in the ground-state of the target nucleus

A_N a proton with momentum $\underline{k}_3$ such that the residual nucleus A_{N-1} (with momentum $-\underline{k}_3$) has an excitation energy $E_s^f - E_s^o$, where E_s^f is the experimental separation energy and E_s^o the one for the least bound proton.

One will find a peak in the energy spectrum only if there is a sufficiently long lived final hole state available. In general the highly excited nucleus with a hole will immediately decay and this will result in a smooth energy spectrum. However, at certain energies the lifetime of the hole may be exceptionally large and give rise to a more or less narrow peak.

The position of such peaks is given by the dynamical properties of the residual nucleus. If these peaks are sharp, as is the case for many upper shell states, they will not be moved by the reaction mechanism. I have heard it said that the rearrangement energy would cause such peaks to slide up and down depending on the degree of adiabaticity of the nucleon removal. This is clearly incorrect. As we saw, the situation may be different for broad energy peaks, where the shape (reflecting the time development of the decaying hole state) may well be influenced by the details of the removal mechanism. However this is a deviation from the impulse approximation and adopting the DWIA one has not anymore this arbitrariness in the shape of the spectrum peaks because the reaction mechanism is fixed to be the sudden one. In fact, one can give a closed formal expression for it, as was first pointed out [7] by Gross and Lipperheide.

I would like to make a few remarks about this approach. For the nuclear structure the essential term in the cross section, neglecting the distortion is

$$\Sigma_f |g_f(\underline{k}_3)|^2 \ \delta(E^f_{A-1} - E)$$

where

$$E = T_0 + E_A - T_1 - T_2 - T_{A-1}$$

$$= E^o_{A-1} + E^f_s - E^o_s$$

E^o_{A-1} being the groundstate energy of the residual nucleus.

In order to avoid normalization problems of the centre of mass wave functions, we normalize all wave functions in an enormous box. The summation is over all final states f of the residual nucleus. In a field theoretical formulation one has $g_f(\underline{k}) = \langle f|a(\underline{k})|i\rangle$, where $a(\underline{k})$ is the proton destruction operator in the Heisenberg picture at t=0 and one has, as is seen by inserting intermediate states,

$$\Sigma_f |g_f(\underline{k}_3)|^2 \ \delta(E^f_{A-1}-E) = \langle i|a^+(\underline{k})\delta(H-E)a(\underline{k})|i\rangle \tag{8}$$

This is the imaginary part of

$$(\pi)^{-1} \langle i|a^+(k) \frac{1}{H-E+i\varepsilon} a(\underline{k})|i\rangle$$

which again equals up to a constant factor the Fourier transform of the retarded hole propagator $\langle i|a^+(\underline{x}=0,t=0)a(x,t)|i\rangle\theta(-t)$; this expression represents the negative time part of the two point Green's function

$$G(x) = -i\langle i|T\{a(x,t)a^+(0,0)\}|i\rangle \quad .$$

Therefore $\Sigma_f |g_f(\underline{k}_3)|^2 \ \delta(E^f_{A-1}-E)$ is just the Kallén-Lehmann density of the proton hole propagator. The proton propagator occurs naturally in the Feynman-Goldstone perturbation theory. One may write down a self consistent Dyson-Schwinger equation for this propagator and in ref.[8]these equations are solved approximately. From Hartree's remark one understands that one will need a strong guidance from experiment to find a useful approximation; comparing the calculated results with experimental ones it seems that much has still to be done in this interesting attempt (see also ref.[9]).

The above established connection between the hole propagator and the overlap integral was made neglecting the distortion. The connection can still be made if one is able to represent the distortion by a constant reduction factor which is the same for all states. In many places, including in one of our reviews, one finds the remark that a more general distortion can be included but that then the off-diagonal expression

$$\langle i|a^+(\underline{k}')\delta(H-E)a(\underline{k})|i\rangle \tag{10}$$

would enter. This remark is incorrect. Clearly, from momentum conservation the above expression vanishes for $\underline{k}' \neq \underline{k}$. Such may not be the case in approximate calculations, but this reflects only a spurious centre of mass movement introduced by the approximations. The expression which one finds for the case of distortion has not anymore a direct connection with the hole propagator.

From the expression (8) sum rules can be derived, for example:

$$\int d^3k \int dE \sum_f \delta(E_f - E)|g_f(\underline{k})|^2 = \int d^3k \int dE \langle i|a^+(\underline{k})\delta(H-E)a(\underline{k})|i\rangle$$

$$= \int \langle i|a^+(\underline{k})a(\underline{k})|i\rangle d^3k = Z \,, \tag{11}$$

because

$$\int a^+(\underline{k})a(\underline{k})d^3k$$

is the proton number operator. This sum rule is also physically understandable. The left hand side is just $\sum_f \int d^3k|g_f(\underline{k})|^2$ assuming pure quasi-free collisions without absorption one can in total knockout Z protons. Of course this sum rule is not fulfilled. Firstly a part of the excitation goes to very high separation energies because of short range correlations. This part of the energy spectrum is not measurable in practice and also the impulse approximation is not valid for it. Furthermore in many cases 90% of the protons is multiple scattered. This falsifies the result completely.

Another sum rule derivable for two body interactions is (Kolthun) [11]

$$\int dE \int d^3k P_E(\underline{k}) \frac{1}{2}\left(\frac{k^2}{2m} - E\right) = E_z = \text{the total proton binding energy.} \tag{12}$$

This sum rule is for the same reason as the number rule invalid in practice. One might perhaps hope that in the ratio of eqs. (11) and (12), which in the ideal case would give the average proton binding energy, the errors will cancel. This is indeed the way in which the sum rule seemed to fit originally. But a little contemplation shows that there is no reason to expect such a cancellation and the rule has for the (e,e'p) case, which is much more favourable than the (p,2p)-one because of the smaller distortion and the more accurate experiments, strongly failed. The nice idea unfortunately does not work in practice.

Looking at the separation energy curves (Fig. 15 of Lecture 1), one sees that there are clear general trends for the separation energies and the widths in dependence of the atomic number. This suggests that it might be useful to parametrize these trends by considering the hole states as being bound in a complex potential. If one would take a purely real potential one would have just the usual single particle approximation. The complex part is supposed to take the finite mean free path of the hole into account and might be connected with many-body theory [11, 8, 9]. Calculations [12] in a very simple model of this type have shown that, depending on its shape, the imaginary part can have a considerable influence on the momentum distribution.

I should at least mention interesting work which has been done by the Maryland group [13] on the off-shellness of the free cross section which already came up in my last talk. This is a very difficult problem. It seems to me that if one is interested in a problem caused by a loss of energy, the mechanism by which this energy is lost must be essential. Therefore I would expect that the off-shell extrapolation necessarily should be strongly dependent on the model used. The Maryland group has in a three body model calculated the matrix element to be used as being a "half-off shell one"; they think that this choice has a more general meaning and do not completely agree with my model dependence remark. Fortunately the uncertainty in the off-shell extrapolation is small in many kinematical situations.

Let me finally come to the use of polarized protons in quasi-free scattering. Again I will only give the physics and mention some open problems.

In general the multiple scattering of the traversing protons in (p,2p) scattering is a disadvantage. The cross sections go down, stray protons cause troubles and the momentum distributions are distorted. However, just this large absorption may have an interesting effect on which our Porto Alegre group has spent some thought [14] and which may be measured at the TRIUMF in the not too distant future.

If one calculates the free cross section which has to be used in the factorized cross-section formula one finds that this in general is one in which the target nucleon is polarized. If the scattering is coplanar, then reflection symmetry through the scattering plane demands that this effective polarization is orthogonal to it. In a symmetrical situation this polarization vanishes because it has then also to be orthogonal to the symmetry plane.

It is easy to see why this effective polarization is in general non-zero. Let us take the asymmetrical geometry of Fig. 1.

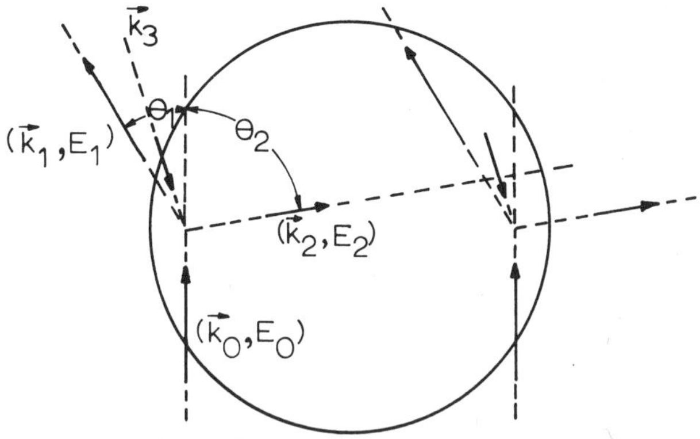

Fig. 1

The protons knocked out at the right will have to travel through less nuclear matter than the ones on the left. Because of the strong absorption mainly the right hand side of the nucleus will contribute to quasi-free processes. On the other hand, because of the spin-orbit coupling there will be a strong correlation between the spin and the momentum at the edge of the nucleus. Selecting by the kinematics a certain momentum means therefore also a spin selection, which means that the knocked out proton was effectively polarized before it was ejected.

There is, however, a very fortunate circumstance. For medium energies just in the angular region needed for the absorption effect, the cross section for protons with parallel spins is much larger than the one for opposite spins. In other words, using incoming protons polarized orthogonal to the scattering plane one can find out what is the effective polarization of the knocked out nuclear proton. If one reverses the polarization direction of the bombarding proton and the effective polarization of the nuclear proton is large, then the cross section can change drastically. The general form of the cross section is given by

$$\frac{d\sigma}{d\Omega} = \frac{d\sigma_0}{d\Omega} \; [1 + (P_1+P_2)P(\theta) + P_1P_2C(\theta)] \tag{13}$$

where the functions P and C are even and odd in θ. $\frac{d\sigma_0}{d\Omega}$ is the unpolarized cross section, $(P_1=P_2=0)$; $P(\theta)$ is responsible for the asymmetry in the scattering of polarized protons on unpolarized ones $(P_2=0)$. C is responsible for the main effect of changing the relative sign of P_1 and P_2 and is therefore the most interesting quantity for us. If it has its maximum value 1 then for 100% oppositely polarized beams the cross-section vanishes.

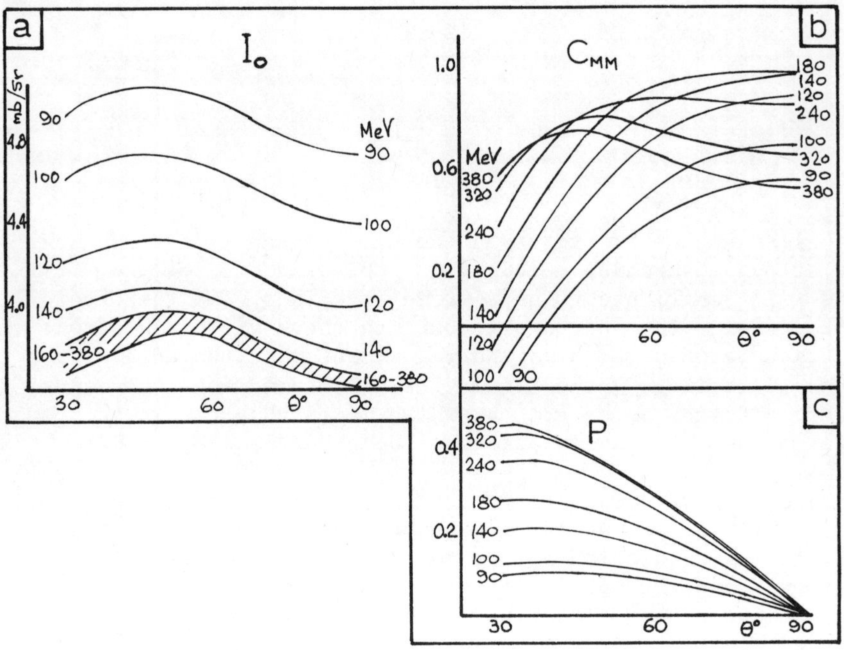

Fig. 2

Fig. 2 shows the functions P and C in the interesting energy domain. Evidently C becomes often near to one!

Results of calculations of quasi-free cross sections for ^{16}O are shown in Fig. 3. We have chosen 320, about 240 and 80 MeV for the incoming and outgoing protons energies.

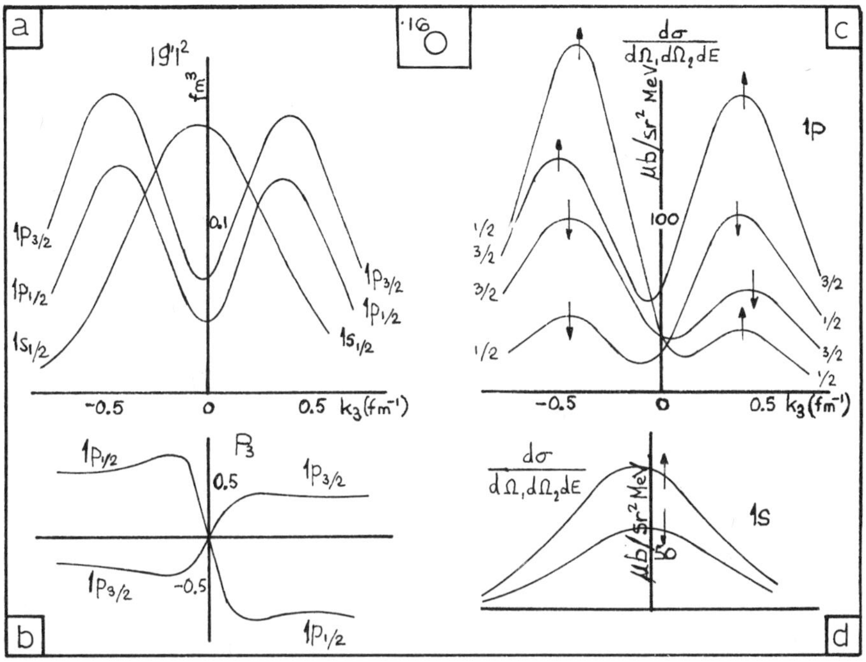

Fig. 3

The only existing quasi-free scattering experiment with polarized protons was recently performed at the Dubna accelerator [14] with 635 MeV incoming protons, as reported by L.I. Lapidus at the Sixth International Conference on High Energy Physics and Nuclear Structure, 1975. The measurements were made on ^{12}C and ^{6}Li, in a not very asymetrical geometry and with a limited energy resolution. As also at 635 MeV the C-function is probably small, it is not surprising that the asymmetries found were about equal to the one of the corresponding free proton-proton scattering.

In the calculations the spin-orbit part of the distorting potential was neglected. It is not quite clear how large its influence will be. In principle it can be included [2], but results in considerable computational complications and also uncertainties, because the factorization fails and the needed phase shifts are not as well known as the free cross sections.

Finally I would like to make a remark on the (p,pn) reaction with polarized protons which, according to some experimentalists may become measurable in the near future. From (p,n) phase shifts one may calculate the spin dependence of the cross section. It appears that C_{np} can become about equal to 0.6, which would make the polarization effects about half as large as in the (p,2p) case.

Acknowledgements: I thank my colleagues Gerhard Jacob, V.E. Herscovitz, C. Schneider and M.R. Teodoro for helpful discussions. The support of the German Agency for Technical Cooperation is gratefully acknowledged.

REFERENCES

[1] D.F. Jackson, preprint, University of Surrey, 1975.
[2] D.F. Jackson and T. Berggren, Nucl. Phys. 62 (1965) 353.
[3] V.V. Balashov, S.J. Griskanova, N.B. Kabachnick, V.M. Kulikov and N.N. Titarenko, Nucl. Phys. A216 (1973) 574.
[4] K.A. Brueckner, H.W. Meldner and J.D. Perez, Phys. Rev. C6, (1972) 773.
[5] S. Pittel and N. Austern, Nucl. Phys. A218 (1974) 221.
[6] S.K. Young and E.F. Redish, Phys. Rev. C, 10 (1974) 498.
[7] D.H.E. Gross and R. Lipperheide, Nucl. Phys. A150 (1970) 449.
[8] V. Wille and R. Lipperheide, Nucl. Phys. A189 (1972) 113.
[9] G.M. Bagradov and V.V. Gorchakov, Sov. J. Particles Nucl. 5 (1974) 122.
[10] D.S. Kolthun, Phys. Rev. Lett. 28 (1972) 182.
[11] H.S. Kohler, Nucl. Phys. 88 (1966) 529.
[12] V.E. Herscovitz, G. Jacob and Th.A.J. Maris, Nucl. Phys. A109 (1968) 478.
V.E. Herscovitz, Nucl. Phys. A161 (1971) 321.
[13] E.F. Redish, G.J. Stephenson and G.M. Lerner, Phys. Rev. C2 (1970) 1665.
[14] G. Jacob, Th.A.J. Maris, C. Schneider and M.R. Teodoro, Phys. Lett. 45B (1973) 181 and preprint.
[15] V.S. Nadezhdin, N.I. Petrov, V.I. Satarov, Contribution to Sixth International Conference on High Energy Physics and Nuclear Structure, Los Alamos, 1975.

Question to Professor Maris from H.G. Pugh

A few years ago you presented a very appealing picture of absorption processes in (p,2p) reactions which suggested that the observed reactions occur principally in the "pole-caps" at the top and bottom of the nucleus as defined by the scattering plane. (This is the basis of your proposed interference experiment in non-coplanar (p,2p) reactions.) Your present picture of absorption, to explain polarization phenomena, is in terms of events occurring on the left and right sides of the nucleus. Is there any conflict with the pole-cap model? If not, how do you reconcile the two pictures?

Answer

I would like to give a twofold answer. In the case of the polarization experiment, we have made a WKB calculation and Miller has made a phase shift calculation which show that the effective polarization effect is a real one and has the expected order of magnitude and sign.

This does not exclude that the pole-caps can contribute considerably, as required for the interference experiment. In ^{12}C because of the negative parity of the p-wave functions the pole-caps tend to contribute little in the scattering plane. But for the d-states in ^{40}Ca, the pole-caps constitute a diluting effect on the polarization. This is, however, taken into account by the distorted wave calculations.

HIGH RESOLUTION MAGNETIC SPECTROMETRY

R. BEURTEY

DPh-N/ME - C.E.N. Saclay

B.P. 2, 91190 Gif-sur-Yvette, France

I. INTRODUCTION

I do not intend to speak about all possible types of spectrometers which are able to perform high resolution experiments at intermediate energies. Our first spectrometer SPESI gives a resolution of 6 10^{-5} in momentum at 1 GeV. It would be difficult to extrapolate the good features and difficulties of our spectrometer to other kinds of spectrometers (e.g. Los Alamos). I will describe two spectrometers which are actually in operation: SPES I, in routine operation since the end of 1972; SPES II in test operations and will be used for taking its first data in July 1975. A few other projects in intermediate energy spectroscopy at Saturne will be described at the end of this lecture (last but not least, the accelerator improvement program). This is necessary to obtain higher quality spectroscopic information. Low-energy nuclear physicists have used magnetic spectrometers for a long time and the "compensation-principles" have been known for more than two decades[A]. The goals were, and still are, to construct a spectrometer system able to:

- analyze as accurately as possible the energy (in fact the *momentum*) of reaction products emitted from a target struck by a primary beam (resolution $\delta p/p$);
- utilize a *large solid angle*, which must be well defined independent of the magnitude of the momentum of the particles to be analyzed (Ω);
- measure *the momenta inside a band* as wide as possible (ΔP_{tot});
- *focus* analyzed particles in both transverse directions on an area as small as possible;

- *be insensitive* to the primary beam characteristics (emittance, momentum dispersion);
- have a large angular range;
- give high angular accuracy.

This set of goals is more difficult to reach at intermediate energies than at the low energies. The large momenta to be analyzed require the construction of large magnets. High mechanical accuracy is necessary. To separate the same levels of a nucleus in a two-body reaction as in the low energy region, a much higher resolution has to be obtained.

In 1967, before the shutdown of the cosmotron, the experiments made at Brookhaven [B)] in nuclear physics opened new research directions. These could be characterized by specific qualities (in comparison to low energy physics).

Finer details on the *spatial structure* of nuclei can be obtained, due to the small wavelength of the projectiles. Alternatively, one should be able to measure *high momenta components* in the wave functions because of the possibility of transferring large "p" in scatterings and reactions.

An almost *instantaneous image* of the nuclei (frozen nuclei), due to the speed of the particles used, can be hoped. It should be possible to knock out or pick up *substructures* deeply and strongly bound, and to create new particles (π, K, ...) utilizing the large c.m. energy.

These are the reasons why Jacques THIRION decided in 1968 to push part of our Nuclear Physics Department in the direction of this Physics, and to build a high resolution magnetic spectrometer able to work near Saturne up to a momentum of 2 GeV/c. The circumstances were favourable: our laboratory had already built and used two magnetic spectrometers ($\sim 2.10^{-4}$ resolution in momentum) for low energy projectiles. Also the accuracy of the particle localization made at this time a quantum jump ... and low energy spectroscopy began to lose its interest.

A large collaboration within our laboratory (physicists, technicians, designers, workmen and the Saturne Department) made it possible to start the project in 1969, to complete the construction of the spectrometers in 1970-71, to install it and the beam line in 1971, and finally in February 1972 to obtain the first spectrum on ^{208}Pb with 200 keV resolution at 1 GeV. A few months later we had reached the design resolution: 6.10^{-5} at 1.04 GeV. Many experiments have been performed since then: (pp) (pp') ($\alpha\alpha'$) (pd) ($p\pi^+$).

My plan will be to describe:
- the principles necessary to achieve a high resolution;
- the first spectrometer "SPESI";
- the second spectrometer "SPESII";
- the future spectrometry around Saturne.

II. THE NECESSARY PRINCIPLES

The basic principles may be classified according to the order of accuracy in the analysis of the magnetic fields.

The high resolution will be obtained by a compromise between:
- the spectrometer *dispersion*;
- the spectrometer horizontal *magnification*, which reproduces the beam spot on the target at the detector;
- the more or less exact *compensation* of the primary beam energy spread, and of the beam angular divergence;
- the reduction of the *intrinsic aberrations*.

IIa. Zero'th Order

We need a spectrometric magnet such that, after a deviation of the particles, the accuracy in localizing these particles at some distance from the exit will give the required resolution for the measured momentum. In the simplified diagram in fig. 1, we see that the dispersion (δx) at a distance (L) from the center of the spectrometer (S) will give a measure of the relative variation of the momentum ($\delta p/p$). Let (α) be the mean rotation angle:

$$\frac{\delta p}{p} \sim \frac{\delta \alpha}{\alpha} \qquad\qquad \delta x \sim L\delta\alpha$$

For a reasonable value of L, say 10 meters, an accuracy of $\sim$ 1 mm for δx, $\delta p/p$ will be 6.10^{-5} for $\alpha \sim 1.5$ rad. We have chosen $\alpha = 97°$ and $B_{MAX} = 17.2$ kGauss. Curvature radius of the mean trajectory: 3.3 meters.

IIb. First Order

The first order effects are very important for the spectrometer and for the *beam line*.

<u>The horizontal focussing</u> of the spectrometer (S) is given by the shape of the entrance pole pieces, and the difference in length of the trajectories on a large radius compared to the small one. It is well known that the radial entrance and exit (fig. 2) gives an automatic focussing for trajectories in the median plane.

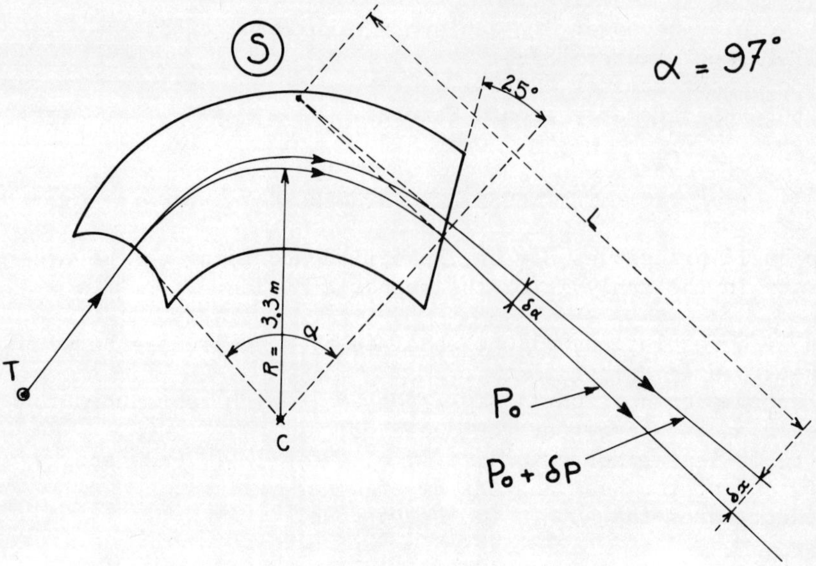

Fig. 1- Schematic view of the Spectrometer and it's Dispersive properties.

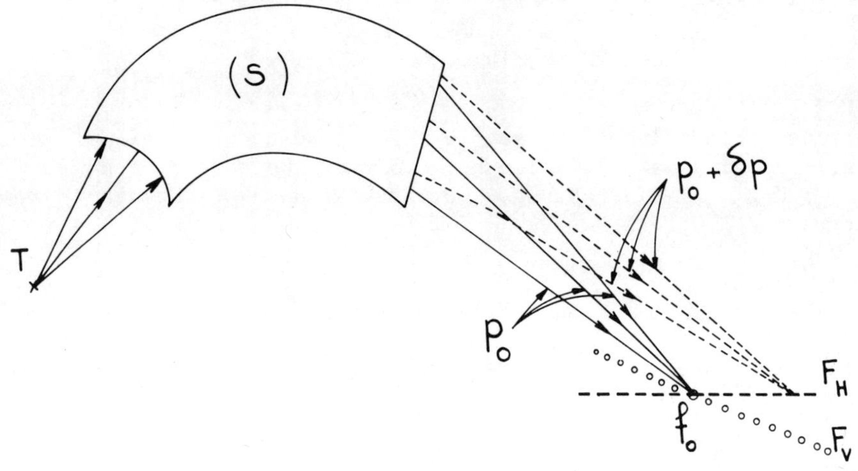

Fig. 2- Focal surfaces of the Spectrometer.

The vertical focussing is made by a quadrupole situated between the target and the (S) magnet, and by the quadrupole effect produced by non-normal exit from the magnet. The quadrupole defocusses horizontally, but the effect of the Q-pole+(S) is chosen such that in the near vicinity of the mean trajectory, both horizontal and vertical focussing give the same focus (f_o). For values of p different from the central value (p_o), one can see that the horizontal and vertical foci are distributed on two different surfaces, cutting f_o practically on the axis of the spectrometer (fig. 2) (Focal Surfaces F_H and F_V). The angles of these surfaces with the axis, and their curvatures, are related to higher order magnetic effects. The angular magnification between the target and the point f_o is $G_H \approx 1$.

Kinematical displacement of the horizontal focal surface (fig. 3). The "x" position of F_H depends to first order on the derivative of the momentum as a function of the scattering (or reaction) angle θ, dp/dθ. Let ($\theta_o \pm \Delta\theta$) be the horizontal aperture of the spectrometer (S) around the mean scattering angle θ. If dp/dθ = 0, all foci are on the "reference focal surface" F_H. If dp/dθ < 0, and $p(\theta_o) = p_o$, then to first order:

$$p(\theta_o + \Delta\theta) \sim p_o - \delta p$$

$$p(\theta_o - \Delta\theta) \sim p_o + \delta_p \quad .$$

The particle emitted at $\theta_o - \Delta\theta$, having a *larger* momentum, will be less deviated than that emitted at θ_o; at ($\theta_o + \Delta\theta$), the momentum being less than p_o, the particle will be *more* curved by S. To first order, the 3 trajectories (and all others) will cut at f'_o instead of f_o.
One sees that the horizontal focal F'_H will go to larger positive X values since (dp/dθ) is more negative as θ_o increases.
For dp/dθ > 0, the focal surface will be nearer to the spectrometer. If (dp/dθ) is very large, the focal F'_H can go to infinity and become virtual. However, the vertical focal F_V remains fixed. The horizontal angular magnification G = Δθ'/Δθ, between the exit and entrance angle diminishes as the focal surface F'_H goes away. G may become much smaller than 1 increasing the error on determining Δθ from a measurement of (Δθ').
In our case, F'_H goes to infinity for $\left|\frac{1}{p_o}\frac{dp}{d\theta}\right| \cong 6.6\ 10^{-4}$/mrd.

First order compensation of the primary beam energy dispersion. Fig. 4 indicates the method used to compensate for the energy dispersion of the beam (at 0°). It necessitates an "analyzer magnet" "A" for the incident beam, which disperses the beam at the target T in such a way that the dispersion at this point is equal to the inverse dispersion of (S) from the focus f_o. In other words, between

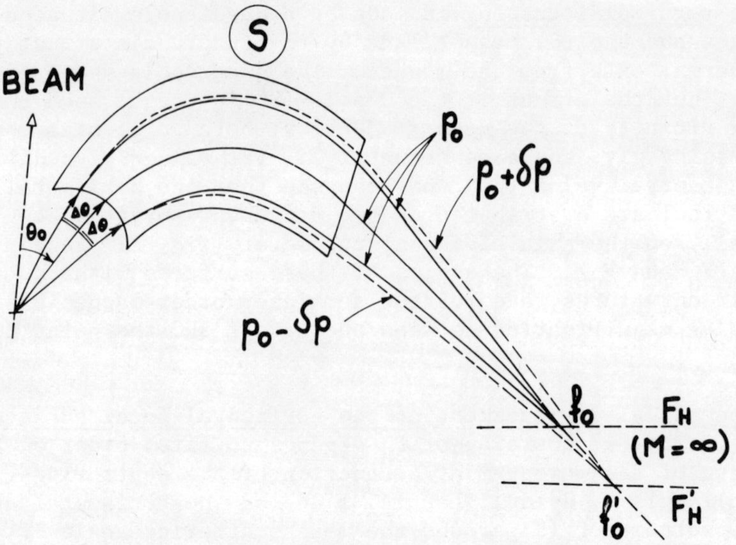

Fig. 3- Kinematical displacement of the horizontal focal surface.

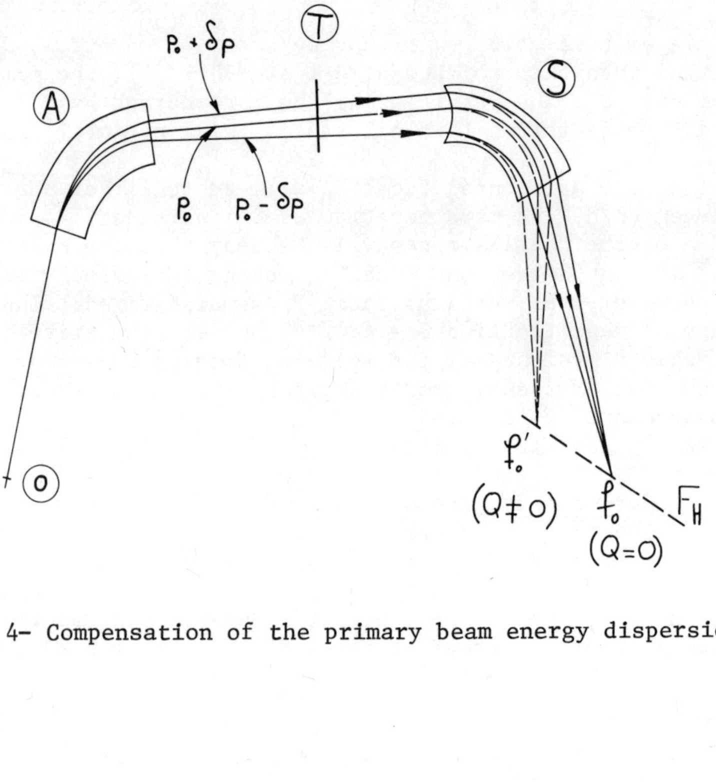

Fig. 4- Compensation of the primary beam energy dispersion.

the object (O) and the final image (f_o), the whole system (A) + (S) must be achromatic. For some energy-loss Q during the traversal of the target, the corresponding particles focus at $f'_o \neq f_o$ and the distance $\overline{f_o f'_o}$ measures Q. That is the reason to call this type of spectrometer a "Q-loss" spectrometer. Q can be due to the energy-loss in a thick target, or to the excitation of the nuclei. For a relative angle θ_o between the axis of (A) and (S) different from zero, and for a given reaction a + A → b + B, a and b being different particles, the analyzed momenta in (A) and (S) are often quite different (for instance, for 400 MeV protons in the reaction $p + {}^6Li \rightarrow \pi^+ + {}^7Li$, the ratio $p_{\pi^+}/p_p \approx 1/3.5$).
In that case, to get a focus f_o independent of the incident momentum, we have to match the dispersion of (A) taking into account the kinematics. For a given θ_0, we have to first order:

$$\left(\frac{dp}{p}\right)_{scatt.} = k\left(\frac{dp}{p}\right)_{inc.}$$

where k is a number depending on the reaction and on θ_o.
Let us call $D^{(A)}$ the dispersion of (A); $D^{(S)}$, that of (S):

$$D^{(A)} = \frac{(dy) \text{ at the target}}{(dp/p) \text{ incident}} \qquad D^{(S)} = \frac{(dy) \text{ at the target}}{(dp/p) \text{ scattered}}$$

Thus we have to realize the condition: $D^{(A)} = k\, D^{(S)}$.

This kinematical matching of the dispersion of (A) to ($kD^{(S)}$) is made using a set of quadrupoles between (A) and the target, which are able to change the linear magnification at T (image of (O)) without changing the focussing.

Compensation of the beam emittance. Of the cross-sectional area of the beam at the target (T), only the *horizontal dimension* is important. The spectrometer has a horizontal magnification $G \approx 1$. To get the minimal spot at the focal surface F_H, this dimension must be smaller than the dimension at F_H corresponding to the desired resolution ($6\ 10^{-5}$ at $F_H \leftrightarrow 1$ mm transverse). It is difficult for high energy beams to get such dimension by using slits (.5 mm). We solved the problem as indicated in fig. 5. Fourteen meters before the entrance of (A) an adjustable slit (O) having a maximum aperture of 8 mm (lead, 20 cm long) is placed. The beam from the machine is almost parallel in (O). This real object (O) is transformed by two successive quadrupoles (Q_1Q_2), both horizontally divergent, into a virtual object (O'). If $G(Q_1Q_2)$ is the angular magnification of (Q_1Q_2), (O') is reduced compared to (O) by a ratio $\sim 1/G(Q_1Q_2)$. The magnitude of G is of the order of 16, (O') has a dimension $\sim$ .5 mm. This object (O') will be reproduced at the target (T) with the same dimension .5 mm (for each value of the beam energy).

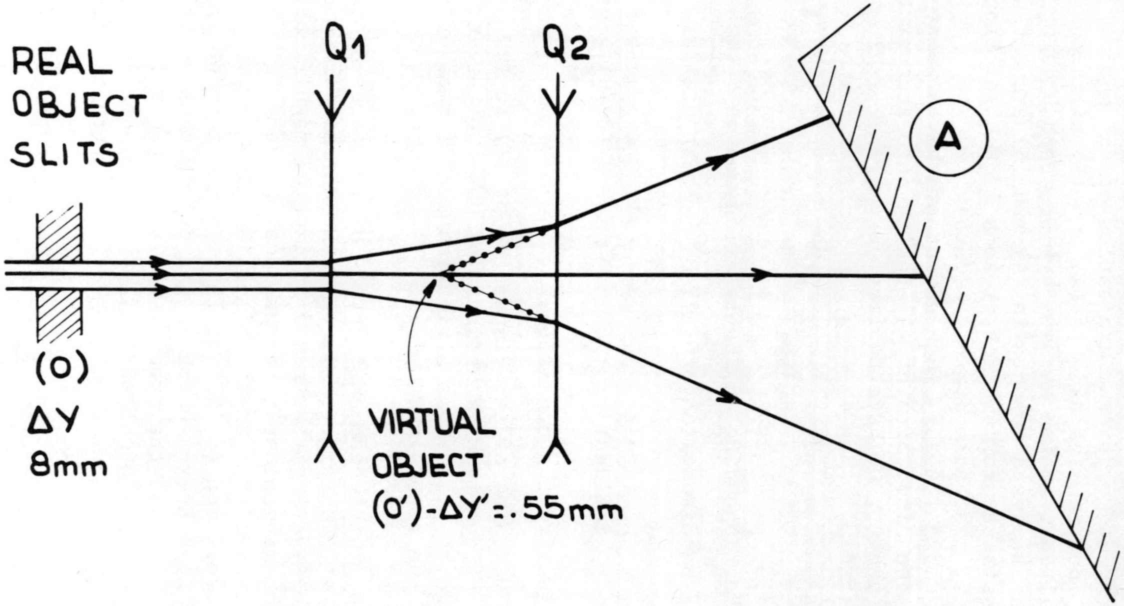

Fig. 5- How to build a small virtual object from a large real one.

The angular divergence of the beam is also important (± 10 mrd max.). If one would focus the beam exactly *on* the target T, a given angle θ_o seen by the spectrometer (θ_o compares to the *mean axis* of the incident beam) would correspond to a *physical* scattering angle:

$$\theta_o - (\delta\theta)_{inc} < \theta < \theta_o + (\delta\theta)_{inc}$$

and due to the kinematics a confusion would result at f_o (or f'_o) related to $\pm \Delta p_{scatt} = \pm (dp/d\theta)\delta\theta$.
It is easy to show how this defect can be compensated to first order in $(\delta\theta)$, by focussing the beam *after* the target T. I indicated in Appendix I a simple geometrical construction explaining how both compensations (Δp_{inc}, $\delta\theta_{inc}$) are obtained for a given kinematics. This focussing will depend on the nature of the reaction and on the scattering angle.
On the other hand, the *vertical* emittance of the incident beam is of little importance (in *our* case). The vertical image will be of the order of 1 cm or smaller, the incident angles $\lesssim \pm 3$ mrd (the *linear* magnification of the spectrometer is, vertically, $G_V \approx 2$).
As a conclusion of this first order study, it is necessary to have between (A) and the target (T) at least *three* quadrupoles in order to:
- vary the horizontal focus position of the beam,
- vary the horizontal magnification (and therefore the *dispersion* of the beam),
- conserve the vertical focussing at the target.

In fact, contrary to fig. 4, it was impossible to get physically (A) and (S) with the same direction of rotation. Fig. 6 indicates the real scheme (A) and (S) rotating in opposite direction; their dispersions are opposite too, and to match the dispersions, that of (A) has to be *reversed*: therefore we make an intermediate image (I) between (A) and the quadrupole triplet ($Q_3Q_4Q_5$) and use this triplet to *reverse* the image (hence the dispersion) at the target. This slight inconvenience is balanced by the possibility of using a cut in (I) (uranium slits) to limit the incident (Δp) (we never used it).

IIc. Second Order and All That...

By construction, such magnets like (A) and (S) having large horizontal and vertical apertures, produce complex aberrations which create at the focussing points (in fact, waists) confusions which need to be corrected, as well for the incident beam (confusion at the target, equal confusion on the focal: $G_S \sim 1$), as for the scattered particles. Those aberrations are parametrized by developing the confusion at the final image into polynomials of the initial variables (at the object): $(y,\theta)(z,\phi)$ and (δp). More generally they can be studied on correlated diagrams (z,ϕ) (y,θ)...

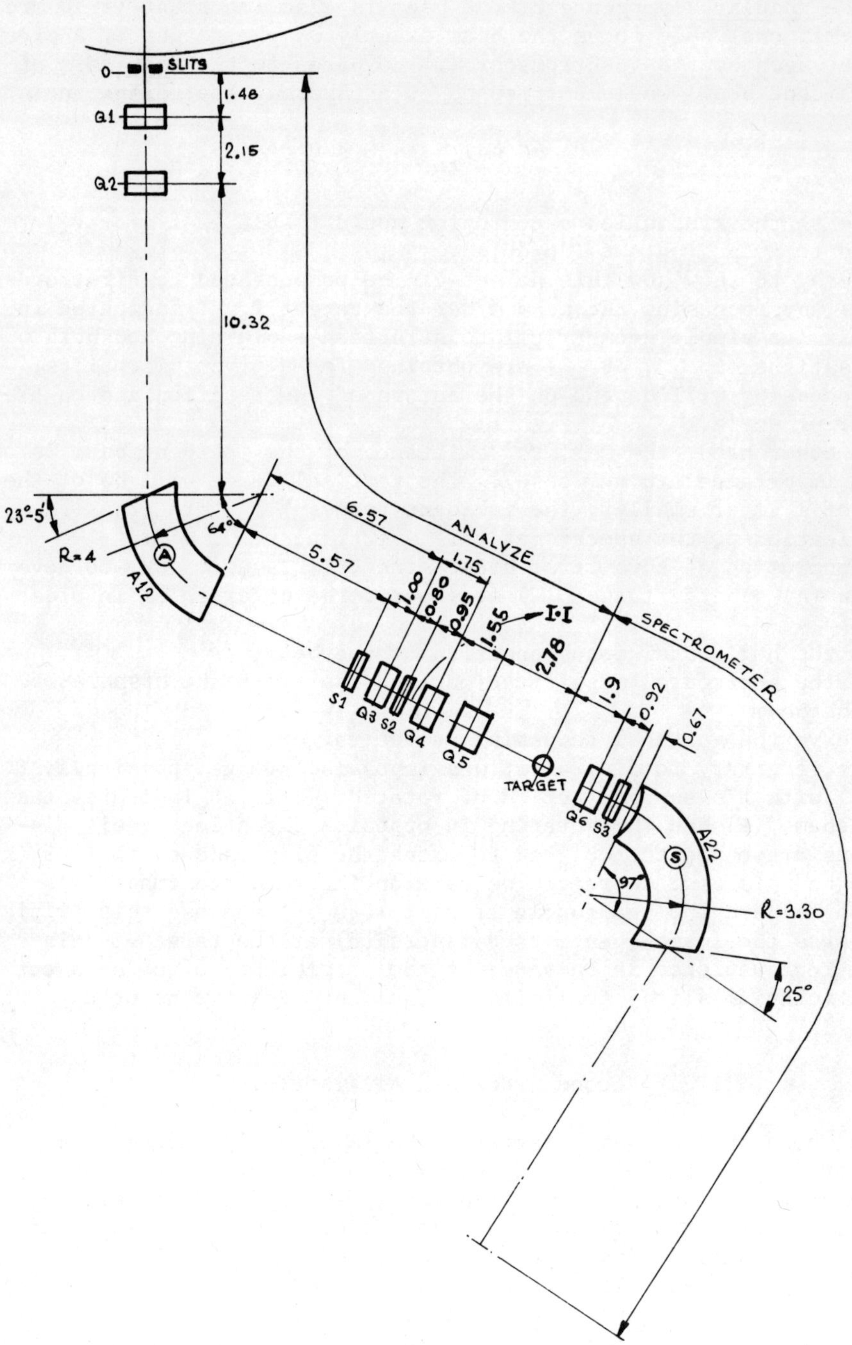

Fig. 6- Layout of the whole Spectrometric Line.

Other effects tend to destroy the resolution, which are not *intrinsic* aberrations of the magnets:
- the non-linear kinematical effects,
- the imperfection of the first order focussing and compensations,
- other physical reasons: finite thickness of targets, intrinsic resolution of the detectors, instabilities of magnetic fields...

Aberrations. How can we find out and then *eliminate* the aberrations? At least, we wish to *reduce* the most important to a level << 1 mm. One proceeds in two steps:
- A *theoretical model* for both magnets (A)(S) allows us to compute a large set of trajectories, hence to determine most influential aberrations, to choose the angles and radii of curvature of the entrance and exit faces, get correct location of the focal planes and their angle with the exit axis, then examine the aberrations to predict where second order corrections will be the most efficient (sextupoles and inner shimming). With a good model for this theoretical work, one is able to reduce greatly the aberrations which cannot be compensated by sextupoles (varying parameters: angles, radii, relative positions...).
- When the magnets are built, one will measure with a great precision the magnetic fields (A, S *and* q-poles). From these maps of the magnetic fields, one makes a very accurate calculation of a set of trajectories, deducing the effective second order intrinsic aberrations. They will be corrected by sextupoles, adjustable shimming inside the magnets (coils), plus a fixed mechanical shimming.
When these new elements are determined a computation of incident and scattered trajectories will allow the optimun values for the currents for each angle for each reaction. For the calculation of a set of orbits from the field mapping, we use a special computer program named "ZGOUBI", much faster (and more accurate) than "ORBIT". Without such a routine, it would have been almost impossible to build and test SPESII: the cost in time and money would have exceeded our possibilities!

Kinematical second order effects will be corrected by a special sextupole placed before the entrance of (S), and by the inner variable shims (electrical shimming). Those shims have to vary *to correct second order effect associated with saturation effects* in (A) and (S).
Residual first order mismatches (also 2nd order!) can be corrected by small variations of the magnet currents at the beginning of an experiment.

III. THE FIRST SPECTROMETER: SPESI

The analyzer (A) and spectrometer (S) magnets were constructed with a great mechanical and magnetic precision. They are of the "window-frame" type, flat pole pieces, turned-up coils at the ends. Vertical focussing for (S) is produced essentially by (Q6) (see fig. 6) and by the exit face angle. Second order effects inside A and S are obtained by punched out iron sheets which are attached at regular intervals on the pole pieces, as shown in fig. 7. Around these sheets there are variable "electric-trimming coils" (which correct 2nd order aberration arising from saturation). A final selection was made after computing their influence on the trajectories. The same figure shows the forms of the entrance and exit edges of the pole pieces.

Magnet "S".
Entrance 0°; Radius 1.4 meter
Exit 25°; Radius 1.4 m (on all figures, this radius has been omitted!)
Useful transverse dimension 44 cm
Pole gap 20 cm
Magnetic field: maximum 19 kGauss
Mean radius of the trajectory 3.3 meters
Total weight 90 tons.

Magnet "A".
Entrance 23°5
Exit 0°; no curvature
Pole gap 14 cm
Useful transverse dimension 44 cm
Mean radius for the particles 4.4 meters
Weight 60 tons.
"Nose pieces" at the entrance-exit of these magnets define a cut-off of the field at definite radii and prevent magnetic field reversal.

Magnetic field measurements were carried out for both magnets with specially designed measurement gear. Hall probes were moved automatically along the magnet at constant radii of curvature. The angular accuracy of the probe positions were exceptionally good. The Hall probes were renormalized continuously with respect to a magnetic resonance probe. Six thousand points in the horizontal plane were measured with an accuracy better than 10^{-4}. The lattice of measurements was used to compute particle trajectories by a computer program. A local expansion of the magnetic field components to 2nd order from 9 points sub-sets was first made. Each trajectory was computed in a moving reference frame along the radius of curvature S, with an expansion to 5th order in the curve element (ds).

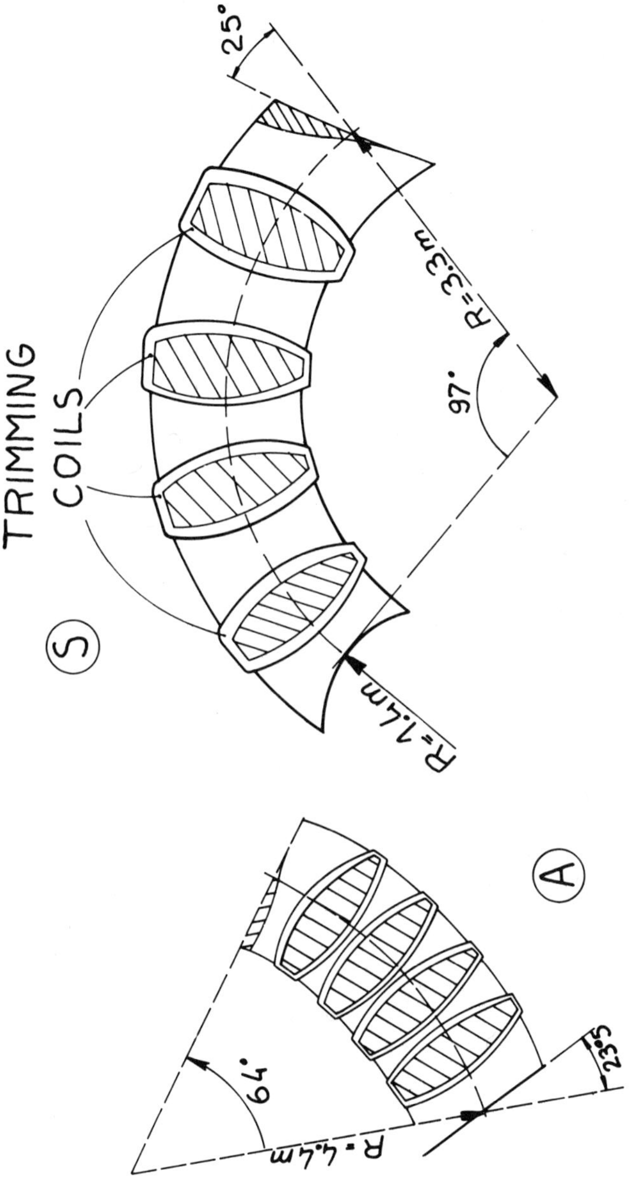

Fig. 7- Internal shims.

Power supply. The stability of the magnetic fields of (A) and (S) are controlled continuously by magnetic resonance probes. The main power supply is common to both (A) and (S). One wanted to benefit by the fact that for many experiments, the final analyzed momentum in (S) is only slightly different from the incident one in (A). Each magnet has three coil-sub-sets $(A_a A_b A_c)(S_a S_b S_c)$. When the main power supply (500 volts, 7000 ampères) is turned on, $(A_a A_b)$ and $(S_a S_b S_c)$ are excited, both magnetic rigidities $(\int_{A,S} B d\ell)$ are the same. For equal momenta on (A) and (S) the focus f_o will *not depend on the stability of this main power supply* (compensation principle!). If $p_f \neq p_i$, A_c is supplied by an independent power supply (which needs to be more stable than the main one!), which can produce a magnetic field in (A) direct or opposite to the main one, for the condition $p_f \gtrless p_i$. For larger differences, the S_a coil can be short circuited.

Other important magnets on the spectrometric line are (fig. 6):
- Two diverging quadrupoles $(Q_1 Q_2)$ before (A), creating the virtual object (O).
- Between (A) and (T), three Q-poles $(Q_3 Q_4 Q_5)$ as was said in IIb.
- Two sextupoles $(S_1 S_2)$ reduce the aberrations due to (A) and the Q-poles (in conjunction with the electric shimming of A).

Those three sextupole corrections are also used to vary the slope of the focal surface of (A) on its exit axis (to match the dispersion of (A) for the momentum compensation).

Between the target and (S) one finds:
- a Q-pole (Q_6) vertically focussing (resp. horizontally defoc.). The field of (Q_6) follows that of (S) to maintain the focussing properties. If $I(Q_6)/I(S)$ varies, one can change slightly the location of the (S) focal plane;
- a sextupole (S_3) correcting 2nd order kinematics and part of the intrinsic aberrations.

The real utilization of the 2nd order coils and magnets is the following:
- Electrical variable shimming inside (A): *horizontal* aberrations)
- (S_1): essentially *vertical* aberrations
- (S_2): mainly the slope of the incident beam focal plane
- (S_3): vertical aberrations of (S) and partly 2nd order kinematical effects
- Electrical shimming of (S): horizontal intrinsic aberrations
- Exit radius of curvature of (S): slope of the horizontal focal plane of (S) - ($\approx$ 45°).

I give in Appendix II the set of magnetic properties of the analyzer-line (between O and T) and of the spectrometric part (between T and F) in the ordinary *matrix form*.

The beam before (O) is prepared to enter the spectrometric line by the following elements:

- The circulating beam inside the machine "Saturne" is extracted by a 2/3 resonance (efficiency $\lesssim$ 30%). The beam is adjusted in position and direction by two of the three extraction magnets in order to be aligned on the axis of the line.
- Two pairs of Q-poles produce *a waist in "O"*, with the following characteristics:
  - horizontally "almost parallel" beam $\Delta y < \pm 4$ mm, $\Delta\alpha_y < \pm 0.75$ mrd
  - vertically $\Delta z \pm 5$ mm, $\Delta\alpha_z \pm 5$ mrd
- Four correction-dipoles are used to delicately align the incident beam on (O).
- In addition to the adjustable slits in (O), *two other upstream slits* can be used to diminish the intensity at the target.

Furthermore, at the entrance of (A), two supplementary slits are used to decrease still more the intensity, or to reduce the incident emittance. When the spectrometer is at small angles, the beam must be stopped in a uranium block.

Position sensitive detectors are insertable along the line before (O) to determine the alignment.

With the reduction due to (O) (and some loss between the machine and (O)), we never got more than 10^{11} protons per burst (more generally 1 to 5 10^{10}). At 1 GeV, the cycle is every 1.8 seconds. Ideally the beam burst is 300 ms long, but time structures of 100, 600, 840 Hz... reduce time effectively!

Most of the beam removal before and at (O) does not enter the experimental room which is separated from the machine area by a 3 m heavy concrete wall between (Q_2) and (A).

The energy stability of the 1 GeV beam can be quite good. (I saw total variations less than 100 keV for one night.) The total energy spread of the incident beam is 300 $\rightarrow$ 500 keV.

IV. DETECTORS AND TECHNICAL DETAILS

The essential measurement for magnetic spectrometry consists in localizing the focal surface (which, due to its small vertical dimension, is almost a *plane*), and to get the crossing point of each trajectory with this focal plane. The position of this point gives the missing mass (or equivalently the excitation energy of the residual nucleus).

IVa. Position Determination

Position is measured by *drift detectors*, the principle of which is shown in fig. 8. A detection chamber is a gas volume where the passage of a particle creates electrons. Under the influence of a transverse electric field, these electrons migrate to the low negative voltage end, where a semi-cylindrical wire proportional counter amplifies the signal. The drift time is proportional to the drift length; this time is measured by a 200 MHz clock, the start of which is given by the trigger signal (passage of the particle through quick detectors), the stop being given by the arrival of the electrons on the wire.

Each of the four actual localization chambers ($H_1H_2H_3H_4$) consists of a *double* counter. The gas is pure methane at atmospheric pressure, drift speed $\sim 10^7$ cm/s (1 mm accuracy = 10 ns; a 50 cm long counter > 5 μs total). The electric field is $\sim$ 1 kV/cm. Two vertical counters of the same type (V_1V_2) are used to verify the vertical dimensions and central position of the scattered particles.

The localization accuracy for each counter varies from .7 mm in the vicinity of the wire to 1.2 mm at 40 cm... (but using four counters reduces the dispersion). I remind you of the fact that 1 mm at 1 GeV $\sim$ 115 keV - 50 cm $\sim$ 55 MeV.

Periodically the drift speed is controlled by a sparker inside the detectors.

IVb. The Trigger

The trigger is obtained by the coincidence of signals produced by the particle to be detected in classical wide scintillators, with eventually 2 or 3 more Cerenkov detectors, depending on the reaction. The coupling of 4 to 6 such detectors reduces the background of the "starts" events and can select the type of particles. Two more small aligned scintillators (jokers: J_1J_2) are positioned in such a way as to cover part of the most intense "peak" (generally elastic!), and are used to control on line the efficiencies of the trigger and of the localization counters.

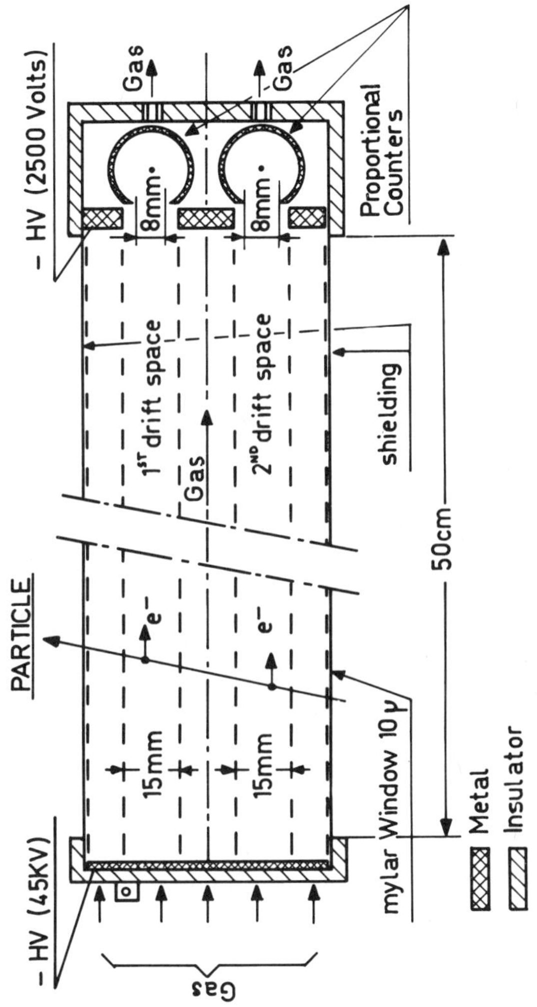

Fig. 8- Double drift counter.

IVc. Monitors

The incident *beam* is monitored by a secondary emission counter. This detector is periodically checked (normalized) by using the reaction:

$$^{12}C(p,pn)^{11}C$$

on a carbon plate in front of it: the measurement of the induced activity allows one to normalize the counter to 5%.

A differential ionization chamber is used to monitor the *alignment* of the beam, and also to measure the total incident current (with right alignment of the beam, the ratio of both countings, secondary emission and ionization, remains constant within a few %).

Two scintillator-telescopes (at 140° and 40°) allow us to monitor the product (BEAM x TARGET).

IVd. Other Equipment

All particles travel in *secondary vacuum* between Saturne and the final window of the spectrometer; there is no multiple scattering, except in this window, the target, and the detectors.

The target chamber can be isolated from the scattering chamber. The target holder can hold 4 or 5 solid targets. Large targets (large compared to the size of the beam) are used for absolute measurements, 5 mm wide strips for relative measurements. A special set-up is necessary for liquid targets (^{3}He, ^{4}He). The reason for using strips is related to the fact that, in case of a variation of the beam position on the target due to a "beam skidding" in energy, even if the resolution is not affected (compensation), the *angular* resolution could be affected: an important change in the beam transverse position causes a change in the mean value of the scattering angle (1 cm displacement $\leftrightarrow$ 0.5° error).

Two television cameras monitor two quartz plates located before and after the target. Before each run, the alignment is checked visually. One checks also the alignment along the incident axis ("theoretical 0° axis") by defocussing the beam on the quartz plates by changing the quadrupole currents ($Q_3Q_4Q_5$). The center of image must remain stable.

Special concrete shielding is installed between (A) and the target. A moveable concrete block (on air cushions) follows the spectrometer motion and partly protects the detectors against the radiation from the uranium block stopping the beam between the target (T) and (S).

V. SPESI: ELECTRONICS, COMPUTER, OPERATIONAL MODE

I would like to give an idea of the operational mode used in a typical 2-body experiment. On the road I shall give, without details, a rough description of the electronics.

We begin generally the experiment by setting the spectrometer on 0°. Using the beam slits and defocussing the beam between Saturne and the object slit (0), one can get a beam of $\sim 10^3$ particles per pulse, and use it to test the counters. One checks:
- the efficiency of the trigger (timings, thresholds...);
- the efficiency of the drift chambers.

For each double counter $(H_{1A}H_{1B})(H_{2A}H_{2B})$ ... $(H_{4A}H_{4B})$, one puts in coincidence the (A) and (B) parts (see fig. 8). Because *real* particles coming from the spectrometer are all inside a horizontal angle $\sim$ 60 mrs, and because (A) and (B) are a very small distance apart, a time gate of $\sim$ 60 ns between both counters rejects an important part of the uncorrelated background: the two electron buckets must arrive within the gate for a real particle.
The four counters $H_1H_2H_3H_4$ are put in coincidence with larger time gates. Residual background will be small, even for a high single counting rate. Finally a gate opened by the trigger (start) requires a signal from the horizontal and vertical coincidences to arrive during the drift times.
Fig. 9 shows a simplified version of the electronics.

The exact drift speeds are determined for all chambers by measuring the drift time between the high voltage plate situated at one end and the wire of the proportional counter. This is done by sending in succession scattered particles at some mean angle onto both ends of each counter: the cuts are clearly seen on each counter spectrum. The speeds for each counter are generally equal within 1%. This normalization is taken into account by the computer. The ratios of the different monitors are checked to be constant for different beam intensities. The computer receives, on line for each event, 4 x 24 bit words comprising $(H_1H_2H_3H_4V_1V_2)$ informations, and reads all important scalers at the end of each burst. Between two successive bursts, an on-line program determines the trajectories.

With the preliminaries out of the way, the experiment begins. For a given θ_o, calculated values are displayed. The values of the magnet setting are set up for this scattering angle. The beam is aligned on the target scintillator.

Two programs are used on-line, utilizing a 4094 channel-block memory. Either the four horizontal counters spectra can be displayed or a set of the reduced data displaying the distribution of the events on the focal surface versus longitudinal position, or the mean angle with the spectrometer axis (corresponding, of course, to the radius of curvature).

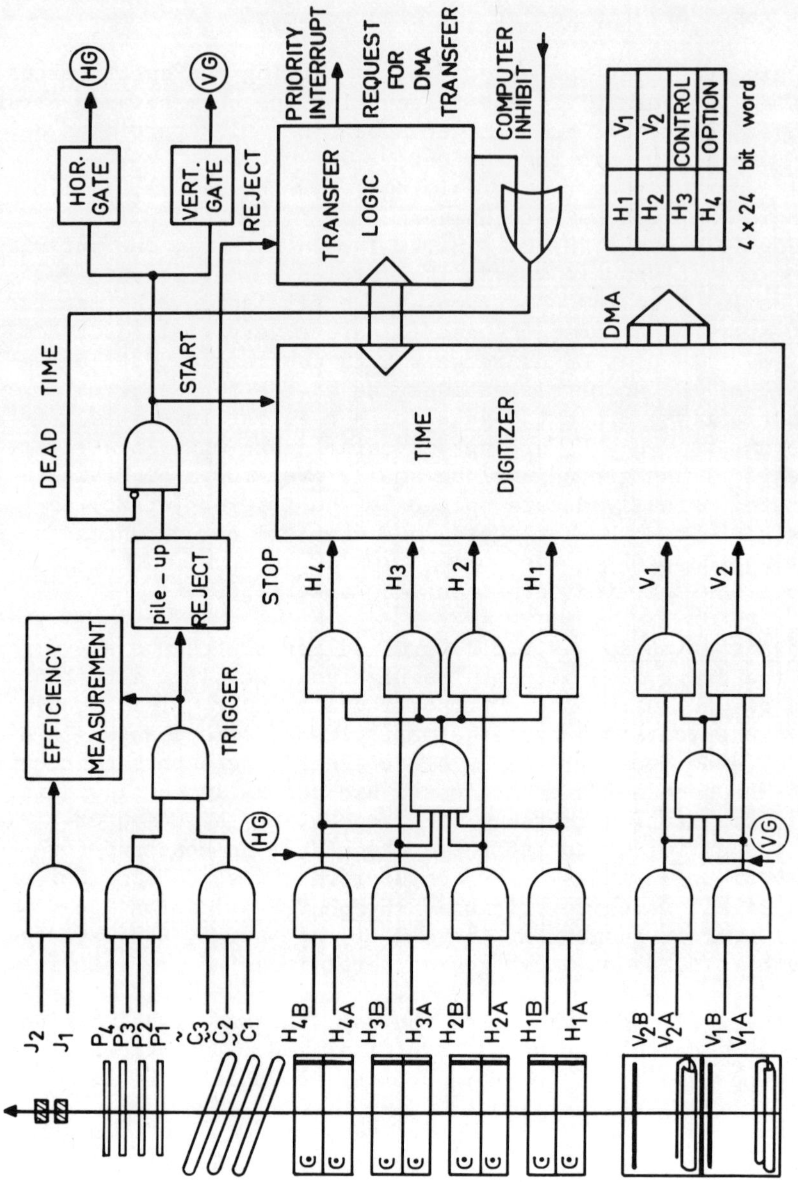

Fig. 9- Electronics of SPESI (simplified).

We can also display: the z (vertical position) and ϕ (vertical angle) spectra; the alignment spectra; a comparison of the transverse position in (H_2H_3) as *measured* directly, compared to the line determined by two events (H_1H_4). The accuracy of this alignment is always $\lesssim$ 1 mm.

Further displays are: *four partial angular distributions* within the horizontal aperture of the spectrometer ($\gtrsim$ 50 mrd). This is done by looking at the measured angles ($\Delta\theta'$) for events arriving on the focal inside some peak representing a final state.

These partial angular distributions are very useful. One can get an accurate relative distribution and verify during the experiment the consistency of measurements at successive angles (overlap method).

An automatic search of the *focal planes position* can be made. For a given peak on the focal plane (fixed by a window), the computer computes the coordinates of the smallest confusion. Doing this for different final states determines the angle of the focal plane. The focal position calculated this way generally differs from the precomputed one by less than 10 cm.

Some correlations can be displayed on line. For example, the correlation between the horizontal angle ($\Delta\theta'$) of the trajectories and the transverse coordinate of their intersection on the focal plane gives a measure of the (θ^2, θ^3...) aberrations. A correction can be made by the sextupole if necessary.

Observation of some scales quantities determine the intensity, the centering of the beam, the dead time, and the efficiencies.

Dead Time and Efficiencies. If the counting rate is high enough ($d\sigma/d\Omega \gtrsim 1$ μb/str) *efficiencies* can be checked on line by comparing the joker counter rates (random subtracted) to corresponding rates of the trigger and drift chamber detectors.

The *dead time* depends strongly on the time structure in the beam. To measure it, we use a standard pulser triggered by a monitor detector to obtain a time distribution which follows the beam structure. The pulse generator signal triggers a fan-out. The outputs are fed to *all* detector preamplifiers, simulating a real event. Comparing the number of sent pulses to the number of such events acquired by the computer gives the total rejection which is essentially due to:

- the dead time of the drift counters ($\sim$ 6 μs);
- the acquisition time of the computer ($\sim$ 10 μs).

The counting rate is tuned such that the dead time be smaller than 20%. Correction is easy.

I remind you of the possibility (we never used it) to compensate on line to some extent residual intrinsic aberrations of the spectrometer ($Q_6 + S_3 + S$). Using a two body reaction which will give good resolution (elastic scattering from a heavy nucleus in thin target), one can construct an "error matrix" experimentally, and look for correlations between (Δy) (transverse confusion at the focal plane) and (ϕz), ($\delta p \phi$)... Such errors with their own signs could be tabulated and subtracted on line.

Finally, the effective zero degree direction angle of the spectrometer is verified in each experiment by the following method (length motions of the ground, due to the weights of the concrete walls, induces smooth variations in time of the axis of the beam).

One measures scattering at small angle on both sides of 0° (generally ± 4°) on a medium A target (e.g. Nickel) and compares, as a function of ($\Delta\theta$) inside the solid angle of the spectrometer, the *ratios* of the elastic cross section to some inelastic one (2^+). Comparing the two angular distributions of this ratio gives the effective zero direction. The accuracy is $\lesssim$ .1° ($\sim$ 1.5 mrd).

Concerning the *solid angle* of the spectrometer, measurements are made with different apertures of the entrance slits. The linearity is better than 1%. Closing those slits ($\sim$ 1 mm) gives an idea of the angular resolution (except for beam and target effects (see fig. 14).

VI. SPESI. QUALITIES AND LIMITATIONS

VIa. Energy (Momentum) Resolution

The resolution, as we saw, depends on so may factors, that I am still amazed that we obtained 95 keV resolution by scattering 1040 MeV protons on ^{208}Pb (fig. 10). In such a case, the resolution was the sum of:
- all intrinsic residual aberrations of the incident beam line, of the spectrometer, and of all residual inaccuracies in the compensation (note that the solid angle of the spectrometer was reduced 1/2 vertically);
- and of the intrinsic resolution of the drift chamber plus some small multiple scattering effects.

The target was thin (50 mg/cm^2) and the kinematical effects were very small.

Other spectra are shown in figs. 11, 12, 13. Note the influence on the energy resolution of a light target kinematics (p/^{12}C), of a thick target and large kinematics (p/^{4}He), of a heavily ionizing projectile (1370 MeV α on ^{40}Ca). Concerning the focal displacement, one has to remember that our drift counters are *fixed* in position. Hence if the focus is displaced by a few meters, even with a mean accuracy $\lesssim$ 1 mm on the detectors, the accuracy of determining the trajectories at the focal plane diminishes quickly as the distance of the focal plane to the counters increases.
A thick target has two effects:
- The target is seen at an angle $\neq 0°$ by the spectrometer, the object has an *apparent width* [e sin θ_0] and the loss of resolution is $\approx$ $(e \sin \theta)_{mm}$ x 115 keV (at 1 GeV) (or $\approx 7\ 10^{-5}$ $(e \sin \theta)_{mm}$ in momentum).
- The multiple coulomb scattering is such that a *physical* angle θ is changed. This change $\delta\theta$ induces an inaccuracy $\delta p^2 = (\frac{dp}{d\theta}^2).d\theta^2$.

VIb. Angular Resolution

It depends not only on the multiple scattering in the target but also on:
- Intrinsic properties of the system: localization accuracy, residual aberrations, multiple scattering in the exit window and detectors.
- Reaction kinematics. The focal plane translation diminishes the angular magnification G_θ. So to obtain the incident angles on the spectrometer from the exit angles ($\Delta\theta'$) measured by the detectors, one has to use:

$$\Delta\theta = \frac{(\Delta\theta')}{G_\theta}$$

and the error $\delta(\Delta\theta) = \frac{\delta(\Delta\theta')}{G}$ increases as G decreases.

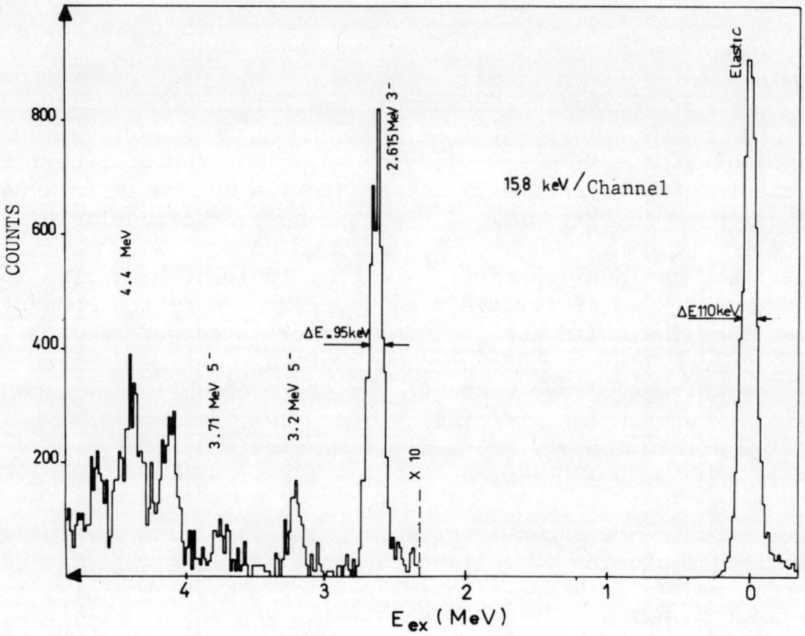

Fig. 10- 1 GeV protons scattered at 10° on ^{208}Pb

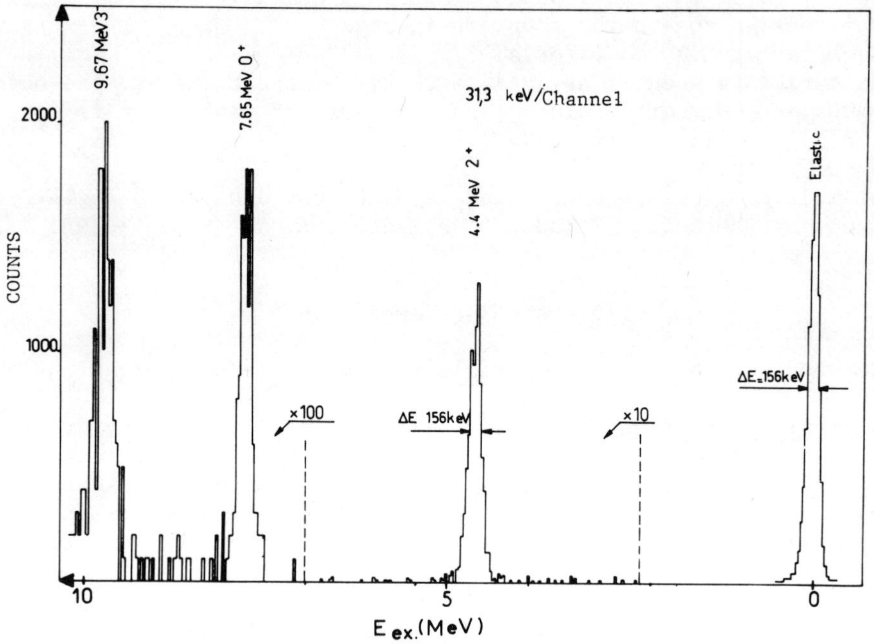

Fig. 11- 1 GeV protons scattered at 6° on ^{12}C

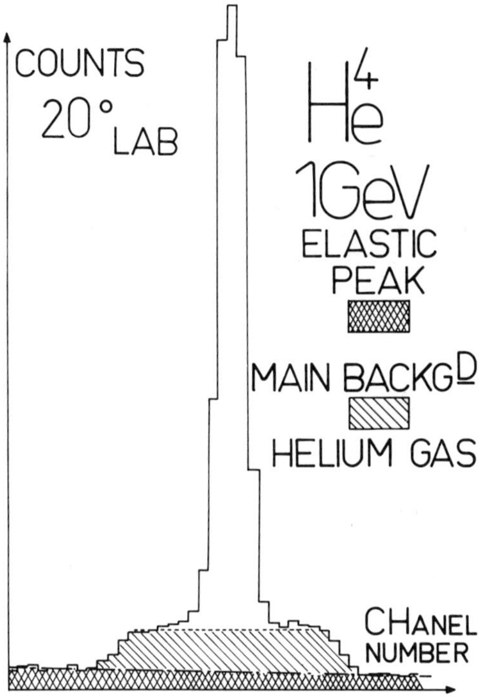

Fig. 12- ^{4}He spectrum

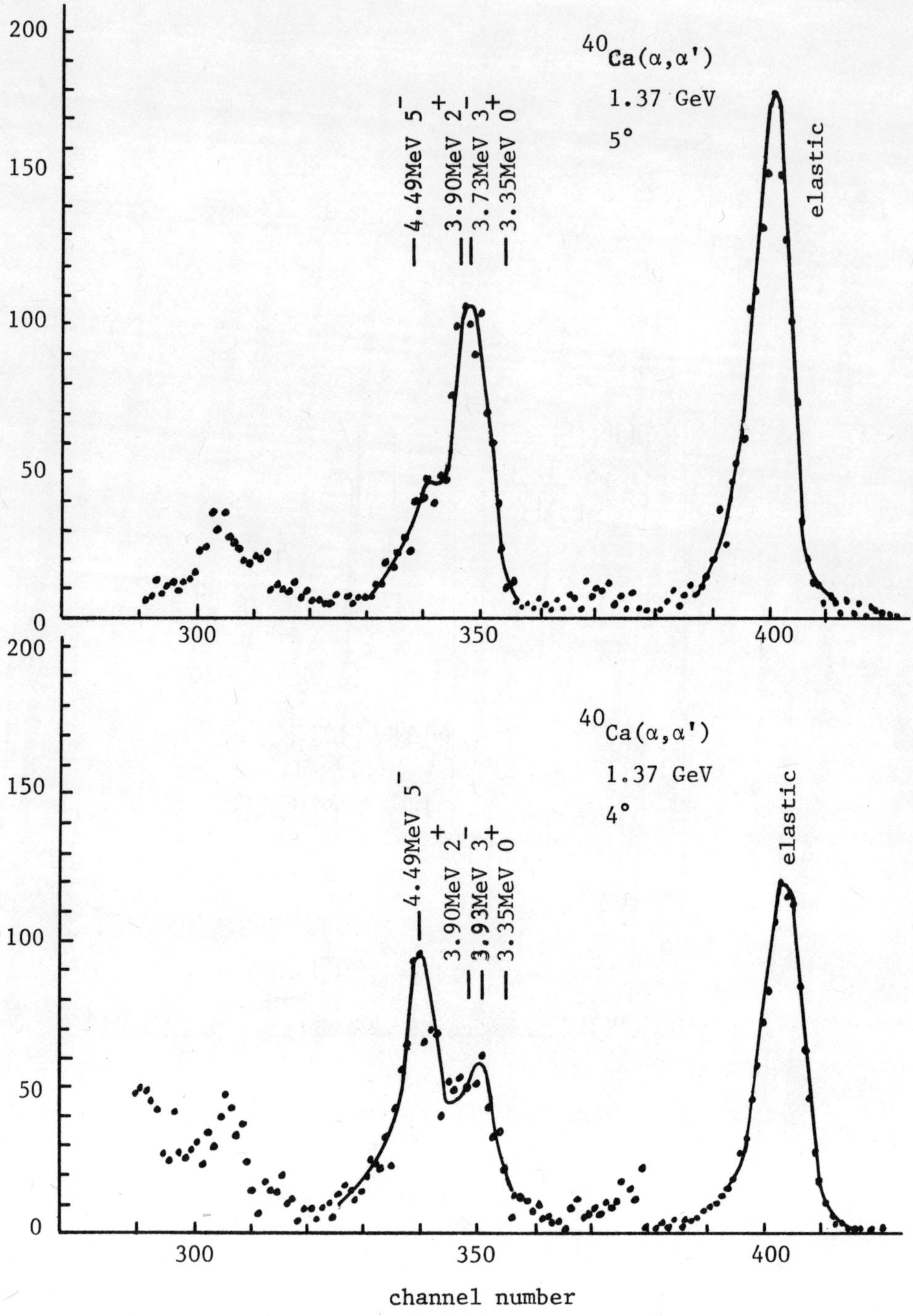

Fig. 13- Excitation energy spectrum of 1370 MeV α particles on ^{40}Ca.

- Width of the target. If the target is large, an energy variation of the incident beam doesn't change the momentum resolution, but affects the scattering angle. Using target strips minimizes this effect.
- Emittance of the beam. Even with the angular compensation (compensation of Δp against the emittance), the full angular width of the incident beam creates an angular confusion. We reduce this effect by closing the entrance horizontal slits of the magnet (A).

Fig. 14 shows the angular resolution obtained with the spectrometer slit narrowed horizontally, for a 5 mm strip of ^{58}Ni. This doesn't take into account the emittance and target width effects.
The angular resolution is of great importance to examine in detail angular distributions oscillating strongly as a function of θ. For instance, with 1370 MeV α's scattered on ^{40}Ca, the diffraction pattern has a periodicity $\sim 2.5°$, i.e. equal to the horizontal aperture of the spectrometer. It would be probably impossible to get a precise pattern for 1370 MeV α's scattered on lead...

VIc. Background Problems

Three kinds of background create errors and difficulties.

The main background in the experimental room is due to stopping the beam on the uranium block (for $\theta_S < 23^\circ_{LAB}$). This background is largely eliminated by a sufficient number of coincidences in the trigger (and the eight-fold coincidences in the four double drift counters). In spite of this, it would be actually difficult to work in the spectrometer room with a beam intensity higher than 10^{11}/burst. The single counting rate on the 60 x 20 cm^2 counters would create fluctuations (especially with the actual time structure).
No problem of this type occurs if cross sections are greater than 1 μb/str: the beam intensity is simply reduced. To go down to 1 nb/str, we need six-fold coincidences and must use thicker targets. This diminishes the momentum resolution, hence the signal to background ratios are worsened. Being patient and substracting residual background could decrease our actual limit to $\sim$ .1 nb/str. To get still lower cross sections would necessitate faster electronics and detectors.

Some physical backgrounds due to other reactions in the target have to be eliminated. For instance:
- in a (p,π^+) reaction, a considerable number of protons of the same momentum as that of the pions are produced due to quasi-elastic scattering (and reactions);
- in elastic and inelastic scatterings at large momentum transfers ($\gtrsim 20^\circ_{LAB}$ for 1 GeV protons on ^{12}C), a copious number of *deuterons* originating from quasi-free reactions:

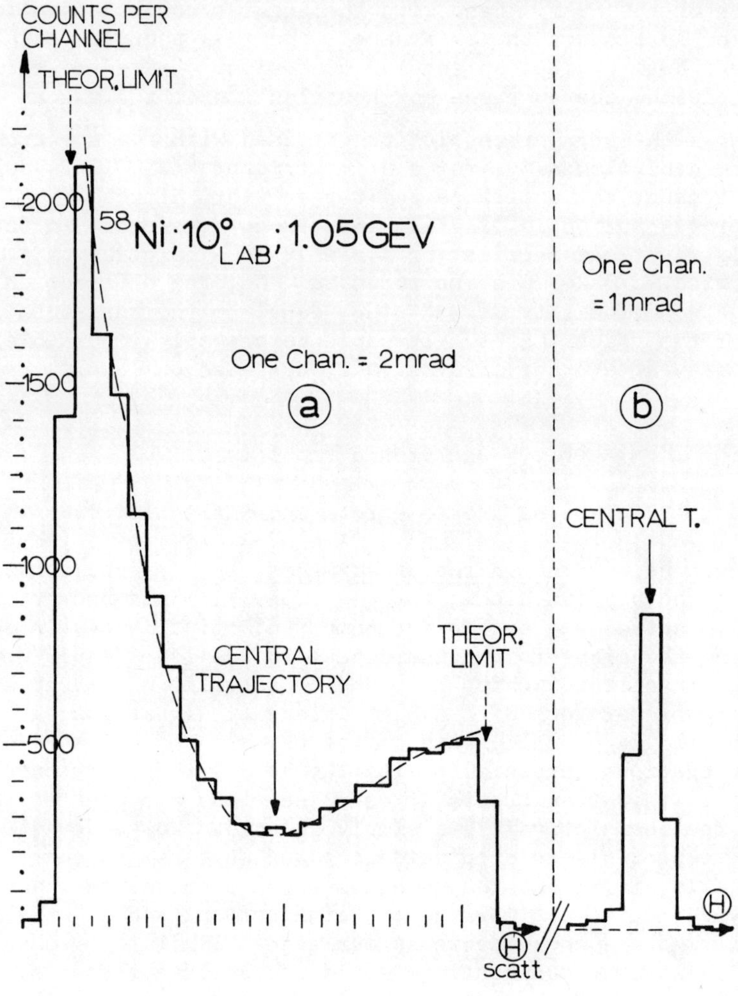

Fig. 14

Angular distribution inside the full spectrometer aperture (a). Angular resolution of (S) + detectors (b). (S) closed.

$$p + (p) \rightarrow D + \pi^+$$

$$p + \binom{p}{n} \rightarrow D + p \text{ (backward scattering on quasi-deuterons)}$$

create very broad bumps superimposed on the 2-body peaks. These reactions are generally less steep (smaller $d/d\theta(d\sigma/d\Omega)$) than the two body reactions or scattering on nuclei. They became more important (relative to the reaction to be observed) as θ increases. We need more efficient rejections against these particles at large angles.
For both the examples discussed in the last paragraph, we used Cerenkov counters to select our particles.

Some peculiar backgrounds produced in the target region can perturb the measurements.
- Small amounts of beam halo may strike the target holders and target chamber and then can perturb the monitors. This small effect is subtracted by counting also with empty target frame.
- The entrance slit of the spectrometer creates lower momenta particles (at 1 GeV, protons lose 110 MeV in passing through the slit). If we need to see very high excitation spectra ($\gtrsim$ 110 MeV), then a coincidence with a thin detector *before* the spectrometer is necessary (but diminishes the resolution).

VId. Problems Concerning Angular Distributions

We generally adjust the beam emittance and the width of the target in such a way as to obtain the desired angular resolutions. This adjustment has to be done because of the more or less rapid changes in the angular distributions as a function of θ. The accuracy of the cross sections decreases:
- for *very high* cross sections ($\gtrsim$ 1 b/str) at small angles,
- for *very small* cross sections ($\lesssim$ 100 nb/str).

We indicated above why the accuracy decreases for small cross sections. For very large cross sections, i.e. generally small angle elastic scattering, the incident beam has to be reduced so much that the *monitor statistics* are poor: constructing a small angle monitor is extremely difficult, a small misalignment of the beam creates large systematic errors.

We use for small angles the *"successive overlap"* method. Each measurement covers an angular aperture of 2.8°, shifting the spectrometer between measurements by 1° (or even .5° at small angles) allows us to extrapolate the cross section, normalizing on the last point having reasonable monitor statistics. Even with a fair monitor statistics we generally use this overlap method, reducing small angular errors. The maximum acquisition rate of the computer ($\sim$ 1000 events per burst) reduces the possibility of measuring inelastic cross sections in the forward direction: as soon as the ratio elastic/inelastic > 10^3, one cannot get enough statistics on inelastic scattering. This is an intrinsic difficulty of the *drift* counters.

Limits of the SPESI system. The physical limits of the system are:
- by *construction* 2.1 GeV/c (p/z) for the incident beam,
  2.0 GeV/c (p/z) for the spectrometer.

Minimum angle in the laboratory $\sim$ 3° (could be decreased for some specific, high cross section problems).
Maximum angle 110° (but now 70° because of "stupid" questions like the location of concrete).
- by the *state of detector possibilities*: cross sections between 100 b/str and 1 nb/str are measured.

Better shielding against background is desirable (uranium block stopping the beam in the room between 3 and 12° only in the near future).
To measure smaller cross sections, part of our laboratory is working on faster detectors, position sensitive chambers, and electronics.
A better beam is needed: that is the reason why we decided to rebuild the accelerator in 1977, not for the specific reason to obtain more beam, but essentially a better emittance, stability, and time structure. I will come back to this point. If we do not want to produce intense secondaries (π), we have the feeling that 10^{12} particles/s is large enough to attack interesting problems down to the level of a few picobarns using high resolution spectrometers of the proper design.

VII. THE PHYSICS: PAST, PRESENT AND FUTURE

The first experiments with SPESI began in February 1972. At the end of 1972, the spectrometer was operational, and data taking started alternating with successive improvements to lower the limit due to background; to improve the rapidity and dependability; and th obtain more reliable normalizations.

During the period between the end of 1972 and the beginning of 1974 we studied elastic and inelastic scattering on separated isotopes (^{6}Li, ^{12}C, ^{40}Ca, ^{58}Ni, ^{208}Pb, ^{16}O).

Elastic scattering has been interpreted within the framework of multiple scattering models (Glauber and KMT)[1,2] using proton density distribution experimentally determined by electron scattering. From these calculations, the conclusion was that any reasonable model predicted the same cross sections at low momentum transfers ($\lesssim 2\ fm^{-1}$) but important differences appeared at large momentum transfers (between 2 and 4.5 fm^{-1}), especially for light nuclei. It must be emphasized also that there are certain ambiguities due to the lack of knowledge of nucleon-nucleon amplitudes beyond 600 MeV. The different parts of the nucleon-nucleon amplitude, summed in a multiple scattering model, do not have the same importance. The spin-orbit part, of small influence for medium and heavy nuclei, should be obtained experimentally in the near future, by scattering of polarized protons. The structure of nuclei used for the calculations was described either by independent nucleons in a deformed or non-deformed potential[3], or by a cluster model[4]. Some studies included short range correlations (SRC), but it is difficult to characterize these without ambiguity, due to uncertainties in the nucleon-nucleon amplitudes. Other small but non-negligible effects due to off-shell effects and approximations in the calculations made SRC difficult to determine. Figs. 15 and 16 show elastic scattering for ^{208}Pb and ^{12}C. The curves are calculated by Dr. Ahmad[6] with the Glauber approach.

Inelastic scattering. Fig. 17 shows the inelastic cross section for the 3^- level at 2.62 MeV of ^{208}Pb. Fig. 18 shows the importance of coulomb scattering. Fig. 19 and 20 present the results for inelastic scattering to the 2^+ and 3^- levels of ^{12}C at 1.04 GeV. The solid curves include coupling and correlations. Fig. 21 shows the inelastic scattering on ^{58}Ni corresponding to the 2^+ level at 1.45 MeV. These calculations were made by Dr. Ahmad, with the Glauber approach, using the vibrational Tassie model.

The study of higher excitation spectra ($Q < 50$ MeV) has shown the existence of:
- particular highly excited states,
- "bumps" corresponding to quasi-elastic processes (see fig. 22).

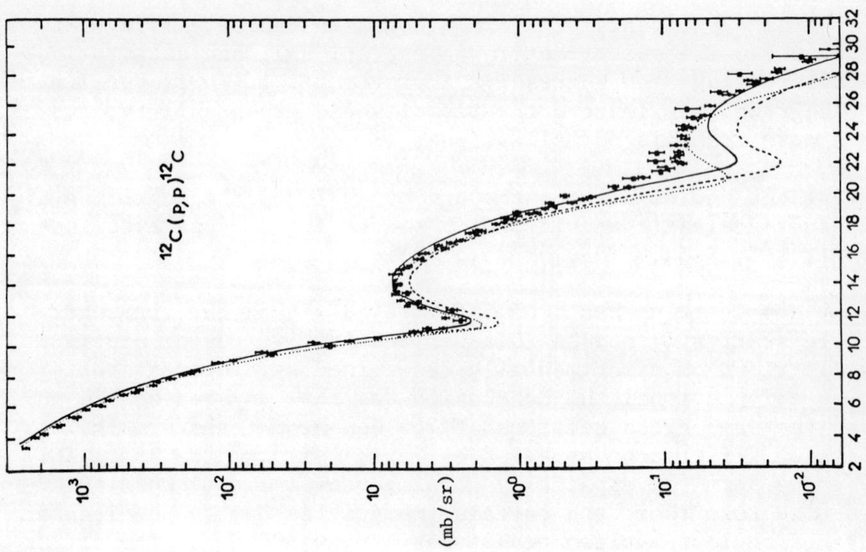

$^{12}C(p,p)^{12}C$
$\theta_{c.m.}$
$\frac{d\sigma}{d\Omega}$ (mb/sr)

Fig. 16

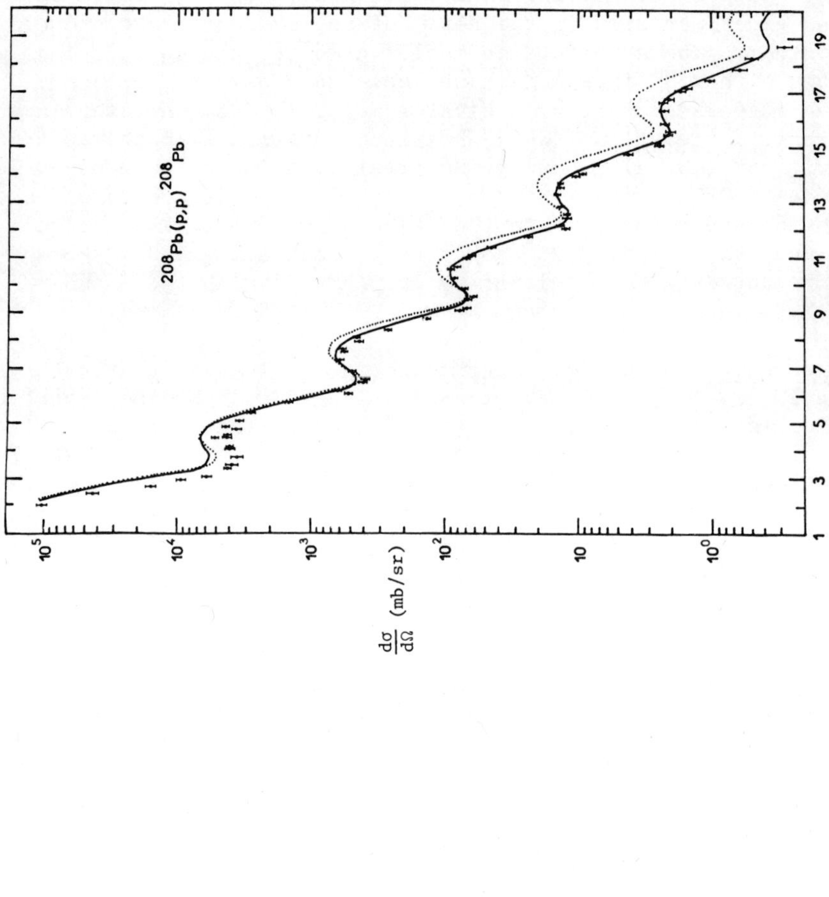

$^{208}Pb(p,p)^{208}Pb$
$\theta_{c.m.}$
$\frac{d\sigma}{d\Omega}$ (mb/sr)

Fig. 15

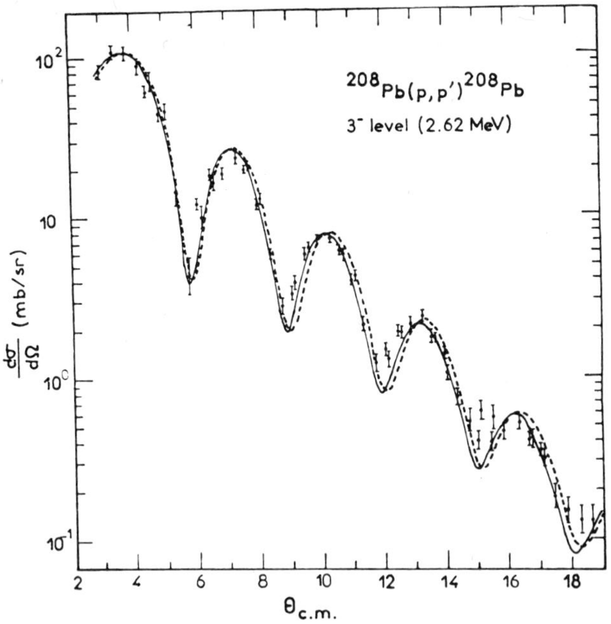

$^{208}Pb(p,p')^{208}Pb$
3⁻ level (2.62 MeV)
dσ/dΩ (mb/sr)
θc.m.

Fig. 17

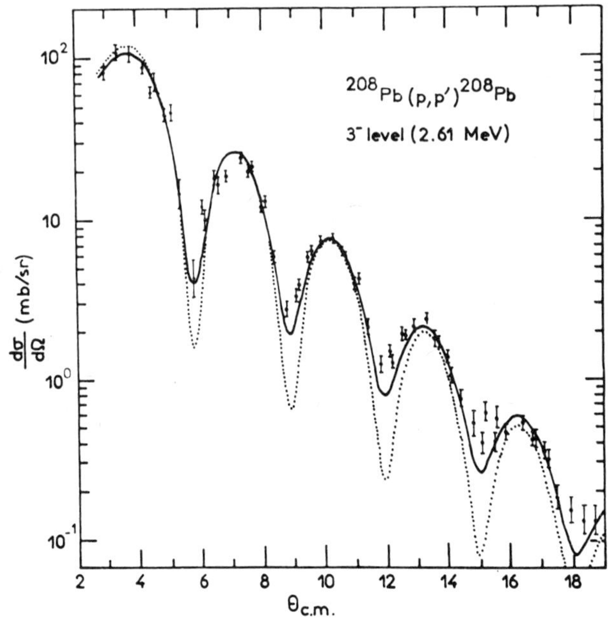

$^{208}Pb(p,p')^{208}Pb$
3⁻ level (2.61 MeV)
dσ/dΩ (mb/sr)
θc.m.

Fig. 18

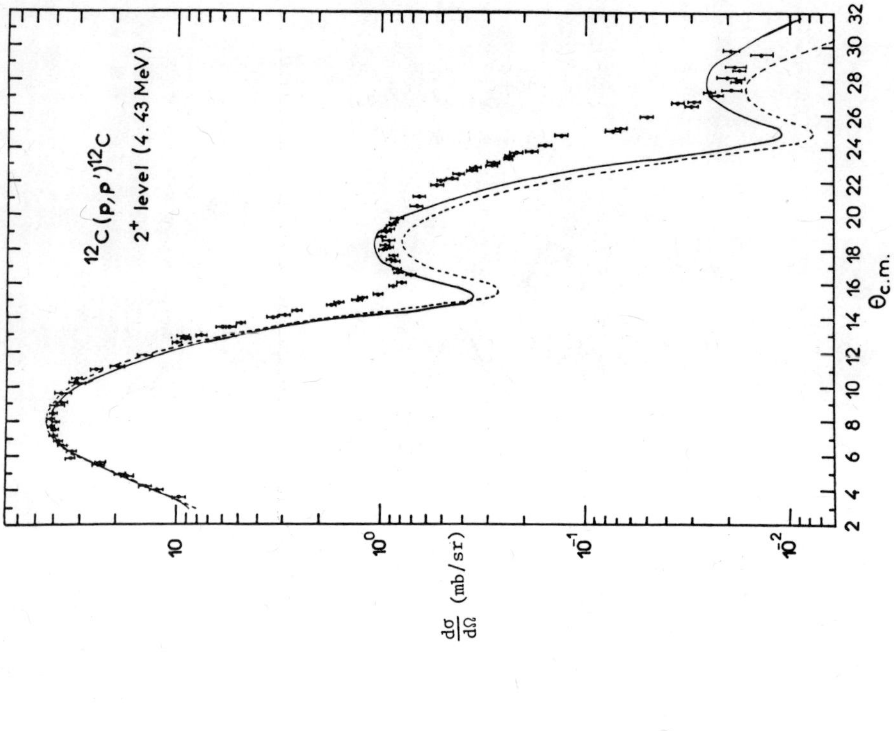

$^{12}C(p,p')^{12}C$
2^+ level (4.43 MeV)
$\frac{d\sigma}{d\Omega}$ (mb/sr)
$\theta_{c.m.}$

Fig. 19

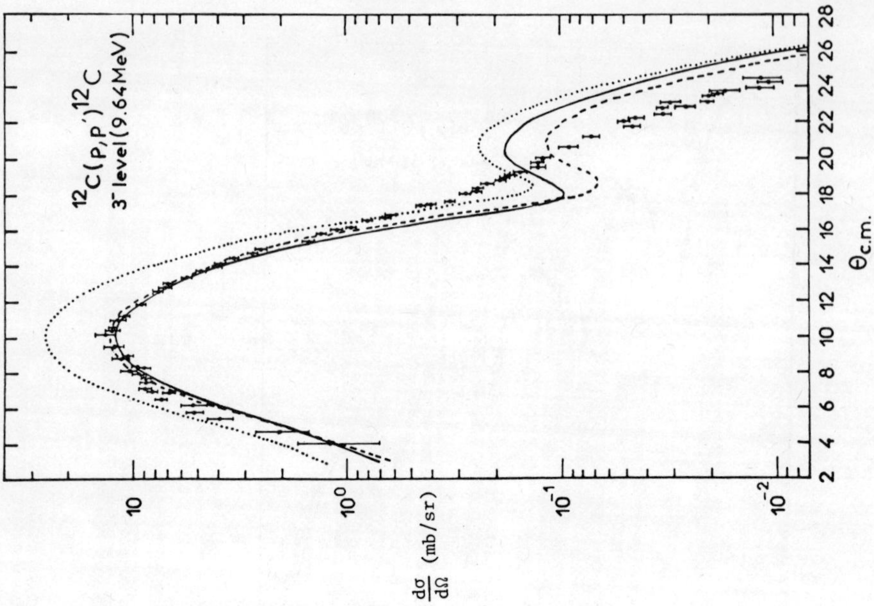

$^{12}C(p,p')^{12}C$
3^- level (9.64 MeV)
$\frac{d\sigma}{d\Omega}$ (mb/sr)
$\theta_{c.m.}$

Fig. 20

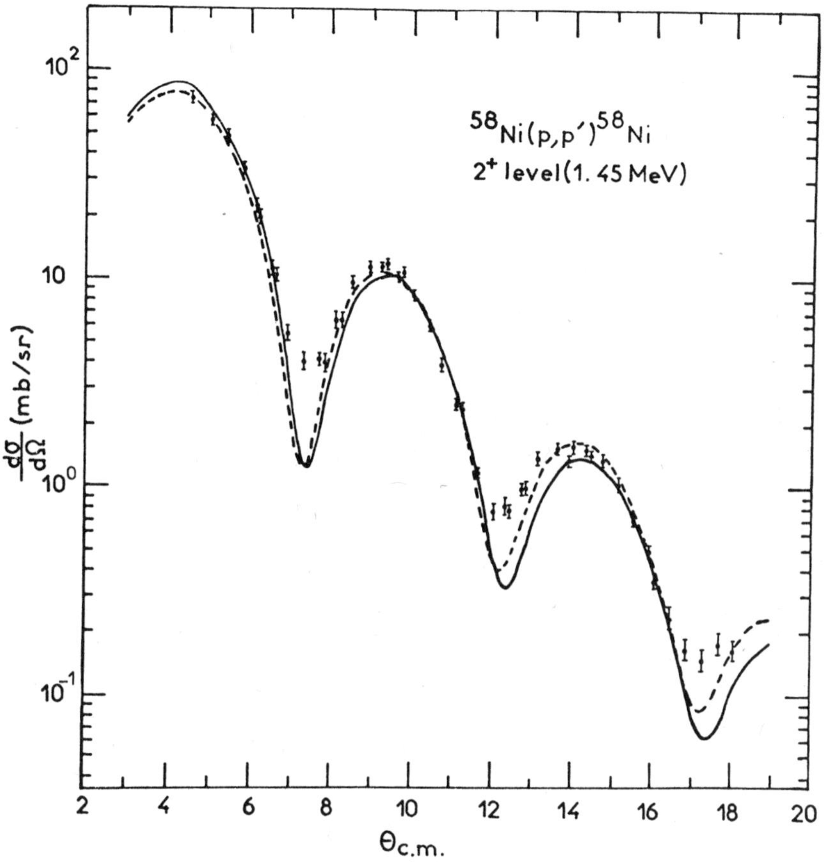

$^{58}Ni(p,p')^{58}Ni$
2^+ level(1.45 MeV)
$\frac{d\sigma}{d\Omega}$ (mb/sr)
10^2
10
10^0
10^{-1}
2
4
6
8
10
12
14
16
18
20
$\Theta_{c.m.}$

Fig. 21

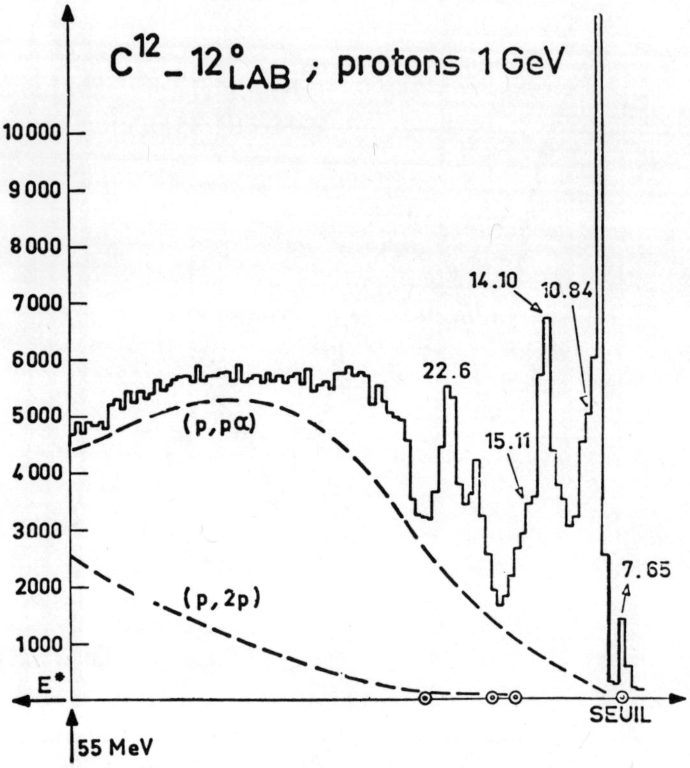

C^{12} - 12°_{LAB} ; protons 1 GeV
10000
9000
8000
7000
6000
5000
4000
3000
2000
1000
E*
55 MeV
SEUIL
(p,pα)
(p,2p)
22.6
14.10
10.84
15.11
7.65

Fig. 22

In particular, on ^{12}C and ^{6}Li one observes a bump, varying in energy as a function of scattering angle, which follows the kinematics of free elastic scattering p/α. The threshold energy is very clearly equal to the energy of separation of an α.

More recently (middle 1973 to the beginning of 1975) we were interested especially in:

- determining with better precision scattering on some medium heavy nuclei. Precise measurements were made for elastic scattering of 1 GeV protons on separated isotopes of Ca and Ti (^{40}Ca, ^{42}Ca, ^{44}Ca, ^{48}Ca, ^{48}Ti) and Ni (^{58}Ni, ^{60}Ni, ^{62}Ni, ^{64}Ni). This experiment was made in collaboration with physicists from Gatchina (USSR) and was used to obtain precisely the variation of nuclear parameters as a function of the number of neutrons. The results for the Ca and Ti are presented on fig. 23[7]. The scattering of 1.37 GeV alphas was also studied on ^{12}C and the Ca isotopes[8] (figs. 24, 25, 26, 27).

- scattering from ^{3}He and ^{4}He (figs. 28 and 29). Fig. 28 shows (p,^{3}He) elastic scattering and some preliminary calculations by R. Frascaria[9] using Glauber formalism, including spin-isospin dependence of the N-N interaction. Three different sets of phase-shifts are used: a) ref. 10 for pp phases and ref. 11 for pn phases, b) and c) ref. 11 for pn phases and ref. 12 for pp phases.
The first measurement at 1050 MeV on ^{4}He had given a contradictory result with the Brookhaven result at 1010 MeV. A systematic study has been made as a function of energy at 350 MeV, 650 MeV, 1050 MeV and 1150 MeV[13].
One sees on fig. 29 that the famous minimum p/He at $t = 0.27$ is relatively pronounced at 650 MeV, non-existent at 350 MeV, and less pronounced at 1 GeV and above. Glauber and KMT type calculations are being done in different laboratories to explain this phenomena. Nevertheless, ^{4}He with its small number of nucleons and total spin zero should be an interesting test nucleus to study the reaction mechanism or possibly structure effects. One can already say that the existence of an "absolute minimum" as a function of (s,t) indicates a very high probability that in the amplitude p/^{4}He, $F = A(t) + C(t)\, \vec{\sigma}.\vec{n}$, the amplitude A has a zero near the real axis in the complex plate (t), and that this zero crosses the real axis at a certain value of incident energy, between 550 and 750 MeV. At this particular point (s_o, t_o), $A_{s_o}(t_o) \equiv 0$. If we suppose, taking into account results from other laboratories (NASA[14], Berkeley[15]), that our distribution at 650 MeV is near that minimum, one can see the great importance of C(t), i.e. of the spin orbit term. It is quite illusory to explain p/^{4}He without introducing a precise and correct spin-orbit term (which we plan to determine later by polarization measurements $\vec{p}$/^{4}He. Lombard[16] at Orsay has shown that, in the vicinity of the minimum, there is equally a very important influence of the coulomb amplitude. Only after taking into account all those effects can we hope to determine detailed structure of ^{4}He.

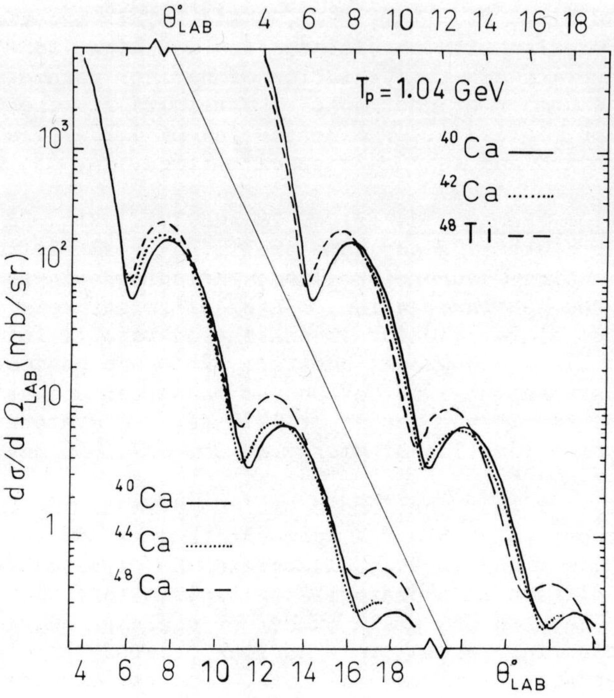

θ°LAB
4 6 8 10 12 14 16 18
$T_p = 1.04$ GeV
40Ca
42Ca
48Ti
10^3
10^2
10
1
dσ/dΩLAB (mb/sr)
40Ca
44Ca
48Ca
4 6 8 10 12 14 16 18
θ°LAB

Fig. 23

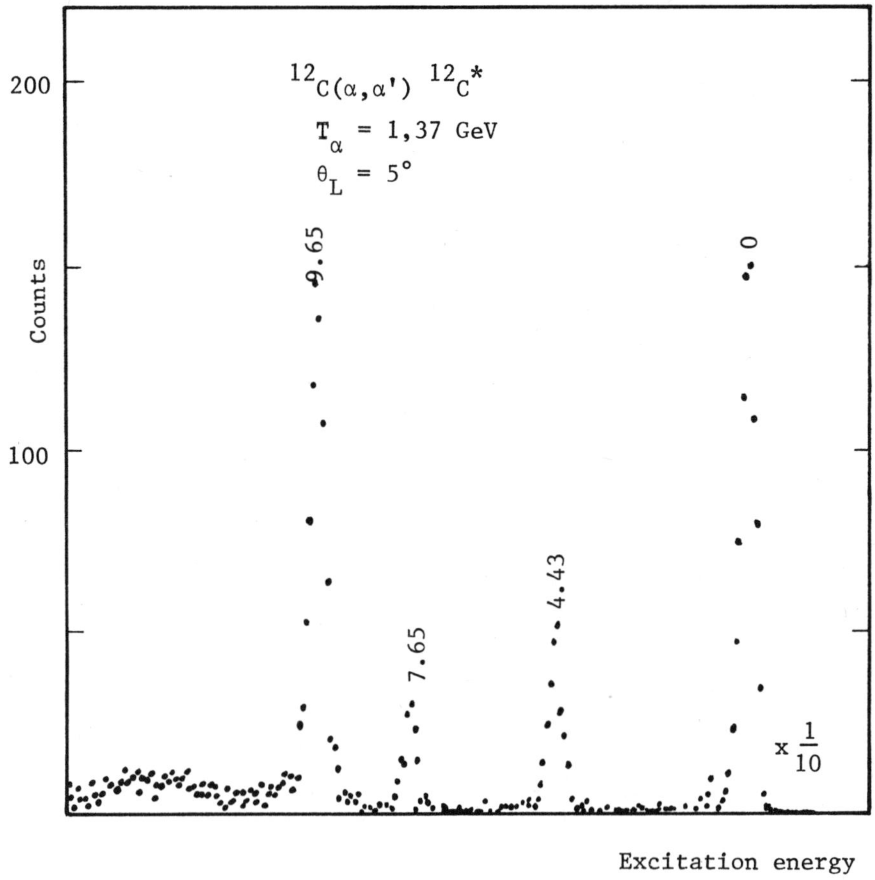

12C(α,α') 12C*
Tα = 1,37 GeV
θL = 5°
200
100
Counts
9.65
7.65
4.43
0
x 1/10
Excitation energy

Fig. 24

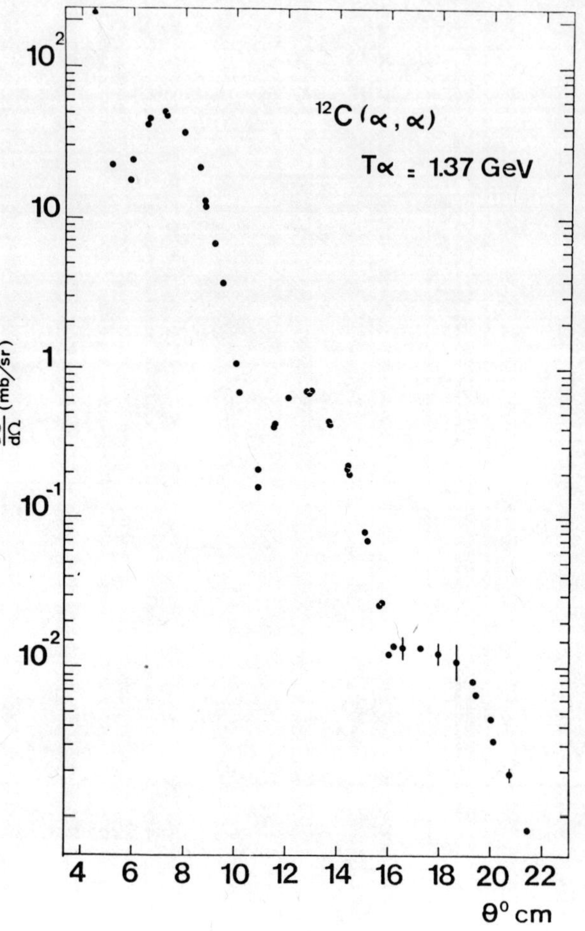

$^{12}C(\alpha,\alpha)$
$T\alpha = 1.37$ GeV
10^2
10
1
10^{-1}
10^{-2}
$\frac{d\sigma}{d\Omega}$ (mb/sr)
4 6 8 10 12 14 16 18 20 22
θ^0 cm

Fig. 25

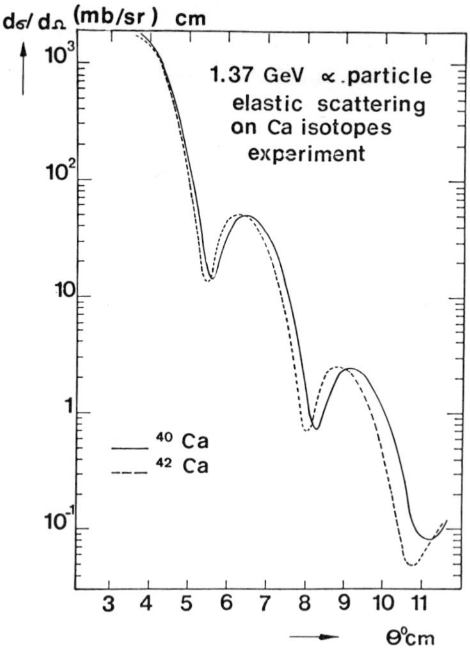
dσ/dΩ (mb/sr) cm
10³
10²
10
1
10⁻¹
1.37 GeV α-particle
elastic scattering
on Ca isotopes
experiment
40 Ca
42 Ca
3 4 5 6 7 8 9 10 11
θ°cm

Fig. 26

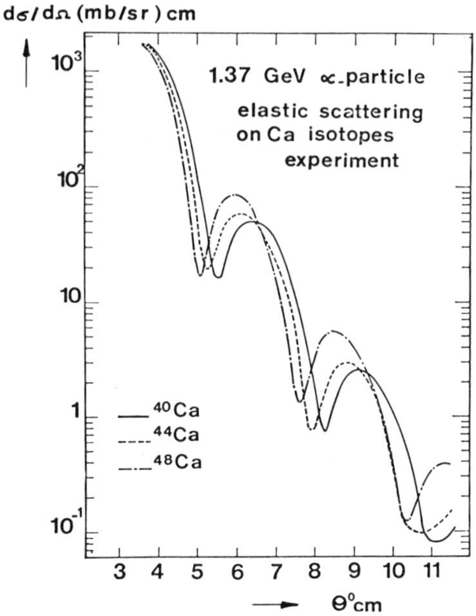
dσ/dΩ (mb/sr) cm
10³
10²
10
1
10⁻¹
1.37 GeV α-particle
elastic scattering
on Ca isotopes
experiment
40Ca
44Ca
48Ca
3 4 5 6 7 8 9 10 11
θ°cm

Fig. 27

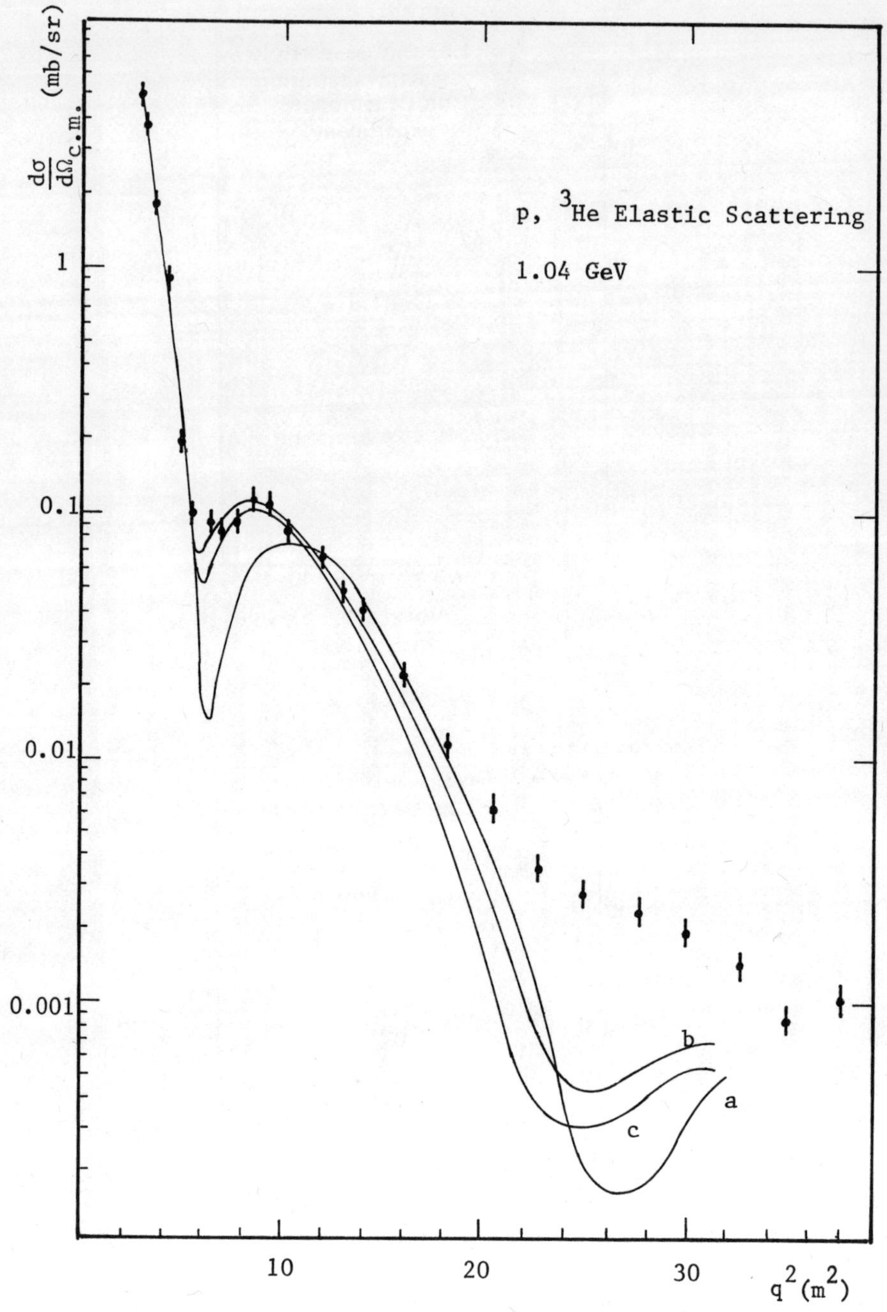

$\frac{d\sigma}{d\Omega_{c.m.}}$ (mb/sr)
p, ^{3}He Elastic Scattering
1.04 GeV
1
0.1
0.01
0.001
b
a
c
10
20
30
$q^2 (m^2)$

Fig. 28

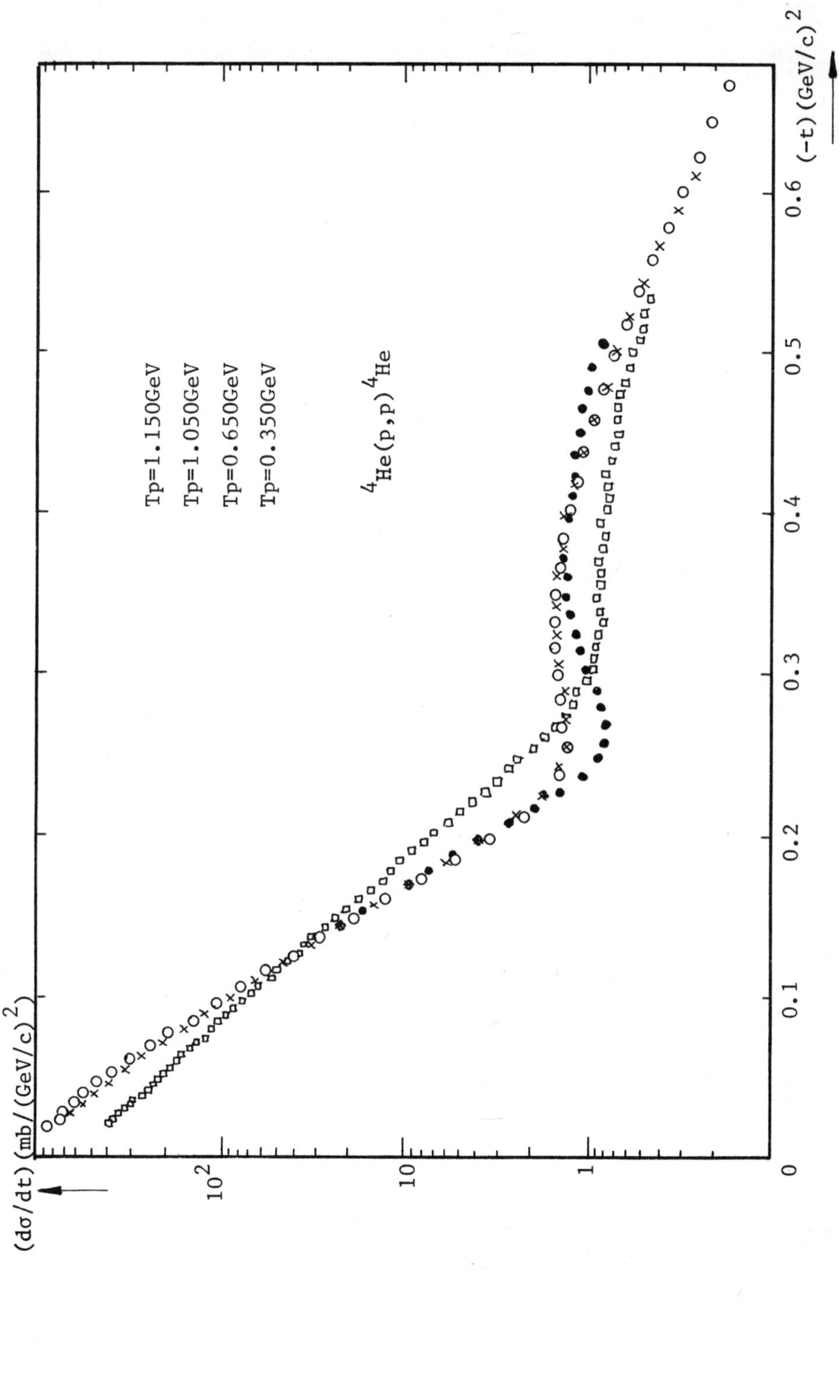
(dσ/dt) (mb/(GeV/c)²)
10²
10
1
0
0.1
0.2
0.3
0.4
0.5
0.6
(-t) (GeV/c)²
Tp=1.150GeV
Tp=1.050GeV
Tp=0.650GeV
Tp=0.350GeV
⁴He(p,p)⁴He

Fig. 29

- initiate high momentum transfer reactions. Two types of reactions have been studied:

  - (p,d) reaction at 700 MeV. Deuteron spectrum observed in the reaction $^{12}C(p,d)^{11}C$ at 700 MeV is presented in fig. 30 for $\theta_{lab} = 10°$. One sees clearly the levels $3/2^-$, $1/2^-$, $5/2^-$, $3/2^-$, $7/2^-$. An analysis by Rost[17] taking into account two-step processes to explain the high spin states ($^{12}C \rightarrow {}^{12}C^*(4.43) \rightarrow {}^{11}C^*(5/2^-$ and $7/2^-$) is shown in fig. 31.

  Other effects should be taken into account for a detailed interpretation:

    - The presence of N* in the ^{12}C wave function (g.s.) should be included in such high momentum transfer reaction. Preliminary calculations by Schaeffer, Rost, Kisslinger[18] show qualitatively this influence. To distinguish between two-step processes and the influence of N*, a study as a function of energy at small angle (in order to reduce two-step processes) seems necessary.
    - The influence of spin effects is shown in fig. 32 from calculations of Ingemarson and Tibell[19] on (p,d) reactions at 185 MeV. The production at Saturne of polarized protons and deuterons will allow in the near future to measure these effects in the entrance and exit channel.
    - More complex effects can play a role such as short range correlations (Tekou[20]).

  - (p,π^+) reaction at 600 MeV. Preliminary results have been obtained on ^6Li (fig. 33) and a spectrum on $^7Li(p,\pi^+)^8Li$. Other measurements are being done on ^9Be, ^{12}C. One still observes the excitation of high spin states which could be connected to phenomena analogous to those obtained in (p,d) reaction: double step processes, effects of N*.

In the near future, we have decided:

- to go on studying high momentum transfer reactions.

In particular:

  - study of $^4He(p,\pi)^3He$ as a function of energy (Orsay-Saclay collaboration);
  - complete (p,π^+) measurements and compare to (d,p);
  - complete measurements on few nucleon systems (p/d; α/α).

- to start polarization measurements.

  - $\vec{p}$/solid targets: ^6Li, ^{12}C, Ca, Ni, Pb to determine phenomenologically the spin-orbit terms without waiting for precise measurements of nucleon-nucleon amplitudes.
  - $\vec{p}/\alpha$ at higher energies to try to clarify particularly this problem.
  - $\vec{d}$/nucleus: for a better understanding of (p,d) or (d,p) reactions.
  - $\vec{d}$/p to know better the deuteron and take out remaining doubts linked to this scattering.

  Still more, a backward scattering experiment $\vec{d}$(p,d)p should give qualitative precious indications on exchange baryons (vasan[21]).

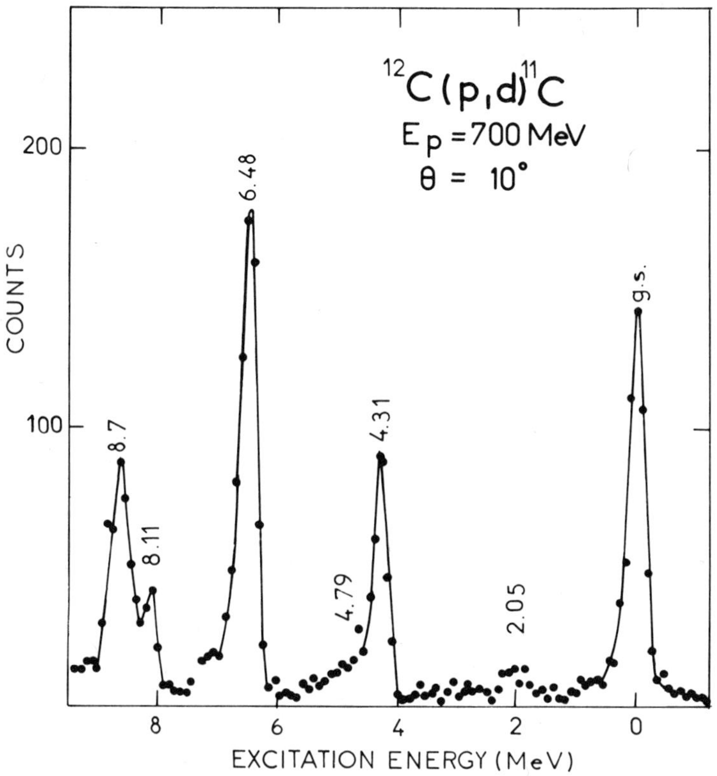

$^{12}C(p,d)^{11}C$
E_p = 700 MeV
θ = 10°
200
100
COUNTS
6.48
g.s.
8.7
8.11
4.31
4.79
2.05
8
6
4
2
0
EXCITATION ENERGY (MeV)

Fig. 30

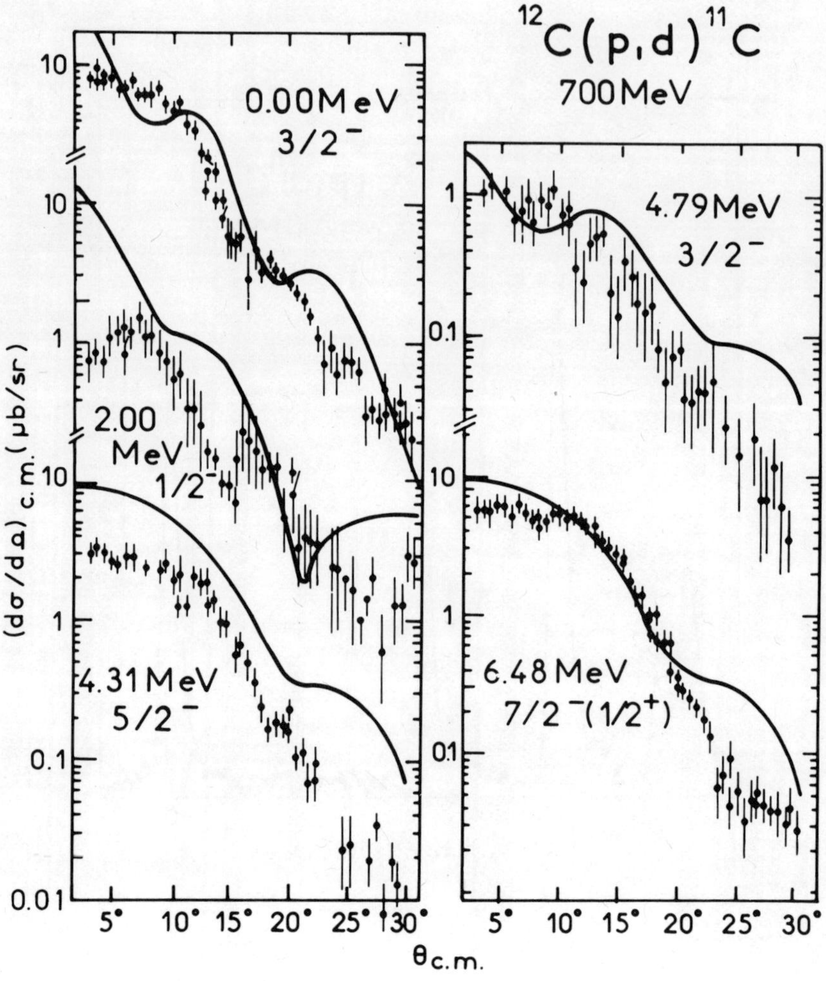

$^{12}C(p,d)^{11}C$
700 MeV
0.00 MeV
3/2⁻
2.00
MeV
1/2⁻
4.31 MeV
5/2⁻
4.79 MeV
3/2⁻
6.48 MeV
7/2⁻(1/2⁺)
(dσ/dΩ) c.m. (μb/sr)
θc.m.
10
1
0.1
0.01
5° 10° 15° 20° 25° 30°

Fig. 31

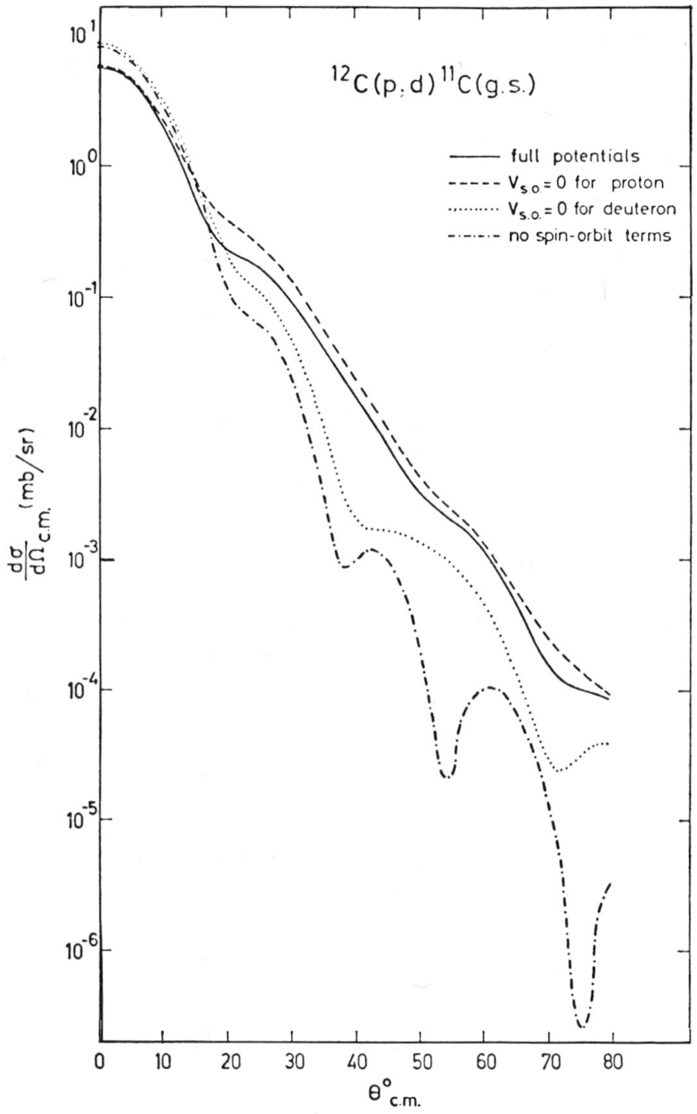
$^{12}C(p,d)^{11}C(g.s.)$
full potentials
$V_{s.o.} = 0$ for proton
$V_{s.o.} = 0$ for deuteron
no spin-orbit terms
10^{1}
10^{0}
10^{-1}
10^{-2}
10^{-3}
10^{-4}
10^{-5}
10^{-6}
$\frac{d\sigma}{d\Omega_{c.m.}}$ (mb/sr)
0
10
20
30
40
50
60
70
80
$\theta^{\circ}_{c.m.}$

Fig. 32

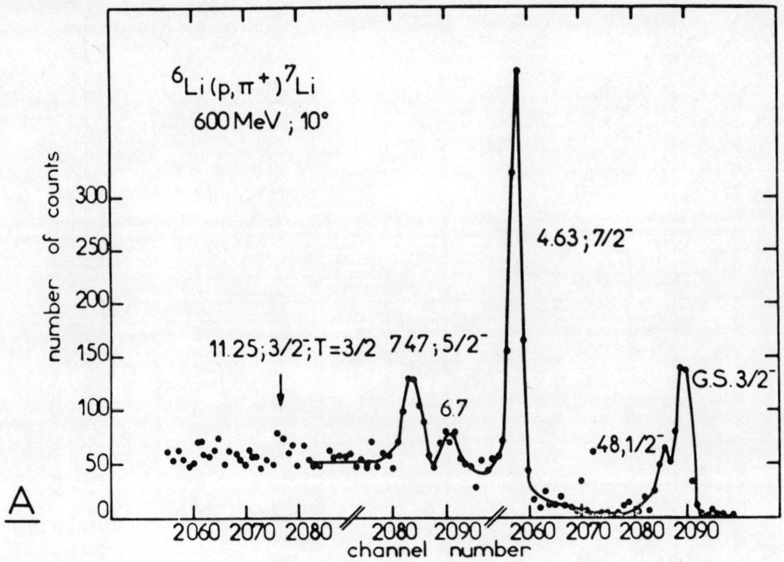

$^6Li(p,\pi^+)^7Li$
600 MeV ; 10°
number of counts
300
250
200
150
100
50
0
11.25;3/2⁻;T=3/2
7.47;5/2⁻
6.7
4.63;7/2⁻
.48,1/2⁻
G.S.3/2⁻
A
2060 2070 2080 2080 2090 2060 2070 2080 2090
channel number

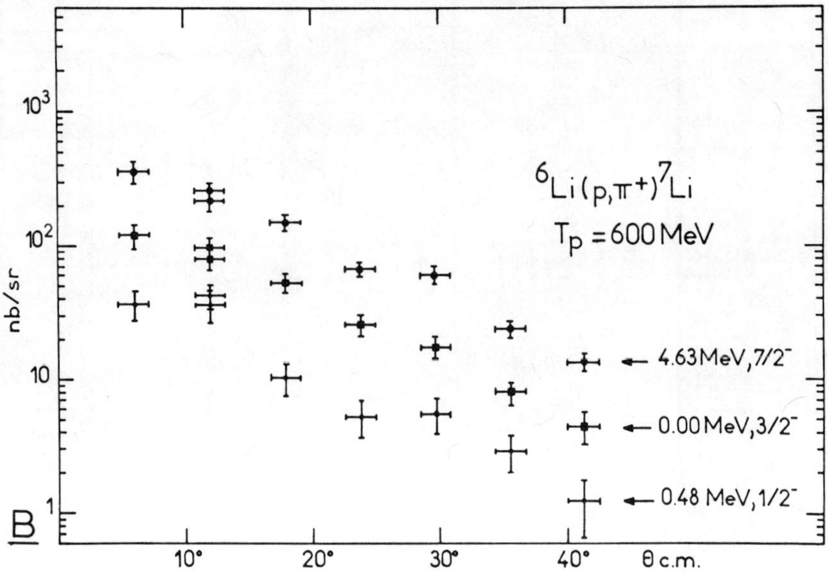

10^3
10^2
10
1
nb/sr
$^6Li(p,\pi^+)^7Li$
$T_p = 600$ MeV
4.63 MeV, 7/2⁻
0.00 MeV, 3/2⁻
0.48 MeV, 1/2⁻
B
10° 20° 30° 40° θ c.m.

Fig. 33

- Eventually very small angle measurements $\vec{p}/p$, $\vec{p}/d$, $\vec{d}/p$.
- to initiate some measurements with 3-body in the final state.

Since three (α,α') spectra have shown the same type of bump as the (p,p') spectra, [attributed to (α,2α)], it should be interesting to couple the spectrometer detecting the energetic α at forward angle with a solid state detector measuring the energy of the recoil α. One could, in this way, examine the excitation spectra of residual nuclei with a resolution between 1 and 2 MeV.

This first study should lead us to a more realistic view of our experimental program after the improvement plan at Saturne is completed. We plan a pair of spectrometers. This will allow us to measure 3-4 body (see below) final states.

Naturally, I did not say anything here about the *experiments* on the 2nd spectrometer SPESII, first tests of which have just been completed, and which should produce physics in July 1975.

VIII. ARGUMENT FOR FUTURE HIGH RESOLUTION SPECTROMETRY AT INTERMEDIATE ENERGIES

The firepower of hadrons (and nuclei as projectiles) is such that a large part of the total cross sections is due to reactions which produce *more than two particles* in the final state. Particularly:

- Between 400 and 1200 MeV per nucleon, the *creation of a pion* leads to a great enhancement of the total cross section. Beyond $\gtrsim$ 1400 MeV, the creation of two pions or more heavy mesons becomes important too and K^+ can be extracted from nuclei.
- The available energy in the nucleon-nucleus C.M. system allows for *important fragmentations* of bombarded nuclei so that many interactions A/B give rise to 3-4-5... body final states. Up to what complexity can we go in studying such few body final state reactions by spectroscopy, and especially magnetic spectrometers? What physical interest? I don't believe that we can answer honestly the second question. Let us rather investigate in a few examples the technical possibilities of high resolution spectroscopy for more than two body reactions, and what kind of information we can hope to extract from it.

VIIIa. THREE-BODY FINAL STATE PHYSICS WITH ONE SPECTROMETER (QUASI-TWO-BODY)

Some of the three-body reactions are practically equivalent to two-body reactions and can be treated almost exactly as two-body problems by a single spectrometer, provided this one has a large solid angle and analyses simultaneously a wide band of momenta. Fig. 34 (αβ) gives examples where the reaction products to be detected are emitted with their tri-dimensional momenta ($\vec{p}_1, \vec{p}_2$) almost equal.

(α) - The charge-exchange (C.E.), single or double, at small angles [$d \rightarrow (2p)$; $^3He \rightarrow (3p)$; $^4He \rightarrow (4p)$...] should have a large cross section (because $\sigma_{np \rightarrow pn}$ is large at small angle). Such a reaction could allow us to compare elastic scatterings to single and double charge exchange on complex nuclei; to compare the excitation of analog states; to test multiple scattering in a more complete frame than pure elastic scattering.

Technically, the (2p)(3p)(4p) packets will appear at intermediate energy as buckshots having relative momenta almost equal to their Fermi-momenta inside the initial projectile (plus some small transfer): they can be analyzed simultaneously by a single spectrometer enough "open" longitudinally (Δp-band) and transversely (ΔΩ). Such a spectrometer is able to analyze accurately each part, reconstruct the invariant mass and momentum of the packet (2p)(3p)(4p), and consequently the invariant missing mass. The simultaneous arrival of the pieces on the focus makes up a good trigger.

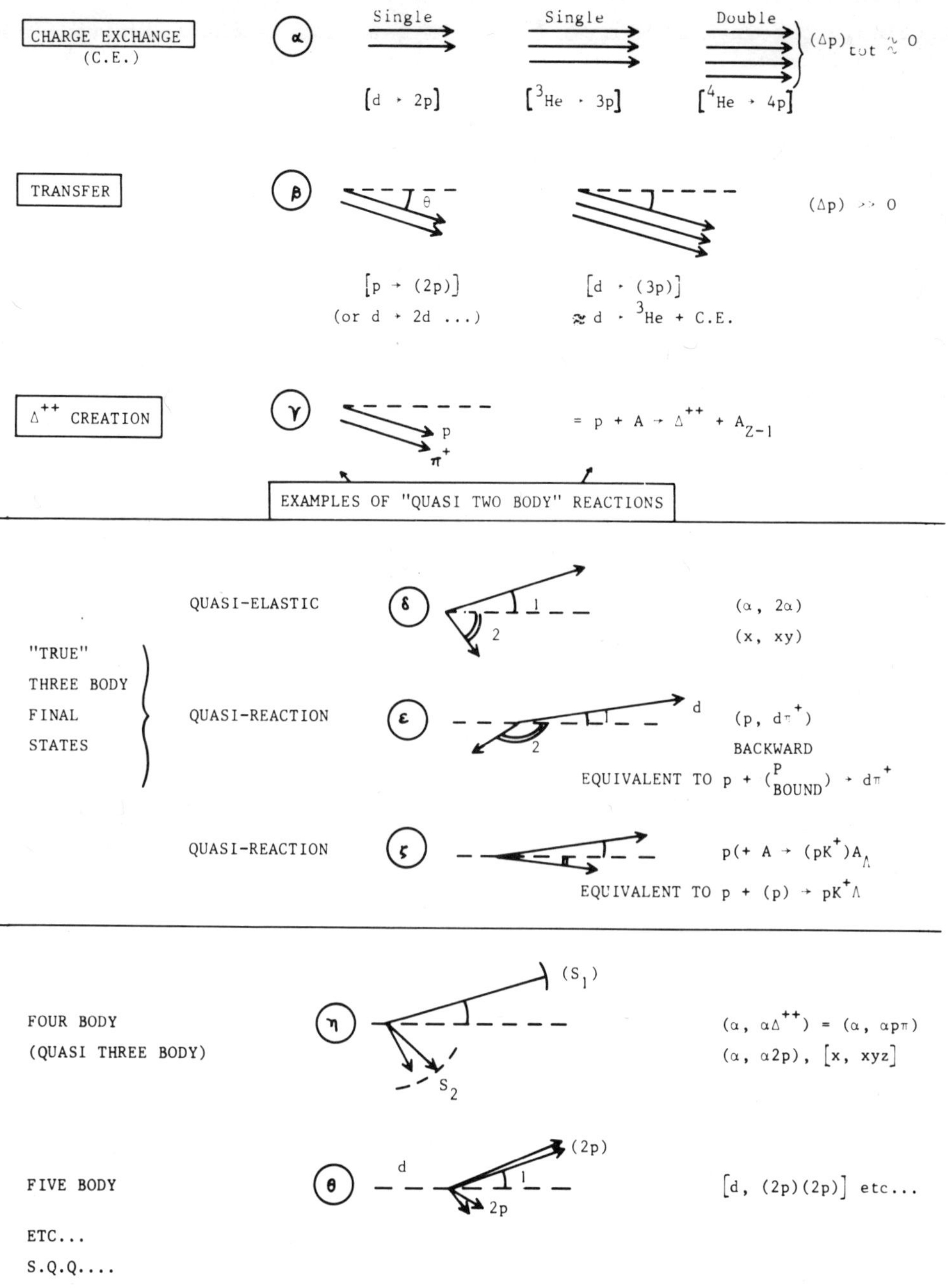
CHARGE EXCHANGE (C.E.)
α
Single
Single
Double
(Δp)tot ≈ 0
[d → 2p]
[³He → 3p]
[⁴He → 4p]
TRANSFER
β
θ
(Δp) >> 0
[p → (2p)]
(or d → 2d ...)
[d → (3p)]
≈ d → ³He + C.E.
Δ++ CREATION
γ
p
π+
= p + A → Δ++ + AZ-1
EXAMPLES OF "QUASI TWO BODY" REACTIONS
"TRUE" THREE BODY FINAL STATES
QUASI-ELASTIC
δ
1
2
(α, 2α)
(x, xy)
QUASI-REACTION
ε
d
1
2
(p, dπ+)
BACKWARD
EQUIVALENT TO p + (P BOUND) → dπ+
QUASI-REACTION
ζ
p(+ A → (pK+)AΛ
EQUIVALENT TO p + (p) → pK+Λ
FOUR BODY
(QUASI THREE BODY)
η
(S1)
S2
(α, αΔ++) = (α, αpπ)
(α, α2p), [x, xyz]
FIVE BODY
θ
d
(2p)
1
2p
[d, (2p)(2p)] etc...
ETC...
S.Q.Q....

Fig. 34

(β) – The proton capture by a proton – the final protons amounting to a "diproton" – is an equivalent reaction to a (p,d) two-body reaction. The invariant mass (2p) $\gtrsim 2M_p$, as far as the relative momentum is small. It would be of interest to compare such a reaction with a (p,d) one. The transferred particle will be different, for instance N* exchange cannot play the same role in p,(2p) as in (p,d).
Reactions such as d → (3p) are more complicated: charge exchange + transfer of a nucleon. Many others can be imagined with heavier projectiles. And so, with a spectrometer, we come gently up to some parts of the heavy ion physics. Let us still enlarge the momentum band to be analyzed at the same time in our spectrometer ($p_{max}/p_{min} \gtrsim 2$. One gets the scheme γ on fig. 34.

(γ) – A proton is transformed in a Δ^{++} by capture of a pion on a nucleus A:

$$p + A \rightarrow \underbrace{(p,\ \pi^+)}_{(\Delta^{++})} + A^*_{Z-1}$$

equivalent to the simple reaction:

$$p + (p) \rightarrow \Delta^{++} + n \qquad \text{on a bound proton.}$$

Within an incident energy range from 400 to 1000 MeV, the forward Δ^{++} disintegration occurs inside a cone, which contains altogether the proton and the pion from disintegration. A part of this cone can be detected by the spectrometer, and the Δ^{++} invariant mass reconstructed. So one does charge exchange physics p → Δ^{++}, where the final nucleus A^*_{Z-1} levels can be reconstructed to 500 keV or 1 MeV more or less, despite the Δ^{++} mass width. One can do some physics on the outgoing channel: scattering of (Δ^{++}) by nuclei.

(γ') – One can generalize (see fig. 34γ) such a measure to treat three-body problems, at forward angles, as long as the center of mass motion concentrates the products of the reaction in a front cone. So, a well defined cutting is done inside three body phase space, taking advantage of the increasing density of the latter near 0°. For example:

"one hole" $pA \rightarrow (d\pi^+)(A-1)^*$ (equivalent to $p(p) \rightarrow d\pi^+$)

"two correlated holes" $pA \rightarrow \left[\binom{t}{{}^3He} \pi^+ \right] (A-2)^*$ (equivalent to

$$\left\{ \begin{array}{l} p(2p) \rightarrow {}^3He\pi^+ \\ p\binom{p}{n} \rightarrow T\pi^+ \end{array} \right\}$$

etc...

In fact, the complete reconstitution of the observed product momenta gives a differential cross section $(d\sigma)^6/d\vec{p}_1\, d\vec{p}_2$ for a well defined final state of the non-observed product; we can now integrate over one or the other variable in the $(\vec{\Omega},p_1,p_2)$ space. The physical content is more rich than in a two body reaction, where the integration is done over wave function overlaps.

These above examples can be generalized: one has to deal with two body quasi-reactions on the substructures of a nucleus, having in free state a well defined kinematics.

On bound structures, the Fermi momenta will cause spread to these kinematics, but a spectrometer with a large (P_{max}/p_{min}) momentum acceptance is able to accept it. Concerning three body reactions (see (ζ) in fig. 34), phase space is much larger, but the forward concentration still holds. A typical reaction is:

$$pA \rightarrow (pK^+)\ A^*$$

equivalent to:

$$p + (p) \rightarrow pK^+\ \Lambda$$

where Λ has been captured by the nucleus (A - 1). One can so reconstruct excited levels of hypernuclei.

VIIIb. THREE BODY AND MORE... PHYSICS

Far from the forward angles, the three body final state physics needs *two* spectrometric tools as soon as one is looking at two particles over three.

Kinematically. One is essentially concerned with two body quasi-reactions: quasi-elastic scattering (x,xy) where x is supposed to be scattered by the substructures y; quasi-reactions (x,yz) where x has reacted with a structure (t) inside the nucleus, along the scheme:

$$x(t) \rightarrow yz \qquad\qquad xA \rightarrow (yz)(A - t)^*$$

Generally, these processes are described by the kinematical schemes (δ) (ε) (ζ) of the preceeding fig. 34.

- In the quasi-elastic scattering, one of the products is emitted forward enough with a large momentum, the other one more backward with a smaller momentum. Exceptional are the events (p,2p) (α,2α) where one is essentially interested by the symmetrical configurations with regard to incident momentum.
- In the quasi-reactions creating a (π,K...), either one of the products goes forward with a large momentum, and the other one backward with a small momentum (see (ε)), or both products go forward together with momenta generally quite different.

In practice, to treat such problems in a general way, one needs to use two coupled spectrometric tools with high analyzing power and large solid angles $(\Omega_1)(\Omega_2)$.

- A partial solution to this problem, in the intermediate energies region, consists of measuring the energy of fast forward particles with one spectrometer and the energy of the slower particle emitted at large angle with a silicon-junction detector. This is what we are going to do in July 1975 at SPESI. We intend to couple two junction detectors $(E,\Delta E)$ to SPESI in order to study the $(\alpha,2\alpha)$ reaction. Of course we shall be limited by the thickness of the detectors to 120 MeV for the kinematic energy of the recoil alphas. The total resolution should be less than 2 MeV.
- Or else, we use two spectrometers having a large solid angle and wide momentum band. Here lies the real future for three body spectrometry. The overall lateral dimensions of a spectrometer restrict the range of their possible relative angles. But preliminary studies show that a minimum relative angle of 45-50° covers most of the cases.
- Of course, once the coupling of two spectrometers is obtained (large $\Delta\Omega$; large Δp), each one can possibly analyze simultaneously two particles or more. The schemes (η) (θ) of fig. 34 show such possibilities.

Amusing though they may be, these possibilities provide a non-negligible physical interest: the possibility to extract from nuclei two correlated substructures (two nucleons, two alphas, ...) is interesting for cluster problems, including, maybe, excited states of nucleons in nuclei (isobars).

Let us remark to conclude that the well known, energetic and angular dispersion compensations of the incident beam can no longer be obtained with two apectrometers. Emittance and dispersion of incident beams will have to be refined, should one desire a very high resolution. This is one of the many reasons to transform entirely Saturne in 1977.

For heavy ion physics, spectrometers with large solid angle and good resolution will be all the more necessary as different reaction products have to be resolved, at least partly, in coincidence.

IX. BRIEF DESCRIPTION OF SPESII SPECTROMETER AT SATURNE SYNCHROTRON (FIRST EXPERIMENT IN JULY 1975)

IXa. Objective

In the beginning, we wanted to build a spectrometer with large solid angle, wide momentum band, good resolution, which could be coupled to SPESI for three body experiments, two particles being detected in coincidence. The maximum momentum necessary did not need to be very large (anyway, increasing it considerably would have been out of our financial possibilities). So we fixed:

- the SPESII parameters: $p_{max} \approx 800$ MeV/c
  Width ($\Delta p/p$) maximum simultaneously analyzable ± 17%
  $\Delta\Omega \sim 2\ 10^{-2}$ str
  Resolution ($\delta p/p$) < $3\ 10^{-4}$
  Length of focal plane: 2 m (perpendicular to optic axis = 1 m with 1 mm $\approx 3\ 10^{-4}\ \delta p/p$)
- general form. The spectrometer SPESII is of QD^2 type (one quadrupole followed by two separated dipoles). We then realized that a lot of interesting problems could be treated, using only SPESII.
- On one hand, some two body reactions, as ($p\pi^{\pm}$, $d\pi^{\pm}$) where the outgoing particle momentum is less than 800 MeV/c for a number of interesting incident energies (with the benefit of the large solid angle).
- One the other hand, some three body reactions as those suggested in (VIIIa.$\gamma\gamma$'). In particular, an experiment producing hypernuclei at 0°, on different targets, was planned in priority.

It should give physical information on excited hypernuclear systems before the end of this year.

Fig. 35 shows the principle scheme of SPESII working at 0°. According to the sign (+) or (-) of the analyzed particles, the incident beam can go either between the two dipoles, or through a hole in the yoke of the first one. The forms of the magnetic poles are complicated in order to partially compensate the aberrations. Due to the large solid angle, it is impossible to get a complete correction of the aberrations for every value (p) of the analyzed momentum. So the resolution will be very good at the center of the focal line (average p_o) and will decrease towards the edges, but a good part of these residual aberrations can be corrected by an error matrix stored in and used on line by the computer.

IXb. Magnetic Properties

The beam line coming from Saturne is finely analyzed - as for SPESI - allowing for an adjustment of incident dispersion on the target, in order to compensate for the energy spread of the beam.

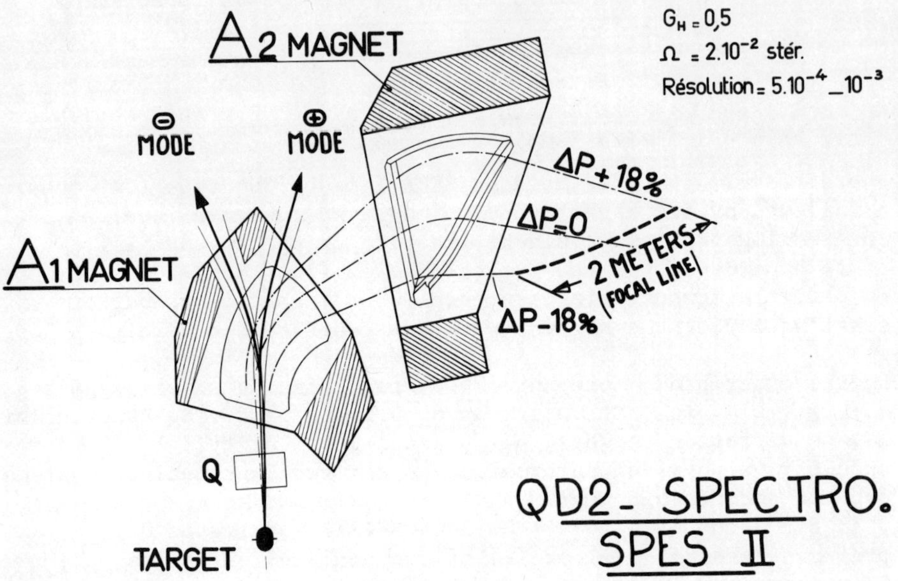
A2 MAGNET
$G_H = 0,5$
$\Omega = 2.10^{-2}$ stér.
Résolution = 5.10^{-4} _ 10^{-3}
⊖ MODE
⊕ MODE
ΔP + 18%
ΔP=0
2 METERS
(FOCAL LINE)
A1 MAGNET
ΔP-18%
Q
TARGET
QD2- SPECTRO.
SPES II

Fig. 35

This is all the more necessary as, for all forecast experiments, the analyzed particle momenta are considerably lower than the incident particle ones (ratio from 1/4 to 1/2): so one has to vary the beam dispersion significantly according to the type of experiment.
In order that the beam goes easily between the two dipoles (see fig. 35) and because the ratio $P_{analyzed}/P_{incident}$ varies for 1/4 to 1/2, one has to vary little the average incident angle of the beam on the target: a special magnet is needed for this purpose.
Concerning the remaining properties (focalization, 2nd order effects on beam, cuttings, alignment detectors and correction dipoles...) the beam line is to some extent similar to that of SPESI.

The flat pole magnets were built, like SPESI, by Rade Koncar in Yogoslavia, according to our drawings (a modified copy was made for SIN laboratory in Zurich).
Following magnetic measurements, it appeared that the main magnetic field was good, but that the leakage field went through some unacceptable reversals. Special magnetic shields were to be constructed to regulate this leakage field.
Having performed trajectory computations based on field mapping, it appears that the desired resolution will be obtained, except on the edges of the focal plane (± 15% in $\Delta p/p$), where corrections, if necessary, will be made by the computer.
It was planned to build a set of multipolar correcting coils between the two dipoles. Those will be constructed only if actual aberrations that can be balanced appear to be larger than the calculated ones.
Let me add a few details. The entrance radius of curvature of the first dipole is chosen in order to correct the vertical aberrations for the *mean* orbit. For extreme orbits ($\frac{\Delta p}{p} \sim 15\%$) those vertical aberrations are *not* corrected (but aberrations of the type $(z\phi)$ do not exist for this type of spectrometer). The relative dispersion was chosen much smaller than that of SPESI ($\sim 1/6$) to avoid a huge detection.

IXc. Detection

Taking into account the larger dimensions of the focal surface compared to that of SPESI, and also the necessity to simultaneously detect two particles, the detection and also the operational mode will be different from that of SPESI.

<u>The localization</u> is performed by two wire chambers perpendicular to the outgoing optical axis of the spectrometer, each chamber being made of five planes of wires, three of which allow for localization along (y + z) (y - z) (z) axis (45°, 90°, 135°) without any ambiguity. For details, see ref. 22.

The trigger will depend on the type of experiment: it will be very simple for a two body experiment, but very elaborate for a three body one, such as the first one forecast:

$$p + A \rightarrow (p\ K^{+}) + (A_{\Lambda}^{Z-1})^{*}$$

for which the predicted cross section is very low. So a very strong rejection is needed against background of parasital reactions giving couples of particles such as (pp) ($p\pi^{+}$), either correlated or not; this rejection is performed, using the fact that the proton and the kaon are emitted at the same time, from the same nucleus of the target. By rebuilding the trajectories, and measuring the relative time of arrival of the two particles "p" and "K^{+}", we are able to require, for the acceptance of an event as a good one, a coincidence from the target, better than one nanosecond. Let us add that specialized Cerenkov detectors will identify the K^{+} and other scintillator detectors will identify and reject correlated pions and protons.

IXd. The Operative Mode

The operative mode also will depend upon the type of experiment. The SPESII facility includes an on line computer (PDP 1145): each candidate (pK^{+}) will be analyzed from:
- trajectories data (y,z);
- relative time of arrival of the two particles.

The missing mass, and consequently the final excited levels of the residual nuclear body, will be constructed with a resolution depending on the target thickness (unequal energy losses for the incident particle and the two outgoing ones). For example, given a reasonable thickness, the hypernuclei experiment should give a total resolution of 600 to 1000 keV (incident beam: 1.25 GeV kinetic energy protons).
During the initial period, a typical two body experiment (such as $pp \rightarrow d\pi^{+}$) will allow us to measure some aberrations of the spectrometer with a nominal magnetic field, and to investigate the possible corrections for kinematical effects of second (and more) order.
Given an acceptable background, i.e. a correct beam transport between the two magnets, one can hope in the two body experiments to measure 10^{-35} cm^{2}/str cross sections.
From now up to the Saturne shutdown (end of 1976) we plan to make, beyond the experiment on the excited levels of hypernuclei, a set of two body, low cross section experiments: ($p\pi^{\pm}$)($d\pi^{\pm}$) and one or two more experiments on three body reactions at zero degree.
Thereafter, we hope to obtain for 1978-79 a third spectrometer (SPESIII), having a large angular and momentum acceptance (see below) to which SPESII could be coupled.

ESSENTIAL CHARACTERISTICS OF SATURNE II

Physical radius	16,8 m
Injection system :	
. injector	Linac 20 MeV
. injection time	400 μs
. injected intensity	10 mA
. injected emittance (normalized)	ε_x = 6 10^{-6} mrd
" " "	ε_z = 6 10^{-6} mrd
. injection magnets	2 magnetics + 1 electrostatic
THE DIPOLES.	
. number	16
. air gap height	0,14 m
. type	H
. radius of curvature	6,3 m
. induction at ejection	0,1 T
. maximum induction	1,95 T
. iron height (for 16 dipoles)	500 t
THE QUADRUPOLES.	
. number	12 focusing + 12 defocusing
. opening radius	0,096 m
. gradient at injection	0,56 T m^{-1}
. maximum gradient	10 T m^{-1}
. maximum inductance on the pole	1 T
. iron weight	75 t
BETATRON OSCILLATION.	ν_x = ν_z = 3,7
LONG STRAIGHT SECTIONS.	
. number	4
. length	8 m
THE RF CAVITIES.	
. number of cavities	2
. harmonic	3
. energy gain per turn	2,8 keV
. RF power	30 kW
BEAM PERFORMANCES.	
. type of particles	H^+ D^+ He^{++}
. maximum energy	2,95 GeV (H^+)
. number extracted particles per second	H^+ : 10^{12} to 1 GeV ; 7 10^{11} to 2 GeV
	D^+ : 5 10^{11} to 1 GeV ;
	He^{++} : 2 10^{11} to 1 GeV
. Normalized emmitance of extraction beam	ε_x = 10^{-5} mrd ; ε_z = 12 10^{-5} mrd
. energy spread of extracted beam	± 5 10^{-4}
. duty cycle at 1 GeV : 40 % ; at 2 GeV : 25 % ; at 2,7 GeV : 15 %	
. Experimental area surface	3 500 m^2
. Number of extraction channel	2
. simultaneous extraction in the 2 channels	Yes, at the same energy
. Total power for experimental areas	30 MVA

Fig. 36

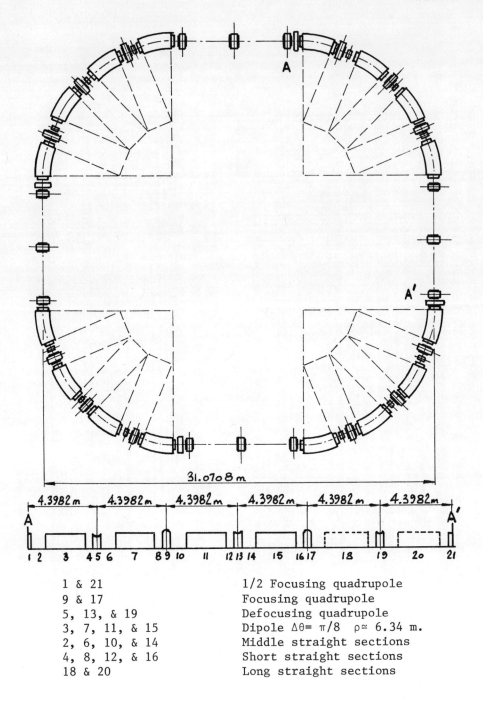

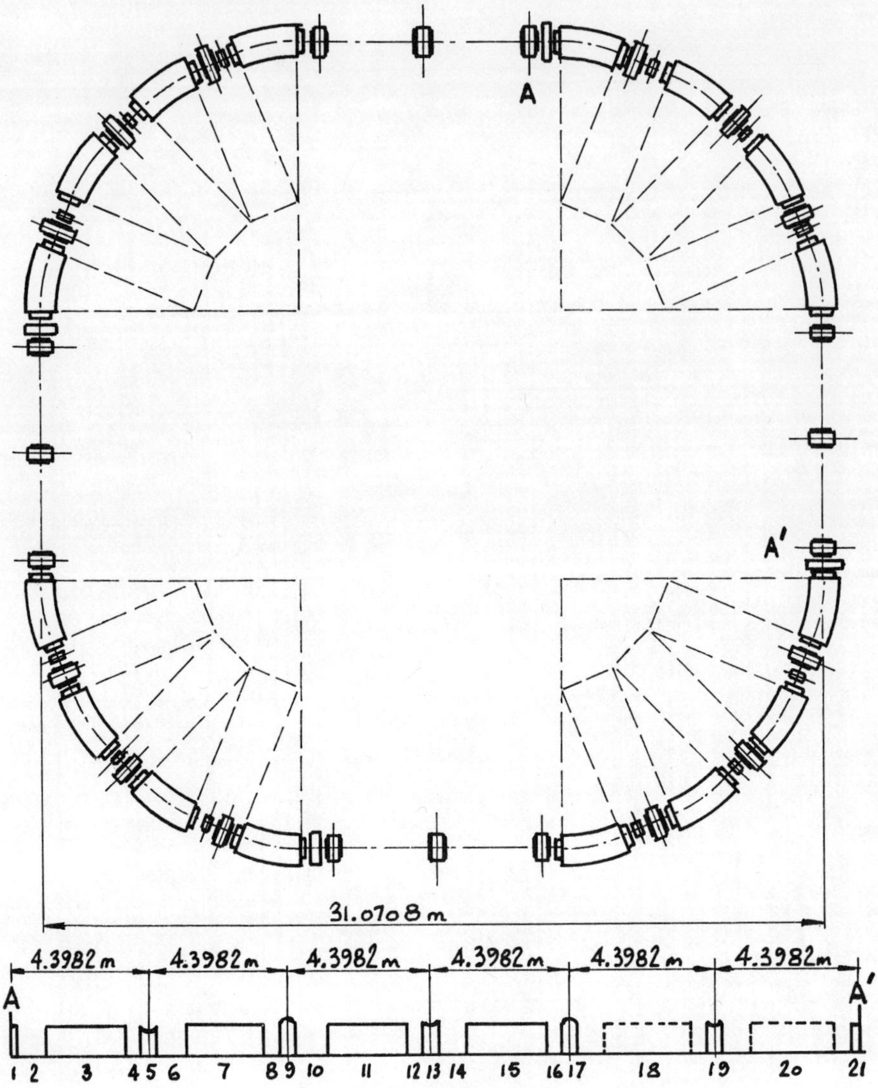

1 & 21	1/2 Focusing quadrupole
9 & 17	Focusing quadrupole
5, 13, & 19	Defocusing quadrupole
3, 7, 11, & 15	Dipole $\Delta\theta = \pi/8$ $\rho \simeq 6.34$ m.
2, 6, 10, & 14	Middle straight sections
4, 8, 12, & 16	Short straight sections
18 & 20	Long straight sections

Fig. 37.- Ring structure of Saturne II.

X. THE FUTURE OF HIGH RESOLUTION SPECTROMETRY AT SATURNE

No fine spectrometry without fine beam! Our operation of Saturne taught us that a machine nearly twenty years old is limited in intensity, quality, and the stability of its extracted beams. The number of breakdowns is hardly admissible and the possibilities to put them right restricted. We therefore decided to transform the whole accelerator.

As the new accelerator is being built in collaboration between the C.E.A. and the university (IN2P3), it has been decided to create simultaneously a "National Laboratory" for nuclear physics at intermediate energies, working with the new machine. For the sum of $\approx$ 40 million francs, we will have:

Xa. A New Accelerator: Saturne II

The main objective being not to produce a large number of pions, it seemed reasonable to increase only slightly the intensity ($\sim$ 2.2 10^{12}/cycle) but rather to get:
- a good emittance (strong focusing by quadrupoles, separated function machine);
- an improved beam extraction system (long straight sections);
- an improved stability and a sufficient flexibility (flexible distribution of beams in *time* and *space*).

The duty cycle will be 20%, the repetition rate will increase. So we can hope to improve (although seldom necessary) by a factor $\lesssim$ 40 the average intensity on *our target* with a better emittance. Fig. 36 gives the expected performances of this new machine.

To realize this objective, the ring will be enlarged and structured as indicated in fig. 37: four-periods structure, eight deviating magnets, sixteen quadrupoles and compensating magnets. The injector will be maintained (20 MeV protons linac). The RF cavity will be entirely new, the two ejection lines (only resonant extraction) will be able to work independently or simultaneously at the same energy. We hope, later, to extract simultaneously on the two lines at *different* energies (money!). Numerous beam position and structure measurements, inside the machine and on the extracted beams will allow for supervision and computer control of the machine. The physical parameters of the machine are indicated on fig. 36.

The extracted beam lines will be also supervised by computer. The main experimental facilities already specified are:
- spectrometry (see below);
- nucleon-nucleon (with polarized beam and polarized target);
- heavy ions;
- test cave using a parasitic beam with possible energetic pion production (300 MeV to 1.5 or 2 GeV);
- one or two detectors testing facilities.

There will be diversified types of beams. Besides the protons, deutons, alpha beams already accelerated, we will produce:
- ^{3}He ions;
- polarized particles $\vec{p}$ and $\vec{d}$ (then $\vec{n}$);
- heavier relativistic ions.
From now on, some tests will be performed with a polarized source. The polarized deuterons will be used before the end of the year, may be also polarized protons (but there is a disturbing depolarizing resonance inside the actual machine).
A new EBIS-type source is being studied at Orsay. The results are promising enough to let us hope to get one in 1978-79:
- heavy ions up to neon (and perhaps even heavier) beam of good intensity (some 10^{10});
- polarized ions beam of greater intensity, by storage inside such a source (the same source should be used).

The status is the following:
- the magnets and the quadrupoles are ordered;
- the shutdown of Saturne I will take place at the beginning of 1977;
- the modification and partial reconstruction of experimental areas will last a year.
The physics experiments will start during spring 1978. They will use to a maximum the specific possibilities of such an accelerator:
- variability in energy (possible frequent energy changes);
- diversified types of particles;
- good emittance and energy definition.

Xb. The Spectrometers

SPESI will be the first facility to be shut down (autumn-winter 1976). It will probably start up again during spring 1978. The beam line will be improved: it will work by itself, or coupled with other low energy detectors, of which some are silicon junction detectors. The localization detection will be improved, made faster, and will enti ely cover the focal plane (actually $\sim$ 1/2).

SPESII will work first by itself:
- either on three body problems around zero degree;
- or two body (or quasi two body) problems with variable angles.
Then it should be coupled with the new spectrometer SPESIII for studying three and four body reactions (if we get the money...).

SPESIII is, today, only a preliminary project. We would like a spectrometer having:
- an average resolution of 5 10^{-4} $(\delta p/p)$;
- a very wide band of momenta simultaneously analyzed ($P_{max}/p_{min} > 2$)
- a solid angle of $\Omega > 10^{-2}$.

The very wide momentum band would allow within the useful incident energy band for the detection of the group ($p\pi^+$) having the Δ^{++} mass, and also some groups of 2 or 3 particles or clusters extracted from a nucleus. Such a tool would be very useful for heavy ions physics, by analyzing several pieces, extracted from an incident ion. The preliminary plan fixes the *maximum analyzed momentum* at $(p/z)_{max} \sim 1.5$ GeV/c. In order to avoid building a monster, we will try to design a set of two magnets, rather similar to SPESII with regard to the form, but supraconducting. To keep the shape of the magnetic field as constant as possible, we will probably work with a *fixed magnetic field* (consequently with a *fixed* mean momentum).

At high energy (between 1.5 and 3.5 GeV/c) a spectrometric facility will be built, working recovering the magnets of Saturne I. Fig. 38 shows such a set. The coils of these magnets will have to be manufactured, and some quadrupoles and sextupoles to be added. There will be two possible ways of working:

- either by using an intermediate vertical image where a detector will allow for a precise time of flight measurement up to the final image;

In that case, the spectrometric characteristics will be as indicated:

- maximum momentum p = 3.5 GeV/c,
- intrinsic resolution $\delta p/p = 1.5\ 10^{-3}$,
- solid angle $\Omega = 3\ 10^{-4}$ str,
- double focusing,
- dispersion at the intermediate image 2.5 cm for $\delta p/p = 1\%$,
- final dispersion 6 cm for $\delta p/p = 1\%$,
- total momentum band 8%.

- or by obtaining the finest possible resolution without an intermediate image, with the following characteristics:
- total momentum band 2%,
- intrinsic resolution $\delta p/p = 1.5\ 10^{-4}$,
- solid angle $\Omega = 3\ 10^{-3}$ str.

There will be no compensation for the incident energy spread. The first sextupole will correct for horizontal aberrations, the last one will get the focal plane perpendicular to the main orbit. Vertical aberrations are not compensated but are very small. The spectrometric set will be fixed. (The total weight and the length of the system would prohibit any rotation around the target.) Therefore it will be necessary to have the incident beam angle varied on the target. A movable dipole plus quadrupole set, acting on the incident beam, will allow us to obtain a physical scattering angle for the analyzed particles in the range 0° to 28° (or -2° to 26°). It will possibly be coupled with a big magnet already existing, and analyzing the slow products of the reaction towards 90°. With such a set of spectrometers, we hope to be well armed in 1978-1979. Light ions (p, d, ^{3}He, ^{4}He) polarized protons and deuterons and heavy ions will be able to give rise to complex reactions needing the finest analysis. We hope to be ready for diverting physics in the eighties.

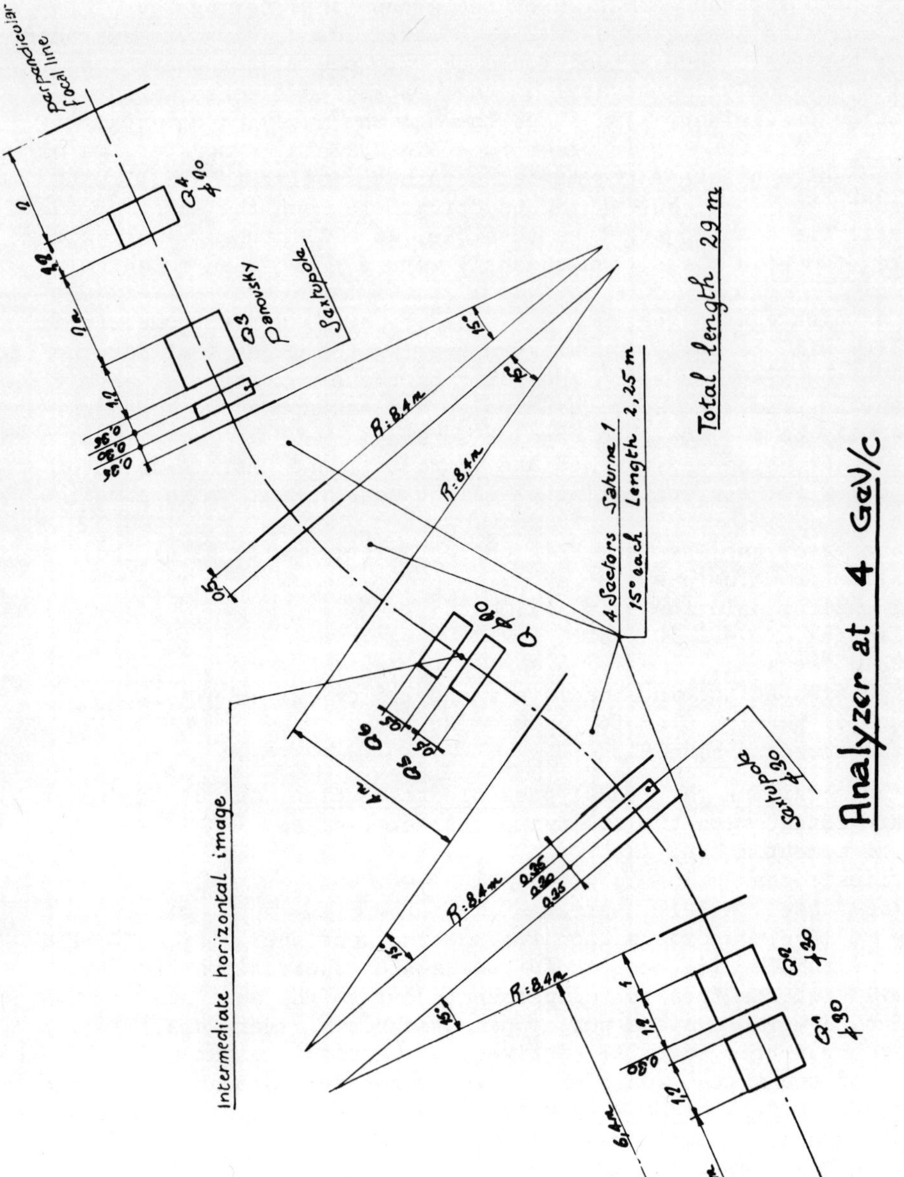

Analyzer at 4 GeV/c
Total length 29 m
4 Sectors Saturne 1
15° each Length 2,25 m
Intermediate horizontal image
perpendicular focal line
Panofsky
TARGET

Fig. 38

For those who would be interested in more precise details (or individual visits) of our laboratories, I give below the list of the physicists participating busily in these spectrometric studies, each with his own specialty:

J. THIRION: comprehensive studies.
J. SAUDINOS: SPESI conception and localization detectors.
D. GARRETA: magnets, corrections, compensations.
J.C. FAIVRE: computers, future detections.
J.M. DURAND: SPESII detectors.
R. BERTINI: physics on SPESII, especially hypernuclei.
R. BIRIEN: building of spectrometers.

REFERENCES

[A] B.L. Cohen, Rev. Sci. Instr., 30 (1959) 415 and BO Sjögren, Nucl. Instr. Methods, 7 (1960) 76.
[B] H. Palevsky et al., Phys. Rev. Letters, 18 (1967) 1200.
[1] R.J. Glauber, Lectures in Theoretical Physics, Boulder 1958.
[2] A.K. Kerman et al., Ann. Phys. (N.Y.), 8 (1959) 551.
[3] V.E. Starodubsky, Nucl. Phys. A219 (1973) 525.
V.E. Starodubsky and O.A. Domchenkov, Phys. Letters, 42B (1972) 319.
[4] I. Ahmad, Phys. Letters, 36B (1971) 301.
[5] E. Lambert and H. Feshbach, Ann. Phys. (N.Y.), 76 (1973) 80.
[6] I. Ahmad, to be published in Nucl. Phys.
[7] G. Alkhazov et al., communication to Santa-Fé Conf., June 1975.
[8] G. Bruge et al., private communication.
[9] R. Frascaria et al., communication to Santa-Fé Conf., June 1975.
[10] N. Hoshisaki, Rev. Modern Phys., 39 (1967) 700.
[11] V. Comparat and A. Willis, private communication.
[12] M. Matsuda and W. Watari, Lett. Nuovo Cimento, 6 (1973) 23.
[13] E. Aslanides et al., communication to Santa-Fé Conf., June 1975.
[14] E.T. Boschitz, Conf. on High Energy and Nuclear Structure (Plenum Press, New York, London, 1970).
[15] G. Igo, contribution to VIth Intern. Conf. on Particle Physics and Nuclear Structure, Santa-Fé, June 1975.
[16] R.J. Lombard, private communication.
[17] S.D. Baker et al., Phys. Letters, 52B (1974) 57.
[18] R. Schaeffer, E. Rost and L.S. Kisslinger, private communication.
[19] A. Ingemarsson and G. Tibell, Physica Scripta, 10 (1974) 159.
[20] A. Tekou, private communication.
[21] S.S. Vasan, Phys. Rev., D8 (1973) 4092.
[22] R. Chaminade et al., Nucl. Instr. Methods, 118 (1974) 477.

APPENDIX I

A SIMPLE GEOMETRICAL CONSTRUCTION TO UNDERSTAND COMPENSATIONS

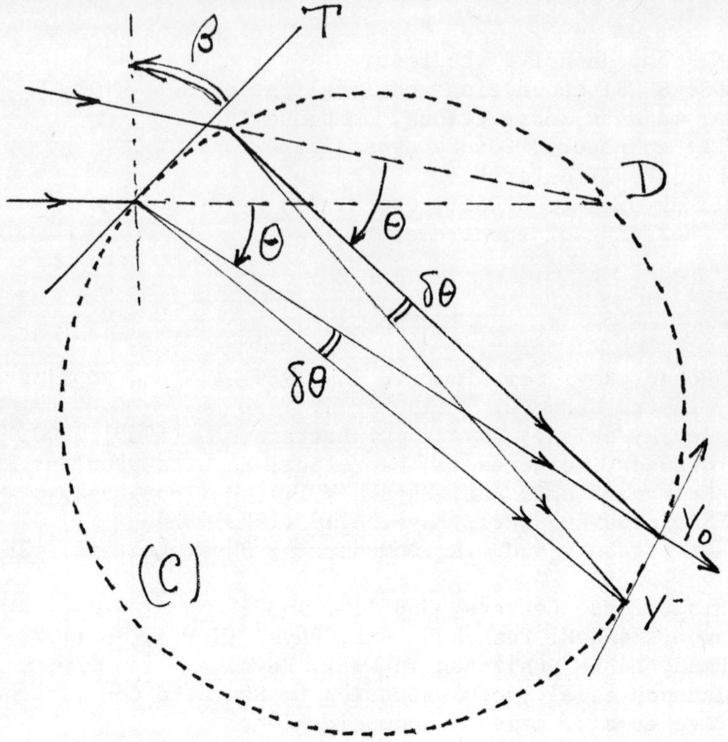

1.- Let us draw a circle (C) tangent to the target T and passing through the focus D of the incident beam. To first order, all particles scattered at the same physical angle θ go through the same point Y_0 on the circle. Particles scattered at a different angle $\theta + \delta\theta$ go through another point (Y^-). Thus to first order on the tangent (Y_0Y) all particles having the same momentum go through a point on Y_0Y, the distance to Y_0 being proportional to $\delta p = \delta\theta . dp/d\theta$. If the distribution of the (Y) points as a function of (δp) is such that (Y_0Y) can be considered to be the "reversed focal plane" of (S), all <u>particles will be focused at the same point after</u> (S). We only have to fix dY/dp at the correct value, and as far as $dp/d\theta$ is fixed, we have to fix the radius of the circle (C)... therefore to fix D for each kinematics and target angle β. <u>This demonstrates the possibility of compensating for the incident angles</u> (in all what precedes, the incident momentum is fixed).

2.- Incident momenta. Let us draw the same circle as in the preceding section 1.

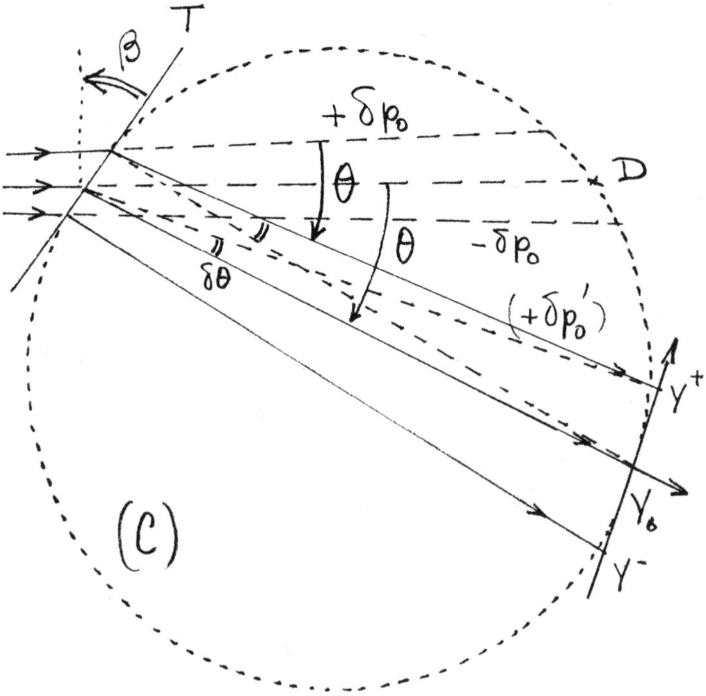

Suppose that the incident beam is dispersed as indicated on the figure as a function of the momentum difference ($\pm\ \delta p_0$) compared to the central value p_0. Then for an incident particle having a momentum ($p_0 + \delta p_0$), the scattered particle at the physical angle θ will have a momentum ($p_0' + \delta p_0'$), the ratio $\delta p_0'/\delta p_0$ being fixed by the kinematics of the reaction.

If ($\delta p_0'$), corresponding to the point (Y^+), is equal to the change in momentum due to the kinematics:

$$\delta p_0' = \left(\frac{dp'}{d\theta}\right).\delta\theta$$

then we can superimpose both figures and say that "all scattered particles having the same momentum after scattering go through the same point Y between the target and the spectrometer, independently of the incident angle and momentum". To realize this, a well definite focussing distance D and dispersion for the incident beam are necessary. They both depend on the target angle (β). This "dispersed object" ($Y^-\ Y_0\ Y^+$) can be considered as an object for the spectrometer, which in turn will focus all particles of the same final state at the same focus.

(See more details on Appendix II)

APPENDIX II

ANALYZER AND SPECTROMETER PROPERTIES IN MATRIX FORM

II.1.- Definitions. Let us define the trivector $\bar{v} = (y, \theta, \delta = \frac{\delta p}{p_o})$ for the trajectories in the horizontal plane.

The vector $\bar{v}$ before and after the analyzer (resp. spectrometer) are related by:

$$\bar{v}^f_A = A\, \bar{v}^i_A \qquad\qquad \bar{v}^f_S = S\, \bar{v}^i_S \tag{1}$$

with

$$A = \begin{vmatrix} A_{11} & 0 & A_{16} \\ A_{21} & A_{22} & A_{26} \\ 0 & 0 & 1 \end{vmatrix} \qquad\qquad S = \begin{vmatrix} S_{11} & 0 & S_{16} \\ S_{21} & S_{22} & S_{26} \\ 0 & 0 & 1 \end{vmatrix} . \tag{2}$$

One defines the kinematics to 1st order by:

$$K(\theta) = \frac{1}{p_S} \frac{dp_S}{d\theta} \qquad\qquad A(\theta) = \frac{\delta_S(\theta)}{\delta_A} . \tag{3}$$

Remark : $A_{11}\, A_{22} = S_{11}\, S_{22} = 1.$

II.2.- Shifting of the focal plane due to the kinematics. $\bar{v}_f = S\, \bar{v}_i$, but $y_i = (\theta, \theta_i, \delta_i = \delta'_i = K\theta_i)$

Hence

$$y_f = S_{16}\, K\theta_i \quad \text{(K given by (3))}$$

$$\theta_f = S_{22}\, \theta_i + S_{26}\, K\theta_i .$$

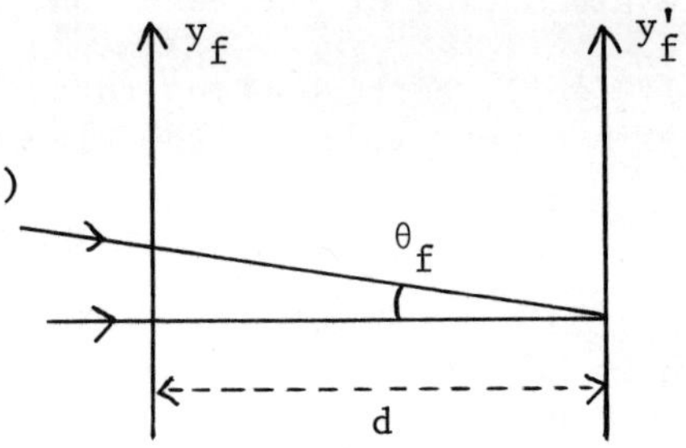

After a translation of length d:

$$y'_f = y_f + d\theta_f = [S_{16} K + d(S_{22} + S_{26} K)].\theta_i$$

$$y'_f = 0 \qquad \text{for} \qquad \boxed{d = -\frac{K\, S_{16}}{S_{22} + K\, S_{26}}} . \tag{4}$$

(4) shows that d = ∞ for $K = -\dfrac{S_{22}}{S_{26}} = .66/\text{rad}.$

$$= 6.6\ 10^{-4}/\text{mrad}.$$

(Example : 46°, pp, at 1 GeV).

II.3.- Compensation for the incident beam emittance (angles).

For particles scattered at the same physical angle θ, the beam being focussed at a distance D from the target, one gets :

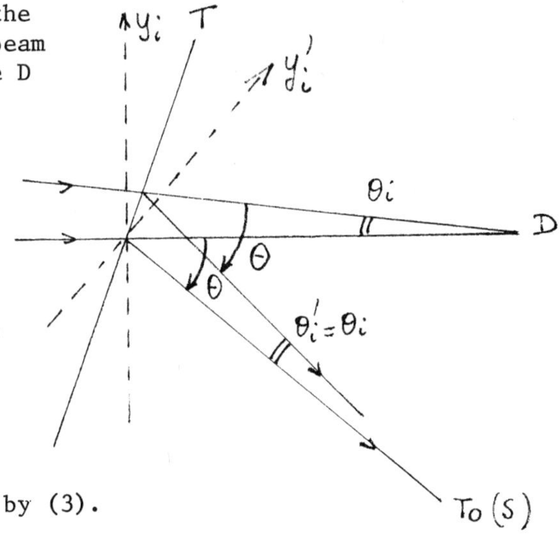

$$\bar{v}'_i = \Delta \bar{v}_i$$

with

$$\Delta = \begin{vmatrix} r & 0 & 0 \\ 0 & 1 & 0 \\ 0 & 0 & A \end{vmatrix} \tag{5}$$

where

$$r = \frac{\cos(\theta-\beta)}{\cos\beta}\ , \text{ A given by (3).}$$

Then $\bar{v}_f$ is related to $\bar{v}_i$ by $\bar{v}_f = |S|.|\Delta|\bar{v}_i = |S'||v_i|$.

$$S' = \begin{vmatrix} r(S_{11} + dS_{21}) & dS_{22} & A(S_{16} + dS_{26}) \\ rS_{21} & S_{22} & AS_{26} \\ 0 & 0 & A \end{vmatrix} \tag{6}$$

Finally one gets for y'_f as function of θ_i :

$$y'_f = [-rD(S_{11} + dS_{21}) + dS_{22}]\ \theta_i \quad . \tag{7}$$

To get $y'_f = 0$ <u>independently of θ_i</u> :

$$D = \frac{d\,S_{22}}{r(S_{11} + dS_{21})} = -\frac{K}{r} \cdot \frac{S_{16}\,S_{22}}{1 + K(S_{11}\,S_{26} - S_{16}\,S_{21})} \qquad (8)$$

(8) gives the incident defocussing to get an image independent of the beam incident angle. (D) depends on (K, r).

II.4.- <u>Compensation for the energy dispersion of the incident beam.</u>

Scattered particles :

$y'_i = r y_i$; $\theta'_i = \theta_i$; $\delta'_i = A\delta_i$.

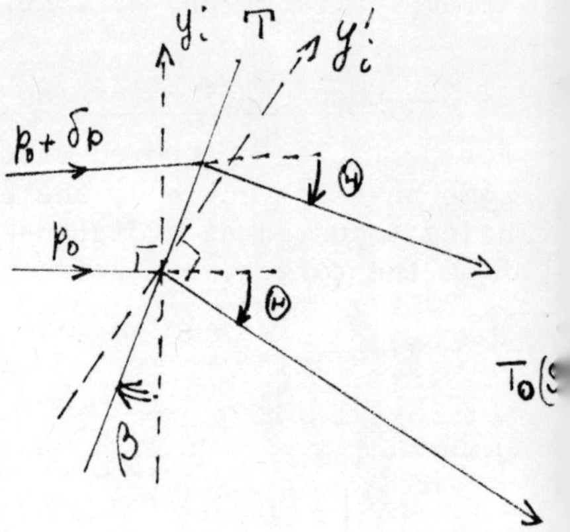

Same matrix as given by (6) for :

$\bar{v}_f = |S'| . \bar{v}_i$.

But for the incident particles :

$$\bar{v}_i = \left\{ y_i = A_{16}\,\delta_i \;;\; \theta_i = A_{26}\,\delta_i \;;\; \delta_i \right\}.$$

Then

$$y'_f = \delta_i \times \left\{ r\,A_{16}(S_{11} + d\,S_{21}) + A_{26}\,d\,S_{22} + A(S_{16} + d\,S_{26}) \right\} \qquad (9)$$

with $D = \dfrac{dS_{22}}{r(S_{11} + d\,S_{21})}$, one gets:

$$y'_f = \delta_i . \left\{ r(S_{11} + d\,S_{21})(A_{16} + D\,A_{26}) - \frac{A}{K}\,d\,S_{22} \right\} \qquad (10)$$

y'_f independent of δ_i and equal to zero if:

$$A_{16} + D\,A_{26} = A\,\frac{D}{K} = -\frac{A}{r} . \frac{S_{16}\,S_{22}}{1 + K(S_{11}\,S_{26} - S_{21}\,S_{16})} .$$

This proves the necessary change in the dispersion of the magnet (A); this is done by the Q-poles ($Q_3\ Q_4\ Q_5$) - (but depends on the target angle β through r).

SOME RECENT EXPERIMENTS PERFORMED AT LAMPF
(± 1 year)

R. Heffner

Los Alamos Scientific Laboratory,
University of California
Los Alamos, New Mexico 87544 (U.S.A.)

I have been asked to discuss a few of the more recent experiments done at LAMPF with the emphasis on the physics to be learned. In the limited time I have I will, therefore, ignore the details of the accelerator and the experimental areas and treat them merely as a source of projectiles. Obviously, too, many very nice experiments must be omitted, so if an experiment in which you have an interest is omitted, you may ask me about it; if it is <u>your</u> experiment, you may <u>tell</u> me about it! In this connection, I would like to make an important disclaimer, namely, that nearly all of the work I will discuss is not my own and so, though I will try to answer your questions, I am clearly not an expert. Perhaps some of the many experts here can help us out.

I am going to start by discussing muon capture experiments, then go to hadron capture and then on to pion physics. Finally, I shall tell you a little about the neutrino program and finish up with some general experiments to be run in the next year or so.

MEASUREMENT OF THE HYPERFINE STRUCTURE (HFS) INTERVAL IN MUONIUM
(Yale, Heidelberg, LASL, University of Bern, University of Wyoming)

Muonium is one of the simplest systems we can think of involving the muon bound with another particle and Professor Primakoff has told us in some detail about the physics of the interaction. Therefore, I shall just say that the experiment provides a beautiful test of the hypothesis that the muon is just a heavy electron; that is, that the electrodynamics of the muon is the same as the electron when scaled by the mass ratio.

Furthermore, the experiment provides a way of measuring fundamental constants such as the muon-electron mass ratio.

Briefly the idea of the experiment is that polarized positive muons are slowed down in an inert gas such as krypton where they capture an electron. Fig. 1 shows a Breit-Rabi diagram for the HFS splitting ($\Delta\nu$) in muonium. The numbers under the heading "No Microwaves" give the initial populations of the magnetic substates and we see that the system has a polarization of 50%. If we observe the positrons emitted from $\mu^+ \rightarrow e^+ + \overline{\nu}_\mu + \nu_e$ decays while in this state of polarization, we shall see a forward-backward asymmetry with respect to the muon spin. Now if microwaves with frequencies near the HFS interval are applied, transitions are induced between various substates and the result is that we can have equal populations of the substates, namely zero polarization. The object of the experiment is then to observe the positrons along the direction of the initial muon polarization (and in the opposite direction) both when the field is on and when the field is off and then to vary the microwave frequency over the resonance region. The signal S is then

$$S \equiv \frac{(e^+/\mu)_{ON} - (e^+/\mu)_{OFF}}{(e^+/\mu)_{OFF}}$$

where (e^+/μ) is the number of observed e^+ per muon stop. S is a maximum ($\pm$) on resonance where the polarization is zero. The location of the resonance peak is then the HFS interval $\Delta\nu$. I should say that the experimenters have evolved much more sophisticated methods than the one which I describe, but this gives the basic idea and was actually used too.

The experiment was performed in a nearly zero magnetic field region and the krypton gas pressure was varied between about 1.6 and 5.3 atmospheres so an extrapolation to zero pressure could be made. The preliminary value of $\Delta\nu$ was measured to be 4463302.2(1.5) kH_z which is a precision of .33 parts per million (ppm). Using the theoretical QED expression for $\Delta\nu$ together with the measured value and the measured values of other fundamental constants which enter, these experimenters deduced the muon (+)- electron mass ratio to $\pm$ 2 ppm, the most accurate measurement of this quantity thus far. One might remark that the mass ratio μ^-/e^- is known only to $\pm$ 100 ppm! In the future this group plans to "repeat" the experiment at 10 kGauss magnetic field where the ratio of the μ^+ to proton magnetic moment can be obtained directly without recourse to the theoretical QED expression.

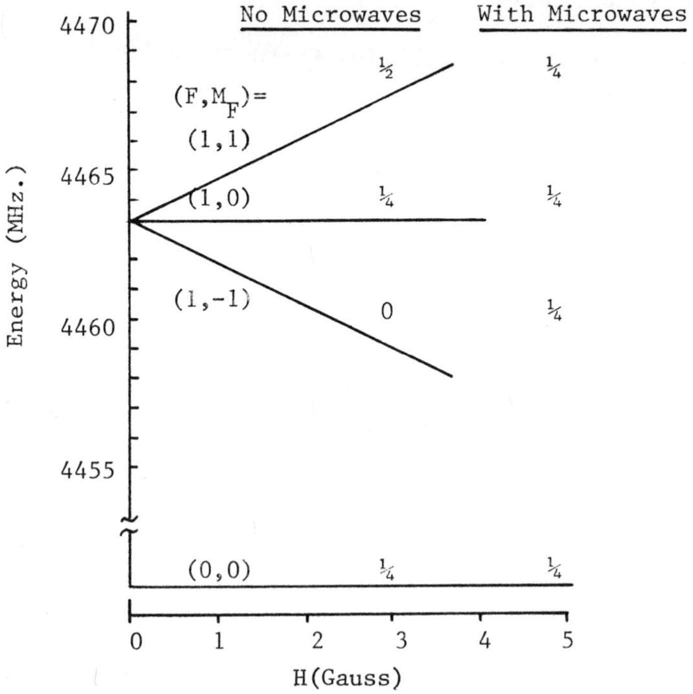

Fig. 1. Breit-Rabi diagram for muonium. The numbers under "No Microwaves" give the relative substate populations without a microwave field and the numbers under "With Microwaves" give the populations with a resonant field.

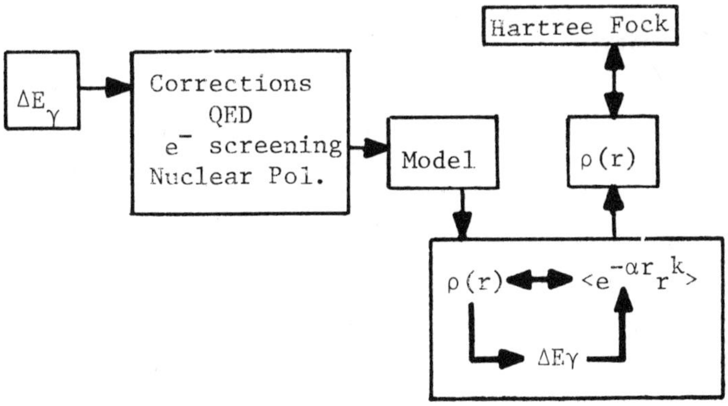

Fig. 2. Schematic diagram showing analysis procedure in obtaining a nuclear monopole charge distribution from measured transition energies.

NUCLEAR CHARGE DISTRIBUTIONS FROM MUONIC ISOTOPE SHIFTS IN THE MASS 60 REGION

(LASL, MAINZ, ERDA, Florida State, Oregon State)

Moving up from the simple muonium system I would like to tell you about some high precision work measuring the changes in the nuclear monopole charge distribution with changing neutron number. Much of what I have to say is taken from a talk by R.B. Perkins at the 1975 Washington APS meeting. Negative muons were stopped in a thick target (A ~ 60) and the 2P → 1S X-rays (~ 1.2 MeV) were observed in a Ge(Li) detector with a precision of about ± 40 eV. Sixteen separated isotopes in the Fe-Zn region were measured.

Fig. 2 is a sketch of the analysis scheme one must perform to obtain the charge distribution, ρ. Very simply, one must correct for all of the other effects (e^- screening, QED, nuclear polarization, etc.) which can cause energy shifts in addition to the effect under study, in this case the finite sized charge distribution. Furthermore, one must have a model for ρ. These experimenters used a two-parameter Fermi distribution, fixing the skin thickness and allowing the radius to vary. The procedure is then to assume a charge distribution and then calculate the transition energies from the Dirac equation putting in all of the corrections discussed above. Now, as you probably know, Ford & Wills [1] and later Barrett [2] found that one can characterize the charge distribution in a relatively model independent way essentially with a single moment, $\langle e^{-\alpha r} r^k \rangle$, where k and α are functions of the nuclear charge and the particular transition. The results here are expressed in terms of an equivalent radius for a uniform charge distribution, R_k. Fig. 3 shows the measured isotope shifts in the Fe-Zn region. Two points are of interest here, one experimental and the other physical. The first is the spectacular precision for ΔR_k, $\pm$ 1 x 10^{-3} Fm. This is to be compared with $\pm$ 10 x 10^{-3} Fm obtained from e^- scattering experiments. The second is the monotonic, almost linear, decrease in ΔR_k as neutrons are added, indicating that the additional neutrons have less and less of an effect on the proton core as more neutrons are added. The surprising thing is that the decrease is nearly linear even as one goes through the region of the closing of the $1f_{\frac{7}{2}}$ proton shell, and this, to my knowledge, is not understood.

This sort of behavior is not new as is shown in Fig. 4 where the ratio of the experimental energy shift to the shift expected if the radius had an $A^{1/3}$ dependence is plotted [3]. The behavior repeats in every major neutron shell, with some deviations in the deformed region.

Negele and Rinker [4] have performed Hartree-Fock calculations using a finite-range density dependent Hartree-Fock theory with a realistic interaction and phenomenological pairing force. They

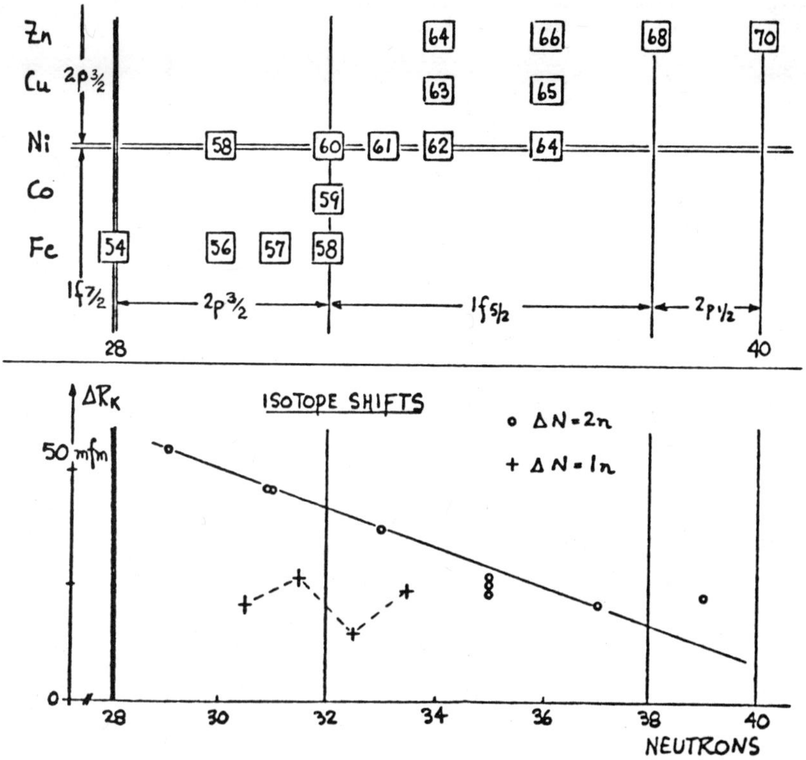

Fig. 3. The bottom frame shows the change in the Barrett equivalent radius as a function of neutron number. The points are plotted half way between the measured isotopes. The top frame shows the nuclear species corresponding to the data below.

found that a spherical solution accounts for only about 1/3 of the experimental shifts in the iron isotopes. They also tried a deformed basis with p-p and n-n pairing and found that the trend toward prolate deformation reproduces much more of the shift. However, their calculations did not show a well defined deformation for these isotopes and suggest that the inclusion of non-axial shapes as well as residual n-p forces may be necessary. It is worth noting that while the isotope shifts are not reproduced, the magnitudes of the charge radii themselves can be calculated to 1%.

STRONG INTERACTION INFORMATION FROM MEASUREMENTS OF HADRONIC ATOM X-RAYS (LASL)

Next I would like to move on to hadronic atom X-rays and to talk about work initiated at LAMPF by Mel Leon [5]. As you know,

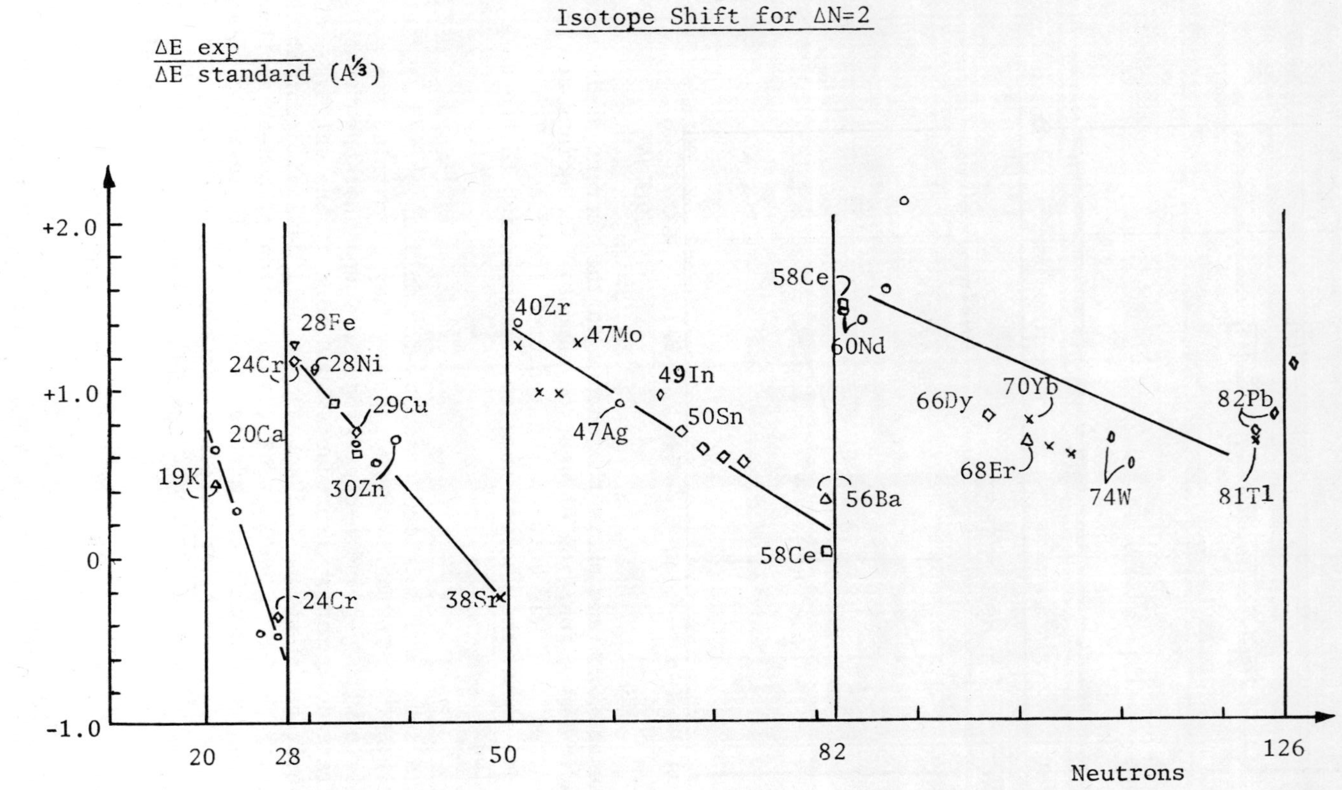

Fig. 4. Compilation of isotope shifts from Engfer et al. [3]. The ratio of experimental shifts (ΔE_{exp}) to shifts assuming an $A^{1/3}$ behavior is plotted as a function of neutron number.

unlike the case for muons the X-rays from the lower lying transitions in hadronic atoms are not observed because the hadron is absorbed relatively quickly from the larger orbits due to the strong interaction. Thus hadrons are not as likely to cause the dynamic nuclear excitations found in muonic atoms. An exception to this can occur, however, when there occurs a hadronic atom transition energy very close to a nuclear excitation energy. (This effect has been known for some time in muonic atoms, too). In such a case the hadron can make a radiationless transition to the lower atomic level while exciting the nucleus to the resonant state. Leon found a particularly good case in ^{112}Cd, illustrated in Fig. 5. Here the 5g → 3d atomic transition is about the same as the excitation energy of the first 2^+ in ^{112}Cd. The wave function for the mixed state can be written as

$$\chi = \sqrt{1 - a^2}\ \phi(5g,0^+) + a\ \phi(3d,2^+)$$

with

$$a = \pm \frac{\langle 3d,2^+ | H_Q | 5g,2^+ \rangle}{E(3d,2^+) - E(5g,0^+)}$$

where

$$\langle H_Q \rangle \ \alpha\ Q_0 \ \langle r^{-3} \rangle \ .$$

Now Q_0 is the nuclear quadrupole moment which can be measured in Coulomb excitation and $\langle r^{-3} \rangle$ is an easily calculated orbital quadrupole strength. The complex energy-difference denominator can be calculated from the Klein-Gordon equation using a phenomenological π-nucleus optical potential.

Referring back to Fig. 5 the strong interaction induced width for the 5g state will cause a depletion of the 5g → 4f and 4f → 3d X-ray yields. Leon did all of these calculations and suggested comparing these transitions in ^{112}Cd to the same transitions in ^{111}Cd where the resonance effect is not so strong. By taking such a ratio the level populations as well as experimental uncertainties tend to drop out. He and his collaborators did the experiment [6] and found good agreement with the theoretical calculations as shown in Table I.

What makes this story even more interesting is the following. Ericson has predicted that the P wave π-nucleus interaction should become repulsive for atoms with $Z \geq 35$ (due to the increasing S-wave π nucleon interaction) and thus the energy shifts should change sign. Dr. Scheck pointed this out to us yesterday. The problem is that 3d → 2p transitions cannot be observed in these heavier nuclei because the pions are absorbed before they reach these low levels. However, Leon has found a resonant condition

TABLE I

OBSERVED AND PREDICTED INTENSITY RATIOS FOR PIONIC CADMIUM

$$R_\alpha \equiv \frac{5\to4}{6\to5},\ {}^{112}\text{Cd} \qquad \frac{5\to4}{6\to5},\ {}^{111}\text{Cd}$$

$$R_\beta \equiv \frac{4\to3}{6\to5},\ {}^{112}\text{Cd} \qquad \frac{4\to3}{6\to5},\ {}^{111}\text{Cd}$$

	Experiment	Theory
R_α	0.65 ± 0.06	$0.72\ {}^{+\,0.07}_{-\,0.13}$
R_β	0.78 ± 0.11	$0.77\ {}^{+\,0.05}_{-\,0.10}$
$R_\alpha\ \frac{111}{110}$	0.69 ± 0.09	0.73 ± 0.05
$R_\beta\ \frac{111}{110}$	0.81 ± 0.10	0.79 ± 0.03

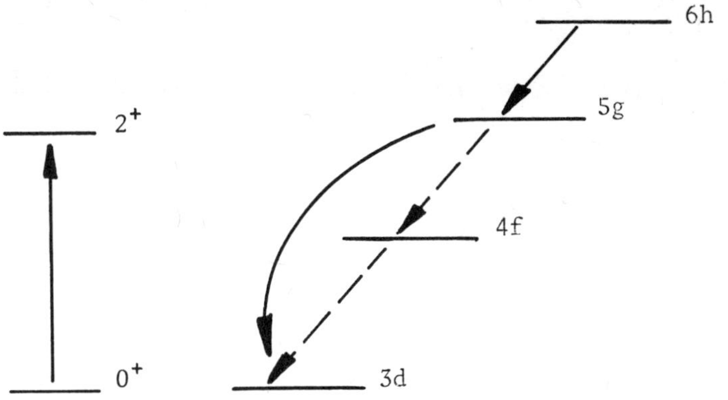

Fig. 5. Schematic diagram for the ^{112}Cd + π^- system showing the resonance between the $0^+ \to 2^+$ transition in ^{112}Cd and the 5g→3d transition in the pionic atom. The dotted line indicates the attenuated X-ray transitions due to the strong interaction induced 5g width.

in ^{110}Pd where the 4f → 3p atomic transition is resonant and thus the 4f → 3d transition should be attenuated (Fig. 6). By measuring this attentuation the energy shift can be determined and hence effect of the π-nucleus P wave force deduced. This experiment will be carried out soon after LAMPF starts up again.

Incidentally the same thing can be tried with more exotic atoms where the strong interaction is less known. Again, Leon has suggested the case of $\bar{p}$ + ^{100}Mo where the resonance effect is expected to be especially strong. His calculation for the experimentally smoothed spectrum is shown in Fig. 7 for various assumptions about the strong interaction. Clearly the calculated spectrum shape is sensitive to the strong coupling and the experiment looks enticing, indeed, though not for LAMPF users!

RADIATIVE PION CAPTURE ON TRITIUM
(LBL, LASL, Lausanne, Case Western Reserve)

I should now like to discuss another pion capture experiment, this time the radiative capture of pions by tritium. The photon spectrum is somewhat sensitive to the three-neutron final state interaction. In particular, the reaction is a good one to search

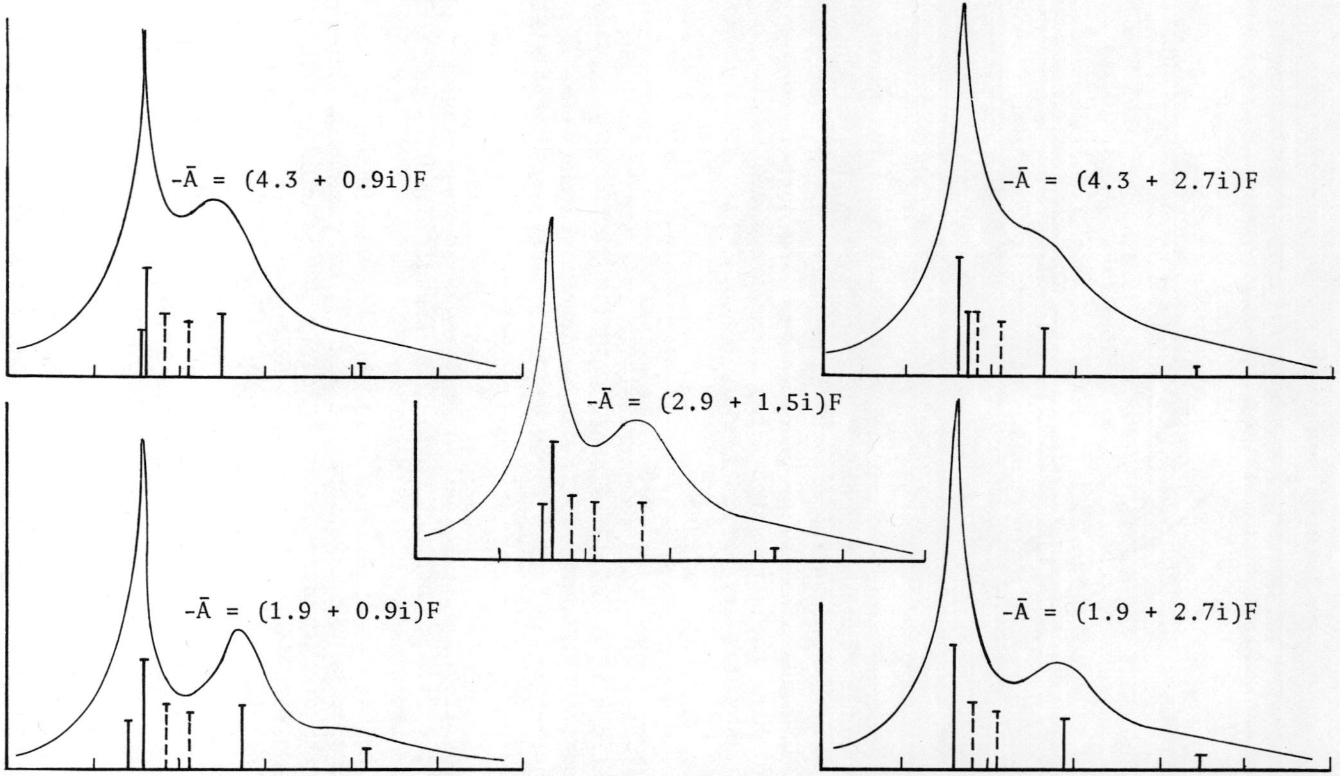

Fig. 7. Calculated line shapes smoothed by an assumed experimental resolution for transitions in the $^{100}Mo + \bar{p}$ system. The different curves show the sensitivity to the strength of the strong interaction.

for a possible bound three neutron state. This is because the gamma ray energy for such a state would be separated from the three body continuum and because the momentum transfer to the three neutron system is small since the photon carries off most of the pion rest mass energy. Furthermore, the reaction is interesting because it is sensitive only to the isospin 3/2 structure in the three nucleon system. Most other feasible reactions produce both 1/2 and 3/2. One can also ask about the possibility for seeing evidence for three-body nuclear forces; that is, can the gamma spectrum be fit by two-body forces alone? Finally, the branching ratio is of interest to the elementary particle picture of nuclei which Dr. Deutsch told us about relating muon capture, β-decay and so on.

Briefly, the negative pions were stopped using a liquid tritium target (60,000 Curies!) and the gamma rays were observed using a pair spectrometer. The overall efficiency of the spectrometer was about 10^{-4} and it had ~ 4 MeV FWHM energy resolution. About one in twenty pions were stopped in the target.

Fig. 8 shows the spectrum obtained after the background from the stainless steel target walls has been subtracted. There is clearly no evidence for a bound tri-neutron in the spectrum. The preliminary measured branching ratio for radiative capture is 4.1

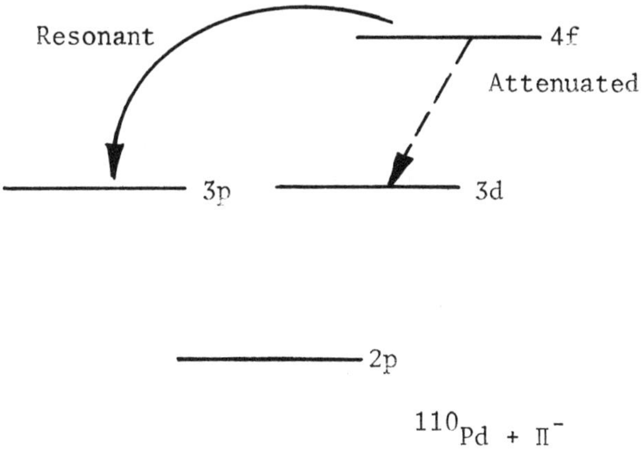

Fig. 6. Schematic diagram for the $^{110}Pd + \pi^-$ system. The notation is the same as for Fig. 5.

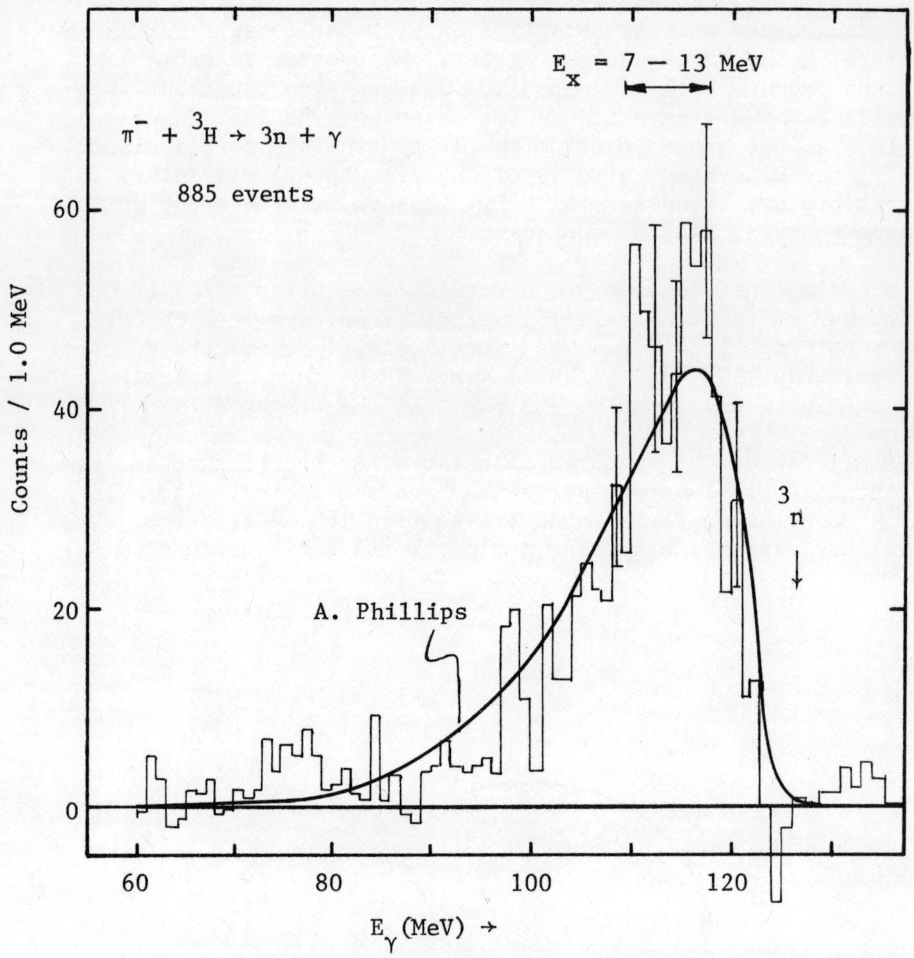

Fig. 8. Measured gamma ray spectrum for $\pi^- + {}^3H \rightarrow 3n + \gamma$ after background subtraction. The notation ^3n indicates the position of a possible bound three neutron system. The solid curve is a calculation by A. Phillips using the Amado model in the impulse approximation for S-wave capture.

±.7%. The solid curve is a calculation of the spectrum shape by Dr. Phillips who will speak to us about his work next Monday. I will just say that, as you can see, he gets good agreement with the measurement without the need for three-body forces. His calculated branching ratio is 6.5 ± 2%. One can also see that there is no evidence for any low-lying resonances which would be manifest as a bump above and beyond the curve calculated by Phillips.

LOW ENERGY $\pi^{\pm}$ p ELASTIC SCATTERING (LASL, Arizona State)

The pion-nucleon system has been well studied with high precision (≃ 1-2%) at incident energies greater than 100 MeV but the published lower energy data is sparse and generally of $\geq$ 10% precision. This situation will be remedied very soon. The interest in this reaction should be clear to all of us now since we have seen so many of those little $t_{\pi N}$ amplitudes in the morning lectures concerning the pion-nucleus optical models. Clearly for detailed studies of pion-nucleus reactions the elementary pion-nucleon amplitudes must be well established.

The pion-nucleon problem has its own intrinsic interest, however, and one of these areas of interest concerns the isosinglet (a_1) and isotriplet (a_3) S-wave scattering lengths. Weinberg [7] has given us a prediction for these quantities using PCAC, current algebra techniques, and the soft pion approximation. He finds $a_1-a_3 = 0.3$ and $a_1+2a_3 = 0$. Most of the previous experimental tests of these predictions have employed the high precision higher energy data (> 100 MeV) and extrapolations to threshold via dispersion relations. In a most recent calculation by Jacob, Hite and Moir [8] using so-called "interior" dispersion relations the values $a_1-a_3 = .265 \pm .003$ and $a_1+2a_3 = -.003 \pm .008$ are obtained.

Another approach is to do an energy dependent phase shift analysis with (as yet non-existent) high precision low energy data and to extrapolate a shorter distance to threshold. Dodder *et al.* [9] have taken such an approach with the published low energy data using an explicitly energy dependent R-matrix analysis. They obtain $a_1-a_3 = .304 \pm .020$ and $a_1+2a_3 = .007 \pm .020$, in agreement with the other values. The larger errors are presumably due to the poorer quality of the data. Hopefully the new low energy data will reduce these uncertainties considerably. It would be quite interesting, for example, to find $a_1+2a_3 = 0$ to a few per cent since the soft pion theorems are expected to be good only to ≃ 10 - 15%. Maybe this would be telling us something!

The experiment in progress at LAMPF has measured $\pi^{\pm}$ p at incident energies of about 30, 40, 50 and 70 MeV over an angular range 50 - 150 degrees in the center of mass. The angular resolution is a few degrees and the expected precision about $\pm$ 5% or better. The data taken last fall is still being analyzed so I have no results for you yet. The group plans to improve their precision with additional running this winter and to start on π^{+} d scattering as well. I should also remark that a group at Saclay has taken some very nice $\pi^{\pm}$ p data in this energy region, too, and this should be coming out soon. There is also a nice measurement at 48 MeV for π^{+}p from a group here at U.B.C. (Auld, *et al.*).

$$\pi^{+} + d \rightarrow P + P$$

(University of S. Carolina, ORNL, Virginia Polytech, LASL)

We have just discussed measuring the π-nucleon elastic amplitudes and now it is appropriate to discuss a reaction which samples the off-shell amplitudes, the π^{+} d $\rightarrow$ pp reaction. We can see the necessity of the off-shell behavior by considering a 50 MeV incident pion which has ~ 130 MeV/c momentum. The two outgoing protons roughly share the incident energy and rest mass and so have about 95 MeV or 430 MeV/c momentum each. The π-nucleon vertices are thus off-shell by roughly half the pion mass.

Koltun and Reitan [10] have shown the importance of considering both pion absorption on a single nucleon and scattering from one nucleon and absorbing on the other one, shown schematically in Fig. 9. Recently, Goplen, Gibbs and Lomon [11] have parameterized the form factors for the off-shell π-nucleon T-matrix as $\sim (k^2+\alpha^2)/(q^2+\alpha^2)$, where α is a free cut-off parameter.

The experiment performed at LAMPF measured the angular distributions to a few per cent precision at 40, 50 and 60 MeV pion energies. The data at 50 MeV is shown plotted versus $\cos^2 \theta_{CM}$ in Fig. 10. Also shown are the calculations of Goplen *et al.*, for different values of the p-wave cut-off parameter α_1. The fit is quite good for α_1 = 340 MeV/c. Fig. 11 shows the sensitivity of the model to the percentage D state in the deuteron for $\alpha_0 = 0$ (s wave). The two curves are for 7.57% and 4.57% D waves and the hatches are the different values for α_1. The data suggest about 4.5% D wave, but this is certainly model dependent. The experimenters have extracted a preliminary fit to this angular distribution which is well described by $A(B + \cos^2 \theta_{CM})$, with $A=1.79 \pm .09$ and $B = .27 \pm .02$ and $\sigma_{TOT} = 6.80 \pm .06$ mb.

I would now like to leave the area of experiments which have already taken data at LAMPF and turn to things which will be done in the next year or so.

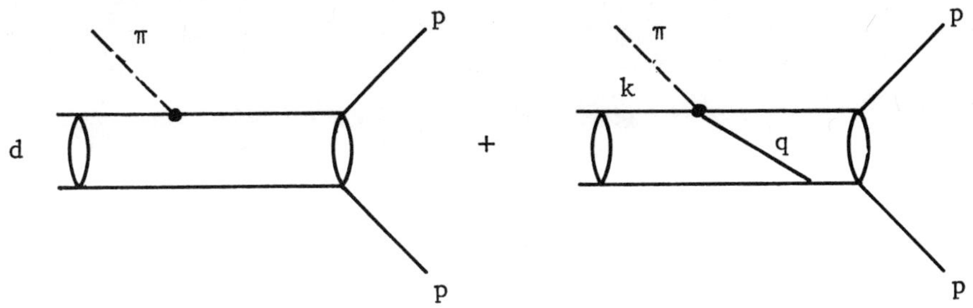

Fig. 9. Schematic diagram for $\pi^+ + d \to p + p$ showing the contributions of absorbtion on a single nucleon and of scattering and then absorbing on the other nucleon. The incident pion momentum is k and intermediate pion momentum is q.

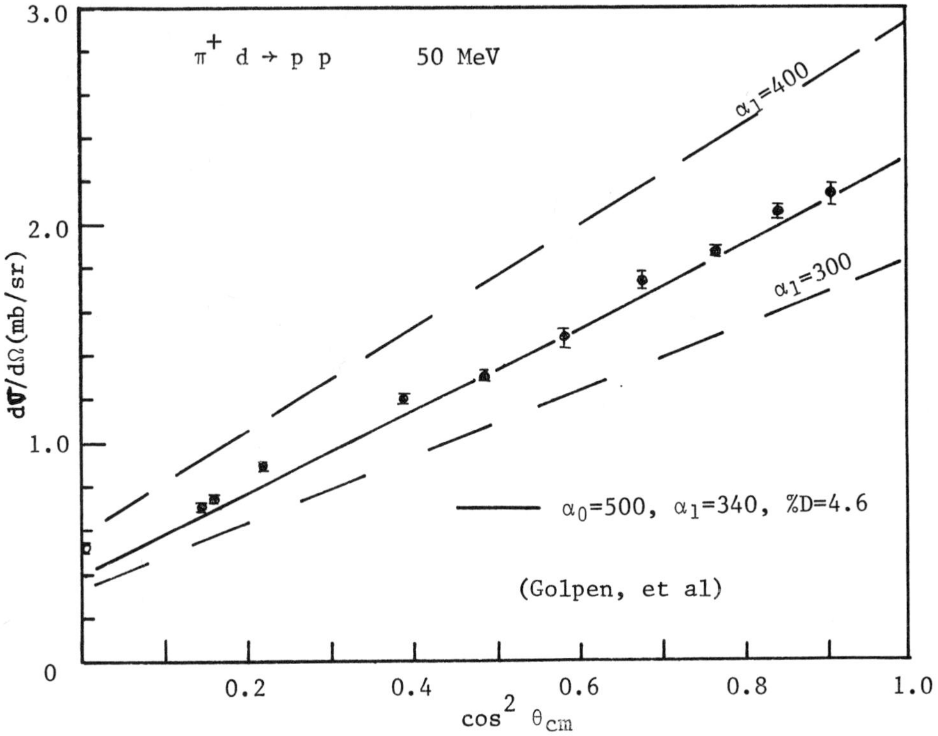

Fig. 10. Data at 50 MeV π energy for $\pi^+ + d \to p + p$ (dark circles). Also shown are the calculations of Goplen et al.

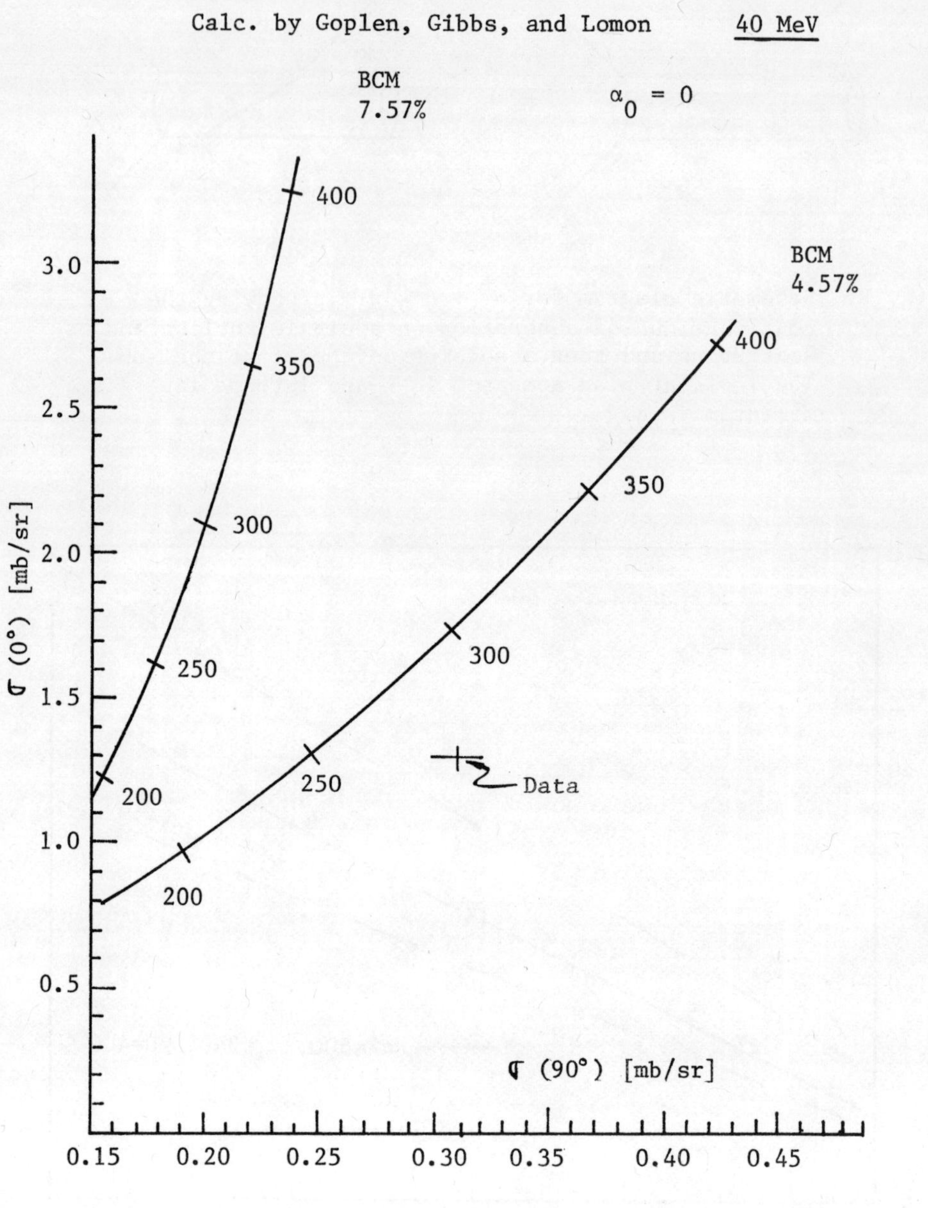

Fig. 11. Calculations of Goplen et al. showing the sensitivity of their model to the percentage deuteron D state (4.57 % vs. 7.57 %). The hatches along the solid curves are for different values of α_1.

NEUTRINO PHYSICS AT LAMPF

I will just briefly describe the neutrino facility and then tell you quickly about one of the first experiments which will run. The facility is conceptually very simple. The 800 MeV proton beam is stopped in a beam stop and adjacent to the beam stop is stacked a lot of shielding. Behind the shielding is the neutrino counting room, so there are no elaborate magnets or things of this sort. The neutrinos come from π^+ and μ^+ decays at rest:

$$\pi^+ \to \mu^+ + \nu_\mu \quad \text{and} \quad \mu^+ \to e^+ + \nu_e + \overline{\nu}_\mu \; .$$

The π^- and μ^- particles created are mostly captured by nuclei before they decay so that the ratio μ^-/μ^+ decays is ~ 1/7000. The maximum neutrino energy is 53 MeV with typical values of ~ 40 MeV. This is to be compared to the sun where the energy is typically < 10 MeV and to a high energy proton synchrotron where the energy is > 500 MeV. For 1 ma proton beams the expected neutrino flux is ~ 10^{14} S^{-1} emitted roughly isotropically.

SEARCH FOR EXOTIC MUON DECAY $\mu^+ \to e^+ \, \overline{\nu}_e \, \nu_\mu$

(Yale, LASL, Saclay, NRC of Canada)

The experiment I shall discuss is concerned with measuring the form of the muon conservation law. You will recall that a muon conservation law was postulated from the non-observation of $\mu^+ \to e^+ + \gamma$ decays and reactions of the form

$$\nu_\mu + Z \to Z' + e,$$

where Z and Z' represent nuclei, for example. All of the known processes involving muons do not as yet uniquely determine the <u>form</u> of the conservation law, however, and in particular we shall consider two forms consistent with the observations, the additive and multiplicative laws. These are:

Additive:
$$\sum_i L_\mu(i) = \text{constant}$$
$$L_\mu = +1 \quad (\mu^-, \nu_\mu)$$
$$= -1 \quad (\mu^+, \overline{\nu}_\mu)$$
$$= \;\; 0 \quad \text{all else}$$

Multiplicative:
$$\prod_i P_\mu(i) = \text{constant}$$
$$P_\mu = -1 \quad (\mu^\pm, \nu_\mu, \overline{\nu}_\mu)$$
$$= +1 \quad \text{all else}$$

One reaction which tests these laws is muonium $\leftrightarrow$ anti-muonium conversion which is allowed by the multiplicative law but forbidden by the additive one. The reaction, incidentally, involves a neutral weak current. Searches for this reaction as well as the equivalent $e^-e^+ \rightarrow \mu^-\mu^+$ have put experimental upper limits on the ratio of the coupling constant for these processes to that for the weak interaction of $\leq$ 5800 and $\leq$ 610, respectively [12]. The "exotic" muon decay, $\mu^+ \rightarrow e^+ \bar{\nu}_e \nu_\mu$, of interest in this experiment, is totally forbidden by the additive law. The multiplicative law allows <u>both</u> $\mu^+ \rightarrow e^+ \bar{\nu}_e \nu_\mu$ <u>and</u> $\mu^+ \rightarrow e^+ \nu_e \bar{\nu}_\mu$ and so the a priori assumption in planning this experiment is that the multiplicative form allows both decays with <u>equal</u> probability.

The object of the experiment is then to search for $\bar{\nu}_e$ coming from the proton beam dump using a large water based Cerenkov counter surrounded by anti-coincidence veto counters to remove charged particle background. The $\bar{\nu}_e\, p \rightarrow e^+ n$ reaction will be the signal for the $\bar{\nu}_e$. The expected positron spectrum from this reaction is sketched in Fig. 12 along with the estimated background from the oxygen in the water. The positron energy resolution is said to be about 10%. At an estimated counting rate of ~ 60/day for a 300 μa proton beam the experimenters hope to be sensitive to a 10% branching ratio for the exotic decays. The experiment will be calibrated with the ν_e electrons incident on heavy (deuterated) water using the reaction $\nu_e\, D \rightarrow ppe^-$.

THE FUTURE (+1 YEAR)

I shall now just take a few more minutes of your time and list a few of what I consider to be some of the more interesting experiments which will have beam time soon after LAMPF comes on again.

1. Double Charge Exchange
   (a) low energy pion channel (LEP)-0°
   (b) EPICS - angular distribution

2. $\pi^0 \rightarrow e^+e^-$: expect ~ 100 events if unitarity branching ratio is true

3. π^0 spectrometer : 1 - 2 MeV resolution (< 1 MeV ?)
   (a) $\pi^- p \rightarrow \pi^0 n$ first experiment < 100 MeV
   $\rightarrow \gamma n$

4. $p + {}^4He$ (HRS)

5. $\pi^\pm$ scattering (EPICS and LEP)

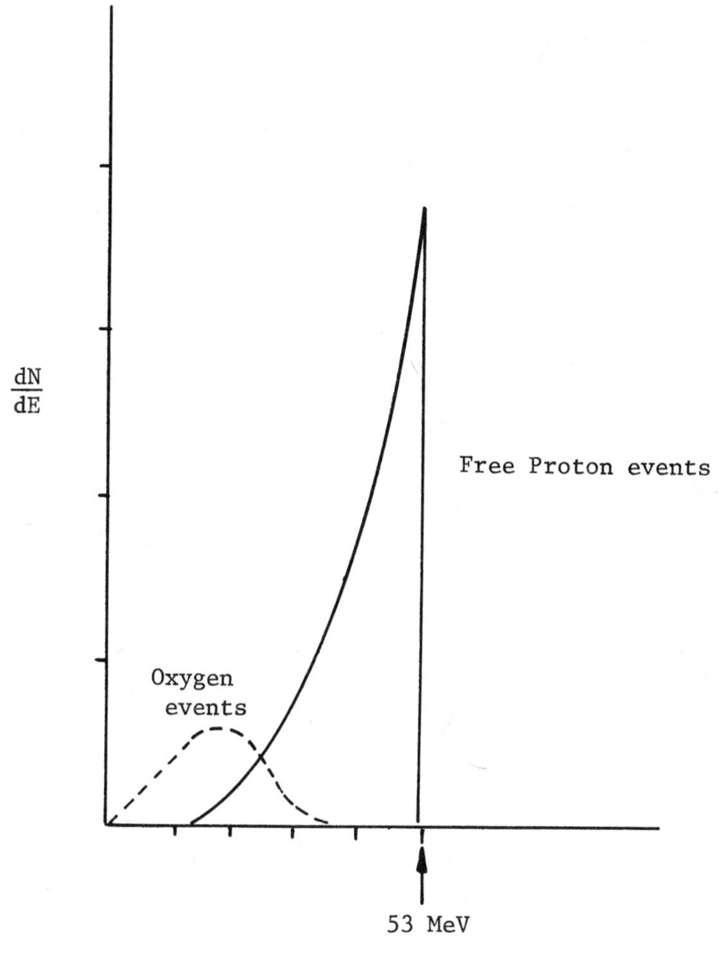

Fig. 12. Calculated positron energy spectrum for the reaction $\bar{\nu}_e p \to e^+ n$. Also shown is the calculated spectrum for an oxygen target.

6. n D scattering

7. Nucleon-nucleon scattering with polarized target

8. $\pi^{\pm}$ D scattering < 100 MeV

9. Continuing : $\pi^{\pm}$ p elastic scattering < 100 MeV

   $\sigma_{TOT}(\pi^{\pm})$ on various nuclei

REFERENCES

[1] Kenneth W. Ford and John G. Wills, PR 185 (1969) 1429.
[2] R.C. Barrett, PL 33B (1970) 388.
[3] The figure is taken from R. Perkins' talk at the 1975 Washington APS meeting. The data are from a compilation by R. Engfer et. al. Nuclear Data Tables 1974.
[4] E.b. Shera et. al. PRL 34 (1975) 535; G.A. Rinker and J.W. Negele Abstract submitted to VI International Conference on High Energy Physics and Nuclear Structure, Santa Fe, New Mexico, 1975.
[5] M. Leon, PL 50B (1974) 425.
M. Leon, PL 53B (1974) 141.
[6] J.N. Bradbury, M. Leon, H. Daniel, and J.J. Reidy, PRL 34 (1975) 303.
[7] S. Weinberg, PRL 17 (1967) 616.
[8] G.E. Hite and R.J. Jacob, PL 53B (1974) 200 and D. Moir, Thesis, Arizona State University (1975).
[9] D.C. Dodder, J.K. Fink, G.M. Hale, and K. Witte, private communication.
[10] See references given in "The Interaction of Pions with Nuclei", Daniel S. Koltun, Advances in Nuclear Physics Vol. 3 (1969) Eds. M. Baranger and E. Vogt.
[11] B. Goplen, W.R. Gibbs and E.L. Lomon, PRL 32 (1974) 1012.
[12] Amato, Crane, Hughes, Morgan, Rothberg and Thompson, PRL 21 (1968) 1709; and Barber, Fittleman, Cheng and O'Neil, PRL 22 (1969) 902.

SOME HIGHLIGHTS OF THE UNIVERSITY OF MARYLAND CYCLOTRON PROGRAM

H.G. Pugh

University of Maryland

Department of Physics, College Park
Maryland 20742, U.S.A.

The Maryland Cyclotron has been operating at full energy for nearly three years now. Thus we have many experimental results published in the literature, and new results which have not yet been published can be found in our Progress Reports. Because of this I will not attempt to cover the entire program but will select some topics which I think may be of special interest to this audience, and on which our cyclotron has produced new results. Clearly, this restriction of topics prevents me from doing full justice to the entire program, so for more information I would like to encourage you to read our annual Progress Report. If anyone would like to be added to the mailing list to receive the current Progress Report and in the future to receive the Progress Report regularly, please send me a post card and I will see to it that this is done.

1. BEAM ENERGIES

Table I shows the particles and maximum beam energies currently available from the cyclotron. The energies given can be exceeded if a special need is demonstrated up to the theoretical limits given. In the case of deuteron, ^{3}He, and ^{4}He, the theoretical limit is provided by the strength of the magnetic field. In the case of protons, the limit is presently given by the radio frequency system and possibly by the electrostatic deflector which equally presents some problems for the high-energy ^{3}He's. Up to the energies listed as regularly scheduled there are no problems due to machine limitations and these beams normally operate extremely reliably, and reproducibly.

TABLE I

BEAM ENERGIES FROM THE UNIVERSITY OF MARYLAND CYCLOTRON

particle	maximum regularly scheduled (MeV)	probable limit (MeV)	limiting factor
p	100	$\geq$ 120	r.f.
d	80	90	field
^{3}He	200	240	field
^{4}He	160	180	field

2. LAYOUT OF THE CYCLOTRON

The layout of the cyclotron and experimental areas is somewhat unusual in that there are two experimental areas vertically one above the other as shown in figure 1. In this figure you can see the beam analysis system which provides an energy resolution of 12 kilovolts in the 100 MeV beam for slits of width 1mm.

Figure 2 shows the layout of the lower experimental areas. Here we have the possibility of doing experiments with moderate resolution either using the entire beam from the cyclotron which has an energy spread of about 200 keV at 100 MeV, or by using the switching magnet to improve the energy resolution to a level of 50 keV or so. The five beam lines shown in this figure are all used for experiments.

Figure 3 shows the layout of the upper experimental area where we eventually expect almost all experiments requiring a high quality beam to be performed. At the moment we have one beam line in operation in the upper area. Because of the extremely good quality of this beam and the very low backgrounds associated with it, it has been especially suitable for the study of low-energy gamma-rays produced by 100 MeV protons. It is also used for any application of nuclear spectroscopy where extremely high quality data are essential. A second line is in the process of being installed in the upper experimental area. This will lead to the spectrometer magnet which is the 40" radius, 180°, n = 1/2 magnet from the old Minnesota linear accelerator. We hope to have this operating next year. This beam also provides the possibility of doing neutron spectroscopy.

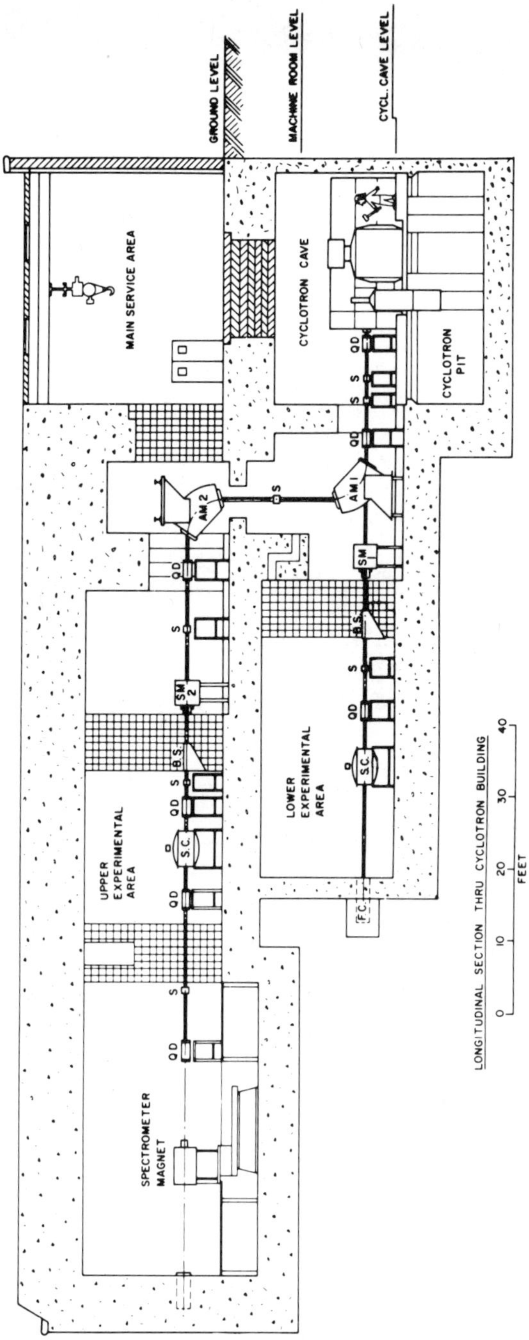
GROUND LEVEL
MACHINE ROOM LEVEL
CYCL. CAVE LEVEL
MAIN SERVICE AREA
CYCLOTRON CAVE
CYCLOTRON PIT
QD
S
AM 1
AM 2
SM 1
SM 2
B.S.
S.C.
FC
UPPER EXPERIMENTAL AREA
LOWER EXPERIMENTAL AREA
SPECTROMETER MAGNET
LONGITUDINAL SECTION THRU CYCLOTRON BUILDING
0
10
20
30
40
FEET

Fig. 1

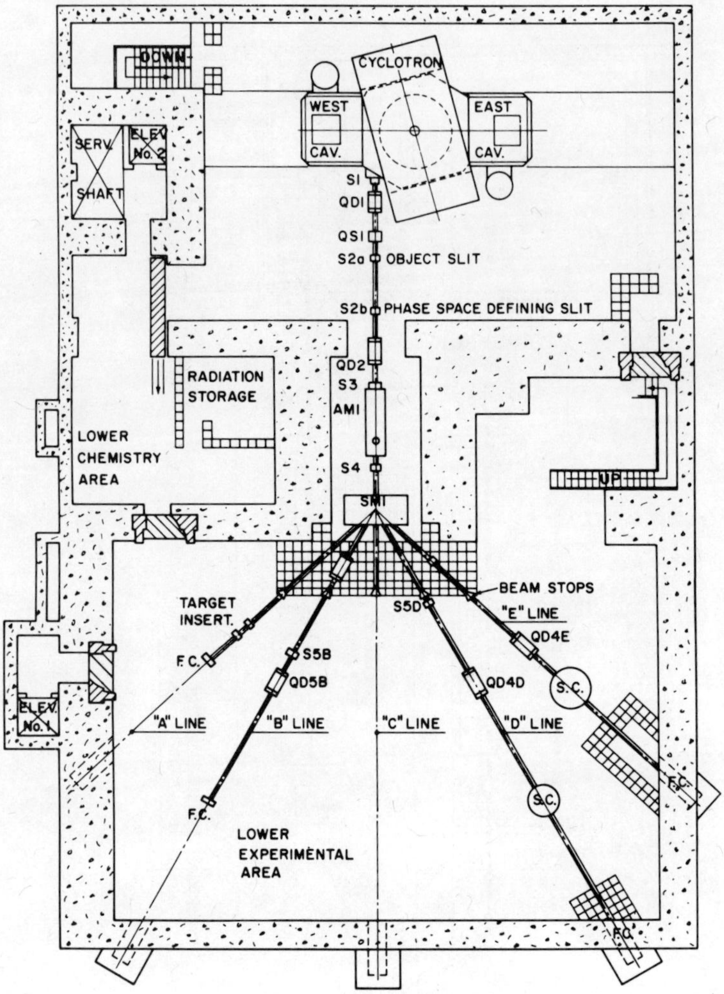

FLOOR PLAN
CYCLOTRON CAVE-LOWER EXPERIMENTAL AREA

Fig. 2

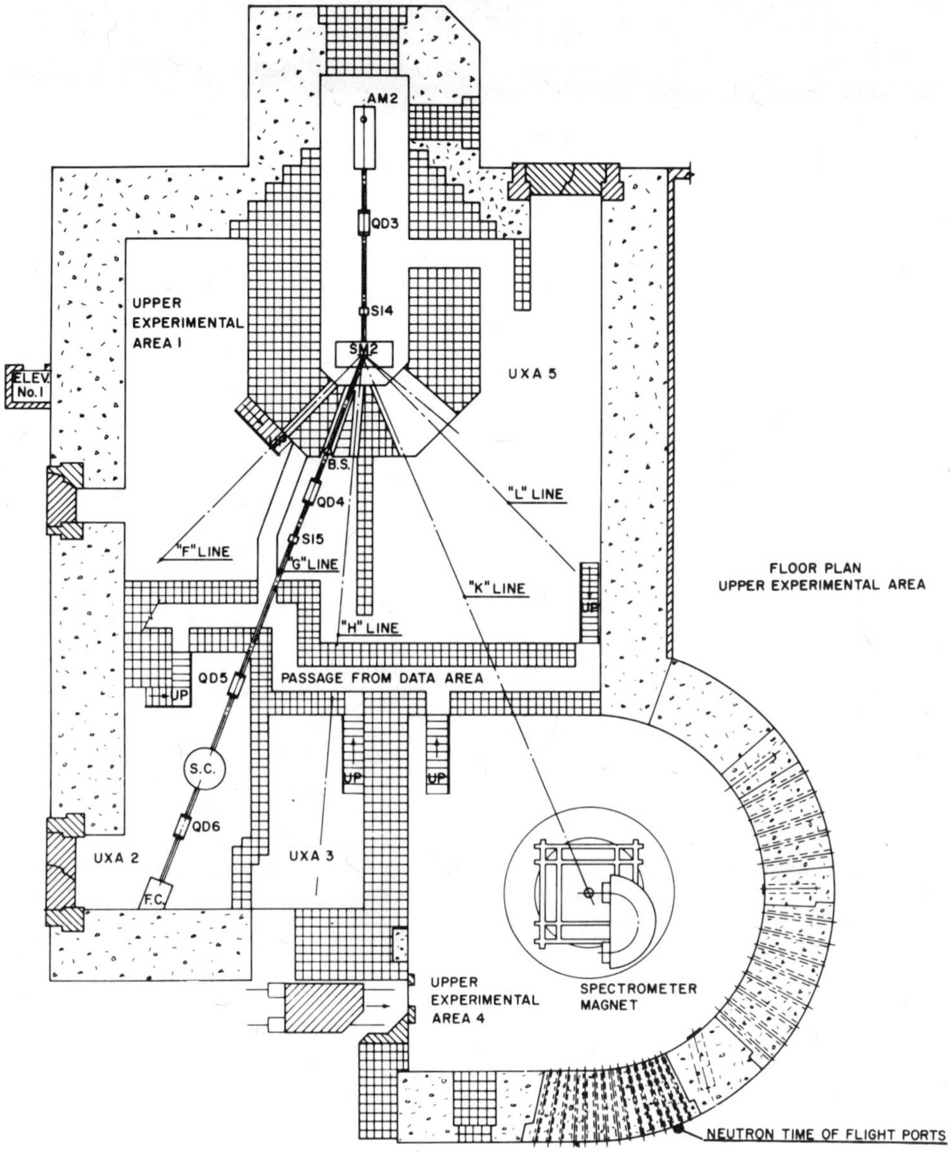

Fig. 3

3. PERSONNEL

Table II gives a list of the Physics and Chemistry faculty who are working with the machine. You will notice that the list is fairly short. The entire facility is approximately one quarter the size of TRIUMF in personnel and funding.

TABLE II

Full-time Faculty (including Research Associates) at the University of Maryland Cyclotron.

PHYSICS

C.C. CHANG	W.F. HORNYAK
N.S. CHANT	B.TH. LEEMANN
P.H. DEBENHAM	H.G. PUGH
J.-P. DIDELEZ (Visitor)	P.G. ROOS
D.A. GOLDBERG	R.I. STEINBERG
H.D. HOLMGREN	N.S. WALL

CHEMISTRY

V.E. VIOLA	W.B. WALTERS
R.A. MOYLE	

4. SOME EXPERIMENTS

I would now like to come to a discussion of a few of the experiments which have been performed and of the conclusions reached from them. In order to be as up to date as possible I will present you with conclusions which can be drawn from even the unpublished results. This, of course, is a very risky procedure and I would like to emphasize that the conclusions are my own version of conversations in which I have engaged at Maryland and therefore cannot be taken as definitive. I would like to refer you to the published literature and the progress reports and to the individuals listed in Table II for the unvarnished truth.

a. Pion Workshop

We do not have a meson factory at Maryland, but we do have a pion workshop. We make our pions very carefully, one at a time. We have made a study of subthreshold pion production and in particular, studied the (^{3}He, π^0) reaction at 200 MeV. The surprising thing about this reaction was not that the cross section was low. It is low, of the order of 10^{-35}cm^2/sr.MeV. But it is much lower than theoretical predictions for coherent production. The experimental result was presented in the Uppsala Conference Proceedings. Banerjee and Wall have recently made some progress towards understanding the process by means of a microscopic treatment of the ^{3}He. At least they have been able to generate theoretical predictions below the experimental observation which is encouraging.

b. Elastic Proton Scattering at 100 MeV

We have obtained some data limited by the availability of high resolution detectors. We have recently obtained 50 kilovolts resolution with lithium drifted germanium detectors and these may enable the program to get well underway. So far, an important conclusion is that polarization data are also necessary. Spin orbit effects in the elastic cross section are comparable to the effects produced by Pauli correlations. This is before we even begin to worry about including short range correlations.

c. Quasi-Free Processes

The (p,2p) reaction has been studied on a variety of light nuclei. It is interesting and significant that the 100 MeV results are qualitatively similar to data at 150 MeV rather than to data at 50 MeV. Analysis of the results in DWIA shows that distortion effects become disastrous only below 100 MeV. This result is important if one considers using (p,2p) reactions as a spectroscopic tool because techniques at 100 MeV are much easier than at higher energies. Of course we cannot study the most deeply bound states, but then this was never expected.

The (p,pα) reaction has been studied also on a variety of light nuclei. A DWIA analysis has been remarkably successful for this process and gives alpha particle parentages which are consistent with shell model predictions. There have been a few measurements made on targets as heavy as calcium but no systematic analysis as yet. We are attempting to increase our counting speeds so that we can extend the measurements to heavier nuclei.

A third kind of process has recently been found to be interesting: that of the quasi-free reaction. The (p,d^{3}He) reaction gives in principle the same kind of information as the (p,pα) reaction. This is made clear by the two diagrams shown below:

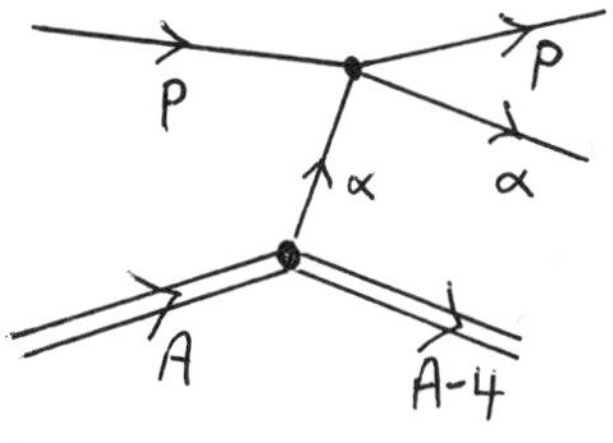

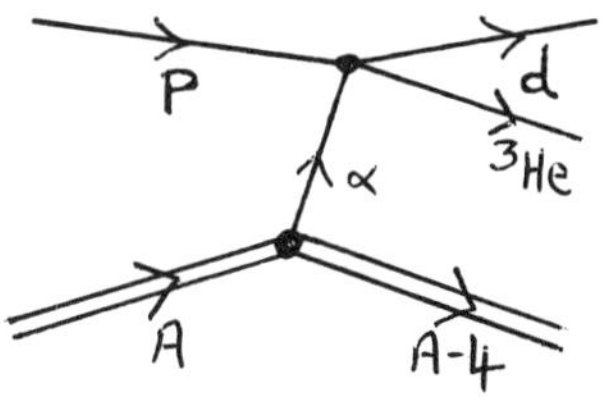

An analysis of the (p,d^3He) data also in DWIA is quite strikingly successful and confirms the information obtained from the $(p,p\alpha)$ reaction. However, some interesting discrepancies remain which will require further investigation. A detailed discussion of these results is given in papers by Chant and Roos at the recent conference on Clustering Phenomena in Nuclei, held at Maryland.

d. Gamma-Rays Following Proton Bombardment

Lithium drifted germanium detectors show a tremendous variety of gamma rays in the region of a few MeV and below. When almost any target is bombarded with any particle very wide-ranging knowledge of nuclei and nuclear energy levels is needed in order to unravel the many different processes that clearly occur. Fortunately the high resolution and high accuracy of the Ge(Li) detectors enable the final nuclei to be identified with few ambiguities. Detailed results have been obtained for proton bombardment of ^{58}Ni, ^{56}Fe, and for ^{4}He bombardment of ^{27}Al.

The final products are generally similar to those obtained from pion absorption except for differences in the initial stages of the cascade. Detailed preequilibrium and evaporation calculations which have been performed are very successful in describing the process. The phenomenon of "multiple alpha emission" seems to be a simple result of the high binding energy of "alpha particle" nuclei. Recent detailed comparisons of pion and proton absorption results seem very interesting and may help to unravel what are the principal processes involved when a pion first interacts with the nucleus.

I cannot resist mentioning the interesting result observed by Hornyak that when ^{232}Th is bombarded with 140 MeV ^{4}He, gamma rays in ^{208}Pb are observed suggesting some catastrophic process in which the whole outer surface of the nucleus is removed.

e. Highly Excited Nuclear States

The high energy beams of the cyclotron make it especially suitable for studying highly excited states. I will give two examples of recent interest.

The giant quadrupole resonance has been observed in inelastic scattering of 70 MeV deuterons. Figures 4 and 5 show spectra for ^{27}Al and ^{58}Ni. The selection of deuterons for projectiles provides interesting comparisons with results obtained using other particles. Firstly, because the deuteron has isospin 0, only isoscalar transitions are observed. Secondly, the results

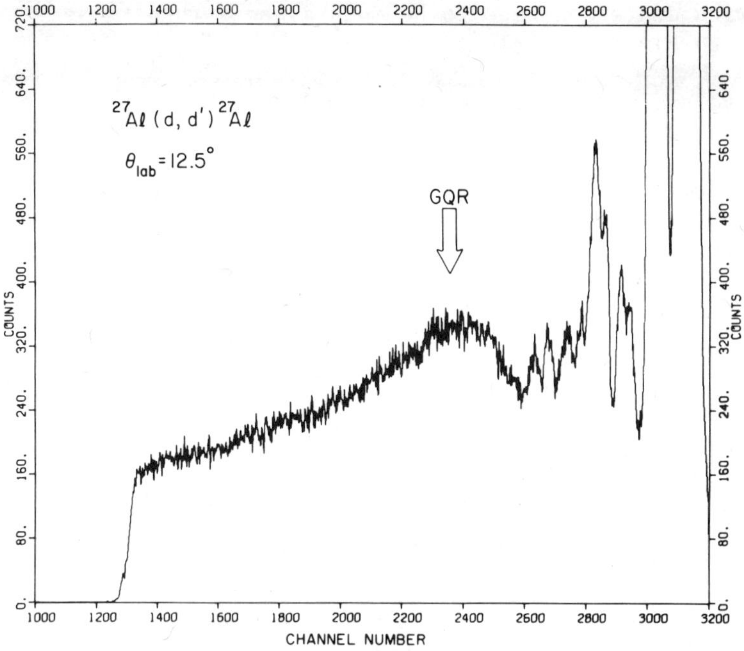
^{27}Aℓ (d, d') ^{27}Aℓ
θ_{lab} = 12.5°
GQR
COUNTS
CHANNEL NUMBER

Fig. 4

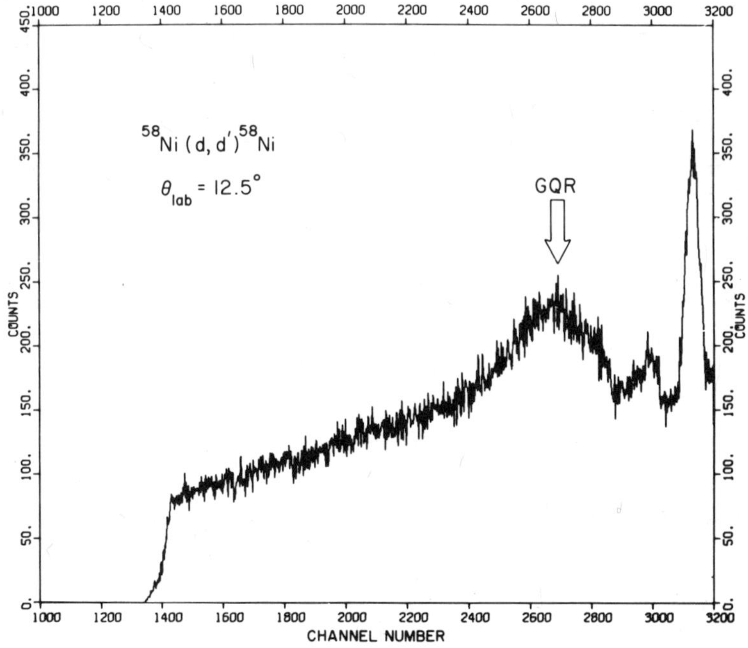
^{58}Ni (d, d') ^{58}Ni
θ_{lab} = 12.5°
GQR
COUNTS
CHANNEL NUMBER

Fig. 5

obtained with deuterons seem to be much cleaner than those obtained with alpha particles.

We have also observed a mysterious new state by proton pickup in the light nuclei. Figures 6, 7 and 8 show spectra for the (d,^{3}He) reaction on ^{6}Li, ^{7}Li and ^{9}Be. All of these show structure at energies corresponding roughly to pickup of 1s protons; however, this does not seem to be the whole story and studies are continuing. Angular distributions for the three targets seem very similar so that the states represent a general property of the light nuclei. These results are correlated with (p,d) neutron pickup results from Uppsala, which in my opinion, are not yet fully understood.

f. Rearrangement Energy Not Observed

There have been many discussions over the years of whether rearrangement energy can be measured. In particular, do spectroscopic results for pickup depend on how quickly the particle is pulled out of the nucleus? A test of this was made by studying the ^{51}V(d,^{3}He)^{50}Ti reaction at 30 and 80 MeV deuteron energies, to complement measurements made at 52 MeV by the Heidelberg group. Transitions to the ground state (0+), 1.55 MeV (2+), 2.67 MeV (4+), and 3.20 MeV (6+) states were studied and accurate angular distributions were obtained. It was found that the spectroscopic factors and ratios of excited states extracted in DWBA were

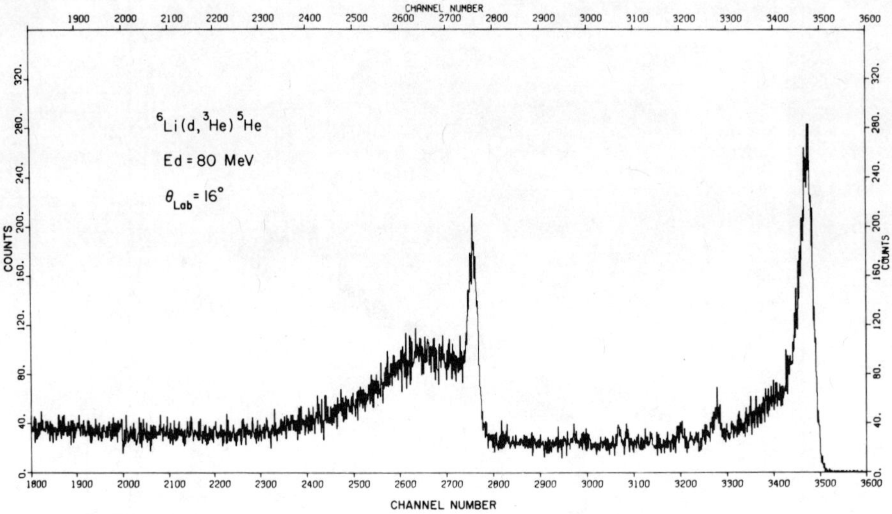

Fig. 6

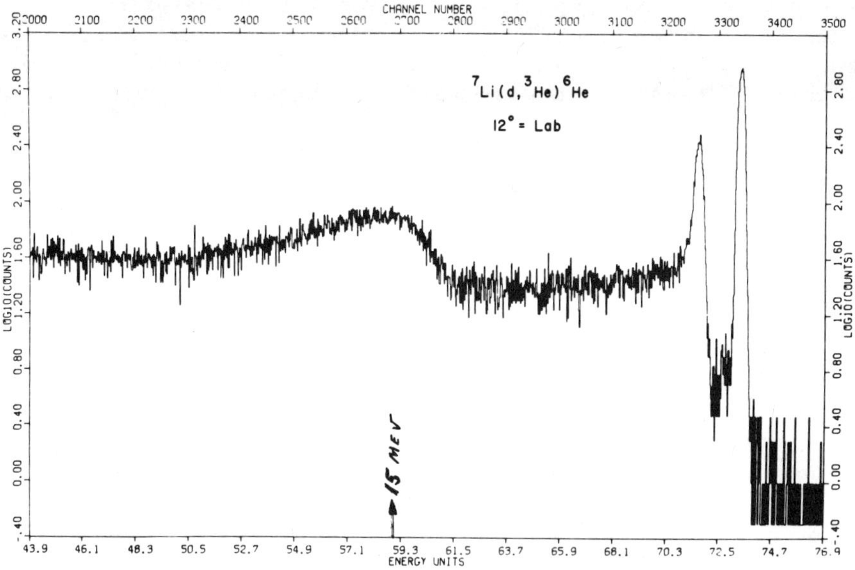
CHANNEL NUMBER
$^{7}Li(d,{}^{3}He)^{6}He$
12° = Lab
15 MEV
LOG10(COUNTS)
ENERGY UNITS

Fig. 7

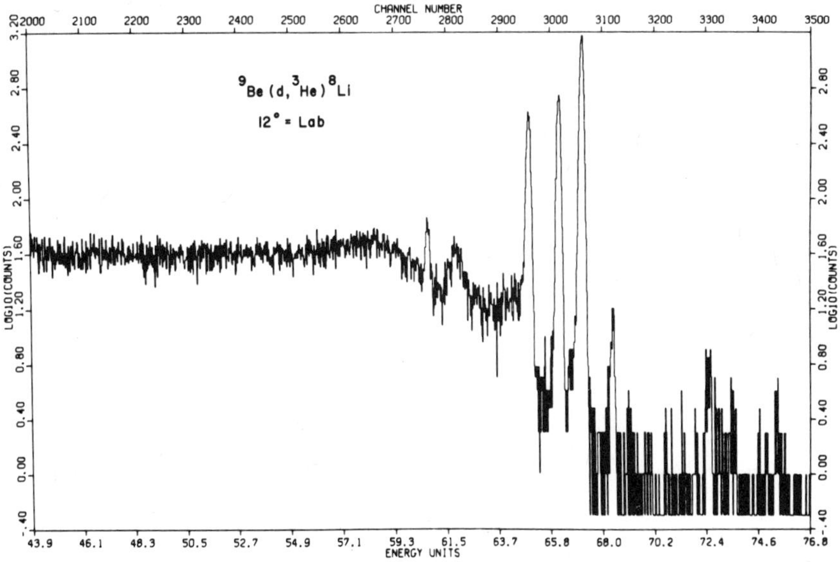
CHANNEL NUMBER
$^{9}Be(d,{}^{3}He)^{8}Li$
12° = Lab
LOG10(COUNTS)
ENERGY UNITS

Fig. 8

independent of bombarding energy. This result holds within an accuracy of 10%. However, it was important to include all the refinements of DWBA to obtain this result. The absence of rearrangement energy in the context of a careful treatment of the reaction mechanism seems to be consistent with the latest discussion of this topic given by Friedman in a recent Physical Review.

4. CONCLUSIONS

The above presents only a very selective and sketchy picture of the cyclotron program. I have omitted all reference to the chemistry program, to few nucleon studies, and to a host of other processes under study. Nevertheless, I have presented enough to show that we relish the prospect of TRIUMF and Indiana coming on the air and of new results to be obtained from those machines (and from LAMPF) in the years to come.

REPORT ON THE STATUS OF THE INDIANA UNIVERSITY CYCLOTRON FACILITY*

AND PREPARATIONS FOR INITIAL EXPERIMENTS

R.D. Bent†

Indiana University

U.S.A.

I. STATUS OF THE FACILITY

The IUCF accelerator consists of three stages (see Fig. 1). The preinjector and injector cyclotron were constructed in the main physics department building and operated there for the first time in May, 1972. They were moved to the new accelerator building in late 1972 and brought into operation again with some modifications and improvements in October, 1973. Assembly of the main cyclotron began in December, 1971 soon after the new accelerator building became available for occupancy and continued in parallel with the reassembly and testing of the small cyclotron. Rough vacuum was first obtained on March 6, 1975, high vacuum on April 2 and first rf voltage on May 15. A beam of 10 MeV protons was injected into the main cyclotron for the first time on June 13, 1975. The present goal is to achieve acceleration in the main cyclotron and extraction during the summer and beam on target in September, 1975. This will allow the first experiments to begin in the late fall of 1975.

---

*Supported by the National Science Foundation and by Indiana University.

†IUCF Scientific Personnel: Faculty Members - R.E. Pollock (Director), G.T. Emery (Associate Director), D.W. Miller (Associate Director), A.D. Bacher, B.M. Bardin, R.D. Bent, P.T. Debevec, D.W. Devins, L.M. Langer, M.E. Rickey, P. Schwandt, P.P. Singh, H.A. Smith and J.G. Wills; Staff Physicists and Research Associates - C.C. Foster, J. Dreisbach, D.L. Friesel, W.P. Jones, R.T. Kouzes, S.A. Lewis and T.E. Ward; Engineers - D.C. Duplantis, J.W. Hicks, W.T. Hoffert, E.A. Kowalski and W.R. Smith.

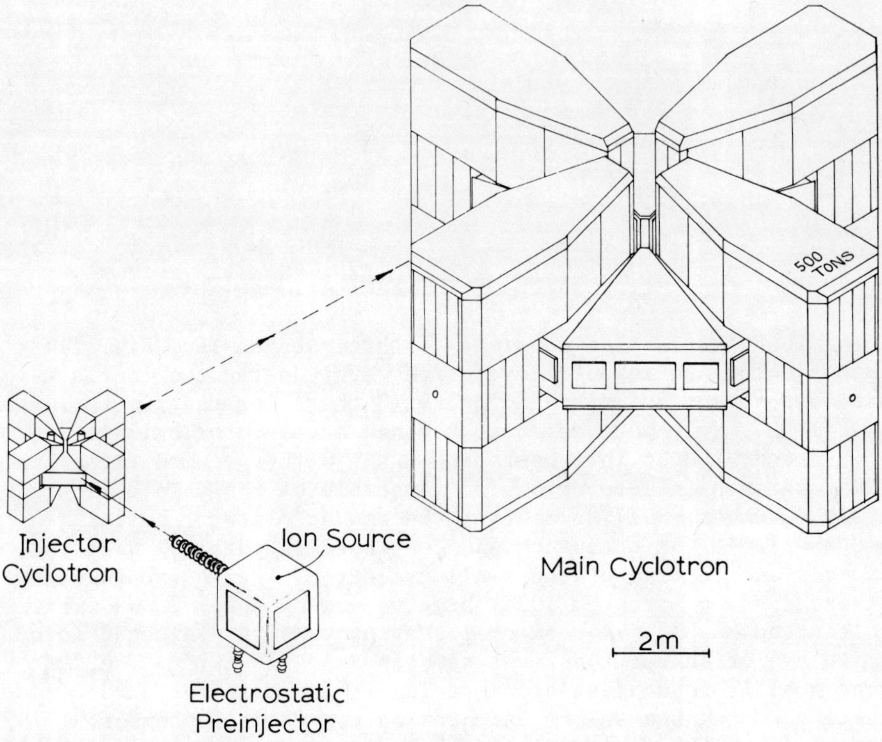

Fig. 1. Approximate scale drawing of the three-stage accelerator

The Indiana cyclotron will produce variable energy beams of light and heavy ions which are well suited for high resolution nuclear structure studies at intermediate energies. The energy ranges at IUCF will overlap at one end with those available at smaller sectored cyclotrons and at the other end with the higher energy meson factories. The beam characteristics anticipated for the first year of operation are shown in Table I. The maximum energies will initially be limited by the terminal voltage limit of the preinjector. A new ion-source terminal and other improvements to bring operation up to the full-energy magnet limit of $(220 \pm 5)Q^2/A$ MeV will be completed by late 1976. The new terminal will stand next to the present one in the large ion-source room and will provide sufficient space and AC power for a polarized ion source and an arc source for multiply-charged ions of Helium and heavier ions.

II. PREPARATIONS FOR INITIAL EXPERIMENTS

The first round of proposals was evaluated by the IUCF Program Advisory Committee (A.D. Bacher, W. Haeberli, D. Hendrie, J. Schiffer, N. Sugarman, L. Wilets, IUCF Associate Director G.T. Emery and IUCF Director R.E. Pollock, Chairman) in February, 1975. Of the 83 different names appearing on the cover pages of the 34 proposals submitted, 62 were from outside IUCF involving 24 different institutions. 6 proposals involved no IUCF participation, 6 proposals involved only IUCF participation and 22 proposals involved collaborations between outside users and IUCF personnel. The total request for beam time was 680 eight-hour shifts. 200 eight-hour shifts were approved for the first 6-month running period. A list of the approved proposals is given in Table II. The fact that no heavy-ion proposals were approved does not accurately reflect the potential of IUCF for accelerating heavy-ions or the current interest in the nuclear physics community for heavy-ion physics. Such proposals were not approved at the present time because beams with $A > 7$ will not be available during the first year of operation.

The layout of the beam lines that will be installed in time for early experiments is shown in Fig. 3. The four main experimental areas are shielded from each other so that set-up can occur in one area while an experiment is in progress in another. It will be possible to operate in this mode since the proposed experiments are well distributed among the different experimental areas. Simultaneous beams in two or more different areas will not be available initially; however, the beam line has been designed so that beam splitting elements can be added at a later time.

TABLE I

ANTICIPATED BEAM CHARACTERISTICS

Beams Available†

Z = 1: $\underline{^{1}H^{+}}$, $\underline{^{1}H_{2}^{+}}$, $^{2}H^{+}$, $^{2}H_{2}^{+}$

Z = 2: $^{3}He^{+}$, $^{3}He^{++}$, $\underline{^{4}He^{+}}$, $\underline{^{4}He^{++}}$

Z = 3: $^{6}Li^{+}$, $^{6}Li^{++}$, $^{6}Li^{3+}$, $\underline{^{7}Li^{+}}$, $\underline{^{7}Li^{++}}$, $^{7}Li^{3+}$

Maximum Beam Energy

Present terminal: (first 12-18 months)	Lesser of 155 MeV or 165 Q^2/A MeV
New terminal: (late 1976)	(220 ± 5) Q^2/A MeV

Energy Resolution

$\Delta E/E \lesssim 0.04\%$ fwhm
(somewhat better for Z = 1)

Time Structure

Macrostructure: 100%

Microstructure: on - 0.2 to 0.4 nanoseconds fwhm
off - typically 30 nanoseconds (rf period)
or by pulse selection up to 250 ns

Beam Intensity

Z = 1: 150nA or (300nA) x (pulse selection fraction)

Z = 2: 100enA (electrical nanoamps)

Z = 3: 30 - 50enA

Intensities on target will be raised by a factor of 3 to 5 with shielding, source and vacuum improvements.

Emittance

Z = 1: 2mm mrad fwhm @ 100 MeV

Z ≥ 2: 4mm mrad fwhm @ 100 MeV

(e.g., 1.0mm x 0.15° fwhm for Z = 1)

†Underlined ions have been accelerated to full radius in the injector cyclotron.

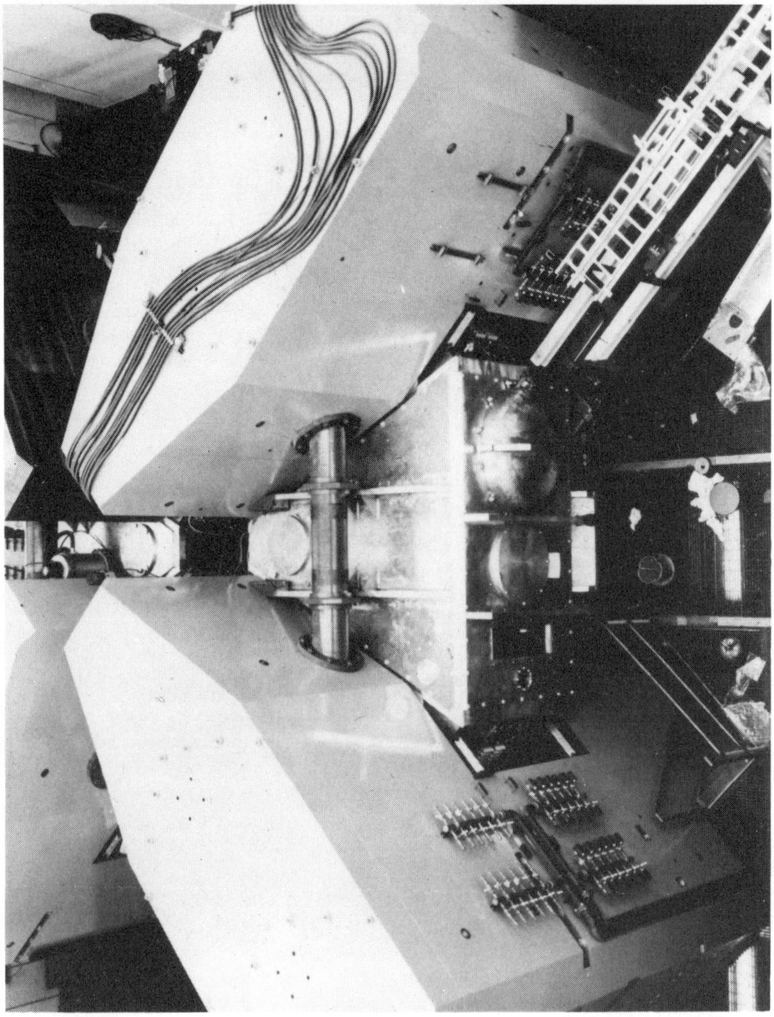

Fig. 2. Photograph taken in March, 1975 of the large cyclotron viewed looking into the north valley. The beam from the injector cyclotron enters from the south, passes through the central region of the main cyclotron and is inflected by magnetic and electrostatic elements into a counterclockwise orbit (as viewed from the top). Acceleration occurs in the east and west valleys and the beam is extracted through the small port visible on the righthand side of the bubble on the north valley vacuum chamber. The beam line to the experimental areas extends off the lower righthand corner of the picture.

TABLE II

RESEARCH PROPOSALS TO IUCF APPROVED FOR FIRST RUNNING PERIOD

(The beam-time unit is an 8-hour shift. The underlined name is that of the spokesman.)

Pion Production Proposals

75-17 *Single charged-pion production in proton-nucleus collisions near thershold;* R.D. Bent, A.S. Broad, P.T. Debevec, R.E. Pollock, IU, L. Gutay, R.P. Scharenberg, Purdue, K.V. Vasavada, Indianapolis; 12 shifts.

75-25 *Studies of pion production near threshold;* G.T. Emery, A.D. Bacher, P.T. Debevec, IU, K. Gotow, D.A. Jenkins, W.C. Lamm, M. Blecher, VPI&U; 6 shifts.

75-31 *Measurement of total (p,π) cross sections through residual activity;* P.P. Singh, M. Sadler, A. Nadasen, D. Friesel, IU; 2 shifts.

Elastic Scattering Proposals

75-20 *Elastic scattering of composite projectiles ^3He and ^6Li from nickel in the energy region around 100 MeV;* P. Schwandt, P.P. Singh, G.S. Adams, A. Nadasen, IU; M.K. Brussel, Illinois; I. Brissaud, Orsay; F.D. Becchetti, Michigan; W.D. Ploughe, Ohio State.

75-29 *Study of ^6Li elastic and inelastic scattering at $E(^6Li)$ = 120 MeV;* F.D. Becchetti, J. Janecke, Michigan; P. Schwandt, P.P. Singh, D.W. Miller, IU.

Collaboration between the authors of these proposals and of proposal 75-02 approved for 25 shifts to study elastic scattering, starting with ^{6}Li at 120 MeV.

Inelastic Scattering Proposals

75-01 *Excitation of giant resonances via inelastic scattering of intermediate energy protons and deuterons;* F.E. Bertrand, D.J. Horen, D.C. Kocher, G.R. Satchler, ORNL; G.T. Emery, D.W. Miller, A.S. Broad, A.D. Bacher, IU; 12 shifts.

75-21 *Excitation of high-spin states by inelastic proton scattering;* G.T. Emery, G.S. Adams, A.S. Broad, D.W. Miller, G.E. Walker, IU; 12 shifts.

Charge Exchange Proposals

75-09 *Search for the giant Gamow-Teller resonance;* P.T. Debevec, P. Schwandt, IU; 12 shifts.

75-10 *Investigation of the (p,n) reaction at intermediate energies;* R.P. Scharenberg, L. Gutay, Purdue; P.T. Debevec, IU; 18 shifts reserved, subject to solution of practical problems.

75-14 (part c), *A study of the isospin makeup of giant resonances;* C.D. Goodman, F.E. Bertrand, Oak Ridge.

75-04 *Neutrons from proton interactions with nuclei;* R. Madey, F.M. Waterman, A.R. Baldwin, Kent State, F.E. Bertrand, Oak Ridge.

8 shifts reserved for collaboration, subject to solution of practical problems, between 75-14 and 75-04.

Transfer and Mass Proposals

75-11 *High-excitation hole states in the (p,d) reaction;* R.E. Pollock, D.W. Devins, D.W. Miller, IU; 7 shifts.

75-24 *Study of (α,^{8}Be) cluster-transfer reactions on light and low-medium mass nuclei;* P.A. Quin, Wisconsin, P. Schwandt, A.D. Bacher, IU; 6 shifts.

75-27 *Study of ^{6}Li-induced α-particle transfer reactions on medium-heavy and heavy nuclei at E(^{6}Li) = 120 MeV;* J. Janecke, F.D. Becchetti, Michigan; 8 shifts.

75-05 *Mass measurements of exotic proton-rich nuclei by (p, ^{6}He) and (^{4}He, ^{8}He) reactions;* A.D. Bacher, R. Kouzes, R.E. Pollock, IU; W. Benenson, E. Kashy, Michigan State; 16 shifts.

Knockout and Quasi-elastic Proposals

75-22 *Studies of knock-out reactions;* D.W. Devins, A.D. Bacher, G.T. Emery, D.L. Friesel, W.P. Jones, H.A. Smith, IU; B.M. Spicer, V.C. Officer, G.G. Shute, Melbourne; 10 shifts.

In-Beam Spectroscopy Proposals

75-15 *Studies of A = 180-195 shape transitional nuclei by in-beam γ-ray spectroscopy;* P.J. Daly, S.K. Saha, Purdue; S.W. Yates, Kentucky; G.T. Emery, T.E. Ward, IU; 9 shifts.

75-03 *In-beam nuclear spectroscopy through (particle, $xn\gamma$) reactions;* P.P. Singh, M. Sadler, G.T. Emery, D.L. Friesel, T.E. Ward, IU; J. Jastrzebski, Z. Sujkowski, Z. Moroz, T. Kozlowski, H. Karwowski, Swierk; 6 shifts.

75-16 *Nuclear structure studies of the cesium isotopes;* R.S. Tickle, W.S. Gray, H.C. Griffin, Michigan.

75-34 *In-beam nuclear spectroscopic studies of the levels of neutron-deficient isotopes of cesium: ^{129}Cs, ^{127}Cs, ^{125}Cs, ^{123}Cs;* S. Jha, P. Boolchand, Cincinnati; G.S. Adams, G.T. Emery, T.E. Ward, H.A. Smith, IU.

9 shifts approved for collaboration between 75-16 and 75-34.

75-08 *Complex nuclear reactions induced by intermediate energy protons;* P.P. Singh, M. Sadler, IU; R.E. Segel, Northwestern; J. Jastrzebski, Z. Sujkowski, Swierk; 4 shifts.

Off-Line Spectroscopy Proposals

75-07 *Studies of neutron-rich nuclei produced by high-energy neutron bombardment;* H.A. Smith, T.E. Ward, G.T. Emery, IU; parasite operation plus 3 shifts.

Applications Proposals: Astrophysics

75-18 *Production of lithium, beryllium, and boron by proton-induced spallation reactions;* A.D. Bacher, R.D. Bent, D.W. Devins, IU; C. Davids, Argonne; 9 shifts.

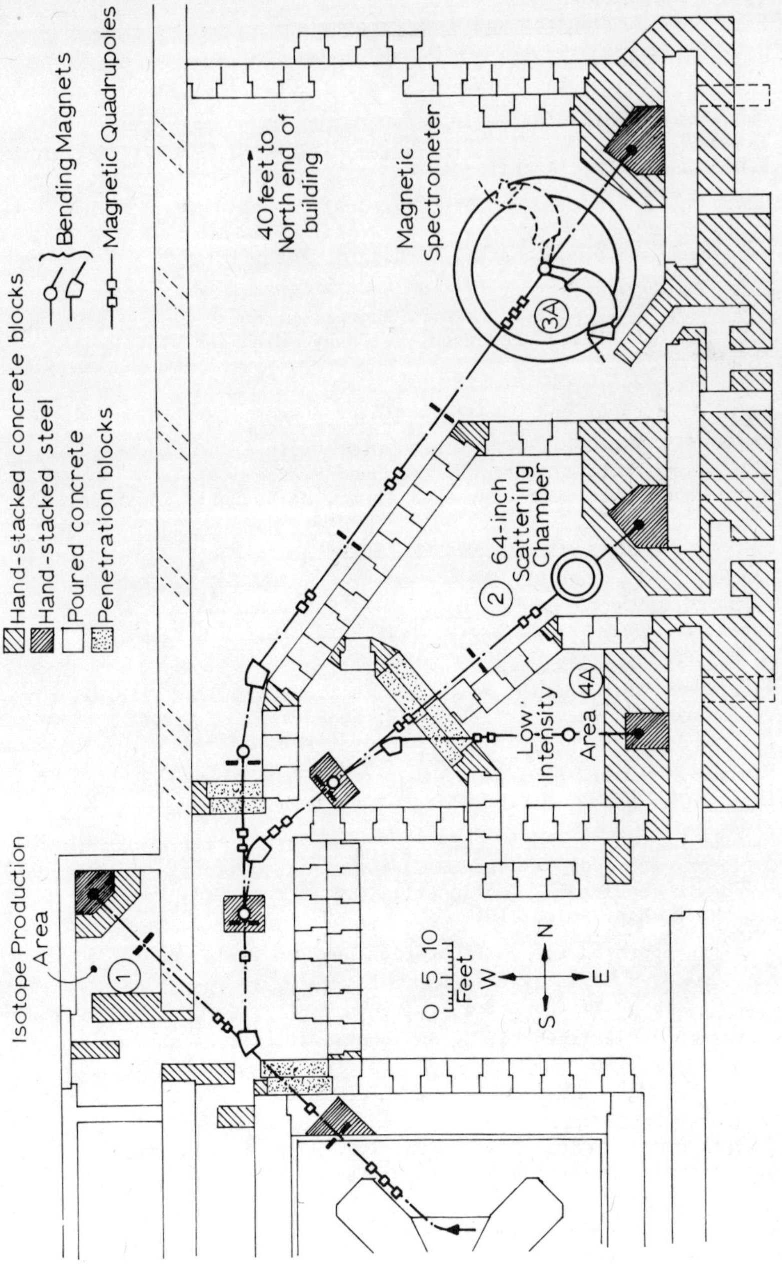

Fig. 3. Experimental area layout

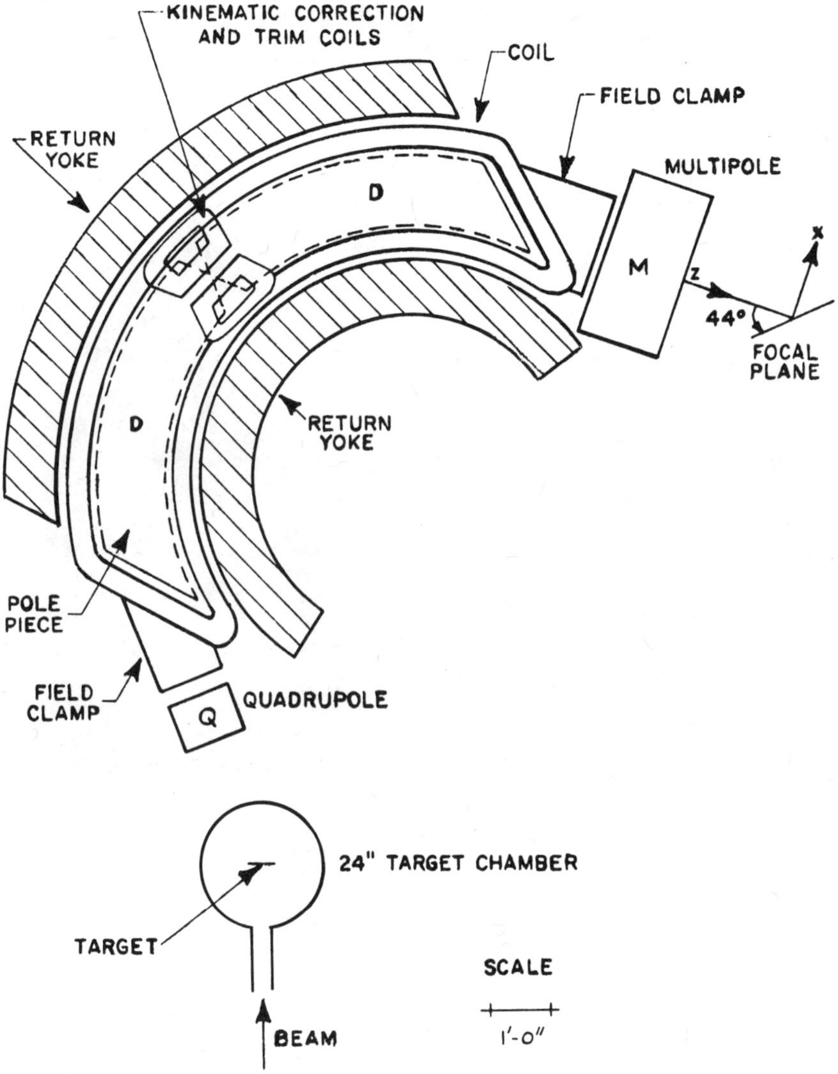

Fig. 4. Layout drawing of the magnetic spectrograph

Fig. 5. Photograph taken in March, 1975 of the experimental areas looking south showing the magnetic spectrograph in the foreground, the 64 inch scattering chamber, and the main cyclotron in the background. The injector cyclotron is under the roof beams south of the main cyclotron, and the preinjector is behind the east wall in the separate ion-source room. The isotope production cave is underground outside the west wall.

A layout drawing of the magnetic spectrograph is shown in Fig. 4. The main magnet yoke and coils are from the spectrometer used by Tyren et al [1] for (p,2p) experiments at the University of Chicago. Following a design study by H. Enge, the Chicago spectrometer was modified to improve its performance by adding quadrupole and multipole elements, new pole pieces with a split in the middle and kinematic and aberration correction coils. The properties of the modified spectrograph are as follows:

Mass-energy product (ME/q^2)	250 AMU-MeV
Maximum solid angle	3.2 msr
Momentum resolution ($\delta p/p$)	2×10^{-4}
Momentum range in the focal plane	3%
Dispersion	15 cm/% in momentum
Flight Path (target to focal plane)	6.2 m

Although this spectrograph was designed for heavy charged-particle spectroscopy, it will be used for initial studies of pion production in proton-nucleus and ^{3}He-nucleus collisions near threshold. In the first year the (p,π^+) reaction will be studied with protons in the 140-160 MeV range to extend the recent 154-MeV Orsay measurements [2] and to test the theoretical predictions of J.G. Wills [3] concerning the cross section behavior very near threshold. Future work will include measurements of $(p,\pi^\pm)$ reactions at higher proton energies ($E_p \lesssim$ 200 MeV) to extend the 185-MeV Uppsala work [4] with improved energy resolution, and studies of the $(^3He, \pi^\pm)$ reaction at $E_{^3He} \approx$ 300 MeV (100 MeV/nucleon) to look for "cooperative" effects and to shed light on the puzzle raised by the Maryland group [5], which reported the $(^3He, \pi^0)$ cross section at $E_{^3He}$ = 200 MeV to be several orders of magnitude smaller than theoretical expectations.

REFERENCES

[1] H. Tyrén, S. Kullander, O. Sundberg, R. Ramachandran and P. Isacsson, Nuc. Phys. 79 (1966) 321.
[2] Y. Le Bornec, B. Tatischeff, L. Bimbot, I. Brissaud, J.P. Garron, H.D. Holmgren, F. Reide and N. Willis, Phys. Lett. 49B (1974) 434.
[3] Private Communication.
[4] See for example S. Dahlgren, P. Grafström, B. Höistad and Å. Asberg in Proceedings of the Fifth International Conference on High-Energy Physics and Nuclear Structure, Uppsala. Ed. G. Tibell (North-Holland, 1973) p. 254.
[5] N.S. Wall, J.N. Craig, R.E. Berg, D. Ezrow and H.D. Holmgren, in Tibell, Op. Cit., p. 279.

SOME RECENT POLARIZED PROTON BEAM EXPERIMENTS AT THE ARGONNE NATIONAL LABORATORY ZERO GRADIENT SYNCHROTRON*

E.C. Swallow

The Enrico Fermi Institute

The University of Chicago, Chicago, Illinois 60637

INTRODUCTION

Let me begin by picking up on the religious reference made by Professor Kabir in his first lecture. In the midwestern part of the United States there are frequent evening "tent meetings" which feature evangelists. These men are charged with inspiring the faithful by recounting the salutary deeds and experiences of others rather than with giving formal sermons. This is rather like my position this evening.

Traditionally, it has also been the case that these men are not professional preachers of the gospel. Again, this corresponds to my position here. My primary areas of interest in particle physics are weak interactions and fundamental symmetries, while most - but not all - of the work I will discuss relates to strong interaction phenomenology. Nonetheless, these interests led me to initiate the second polarized proton beam proposal submitted at the Argonne Zero Gradient Synchrotron (ZGS), so I am fairly well acquainted with the program there.

I hope to convince you that investigations with high energy polarized protons are indeed taking place at the ZGS, and that they are yielding interesting new results. I think you will also see that this line of research holds considerable promise for the future, including experiments on weak interactions and basic

*Work supported in part by the United States National Science Foundation under Grant NSF GP 32534 A-2.

symmetries, both at the ZGS and at new intermediate-energy machines like TRIUMF.

A polarized proton beam opens up several important new physics possibilities. It can be used in conjunction with a polarized target to study proton-proton interactions in pure initial spin states [1]. It can also be used to study spin effects in elastic scattering at very small angles and in inelastic processes. Such experiments are extremely difficult - perhaps impossible - with polarized targets because of the substantial backgrounds which arise from interactions involving bound nucleons. There are also many technical benefits such as the ability to do rapid polarization reversal and to use a wide variety of detector configurations, including bubble chambers. The price one pays, of course, is that the variety of possible projectiles and targets is quite limited and that beam intensities are two to three orders of magnitude lower than for unpolarized protons.

ZGS POLARIZED BEAM

The first acceleration of a polarized proton beam to multi-GeV energies took place at the Argonne National Laboratory Zero Gradient Synchrotron in July, 1973 - nearly two years ago. Polarized beam momenta as high as 8.5 GeV/c have since been reached [2], with normal physics operation at 6 GeV/c and below. Circulating intensities as high as $3\text{-}6\times10^{9}$ protons per pulse have been obtained with a 2.6 sec pulse period at 6 GeV/c. In intermediate energy physics units, this is a feeble time averaged beam current of about 0.3 nA. Approximately 20% of the circulating beam can be extracted into one of the external proton beam lines where two to four experiments typically operate simulataneously. The extracted beam is transversely polarized in the vertical direction. During normal operation, the magnitude of the polarization is now 65-75%. Beam will be accelerated to full ZGS energy (12 GeV/c) at some time in the near future.

The polarized proton ion source [2] is a ground state atomic beam source designed and built by Auckland Nuclear Accessory Co., Ltd. of Auckland, New Zealand. When operated in a pulsed mode, it produces a pulsed current of 30 μA with a polarization of 75-80%.

Beam is injected into the main ring of the ZGS by a 50 MeV proton linac, so polarized protons at this energy could also be made available for experiments. The linac output current is about 15 μA.

EXPERIMENTAL PROGRAM

The ZGS polarized proton beam experimental program which has developed over the last few years is summarized in Table I. Future experiments which are presently under consideration are shown in Table II.

In many cases, listed experiments are part of a coherent program by a particular research group. The Michigan-ANL-St. Louis collaboration, under the leadership of A.D. Kirsch, deserves special mention because they provided much of the impetus for the development of the polarized beam facility [3]. Their primary interest has been in using the polarized beam in conjunction with a polarized target to study p-p interactions in pure initial spin states (E-324, E-366, E-381). They have recently observed [4] a dramatic momentum dependence of the p-p total cross section difference for spins parallel and anti-parallel. The difference is small above about 3 GeV/c and appears to rise rapidly below that momentum. The Rice University group proposes (P-395) to investigate this further in the 1-3 GeV/c region.

The ANL-Northwestern collaboration is pursuing a related program (E-372, E-385, P-401, P-402) aimed at obtaining a full amplitude analysis for p-p elastic scattering [5]. Initial results from this ambitious project are now available [6], but a great deal of work remains to be done before the amplitudes can be completely determined.

In the area of inelastic processes, the Argonne Effective Mass Spectrometer group has undertaken a systematic investigation (E-339, E-391) of N^* and Δ production occurring at the polarized proton vertex [7]. Several studies of inclusive reactions have also been performed (E-336) or proposed (P-393, P-394, P-399), including one quite extensive systematic effort proposed by the Indiana group (P-399).

Having given this quick overview of the ZGS polarized proton beam program, I will now present brief discussions of several experiments which I find particularly interesting. Needless to say, my selection is biased in favor of my own work, so I will try to offset this by starting with experiments performed by others.

POLARIZATION PARAMETER IN p-n ELASTIC SCATTERING

Using the Effective Mass Spectrometer, R. Diebold and co-workers have determined the polarization parameter (strictly speaking, the analyzing power) for p-n elastic scattering [8] at 2, 3, 4, and 6

TABLE I

Current Polarized Proton Beam Experiments at the ZGS

Experiment No.	Title, Spokesman, Institutions	Detectors	Status
E-324	Total p-p Cross Sections Using a Polarized Target and a Polarized Beam, A.D. Kirsch, (U. of Michigan, ANL, St. Louis U.)	PPT Ctrs	Complete [1]
E-336	Measurement of Λ's Produced by a Polarized Proton Beam, R. Winston, (U. of Chicago, ANL, Ohio State U.)	EMS	Complete [22]
E-339	Study of Resonance Production with a Polarized Proton Beam Using the Effective Mass Spectrometer, A.B. Wicklund, (ANL)	EMS	Complete
E-354	Parity Violation in Proton Scattering Processes, D.E. Nagle, (Los Alamos, U. of Chicago, U. of Illinois)	Ctrs	Complete [11,13]
E-364	Measurement of the Polarization Parameter at Small Angles in p-p Elastic Scattering, D.R. Rust, (Indiana U. ANL, U. of Chicago, Ohio State U.)	EMS	Complete [15]
E-366	Feasibility Study for Measurement of the Recoil Spin in p-p Elastic Scattering with a Polarized Beam and Polarized Target, A.D. Kirsch, (U. of Michigan, ANL, St. Louis U.)	PPT Ctrs	In progress [23]
E-367	Study of 6 and 12 GeV/c p-p Interactions Using the ZGS Polarized Beam and the 12-Foot Bubble Chamber, D.K. Robinson, (Case Western Reserve U., Carnegie-Mellon U., U. of Michigan)	12' HBC	In progress
E-372	Proton-Proton Elastic Scattering with Polarized Proton Beam and Determination of p-p Scattering Amplitudes, D. Miller, (Northwestern U., ANL)	PPT MWPC Ctrs	Complete [6]
E-376	Measurement of the Polarization Parameter for p-n Elastic Scattering from 2 to 6 GeV/c, R.E. Diebold, (ANL)	EMS	Complete [8]
E-381	Elastic p-p Cross Sections Using a Polarized Target and a Polarized Beam, A.D. Kirsch, (U. of Michigan, ANL, St. Louis, U.)	PPT Ctrs	Complete [1,4]
E-385	Proton-Proton Scattering with S-Type Polarized-Proton Beams, N-Type Polarized Target and a Spin Analyzer for Recoil Protons, B. Ŝandler, (ANL, Northwestern U.)	PPT MWPC Ctrs	In Progress [6]
E-391	Measurement of Polarization Effects Using the Effective Mass Spectrometer and Polarized Beam at 9 and 12 GeV/c, S. Kramer, (ANL)	EMS	Approved

TABLE II

Future Polarized Proton Beam Experiments at the ZGS

Experiment No.	Title, Spokesman, Institutions	Detectors	Status
P-393	Measurement of the Asymmetry in Inclusive p-p Scattering, M. Marshak, (U. of Minnesota and Columbia U.)	Ctrs	Proposed
P-394	Measurement of Spin Rotation Parameters in Λ Production at 12 GeV/c Using the Polarized Proton Beam and the Effective Mass Spectrometer, A. Lesnik, (Ohio State U., ANL, U. of Chicago)	EMS	Proposed
P-395	Measurement of the Total Cross Section for Proton-Proton Scattering in Pure Initial Transverse Spin States in the 1-3 GeV/c Region, G. Phillips, (Rice U.)	PPT MWPC Ctrs	Proposed
P-398	Measurement of Polarization in the Coulomb Interference Region in p-p Scattering, D. Rust, (Indiana U.)	MWPC Ctrs	Proposed
P-399	Study of Inclusive Reactions Using the ZGS Polarized Beam, S. Gray, (Indiana U.)	MWPC Ctrs	Proposed
P-401	Measurement of Observables (N,S; 0,S), (0,S; 0,S), and (N,0; 0,N) at 6 GeV/c, D. Miller, (Northwestern U., ANL)	PPT MWPC Ctrs	Proposed
P-402	Measurement of Observable (S,S; 0,0) in Proton-Proton Elastic Scattering at 2, 3, 4 and 6 GeV/c, K. Nield, (ANL, Northwestern U.)	PPT MWPC Ctrs	Proposed
P-403	Search for Parity Violation in p-Nucleus Scattering, D. Nagle, (Los Alamos, U. of Chicago, U. of Illinois)	Ctrs	Proposed

GeV/c (E-376). This was done by bombarding a liquid deuterium target with polarized protons. The left-right scattering asymmetry was measured by observing the angle and momentum of fast (forward) scattered protons. Both p-n and p-p events were recorded, and they were distinguished by the absence (or presence for p-p) of a recoil proton. The p-p events provide a check on the purity of the sample and the accuracy of the deuterium corrections.

Polarization effects in p-p and p-n elastic scattering are intimately related to one another, and they can be used to separate the I=0 and I=1 t-channel exchange contributions to the spin-flip amplitude. One might reasonably espouse either of two simple qualitative expectations. Pure I=0 exchange, as might be expected in an optical model [9], gives rise to equal polarizations in the two reactions. On the other hand, the I=1 amplitudes have opposite signs for the two reactions. Thus a single-flip amplitude arising from pure I=1 exchange would give mirror symmetry similar to that observed in $\pi^{\pm}p$ elastic scattering.

The experimental results, shown in Figure 1, are fairly close to the optical model expectation at 2 GeV/c, but this becomes less true as the momentum increases. One is tempted to speculate that they might be moving toward a mirror-symmetric configuration at higher energies. This group plans to extend their measurements to 9 and 12 GeV/c as part of the approved experiment E-391.

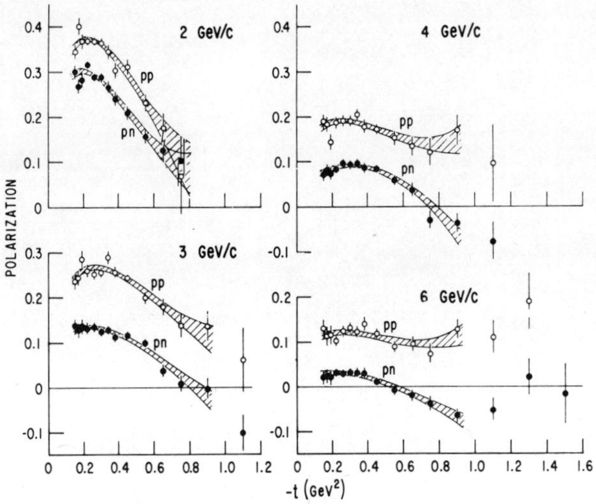

Fig. 1. Polarization parameter for p-p and p-n elastic scattering at four momenta. The errors are statistical only and do not include the ± 6% scale uncertainty from the beam polarization. (Taken from Ref. 8.)

PARITY VIOLATION IN PROTON SCATTERING

One of the experiments which I find most intriguing is the search for parity violation in proton scattering at 6 GeV/c performed by a Los Alamos-Chicago-Illinois collaboration (E-354). Their approach is to look for a dependence of the total interaction cross section on the helicity of the incident protons. This is done by using integral counting techniques to perform a transmission measurement. Careful monitoring of beam characteristics, highly symmetric experimental apparatus, and rapid polarization reversal are essential to achieve the precision required for a really interesting result. A longitudinally polarized beam is obtained by bending the vertically polarized 6 GeV/c beam downward through an angle of 7.75°.

Any (real) dependence of the interaction cross section on the helicity of the incident protons is clear evidence for a parity violating contribution to the nucleon-nucleon interaction. Such an effect is, in fact, predicted [10] by the conventional current-current model for the weak interactions. Since it arises from an interference between the strong and weak amplitudes, the effect is expected to be first order in the weak coupling constant and thus of order 10^{-6} or 10^{-7}. Contributions from hadronic weak neutral currents could increase this by one or two orders of magnitude under a variety of circumstances.

The particular appeal of this type of experiment is that it offers the possibility of exploring hadronic weak interactions in some detail. Once such helicity dependence is observed, it should be possible to explore its energy dependence and isospin properties, and perhaps even to look for it in specific reaction channels.

Measurements [11] by this collaboration have given a value for $A \equiv (\sigma^+ - \sigma^-)/(\sigma^+ + \sigma^-)$ of $A = (5 \pm 9) \times 10^{-6}$ for p-Be scattering at 6 GeV/c. An earlier investigation [12] of p-p scattering at 15 MeV by the Los Alamos and Illinois contingents also gave a negative result, $A = (1 \pm 4) \times 10^{-7}$. In their most recent 6 GeV/c run at ANL, they used a water target to minimize noise arising from target thickness inhomogeneity and observed a value of $A = (15 \pm 2) \times 10^{-6}$, a clear parity violating effect [13].

Unfortunately, there is a parity violating background which prevents any straightforward interpretation of this result. Normal strong interaction processes transfer at least some of the (longitudinal) beam polarization to hyperons produced in the target [14]. The hyperons subsequently decay *via* the weak interaction. To the extent that parity is violated in these decays, and to the extent that the hyperons are longitudinally polarized, there will be a forward-backward asymmetry in the emission of the decay products in the hyperon rest frame. (This asymmetry is particularly large for Σ^+ and Λ^0 hyperons.) The laboratory angular distribution of the decay products thus depends on the hyperon polarization, and therefore on the beam polarization. Since some of the charged decay products strike the downstream transmission counters, it appears to the apparatus that the interaction rate in the target depends on the (longitudinal) beam polarization, which is the signature of parity violation in this experiment. An estimate of this effect indicates that it can indeed give rise to A values like 2×10^{-5}.

It might be possible to attack the hyperon background problem by varying the size of the transmission counters and extrapolating to zero acceptance, but it seems doubtful that the required precision could be achieved in this way. The Los Alamos-Chicago-Illinois collaboration proposes to deal with the problem in a different manner (P-403). They want to place an appropriate set of magnets after the target to sweep out the hyperon decay products so that they cannot hit the transmission counters. Using this approach, they hope to reach an accuracy of a few parts in 10^7 at 6 GeV/c.

POLARIZATION PARAMETER IN SMALL-ANGLE p-p ELASTIC SCATTERING

An Indiana-ANL-Ohio State-Chicago collaboration, of which I am a member, has measured [15] the polarization parameter P (analyzing power) in p-p elastic scattering at 6 GeV/c with fine t (four-momentum transfer squared) resolution over the interval $0.02 < -t < 0.5$ GeV^2. Previous experiments which measured P near 6 GeV/c were limited to the momentum transfer range $-t > 0.1$ GeV^2 because they used a polarized target as their source of polarized protons. They also had rather coarse t resolution. Any rapid variation in P near $-t = 0.13$ GeV^2, where there is a break in the exponential dependence of the differential cross-section at high energy [16], would thus have been missed.

The Argonne Effective Mass Spectrometer (EMS) [17] was used for this experiment (see Fig. 2). A beam of transversely polarized 6 GeV/c protons was incident on a 50 cm liquid hydrogen target. The magnitude of the polarization was typically 56%, and the sign of the (vertical) polarization was reversed approximately every two hours. Forward scattered particles were measured in five sets of magnetostrictive wire spark chambers placed before, within, and behind a magnet with large aperture. The mean resolution in $-t$ was ± 0.006 GeV^2.

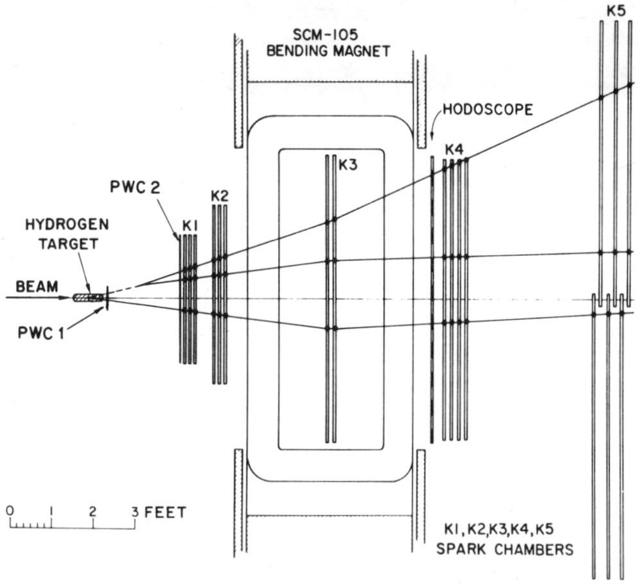

Fig. 2. Plan view drawing of the Argonne Effective Mass Spectrometer, including proportional wire chambers added for experiment E-336. Recoil counters used in the elastic scattering trigger are not shown.

Two different triggers were employed during the course of the experiment. Both used a large array of scintillation counters downstream of the magnet to detect forward scattered particles and a veto counter located on the deflected beam line, just behind this array, to reject unscattered beam protons. Scintillation counters were also placed on both sides of the hydrogen target to detect recoil protons. The first trigger, for scattering events with $-t > 0.04$ GeV^2, required a beam particle signal in coincidence with a recoil particle signal, a forward particle signal, and no signal from the beam veto counter. For $-t < 0.04$ GeV^2 elastic recoil protons stop in the target. Thus, the second trigger, which provided our sample of these very-small-angle events, did not require a signal from either of the recoil counters.

In order to obtain a pure sample of elastic events, we applied several cuts to the data. The most important of these were a cut on the chi-squared of the fit to the measured trajectory and a cut on the missing mass calculated assuming the forward particle was a proton. Our residual inelastic contamination is estimated to be less than 2%.

Denoting the beam polarization by P_B, the left-right scattering asymmetry is given by

$$a(t) = P_B P(t) = K \frac{N_L(t) - N_R(t)}{N_L(t) + N_R(t)} \quad ,$$

where N_L and N_R are the numbers of left and right scatters respectively, and K is a constant determined by the range of azimuthal angles accepted. The recoil counters limited the azimuthal acceptance to $\pm 24^\circ$ from the horizontal, yielding K = 1.029. For data taken without the recoil counters in the trigger, the full 2π range was accepted. Each of these events was weighted by the absolute value of the cosine of the angle between the scattering plane and the horizontal before it was added to N_L or N_R. In this case $K = 4/\pi$. The asymmetries for the two signs of beam polarization were averaged to remove time-independent instrumental biases.

The beam polarization was expected *a priori* to be between 50% and 70%. A more precise value was obtained by normalizing our data to agree on the average with the results of Ref. 18 in the interval $0.1 < -t < 0.5$ GeV^2. This gave a mean beam polarization of (56 ± 4%), where the dominant uncertainty is the absolute normalization in Ref. 18.

The final results for P(t), determined from 3×10^5 p-p elastic scatters at 6 GeV/c, are shown in Fig. 3. The indicated

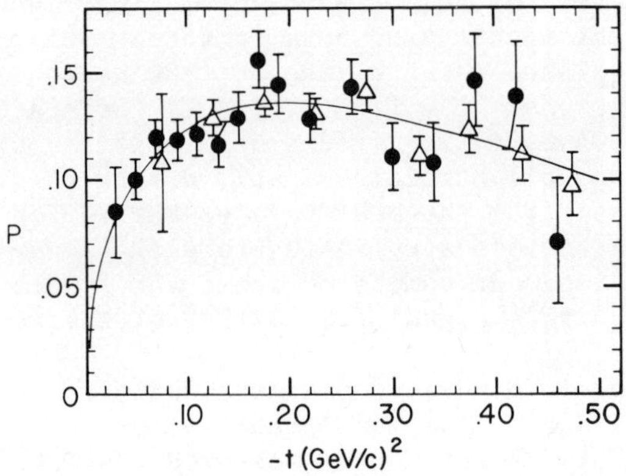

Fig. 3. The polarization parameter P for p-p elastic scattering at 6 GeV/c as a function of -t. Results from this experiment (solid circles) are shown along with those of Ref. 18 (open triangles). The curve is the empirical formula discussed in the text.

errors are purely statistical. There is, in addition, a systematic scale uncertainty of ±7% in all values of P(t) due to the uncertainty in the beam polarization as discussed above. Other sources of systematic error are not expected to contribute significantly.

Our results, combined with those of Ref. 18, are well fitted (χ^2 = 17 for 22 degrees of freedom) by the empirical formula

$$P(t) = (0.506 \pm 0.021) \sqrt{-t} \ \exp[(2.52 \pm 0.18)t]$$

over the interval -t = 0.02 to 0.5 GeV^2, as shown by the curve in Fig. 3. There is clearly no significant evidence of structure near -t = 0.13.

Enough data has been taken to increase our sample size by at least a factor of 3, so a more accurate result will be available when this is completely analyzed. In addition, the Indiana group has plans for a precise study of P(t) in the Coulomb interference region (P-398).

POLARIZATION EFFECTS IN INCLUSIVE Λ° PRODUCTION

At about the time practical plans were begun for the polarized beam at the ZGS, the enthusiasm of Larry Ratner, Alan Krisch, and our friend Gary Marmer also infected some of us in the Argonne-Chicago-Ohio State collaboration. We soon came to recognize the unique value of the polarized beam--as opposed to polarized targets--for investigating spin effects in *inelastic* processes.

Combining this realization with our interest in polarized hyperon beta decay experiments caused us to suggest that this facility might provide a copious source of polarized high-energy hyperons. More specifically, we offered the intuitive speculation [14] that a fast forward hyperon, being in a sense the persistent image of the incident proton, "should" retain the beam's polarization along with its baryon number. To study this interesting possibility in a direct and concrete fashion, our collaboration undertook ZGS experiment E-336, an investigation of inclusive Λ^0 production with the 6 GeV/c polarized proton beam.

If the polarized beam should prove to be a copious source of high-energy polarized hyperons, it would be possible to adapt many of the techniques developed for investigating CP phenomenology in the K^0 system to the study of hyperon decays. Among the new possibilities would be a polarized $\Lambda^0 \to p + e + \nu$ experiment with an order of magnitude increase in sample size [19]. One could, of course, also perform precision magnetic moment measurements. While we are dreaming, we might, in addition, hope for an accurate determination of the decay parameters for $\Xi \rightarrow \Lambda\pi$ --with attendant tests of time reversal invariance and unitarity [20] --and for polarized hyperon scattering experiments. In any case, we can at least expect to gain some new insight into the spin structure of multiparticle production amplitudes by studying the new class of observables which is accessible in hyperon production experiments.

Parity conserving inclusive processes of the form spin ½ (polarized) + unpolarized → spin ½ (analyzed) + anything can be described [14] by observables which are a generalization of the Wolfenstein spin rotation parameters for proton-proton elastic scattering. The polarization (P+P'), the analyzing power (P-P'), the depolarization parameter D, and the differential cross section σ are related as follows:

$$\sigma = \sigma_o[1 + (P-P')\,\vec{e}\cdot\hat{n}]$$

$$\sigma\,\vec{f}\cdot\hat{n} = \sigma_o[(P+P') + D\,\vec{e}\cdot\hat{n}].$$

Here $\vec{e}$ is the initial (beam) polarization vector; $\vec{f}$ is the final (Λ^o) polarization vector; $\hat{n}$ is the production plane unit normal (along $\vec{\text{beam}} \times \vec{\Lambda}$); and σ_o is the cross section for unpolarized protons. (P+P') measures the component of the final-state baryon polarization along the production plane normal that is produced independently of the beam polarization. (P-P') measures the left-right production asymmetry of the final-state baryon when the incident beam is 100% vertically polarized. D measures the component of the beam polarization along the production plane normal that is retained by the final-state baryon. For *elastic scattering*, time reversal invariance constrains the polarization to be equal to the analyzing power. However, for *inelastic processes*, like inclusive Λ^o production, the two parameters are *not* required to be equal. Differences between polarization and analyzing power have previously been observed only in low energy nuclear reactions [21]. Our results from E-336 contain the first observation of this effect at high energy [22].

We have determined (P-P'), (P+P'), and D for lambdas produced in the forward direction by transversely polarized 6 GeV/c protons incident on a 20 cm liquid hydrogen target. This experiment, like the p-p and p-n elastic polarization measurements described earlier, used the Effective Mass Spectrometer which is shown in

Fig. 2. Two multiwire proportional chambers (PWC's) were incorporated into the trigger logic as multiplicity sensing hodoscopes; the trigger requirement was that there be at least two more charged particles at PWC2 than at PWC1. This requirement selected $\Lambda^o \rightarrow p\pi^-$ decays which occurred between the two PWC's. Chamber sizes were chosen to minimize triggering biases which might depend on event multiplicity. We also required at least two particles in the counter hodoscope behind the spectrometer magnet. Non-interacting beam particles were suppressed by a small veto counter just behind this hodoscope. Thus our trigger was

$$\text{Trigger} = \text{Beam} \cdot \overline{\text{Beam Veto}} \cdot \text{PWC}(\Delta N \geq 2) \cdot H(\geq 2).$$

During data taking, the sign of the beam polarization was reversed approximately every two hours. The analyzing magnet polarity was also reversed several times.

In our data analysis, the $p\pi^-$ invariant mass was calculated for all plus-minus charge combinations and compared with the known Λ^o mass. Cuts on the quality and position of the production and decay vertices reduced the background under the Λ^o mass peak to less than 1%. The RMS mass resolution was about 1 MeV. Our surviving data sample contained 15281 Λ^o events.

The analyzing power (P-P') gives rise to a left-right production asymmetry. By counting the number of lambdas produced left and right, for the beam polarization up or down, and for opposite polarities of the spectrometer magnet, we can express the analyzing power as a pure ratio,

$$P-P' = \frac{1}{e'} \cdot \frac{G_L - G_R}{G_L + G_R} \quad .$$

Here G_L (Geometric left) $\equiv \{(L\uparrow+)\ (L\uparrow-)\ (R\downarrow+)\ (R\downarrow-)\}^{\frac{1}{4}}$, G_R (Geometric right) $\equiv \{(R\uparrow+)\ (R\uparrow-)\ (L\downarrow+)\ (L\downarrow-)\}^{\frac{1}{4}}$, e' is the magnitude of the mean beam polarization along the production plane normal, and for example, $(L\uparrow+)$ denotes the number of detected lambdas produced to the left when the beam polarization was up and the spectrometer magnet polarity was positive.

To extract the parameters (P+P') and D from the data, we made use of the polarization-analyzing properties of the $\Lambda^o \rightarrow p\pi^-$ decay. Thus, each category like $(L\uparrow+)$ was further subdivided according to whether the decay proton was emitted above or below the production plane. Similar ratios were then constructed, yielding the parameters (P+P') and D. All dependence on beam intensity, target density, geometric detection

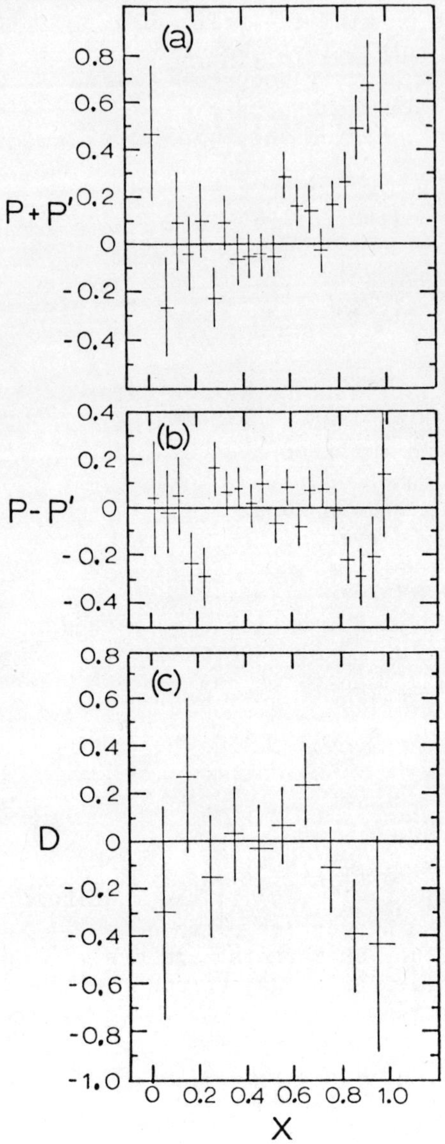

Fig. 4. Values for the spin rotation parameters versus the Feynman scaling variable x: (a) polarization (P + P'); (b) analyzing power (P - P'); (c) depolarization parameter (D).

efficiency, and production cross section cancels in these ratios.

In order to monitor the beam polarization, we continuously recorded p-p elastic scattering events along with our Λ^0 data. By taking the beam polarization to be (60 ± 3)%, we obtain values for the elastic analyzing power which agree very well with those obtained from polarized target experiments [18].

Our results for the polarization (P+P'), the analyzing power (P-P') and the depolarization parameter D are shown as a function of the Feynman scaling variable x in Figure 4(a), 4(b), and 4(c), respectively. There is structure in all three parameters for x greater than 0.8. In particular, I call your attention to the disparity between (P+P') and (P-P'); the integrated difference for x >0.8 is 0.465 ± 0.082.

This structure in the spin parameters can be qualitatively understood in terms of t-channel exchange properties of the production amplitudes. The fact that P' is non-zero and D is negative implies that unnatural parity exchange contributions are dominant [14]. However, P is also finite, and D is certainly not -1, indicating that natural parity exchanges also contribute.

Our analysis of this data is continuing, and we intend to evaluate the spin rotation parameters which characterize Λ^0 polarization in the production plane. When the full-energy polarized beam becomes available, we also plan to extend these measurements to 12 GeV/c (P-394). Once that is done, we will be able to evaluate the prospects for polarized hyperon beams at the ZGS more precisely.

CONCLUSION

I have reviewed the polarized proton beam program which has developed at the ZGS as well as some of the plans for the future. I have also given brief descriptions of several of the experiments and discussed their results. It is my hope that this has convinced you that polarized beam physics is "for real" and that it can be interesting - perhaps even exciting. So I look forward to hearing about the first polarized beam results from TRIUMF in the near future.

REFERENCES

[1] E.F. Parker, et al., Phys. Rev. Letters 31, (1973) 783 and J.R. O'Fallon, et al., Phys. Rev. Letters 32, (1974) 77.

[2] T. Khoe, et al., Particle Accelerators (to be published).

[3] A.D. Krisch, "Experiments with Polarized Beams and Targets", in *Proceedings of the 1972 Canadian Institute of Particle Physics Summer School*, edited by R. Henzi and B. Margolis (McGill University, Montreal, 1972) p. 209.

[4] L.G. Ratner, private communication.

[5] A. Yokosawa, "An Experimental Program for p-p Scattering Amplitude Measurements", in *Proceedings of the Argonne Symposium on High Energy Physics with Polarized Beams*, edited by G. Thomas and A. Yokosawa, Argonne Report ANL/HEP-7440 (1974) p. XIII-1.

[6] G. Hicks, et al., Argonne preprint ANL-HEP-75-01, submitted to Phys. Rev. Letters.

[7] A.B. Wicklund, "Inelastic Polarized Proton Interactions at 6 GeV/c", in *Proceedings of the Argonne Symposium on High Energy Physics with Polarized Beams*, edited by G. Thomas and A. Yokosawa, Argonne Report ANL/HEP-7440 (1974) p. XV-1.

[8] R. Diebold, et al., Argonne preprint ANL-HEP-PR-75-25, submitted to Phys. Rev. Letters.

[9] A.W. Hendry and G.W. Abshire, Phys. Rev. D10, (1974) 3662; and F. Halzen, "Theoretical Interpretation of Experiments with Polarized Proton Beams", in *Proceedings of the Argonne Summer Study on High Energy Physics with Polarized Beams*, edited by J.B. Roberts, Argonne Report ANL/HEP-75-02 (1974) p. XXIV-1.

[10] E.M. Henley, "Possible Symmetry Violations in Hadron Scattering", op. cit., p. XII-1.

[11] J.D. Bowman, et al., Phys. Rev. Letters 34,(1975) 1184.

[12] J.M. Potter, et al., Phys. Rev. Letters 33, (1974) 1307.

[13] Preliminary results reported at the Santa Fe Conference; also private communication with J.D. Bowman and H.L. Anderson and ZGS Proposal P-403.

[14] E.C. Swallow, Phys. Letters 49B, (1974) 91 and references therein.

[15] D.R. Rust, et al., Enrico Fermi Institute preprint EF1-75-4, to be published in Physics Letters.

[16] G. Barbiellini, et al., Phys. Letters 39B, (1972) 663; also R.A. Carrigan, Jr., Phys. Rev. Letters 24, (1970) 168.

[17] See I. Ambats, et al., Phys. Rev. D9, (1974) 1179, and references therein.

[18] M. Borghini, et al., Phys. Letters 31B, (1970) 405.

[19] See R. Abrams, et al., in *Planning for the Future: ZGS Workshops Summer 1971*, Argonne Report ANL/HEP-7208 (1971), Vol. I, p. 369.

[20] See P. Eberhard, CERN Report CERN 72-1 (1972).
[21] R.N. Boyd, et al., Phys. Rev. Letters 29, (1972) 955.
[22] A. Lesnik, et al., Enrico Fermi Institute preprint EFI-75-30, submitted to Phys. Rev. Letters.
[23] R.C. Fernow, et al., Phys. Letters 52B, (1974) 243.

PROTON PHYSICS AT SIN

L.G. Greeniaus

DPNC, University of Geneva

Geneva, Switzerland

The SIN 600 MeV sector focussed cyclotron at Villigen has been in operation for about 18 months. However, the final beam lines and experimental facilities are just now in the final stages of completion. These include four areas for pion nuclear physics, a biomedical pion beam, a superconducting muon channel with three experimental areas, a neutron beam line, an area for physics with polarized protons and a low energy proton area.

Unlike TRIUMF and LAMPF which have extensive experimental programs for both mesons and nucleons, at SIN a very heavy emphasis has been placed on meson physics. This is readily seen if one compares the number of meson and nucleon experimental zones. About a year ago, it became apparent that there would be a severe lack of space in the main experimental hall. This led to major changes in the proton physics area and it was decided to build the zone outside of the main building. Construction of the proton area was delayed because of this and a combination of other factors, but the final result has been the design of a very versatile and useful polarized proton beam. Only one proton experimental area has been planned with a small size of 12m x 12m.

The main constraint on use of a proton beam for physics (as opposed to meson production) is that the meson physics program must not be disturbed. This effectively excludes using the primary accelerated beam for long periods of time. To avoid this problem, a parasitic polarized proton beam will normally be obtained by small angle scattering from the first π-production target (the M-target). (To see the layout of the main hall refer to the talk presented by Dr. Scheck). This is a thin (∿1 cm) carbon or beryllium target used to produce low intensity pion beams for the πM1, πM2 and πM3 areas. It will also be possible to use a 50 na accelerated polarized proton beam in a main user mode.

The main characteristics of the two proton (pM1) beams are summarized in Table I. The general layout around the M-target

TABLE I

PROPERTIES OF THE pM1 PROTON BEAMS

		Parasitic Mode	Main User Mode
Origin		8° scattering from the M-target	Accelerated beam from Polarized Ion source
Beam energy *		300 - 600 MeV	300 - 600 MeV
Momentum Resolution		0.2%	0.2%
Intensity:	300 MeV	$< 10^8$ p/s **	$\lesssim$ 0.5 na
	600 MeV	$\lesssim 10^{10}$ p/s **	~ 50 na
Polarization		0.38 ± 0.02	0.80
Transverse		Yes	Yes
Longitudinal		Yes	Yes
Unpolarized		No	Yes
Availability		On request	< 15 days/year

*With a variable length Cu degrader.
**For 100 μa primary beam and 1 cm carbon for M-target.

station and the pM1 beam transport is shown in figure 1. For the parasitic beam, protons scattered at 8° from the M-target are focussed, deflected through 82° by a pair of analyzing magnets and directed along a 35 meter transfer line to the pM1 experimental area just outside the main hall. Measurements of the p-C analyzing power [1] by our group at the CERN-SC show that the beam polarization will be $P_o \simeq 0.38 \pm 0.02$. A 600 MeV beam in the experimental area should have an intensity of $\sim 10^{10}$ p/s for a 1 cm thick target and 100 μa in the primary beam. The accelerated polarized beam could have a polarization as large as 0.80 and an intensity of 50 na. Initial tests show a polarization around 0.6. Except that it is deflected magnetically into the pM1 line instead of scattered, the beam transport is unchanged.

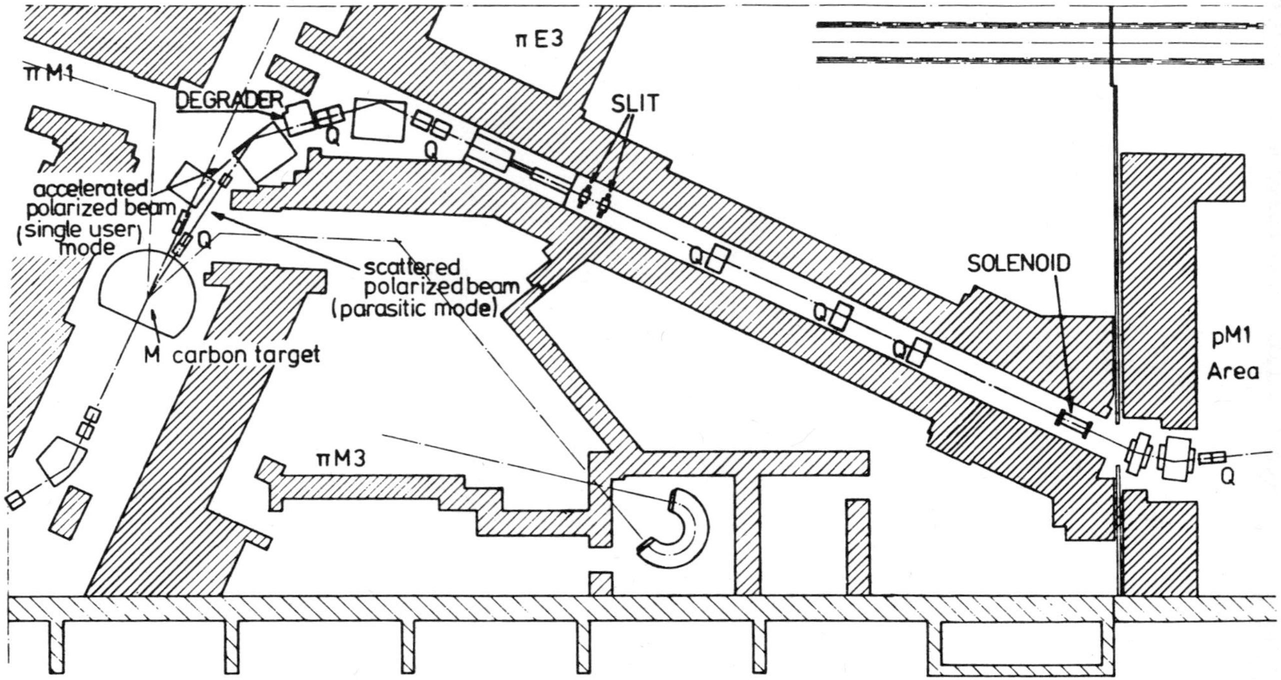

Fig. 1. Layout of the beam transport system near the M-target station. The pM1 beam is produced by scattering in the M-target (Carbon or Be) if the high intensity primary beam is being used. The low intensity accelerated beam is deflected magnetically into the pM1 line. The transfer tunnel from the SLIT to the pM1 experimental area is about 20 meters.

Energies for the pM1 beam in the range between 300 and 600 MeV can be obtained. A variable length Cu degrader has been constructed and installed between the first two analyzing magnets. Calculations indicate that degrading the energy to 300 MeV will reduce the beam intensity by at least a factor of 100. Slits in the degrader and after the beam blocker will allow a momentum resolution of 0.2% to be obtained. A superconducting solenoid in the transfer line can be used to rotate the polarization vector to any desired angle in the plane transverse to the beam direction. For vertical polarization a transverse polarized beam is obtained. When the spin is in a horizontal plane, the final magnetic deflection into the pM1 area, 30°-35° depending on the beam energy, results in a longitudinally polarized beam for experiments.

The two types of pM1 beam complement one other very well. The main advantage of the scattered beam is that it is truly parasitic and will be available whenever the main beam is being used. The accelerated polarized beam will be scheduled less than 15 days/year due to the heavy pressure on machine time. While the scattered beam is perfectly respectable for many applications, there are certain experimental advantages in using a beam with higher polarization and intensity - for example, statistics and reduction of systematic error. Since a large fraction of the time spent on an experiment involves setup, debugging, testing and preliminary measurements, etc., the existence of the two types of beams is advantageous. Careful preparation of a measurement can be done in the parasitic mode, while the hard-to-get main user time can be dedicated to the actual data collection.

For nucleon physics at medium energies, μa beams are too intense to be generally useful. Even one hundred na is a fairly intense beam. For many types of experiments 0.1 na is sufficient and in particular cases even this is too much. Except for its relatively low availability, the SIN accelerated polarized beam is comparable to those at TRIUMF and LAMPF. However, the pM1 scattered beam is adequate for the majority of the proton experiments already proposed or under consideration.

At present, only our group [2,9] has proposed experiments for the pM1 beam. From our point of view, this is a very enviable position. In the medium energy range there is a lack of varied and precise experimental data - especially for polarization phenomena. Given the uncertainty in describing the nucleon-nucleon interaction and its importance in understanding many other nuclear processes, continued activity in this field is amply warranted. Our experimental program has as its ultimate goal a complete experiment (10-14 separate measurements) at several angles and energies for the p-p system. Accurate p-p elastic scattering data of chosen polarization phenomena will be obtained for all angles. This includes the Coulomb-nuclear interference region which until the present has remained almost inaccessible to experimentalists (except for $d\sigma/d\Omega$). Both polarized beams and targets will be used. An effort will also be made to study in detail the inelastic reaction $pp \rightarrow \pi^+ d$.

Two proposals [3,4] have been accepted by the SIN program committee. Together they propose measurement of the Wolfenstein parameters P, D, R, A (and possibly A' and R') at all angles. Statistical errors of about ±0.02 for P and ±0.05 for D, R and A are expected. In figures 2 to 5, calculations of P, D, R and A for the Livermore phase shifts [5] are shown. The small and large angle regions are indicated. The rapid variations with both angle and energy are interesting. The error corridor from a single energy analysis would be about 10-20% wide (relative error). The small angle experiment will probably begin in October 1975 and is a logical continuation of our small angle $d\sigma/d\Omega$ and P measurements at CERN [6,7]. Much of the equipment for the new experiment has already been tested this summer at the CERN-SC. The large angle measurements will probably start in the summer 1976.

The equipment available includes: multiwire proportional chambers with a total of 7424 wires, a spectrometer magnet, PDP 11/20 with disk and 2 tape drives, liquid hydrogen target and a polarized proton target (under construction), and a turntable capable of supporting movable spectrometer and polarimeter arms. The layout of the pM1 area and the configuration of the apparatus to be used in the large angle experiment are shown in figure 6. The polarimeter and spectrometer can be positioned independently in the range $-90^{\circ}<\theta<90^{\circ}$. For the small angle measurements the polarimeter will be rotated into the polarized beam (intensity $\sim 2.10^5$ p/s). The large angle experiment will require 2.10^8 p/s. Both intensities are attainable with the scattered pM1 beam. The main characteristics of the experiment are summarized in table II. Mutliwire proportional chambers will be used to observe individual particle tracks before and after scattering from a liquid hydrogen target and the carbon analysing target of the polarimeter. A fast decision system [8] will reject within 2-3 μs all events that do not undergo a double scattering. A detailed experimental layout is given in figure 7. It will be necessary to obtain $1\text{-}2.10^5$ good events at each scattering angle to obtain the desired statistical errors. This experiment should provide the first detailed measurements of triple-scattering parameters in the Coulomb-nuclear interference region.

Figs. 2-5. Values of the Wolfenstein parameters P, D, R and A calculated from the Livermore phase shifts. The region for the small angle experiment is not cross-hatched. An error coridor for each curve is about 10%-20% (relative error).

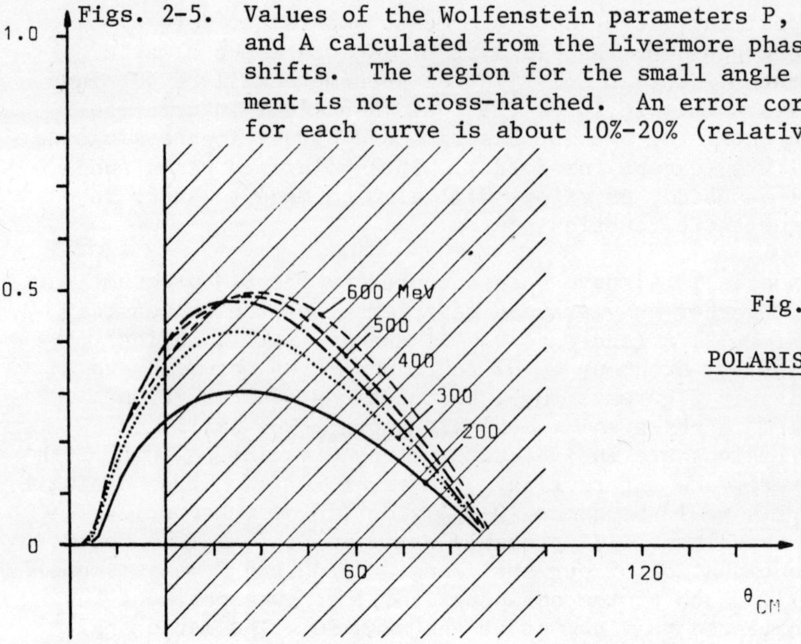

Fig. 2

POLARISATION

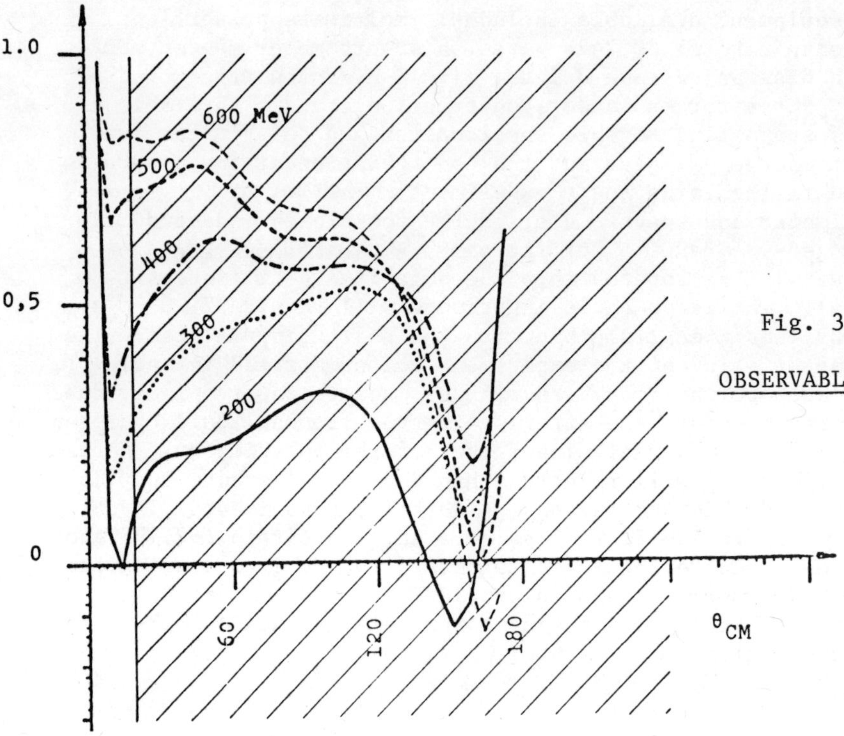

Fig. 3

OBSERVABLE D

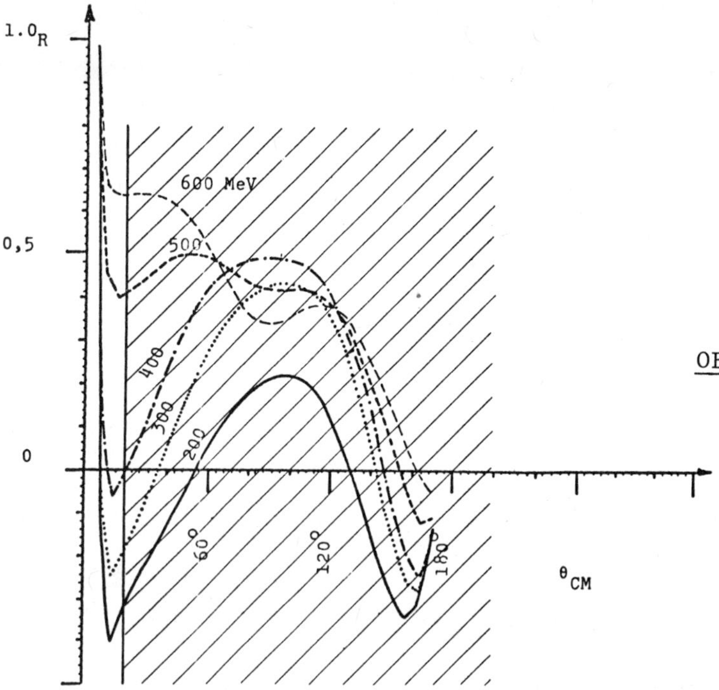

Fig. 4

OBSERVABLE R

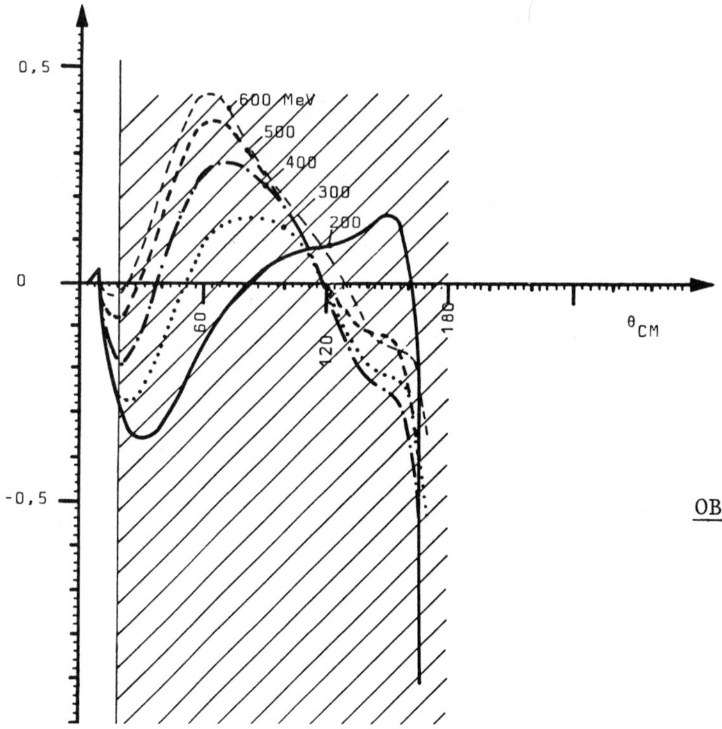

Fig. 5

OBSERVABLE A

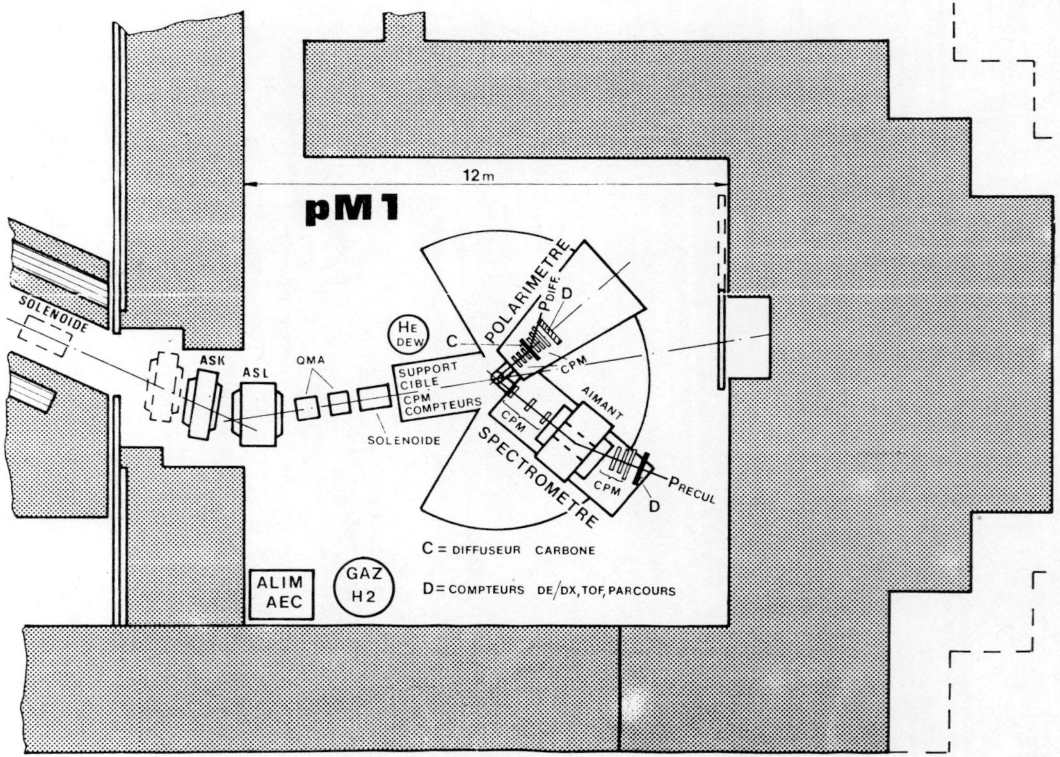

Fig. 6. Layout of the pM1 experimental area. The turntable supports a spectrometer and a polarimeter, shown in a configuration to be used for the large angle scattering experiments. The support can hold wire chambers, counters, and a LH_2 target and cryostat or polarized target. The deflecting magnets ASK and ASL can be moved so that the deflection angle into the zone can be varied in the range 30^0 - 35^0. (To obtain longitudinally polarized beams at different energies.)

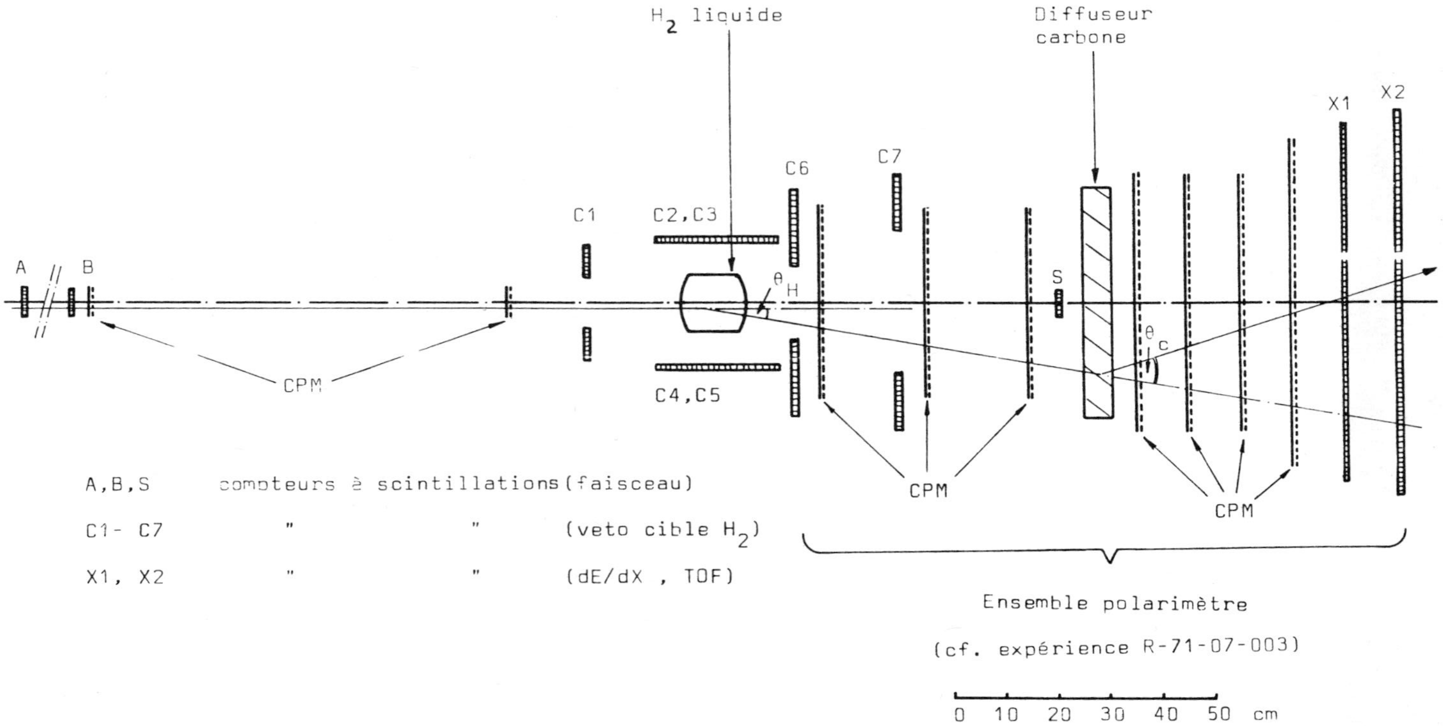

Fig. 7. Schematic of the experimental setup for the small angle scattering experiment. Particle tracks before and after each scattering will be observed. Most inelastic events will be vetoed by counters around the LH_2 target. Time of flight and de/dX for the outgoing particles will be measured.

TABLE II

Summary of Proposed Experiment to Measure P, D, R and A for Small Angle pp Scattering

Energy Range		400 to 600 MeV
	* * *	
Angular Range: LAB		1° - 10°
CM		2° - 25°
	* * *	
# of data points		10 at each energy
		10^{5} good events/point
	* * *	
Relative error between data points		dependent on beam polarization and number of events collected. See Ref. 4 for details.
	* * *	
Normalization Error (due to error in P_{o} and carbon analyzing power)		3% - 5%
	* * *	
<u>System Properties</u>		
Acceptances		$0<\theta_{H}\lesssim 10^{o}$
		$\theta_{c}\lesssim 15^{o}$ for $\theta_{H}<7^{o}$
	* * *	$\theta_{c}\lesssim 10^{o}$ for $\theta_{H}>7^{o}$
Angular Resolutions		$\Delta\theta_{H}\sim \pm 0.3^{o}$
	* * *	$\Delta\theta_{c}\sim \pm 0.6^{o}$
Systematic Errors	* * *	$\delta\theta_{H} \simeq \delta\theta_{c} < 0.01^{o}$
Inelastic Background	* * *	$\sim 1\%$
Carbon Analyzing Power		$0.30 \rightarrow 0.40$
	* * *	600 MeV $\rightarrow$ 400 MeV
Beam Intensity	* * *	2.10^{5} p/s
Good Event Rate	* * *	10 ev/s
Electronic Decision (in 2-3 μs)		$\theta_{H} \gtrsim 1^{o}$, $\theta_{c} \gtrsim 3^{o}$
	* * *	1 track only
Efficiency of beam Veto		$\sim 90\%$

It will be necessary to measure the effective analyzing power of carbon in the energy range 200-400 MeV for the large angle experiment. This will be done with the polarimeter. A configuration similar to the polarimeter, but with large chambers placed close to a CH_2 target should be ideal for precise measurements of $d\sigma/d\Omega$ and asymmetry for the reaction $pp \to \pi d$. A polarized target is being constructed and will be ready for use in early 1977 for the more complicated measurements required by the complete experiment.

The SIN proton physics beam does not have as large an experimental program or all the facilities available at LAMPF and TRIUMF. However, its variable energy and continuous availability with either transverse or longitudinal polarization make it an extremely useful facility. Our group's program for the next 5 years represents a considerable effort in the nucleon-nucleon field and will make very effective use of this beam.

REFERENCES

1 D. Aebischer et al., Nucl. Instr. and Meth. 124 (1975) 49. "Measurements of the p-C Analyzing Power with a Small Angle Scattering System".

2 The people actively participating in our SIN experiments are: D. Besset, B. Favier, G. Greeniaus, R. Hess (group leader), C. Lechanoine, J.C. Niklès, D. Rapin and D.W. Werren from the University of Geneva, Ch. Weddigen from the Kernforschungszentrum, Karlsruhe, and A. Junod from the ETHZ.

3 SIN experimental proposal R-71-07.2, Jan. 1973 "Diffusion Elastique Proton-Proton entre 400 et 600 MeV. I. Mesure des Paramètres D et R".

4 SIN experimental proposal R-71-07.3, June 1974 "Diffusion Elastique Proton-Proton entre 400 et 600 MeV. II. Mesure des paramètres P, D, R et A à petits Angles".

5 M.M. Macgregor et al., Phys. Rev. 182 (1969) 1714. R.A. Arndt et al., Phys. Rev. C9 (1974) 555.

6 D. Aebischer et al., Helv. Phys. Acta 47 (1974) 72. "Diffusion Elastique Proton-Proton dans la région d'interférence nucléaire coulombiene entre 300 et 600 MeV".

7 D. Aebischer et al., Helv. Phys. Acta (to be published). "Measurement of Polarization for Small Angle p-p Elastic Scattering at 399 MeV".

8 D. Aebischer et al., Nucl. Instr. and Meth. 117 (1974) 131. "A Fast On-Line Selection System for Particle Trajectories detected by Multiwire Proportional Chambers".

9 In August 1975, a proposal to use the pM1 beam for proton-tomography was submitted (SIN experimental proposal A-75-02.1). The spokesman is L. Dubal.

PARTICIPANTS

J. H. ALEXANDER	University of British Columbia, Canada
B. ALLARDYCE	CERN, Switzerland
J. ASAI	University of Laval, Canada
D. M. ASBURY	University of Surrey, England
R. E. AZUMA	University of Toronto, Canada
G. A. BEER	University of Victoria, Canada
R. D. BENT	Indiana University, U.S.A.
R. BEURTEY	C.E.N. Saclay, France
J. L. BEVERIDGE	TRIUMF, Univ. of British Columbia, Canada
E. BLACKMORE	TRIUMF, Univ. of British Columbia, Canada
J. BREWER	University of British Columbia, Canada
R. C. BROWN	TRIUMF/Rutherford Laboratory, England
E. T. BOSCHITZ	SIN, Villigen, Switzerland
D. A. BRYMAN	University of Victoria, Canada
J. CAMERON	University of Alberta, Canada
H. S. CAPLAN	University of Saskatchewan, Canada
A. Z. CAPRI	University of Alberta, Canada
A. A. CONE	Vancouver City College, Canada
S. COSTA	University of Torino, Italy
M. CURRIE-JOHNSON	University of British Columbia, Canada
J. DEUTSCH	Université de Louvain, Belgium
M. DILLIG	Universität Erlangen-Nürnberg, Germany
M. S. DIXIT	Carleton University, Canada
E. T. DRESSLER	University of Saskatchewan, Canada
D. DUPLAIN	University of Montreal, Canada
E. D. EARLE	Atomic Energy of Canada, Chalk River
R. A. EISENSTEIN	Carnegie-Mellon University, U.S.A.
G. T. EWAN	Queen's University, Canada
L. FELAWKA	University of British Columbia, Canada
A. C. FONSECA	Notre Dame University, U.S.A.
D. M. GARNER	University of British Columbia, Canada
R. GREEN	Simon Fraser University, Canada
G. GREENIAUS	CERN, Switzerland
F. N. GYGAX	ETH/SIN, Switzerland
I. HALPERN	University of Washington, U.S.A.
M. HASINOFF	University of British Columbia, Canada

R. S. HAYANO	Univ. of Tokyo/Univ. of British Columbia
R. H. HEFFNER	Los Alamos Scientific Laboratory, U.S.A.
E. M. HENLEY	University of Washington, U.S.A.
S. JACCARD	University of British Columbia, Canada
K. P. JACKSON	University of Toronto, Canada
B. K. JAIN	University of Manitoba, Canada
P. W. JAMES	University of Victoria, Canada
R. R. JOHNSON	University of British Columbia, Canada
M. DE JONG	University of Manitoba, Canada
P. K. KABIR	University of Virginia, U.S.A.
J. R. KANE	College of William and Mary, U.S.A.
A. N. KAMAL	University of Alberta, Canada
S. K. KIM	University of Victoria, Canada
H. KOCH	CERN, Switzerland
K. S. KRANE	University of Oregon, U.S.A.
J. K. P. LEE	McGill University, Canada
D. LIM	University of Victoria, Canada
K. F. LIU	C.E.N. Saclay, France
R. MACDONALD	University of British Columbia, Canada
H. B. MAK	Queen's University, Canada
K. MALTMAN	University of British Columbia, Canada
Th. A. J. MARIS	Universidade do Rio Grande do Sul, Brazil
G. MARSHALL	University of British Columbia, Canada
G. R. MASON	University of Victoria, Canada
T. MASTERSON	University of British Columbia, Canada
E. L. MATHIE	University of Victoria, Canada
R. H. McCAMIS	University of Alberta, Canada
D. K. McDANIELS	University of Oregon, U.S.A.
D. F. MEASDAY	University of British Columbia, Canada
S. L. MEYER	National Science Foundation, U.S.A.
A. MILLER	University of Alberta, Canada
B. T. MURDOCH	University of Manitoba, Canada
G. C. NEILSON	University of Alberta, Canada
A. OLIN	University of Victoria, Canada
D. PAI	University of British Columbia, Canada
P. PAVLOPOWLOS	CERN, Switzerland
M. PEARCE	University of Victoria, Canada
J. E. D. PEARSON	University of British Columbia, Canada
C. PERRIN	SIN, Switzerland
A. C. PHILLIPS	University of Manchester, England
P. R. POFFENBERGER	University of British Columbia, Canada
J. M. POUTISSOU	Univ. of B.C./Univ. of Montreal, Canada
H. PRIMAKOFF	University of Pennsylvania, U.S.A.
H. G. PUGH	University of Maryland, U.S.A.
A. REITAN	Univ. of B.C./Univ. of Trondheim, Norway
J. REYES	University of Victoria, Canada
J. R. RICHARDSON	TRIUMF, Canada
P. ROBERSON	University of Wyoming, U.S.A.
B. C. ROBERTSON	Queen's University, Canada

M. SALOMON	University of British Columbia, Canada
J. T. SAMPLE	University of Alberta, Canada
F. SCHECK	SIN, Switzerland
A. SCHENCK	E.T.H./SIN, Switzerland
K. I. H. H. SHAKARTCHI	TRIUMF, Canada
D. SHEPPARD	University of Alberta, Canada
R. SLOBODA	University of Alberta, Canada
A. W. STETZ	University of Alberta, Canada
A. STAMP	University of British Columbia/New Zealand
E. C. SWALLOW	Enrico Fermi Institute, U.S.A.
L. W. SWENSON	Oregon State University, U.S.A.
A. SZYJEWICZ	University of Alberta, Canada
A. W. THOMAS	University of British Columbia/CERN
J. VAVRA	University of British Columbia, Canada
J. VINCENT	University of Victoria, Canada
P. WALDEN	University of British Columbia, Canada
J. B. WARREN	University of British Columbia, Canada
W. J. WIESEHAHN	Simon Fraser University, Canada
J. G. WILLS	University of Indiana, U.S.A.
R. WOLOSHYN	Massachusetts Institute of Technology, U.S.
T. YAMAZAKI	Univ. of Tokyo/Univ. of British Columbia

SUBJECT INDEX